T0262897

Wireless Sensor Networks

Principles and Practice

OTHER TELECOMMUNICATIONS BOOKS FROM AUERBACH

Wireless Sensor Networks

Principles and Practice

FEI HU • XIAOJUN CAO

CRC Press is an imprint of the
Taylor & Francis Group, an informa business
AN AUERBACH BOOK

Auerbach Publications
Taylor & Francis Group
6000 Broken Sound Parkway NW, Suite 300
Boca Raton, FL 33487-2742

International Standard Book Number: 978-1-4200-9215-8 (Hardback)

Library of Congress Cataloging-in-Publication Data

Hu, Fei, 1972-
Wireless sensor networks : principles and practice / Fei Hu and Xiaojun Cao.
p. cm.
Includes bibliographical references and index.
ISBN 978-1-4200-9215-8 (alk. paper)
1. Wireless sensor networks. I. Cao, Xiaojun, 1972- II. Title.

TK7872.D48H82 2009
681'.2--dc22 2009044545

Visit the Taylor & Francis Web site at
http://www.taylorandfrancis.com

and the Auerbach Web site at
http://www.auerbach-publications.com

To Fei's family—*Fang* and *Gloria* ……

To Matt's family ……

Contents

Preface

Wireless sensor networks (WSNs) are one of the hottest topics of research in today's era of information explosion. The latest advances in science and engineering have paved the way for employing the low-power, low-cost WSN, which provides a high order of spatial and effective resolution for an ever-increasing number of applications, such as infrastructure protection and security, surveillance, healthcare, habitat/environment monitoring, food safety, and smart energy. Although the use of WSNs has many advantages over traditional networking techniques and sensing methods in the aforementioned applications, it also poses many challenging issues and optimization problems in the design of network architectures, protocols, and algorithms. To resolve these challenges despite constraints on available energy, bandwidth, memory space and computing power, high rates of node failure and data loss, adverse communication environments, and unique application requirements, significant efforts have been made in the academia and industry.

Although some books on sensor networks have been published recently, most (if not all) of them are not suitable to be used as textbooks due to the limited number of topics covered or to the nature of editorial collections. This textbook attempts to comprehensively discuss all the major technologies, standards, topics, and developments in sensor networks. It covers almost all aspects that readers need to know to enter this burgeoning field, including hardware design, medium access control, routing schemes, transport protocols, OS support, middleware, data management, localization, synchronization, security, actuator/underwater/video sensor networking, power control, and sensor simulations.

This textbook makes complicated concepts easy to understand through interesting examples and WSN applications. In addition, it has exercises, assignments, and detailed case studies that help readers understand the contents and then apply their knowledge in designing their own applications or for solving real-world problems. We have also included some practical sensor network design cases such as medical applications.

Targeted Audiences

This book is ideal for senior college students or first-year graduate students who are majoring in computer engineering, electrical engineering, or computer science. It is also an excellent reference book for sensor network designers, researchers, and engineers who wish to fully exploit WSN technology, and for government employees who wish to use WSNs to enhance homeland security.

Scope of This Book

Because we target both engineering and science students, we have covered both hardware and software topics in this textbook. This book, which consists of 18 chapters, is organized in the following manner:

	Chapter 1. Introduction (WSN overview; basic network concepts)
Computer engineering knowledge	**Chapter 2.** Hardware (micro-sensors with microcontrollers and radio)
Network protocol stack	**Chapter 3.** MAC layer (neighborhood wireless transmission)
	Chapter 4. Routing layer (find an optimal source-to-end path)
	Chapter 5. Transport layer (loss recovery, congestion control)
Computer science knowledge	**Chapter 6.** Operating system (such as TinyOS)
	Chapter 7. Middleware (hide networking details for programmers)
	Chapter 8. Sensor data management
Advanced WSN topics	**Chapter 9.** Localization (also called calibration; very useful)
	Chapter 10. Clock synchronization (correct clock drifts in sensors)
	Chapter 11. Security (countermeasure WSN attacks)

(continued)

Special sensor networks	**Chapter 12.** Wireless actor and sensor networks (with mobile actors)
	Chapter 13. Underwater sensor networks (using acoustic; not RF)
	Chapter 14. Video sensor networks
Miscellaneous	**Chapter 15.** Energy models and low-energy design
	Chapter 16. WSN simulators
Case studies	**Chapter 17.** WSN for tele-healthcare applications
	Chapter 18. WSN for light control

What Can Be Covered in a Course

For a One-Semester (15 Weeks) Course: The following table is our suggested time allocation plan among different topics. Instructors should adjust their teaching plan based on students' feedback and learning practices.

Time Length	*Teaching Topics*	*Chapters*
2 weeks	WSN basics; sensor hardware (for computer science major, the hardware part can be shortened)	Chapters 1 and 2
2 weeks	MAC layer (teach at least two MAC schemes, emphasizing "energy-saving" design)	Chapter 3
2.5 weeks	Routing layer (teach proactive/reactive routing schemes, emphasizing "scalable" design)	Chapter 4
1.5 weeks	Transport layer (teach both "reliable end-to-end transmission" and "congestion control")	Chapter 5
1 week	Operating system; middleware (for computer science major, 2 weeks may be used)	Chapters 6 and 7
1 week	Sensor data management (for computer science major, 2 weeks may be used)	Chapter 8
1 week	Sensor localization; time synchronization (for PhD/MS students, 2–3 weeks may be used)	Chapters 9 and 10

(continued)

(continued)

Time Length	*Teaching Topics*	*Chapters*
1 week	WSN security (Teach µTESLA, Key pre-distribution)	Chapter 11
1.5 weeks	Special sensor networks (especially underwater WSN)	Chapters 12 through 14
0.5 week	Energy models; WSN simulators	Chapters 15 and 16
1 week	Case studies	Chapters 17 and 18
Total: 15 weeks	In each chapter, teach both math principles and concrete design cases. Leave some topics for after-class reading assignments.	

Note that some time should also be allocated for class labs.

For a One-Quarter (10 Weeks) Course:

Time Length	*Teaching Topics*	*Chapters*
1.5 weeks	WSN basics; sensor hardware	Chapters 1 and 2
1 week	MAC layer (emphasizing "energy-saving" design)	Chapter 3
1.5 weeks	Routing layer (teach proactive/reactive routing schemes; emphasizing "scalable" design)	Chapter 4
1 week	Transport layer (teach both "reliable end-to-end transmission" and "congestion control")	Chapter 5
0.5 week	Operating system; middleware (for computer science major, 1.5 weeks may be used)	Chapters 6 and 7
0.5 week	Sensor data management (for computer science major, 1.5 weeks may be used)	Chapter 8
1 week	Sensor localization; time synchronization (for PhD/MS students, 2–3 weeks may be used)	Chapters 9 and 10

(continued)

Time Length	*Teaching Topics*	*Chapters*
0.5 week	WSN security (Teach μTESLA, Key pre-distribution)	Chapter 11
1 week	Special sensor networks (especially underwater WSN)	Chapters 12 through 14
0.5 week	Energy models; WSN simulators	Chapters 15 and 16
1 week	Case studies	Chapters 17 and 18
Total: 10 weeks	In each chapter, teach both math principles and concrete design cases. Leave some topics for after-class reading assignments.	

For computer engineering majors, Chapter 2 (sensor hardware) is important. This chapter may be allocated more time as it would require a detailed study. For computer science majors, Chapters 6 through 8 (OS, data management) should be covered in more details.

Some chapters, such as Chapters 8 through 10 (localization, synchronization, and security), may be assigned to PhD/MS students as term paper topics (i.e., the students are required to explore this topic in more detail and submit a research paper based on their investigations). Chapters 17 and 18 could be used as projects for senior students.

While teaching, the use of survey-like PowerPoint® slides is not recommended in class as this book covers WSN topics in detail. Instructors should select good design examples to elaborate certain concepts. For instance, when studying MAC layers, at least one of the MAC schemes (such as S-MAC) should be taught in detail.

Math principles are extremely important to WSN design. Therefore, if any chapter has some good math models, they should be studied carefully. These math principles should especially be emphasized for PhD/MS students.

MATLAB® is a registered trademark of The MathWorks, Inc. For product information, please contact:

The MathWorks, Inc.
3 Apple Hill Drive
Natick, MA 01760-2098 USA
Tel: 508 647 7000
Fax: 508-647-7001
E-mail: info@mathworks.com
Web: www.mathworks.com

Acknowledgments

Dr. Xiaojun Cao has contributed about one fifth of the contents of this book, and the remaining text has been written by Dr. Fei Hu. We wish to express our gratitude to all those who have helped us during the preparation of this book. Students in the Department of Electrical and Computer Engineering at the University of Alabama, Tuscaloosa, have helped to edit some figures and math equations. They have also helped to edit some of the content. We are grateful to Rahul Mallampati for helping us revise the content, edit some figures, and format the text in Microsoft Word®. Thanks also to Barnali Chakrabarty and the staff at Auerbach Publications who have provided us continuous support during the writing effort.

A final note: Much of the content and concepts of the textbook are based on existing research efforts from the literature that we cannot list here specifically. We would like to particularly express our appreciation to those authors who have published excellent materials on WSNs.

Disclaimer

As the purpose of this textbook is to explain the latest concepts on wireless sensor network (WSN) design in a textbook format to train students and engineers, we would like to make it clear that this book is not meant for publishing innovative research ideas. Although we have tried our best to provide due credit to all publications cited in this book, there could still be some errors. We would like to sincerely thank all the authors who have published WSN materials that have been cited by us. Any errors or questions on the content of this book can be addressed to us (Fei Hu: fei.hu@ieee.org or Matt Cao: cao@cs.gsu.edu); we will correct the errors and thus improve this textbook in future editions.

Authors

Fei Hu is currently an associate professor in the Department of Electrical and Computer Engineering at the University of Alabama (main campus), Tuscaloosa. His research interests include sensor networks, wireless networks, network security, and their applications in biomedicine. His research has been supported by the U.S. National Science Foundation, Cisco, Sprint, and other sources. He received his PhD in 1999 in signal processing from Tongji University, Shanghai, China, and in 2002 in electrical and computer engineering from Clarkson University, New York. He received his BS and MS in telecommunication engineering from Shanghai Tiedao University, China, in 1993 and 1996, respectively.

Xiaojun Cao is currently an assistant professor in the Department of Computer Science at Georgia State University, Atlanta. He received his BS from Tsinghua University, Beijing, China, and his MS from the Chinese Academy of Sciences, Beijing, China. In June 2004, he received his PhD in computer science and engineering from the State University of New York at Buffalo. Dr. Cao's research has been sponsored by the U.S. National Science Foundation, IBM, and Cisco's University Research Program. He is a recipient of the NSF CAREER Award, 2006–2011. His research interests include modeling, analysis, and protocol/algorithm design of communication networks. Important among these are optical networking, waveband switching, optical burst switching, mobile ad hoc networks, sensor networks and security, and optical wireless communications.

BASICS

I

Chapter 1

Introduction

1.1 Basics

One may have seen "sensors" on many occasions. This book targets tiny sensors that have RF (radio frequency) communication capabilities. These sensors could form a wireless network, called the *wireless sensor network* (*WSN*). Then a natural question arises: "Why do WSN technologies advance so quickly?"

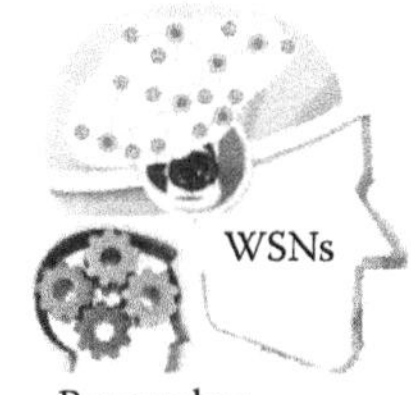

WSNs become a reality because of the integration of three technologies: (1) microelectromechanical systems (MEMS), which could make sensors' mechanical parts fit into a very tiny chip (even less than a quarter!); (2) digital electronics, which make a tiny chip (with a microcontroller) powerful enough to handle the incoming sensor data (such as data compression, data fusion, and networking operations); and (3) wireless (RF) communications, which relay the sensor data among many sensors.

As shown in Figure 1.1, a WSN sensor typically includes *an analog sensing chip* to sense environmental parameters (such as temperature and light), a *microcontroller* to execute local data processing (such as video compression) and networking operations (such as performing a routing protocol with a neighbor sensor), and a *radio transceiver* to send/receive sensed data through a wireless medium. The entire sensor can be powered by batteries or other power sources (such as solar energy) with a lifetime of several months to a few years.

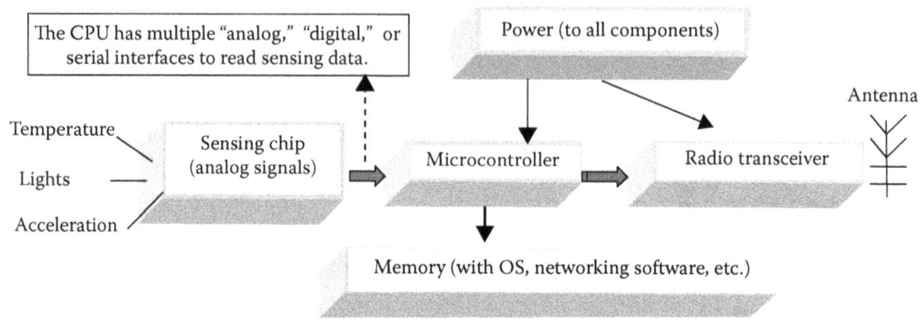

Figure 1.1 WSN sensor hardware components.

In Chapter 2, we will discuss each WSN sensor component in detail. In this chapter, the following few important points are covered:

1. Figure 1.1 lists only the most important components of a WSN sensor. There could be other circuit parts depending on practical application requirements. For instance, we may plant a GPS receiver in a sensor to keep track of accurate positions. A solar panel that avoids the use of AA batteries could be used to absorb solar energy.
2. The following facts explain that not all devices that can sense environmental parameters can be termed as a "WSN sensor":

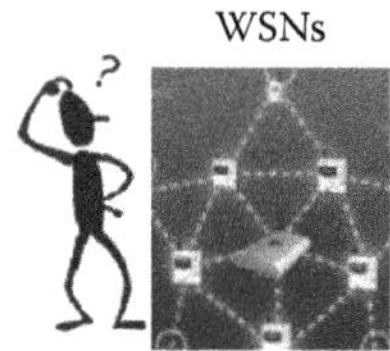

Analog sensor, digital sensor, and WSN sensor: An *analog sensor* detects environmental parameters and, accordingly, changes its voltage level or other signals. Its output is a continuous, weak, and noisy analog signal. A *digital sensor* has an internal ADC (analog-to-digital converter) and a low-capacity CPU (also called a microcontroller). It can interface with a computer to display the sensed data. A *WSN sensor* adds the RF communication capability to a digital sensor. Its CPU runs wireless network protocols, such as hop-to-hop routing protocols. Moreover, the design of a WSN sensor emphasizes the tiny size, low cost, and low energy consumption.

To build a practical WSN application, a WSN sensor should have the following features: tiny size, low cost, and low energy consumption.

1. *Tiny size*: A WSN sensor should be portable to achieve a large-scale, convenient deployment. For instance, we may ask each patient in a nursing home to carry a few medical sensors to achieve anytime, anyplace monitoring. If medical sensors are large (say, larger than a cell phone), it is not convenient for a patient to carry

them. As another example, if we want to achieve environmental surveillance in a large city, many tiny sensors could be dropped from a plane. If the sensors are large, they may not be easily deployed. Moreover, it is better for the sensors to look more "hidden" to achieve a safe, "clean" sensing in the environment.

2. *Low cost*: A WSN should be able to operate well even though there are numerous sensors (greater than thousands) in the network. Therefore, each sensor should have low cost for popular applications. In the future, the unit price of sensors could be less than $1 each [Akyildiz02].
3. *Low energy consumption*: Because each sensor is designed to be disposable, we need not replace every sensor's batteries, especially in a large-scale network. If we wish a WSN to operate for a long time, it should have low energy consumption.

In the future discussion, unless we specify the type of sensor (analog or digital), when we use "sensors," we mean "WSN sensors." A "WSN sensor" is often called a "mote."

WSNs have a wide range of applications in health, military, homeland security, and other domains. For example, the physiological data about a patient can be monitored remotely by a doctor through a medical sensor network. While this arrangement is very convenient to the patient, remote doctors can monitor the patient's condition 24/7. WSNs can also be used to detect chemical pollutions, such as *E. coli* inside drinking water. A well-designed WSN can quickly find out pollutants' names and locations [John06].

Mobile ad hoc networks (MANETs) [CPERKINS00] have attracted much attention. A typical example is the wireless network among some laptops carried by people who are on the move. Because of the nodes' mobility, the design objective of a MANET is to make the routing protocols adapt to quickly changing network topology.

So, what are the differences between WSNs and MANETs?

WSNs

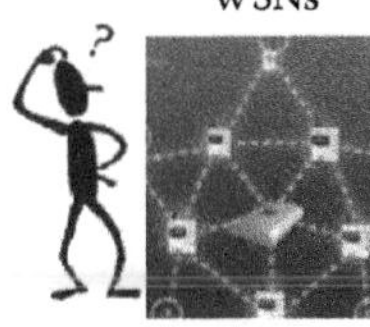

Difference

WSNs and MANETs [Akyildiz02]:

- The number of sensors in a WSN can be several orders of magnitude higher than the nodes in a MANET. Therefore, sensors are more densely deployed.
- Due to the low-cost design objective, sensors are prone to failures. But the MANET nodes (such as laptops) could be designed to have strong calculation capability.
- Most typical WSN applications do not require mobility, that is, the sensors are stationary. But a MANET node has high mobility.
- Sensors are limited in power (typically, battery driven), computational capacity (their CPUs have slow operation frequency), and memory (typically <100 kB).

Similar to a MANET, in a WSN, hop-to-hop communications are necessary because of the limited RF communication distance between each sensor. For instance, most of the current WSN sensors can only transmit data to a distance less than 300 ft. It is not possible to ask a remote sensor to directly (i.e., using a single hop) communicate with a server.

Besides the limited wireless signal-broadcasting distance, from the energy consumption viewpoint, a multi-hop approach is better than a single-hop one. This is because the signal energy level fades away quickly as the distance increases:

$$\text{RSS} \propto \frac{1}{d^{\alpha}} \tag{1.1}$$

where

RSS is the *received signal strength* in a receiver
d is the radio signal propagation distance between a sender and a receiver
α is the *path loss ratio*, whose value is typically between 2 and 5

The larger the path loss ratio, the smaller the RSS. The path loss ratio, α, varies in different radio propagation terrains and weather conditions.

If $\alpha = 2$ and if we increase the distance, d, by 10 times, the RSS will be 100 times weaker than the original value. Therefore, we can assume the following:

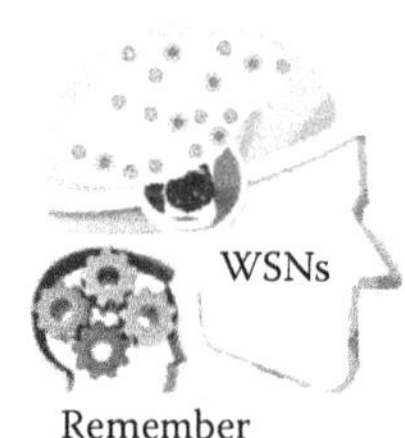

In WSNs, multi-hop, relay-based data communications can generally save more energy (and make the received signal stronger) than direct sender–receiver (1-hop) communications. Also remember: In WSNs, one of the top concerns is energy consumption. That is why so many WSN communication mechanisms are proposed with energy efficiency as the main target.

In WSNs, different sensors that can measure mechanical, thermal, biological, chemical, optical, and magnetic parameters may be attached to WSN nodes to measure the parameters of the environment. In some cases, actuators may be used in WSNs to perform some responses based on the sensors' inputs. However, if the sensors communicate through other strong hardware components, such as actuators, the common WSN design concerns (such as low power, low cost, and short communication range) may not exist anymore. This book focuses on common WSNs with these resource constraints [Jennifer08].

Remember: WSNs have very serious *design and resource constraints* [Akyildiz02, CPERKINS00]. Based on the definition in [Jennifer08], *resource constraints* mean that each sensor has very limited power supply, short wireless communication range, low network bandwidth, and limited CPU processing and memory storage; *design constraints* emerge from environmental conditions and application requirements. For instance, an indoor environment has lots of obstacles, which means the wireless communication quality for it could be lower than that for outdoor applications.

Case study

Can we call a *multi-camera network* (using wireless signals) a WSN, as each camera uses light sensor(s) to capture picture pixels? If a camera does not have serious resource constraints (for instance, its memory could be over 1 GB and its CPU could operate with more than 16 bits of bus width), we normally call such a network an ad hoc network or a general wireless network (instead of calling it a WSN). However, if each camera has serious resource constraints (for instance, it has an 8 bit CPU, <100 kB storage, and limited radio communication distance), we could call it a WSN. Recently, the concept of video sensor network (VSN), which consists of many low-cost video sensors, has been proposed. The VSN is a special type of WSN. Remember: The design of a WSN is so challenging because of such serious resource constraints. If these constraints were absent, we could easily borrow the network design ideas from traditional wireless networks.

Case study

Can we call a *multi-robot system* a WSN, as each robot could have one or multiple sensors? Normally, we do not call a multi-robot network a WSN due to the following reasons: Although each robot has tiny, low-memory, slow-CPU sensor(s), it also has other circuit components that may achieve powerful CPU calculations or long-distance RF communications. Therefore, we cannot say that all wireless networking functionalities are achieved only by the tiny sensors. In this case, we would call it a MANET.

Case study

Can we call a *multi-vehicle network* a WSN, as each vehicle has hundreds of tiny sensors? If we limit our research to the wireless networking achieved by the tiny, RF-capable sensors in different vehicles, we may call such a network a WSN. Although vehicles normally use strong RF antennas to maintain their communications, the main challenge is the highly dynamic network topology due to vehicles' mobility. Therefore, we call such an assembly a vehicle ad hoc network (VANET) instead of a WSN.

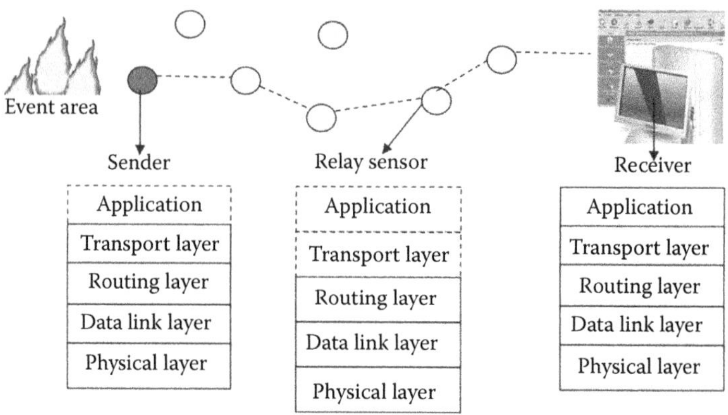

Figure 1.2 WSN network protocol stack.

After we understand general WSN concepts, a natural question that arises is "how challenging is it to design networking protocols among tiny sensors with serious resource constraints?" To answer this question, first, we simply review the network protocol concepts. Later on, we will explain the design challenges in each protocol layer. For more details on protocols, please refer to other course materials such as Computer Networks, Wireless Networks, Digital/Data Communications, and so on.

As shown in Figure 1.2, assume that a sender (sensor) reports the event data (such as fire) to a remote server (receiver). The sender needs to use multi-hop communications to relay its data through some intermediate sensors. Based on the Open Systems Interconnection (OSI) standard, we have seven layers of network protocols, that is, application layer, session layer, presentation layer, transport layer, routing layer, data link layer, and physical layer. However, in WSNs, typically we do not need the session layer and the presentation layer. But, as shown in Figure 1.2, on the receiver side, we do need the other five layers to achieve successful sensed-data collections:

1. *Application layer*: The receiver needs to display the data on the screen. The application layer defines the sensor data display format and performs sensor database management. If the sensor data has to be displayed on Internet Web pages, the application layer needs to understand Internet application layer protocols, such as HTTP.
2. *Transport layer*: The TCP is a typical transport layer protocol. The main functionalities of the transport layer are to (a) achieve "end-to-end" (E2E) reliable data transmission and (b) reduce network congestion. The TCP achieves E2E reliable transmission through packet retransmission and time-out check. The TCP also reduces network congestion through data-rate control. However, in WSNs, the TCP is not a suitable transport layer protocol due to its high overhead. In Chapter 5, we will discuss transport layer in WSNs in detail.

3. *Routing layer*: It achieves hop-to-hop data forwarding among numerous sensors. It searches the optimal path that has low energy consumption, or low delay, or other good features. Once the optimal path is established, the sensed data can be relayed one-by-one by sensors. The routing layer also maintains the route in case the network conditions change from time to time (for instance, a sensor in the path may drain its batteries).
4. *Data link layer*: While the transport layer is responsible for E2E transmission control, the data link layer only handles neighboring (1-hop away) nodes' communication issues. For instance, a sensor may determine whether or not it should adjust its sending rate, based on its upstream and downstream sensors' buffer setups. Sometimes the data link layer is called the *Medium Access Control* (*MAC*) layer. Actually, MAC is part of data-link-layer tasks because MAC only takes care of the wireless-medium-sharing issues among 1-hop neighbors. MAC ensures that all neighboring sensors do not cause signal transmission conflicts, while a data link layer may handle error detection, data framing, and other tasks.
5. *Physical layer*: It converts meaningful data to wireless signals through encoding/modulation and other wireless communication modules. As this layer only sees "signals" such as voltage levels ("0" or "1"), it cannot understand any higher-layer issues (such as routing, data content, and reliability).

Just like the Internet network, WSNs also need the above five layers of protocols. Note: Typically, the sensors do not run application layer protocols, because it is the server's task to correctly display the sensor data. The relay sensors between a sender and a receiver should not run transport layer protocols, because the transport layer only exists between two ends (i.e., the source and the destination). Figure 1.2 uses the dashed boxes to represent nonexistent layers.

In the following discussion, we provide an overview of the design issues in different WSN layers. Akyildiz et al. [Akyildiz02] present a more comprehensive review on these layers. We also cover other important issues such as sensor localization.

1.2 MAC Layer [John06]

A MAC protocol coordinates signal transmissions over a shared RF channel. When a group of sensors communicate using an RF, the MAC protocol determines the communication schedules and rules, because at any time only one pair of users can use the frequency to send out data to each other. The MAC protocol determines the wireless-channel-occupancy durations and many other things.

The most popularly used channel-sharing solution is the *contention-based* scheme. In such a scheme, a sensor having messages to transmit will first listen to the channel to see whether or not it is idle (i.e., not busy). If the RF channel is idle, it immediately transmits data. If the RF channel is busy, it waits (sometimes uses exponential backoff) and tries again later.

In many wireless network MAC protocols, sensors that do not send or receive a packet in a given time frame go into *sleep* mode to save energy. Some variation schemes exist for such a sleep-based protocol. The main point is that a WSN MAC scheme should be energy efficient and collision free, have low-complexity schedule control and low memory requirements, and be able to adapt to changing RF and networking conditions [John06].

1.3 Routing

WSNs use multi-hop routing paths to forward data. Traditional routing schemes, such as the IP (Internet Protocol), do not perform well in WSNs. For instance, the IP assumes highly reliable wired connections (such as fiber optics or cables), where packet errors are rare; this is not true for WSNs, because wireless links have high bit error rates. Many MANET routing solutions are also not suitable to WSNs, because they are often optimized for highly mobile nodes, and they often assume *symmetric* links between neighbors (i.e., if node A can reliably reach node B, then node B can reach node A). Such an assumption may not be suitable for WSNs where nodes are typically stationery. Therefore, new routing schemes are needed in WSNs [John06].

For WSNs, which are often deployed in an ad hoc (random) fashion, the routing protocol typically begins with neighbor discovery. Sensors send out rounds of HELLO messages (packets) and build local neighbor tables. The tables typically have part of the following information: each neighbor's ID, location, remaining energy, delay via that sensor, and an estimate of link quality [John06].

1.4 Other Communication Issues [John06]

There could be some other communication issues besides the above basic protocols:

1. *Reliability*: Remember that each wireless link has a high packet error rate due to the unreliable nature of wireless communications. Its error rate could be 1/100 (i.e., one out of 100 packets could get damaged due to wireless interference). How do we measure link quality? For this, we could use packet drop

rate, received signal strength, etc. Another reason that causes link unreliability is that WSN links are typically *asymmetric*, that is, even though sensor A can successfully send a packet to sensor B, the reverse link from sensor B to sensor A may not be reliable [John06].

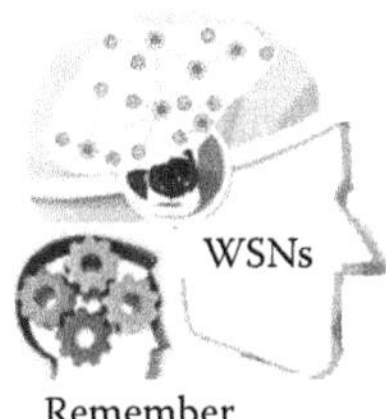

Typically, reliable data transmission is achieved in the transport layer. The sender could invoke a timer after sending a packet. When the acknowledged packet for such a sent packet cannot be received before the timer expires, the sender will retransmit the packet.

On the other hand, in WSNs, E2E retransmission does not work well due to the large amount of relay sensors having unreliable wireless links among them. Therefore, it is better to use retransmissions in each hop instead of waiting for the destination sensor's feedback. In this case, we could say that the reliability is actually achieved in the data link layer instead of the transport layer.

2. *Designing proper wake/sleep schedules*: The best way to save energy is to put a sensor into sleep mode. However, it is challenging to determine the wake/sleep schedule for a group of neighboring sensors, based on the practical data transmission timing conditions.
3. *Unicast, multicast, and anycast semantics*: In some cases, a WSN server routes messages to a geographic destination area with a group of sensors. Should the server talk with one specific sensor in that area or all sensors in that area? There are a few choices. First, to achieve *unicast* communication, we can include a specific destination address (sensor ID) in the sensor's message. Second, we can specify that a message should be sent to a few sensor IDs. This is called *multicast* communications. Third, we can just specify an event area and send the message to any node in that area. This is called *anycast* communication. Sometimes we do not specify any destination, and just simply flood (*broadcast*) a command message to the entire network. Many WSN routing schemes support the above unicast, multicast, anycast, or broadcast communications [John06].
4. *Real time*: In many WSN applications, it is important to deliver the data to a destination within a delay threshold. For instance, if a patient has a heart attack, the electrocardiogram (EKG) data should be transmitted back to a doctor within one second.
5. *Mobility*: In most WSN applications, sensors are stationery. If sensors are mobile, it is very challenging to design protocols that can adapt to a large-scale, mobile network topology.

6. *Broken links*: As WSN sensors have limited RF transmission range, it is possible that there are no forwarding sensors in the path where a message is supposed to travel. Or, the sensors may drain their batteries and not work anymore. The routing protocols should be able to deal with such broken wireless links.
7. *Security*: A malicious user can conduct a wide variety of attacks on WSN protocols. For instance, an attacker may itself become a valid relay node and then intentionally drop the packets. Security is an important area in any wireless network due to the unreliable, broadcast-based radio signal transmissions.
8. *Congestion*: A WSN could have high traffic density in some areas due to frequent events in that area. A good routing protocol should try to detour to escape from such congested areas. How to detect congestion areas and how to avoid them are two challenging issues.

1.5 Sensor Localization

Node localization: We often need to detect the exact location of a sensor in a WSN. If an event is detected, we need to know the exact sensor location. Many issues need to be considered in node localization. For example, how do we efficiently use beacons (nodes that know their locations) to find other nodes' locations? If using beacon nodes, how do we determine their communication ranges? Depending on different localization accuracy requirements (for instance, <5 or <1 m), we need different algorithms. Is the system indoors or outdoors? Is it a two-dimensional (2D) or a three-dimensional (3D) localization problem? What is the algorithm's communication overhead (or how many command messages does it use in unit time)? How long should it take to localize a sensor? And many other issues should be considered [John06].

In outdoor applications, we may equip each node with a GPS. This solution seems simple. However, the sensor's cost will rise, and thus such a scheme becomes unacceptable in most WSN applications. Localization schemes can be classified as either *range based* or *range free*. In *range-based* schemes, we first determine the range, that is, the distance between nodes. Then we can use geometric principles to calculate the exact locations. An example of this is as follows: We can use some special hardware or circuit to detect the time difference between arrivals of sound and radio waves. Then, such a difference can be converted into a distance measurement. In *range-free* schemes, we do not determine distances directly. Instead, the number of hops is used. Once the hop counts are determined, we can estimate the distances between nodes using an average distance per hop. Obviously, *range-free* solutions may not be as accurate as *range-based* solutions. But they do not require extra hardware in the sensors [John06].

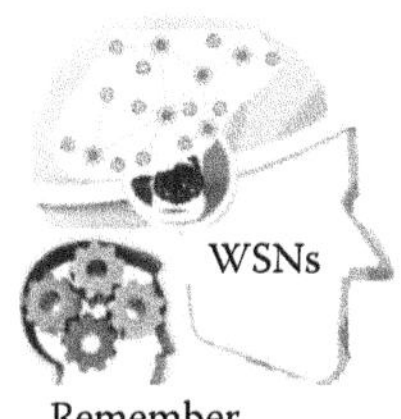

When proposing a new WSN protocol, always keep in mind the low-cost requirements of sensors/systems. For example, adding a GPS to each sensor could solve many problems easily. However, a GPS requires expensive satellite communication systems to receive timing/position information. Currently, many commercialized sensors still cost more than $100 per unit. However, our long-term goal is to make each sensor cheaper than $1. Thus, large-scale deployment becomes feasible.

1.6 Clock Synchronization [John06]

The clocks of each node in a WSN should have the same time control scheme. In many cases, we need to know at what time the event occurred. We also need to use accurate time to achieve some network tasks. For example, when sensors prepare sleep/wake-up schedules, they need to know what time to go to sleep and wake up. In some localization algorithms, we require to measure time difference.

Because the internal clock control hardware/software in a tiny sensor could have a clock drift from time to time, it is necessary to synchronize its clock readings periodically.

Traditional Internet uses the NTP (Network Time Protocol) [DLM91] to synchronize clocks in different Internet hosts. But NTP is too complex (thus, needs extensive memory and calculation overhead) for WSNs, because they require frequent message exchanges. GPS is too expensive. Some good clock synchronization protocols have been proposed, such as RBS [JElson02], TPSN [SGaneriwal03], and FTSP [MMaroti04]. We will discuss this in detail in Chapter 10.

1.7 Power Management [John06]

Today, most commercial WSN nodes (such as Mica2 and MicaZ [Crossbow08]) run on two AA batteries. If we continuously run sensing functions in the sensors without a good power control, these sensors can drain the batteries in a few days. However, most WSN applications prefer the power lifetime to be larger than a few months or even one year. Therefore, it is important to perform power management in sensors.

Today, renewable power sources are under active research. This means that we could use solar cells in sensors or just utilize energy from a sensor's movement or wind. For example, underwater sensors can store energy from water current. The efficiencies of batteries and low-power circuits are improving each year. Many sensor products can set up multiple power-saving states (off, idle, on) for each component of the sensor (such as the analog sensor chip, the RF transceiver, and the microcontroller). These components will be active only when they have tasks.

Other ways to save power include the following few aspects: We could put sensors in a complete sleep state, as listening to messages also consumes energy; and carefully design a good wake-up/sleep schedule, so that a sensor wakes up only when it needs to help relay data.

To save power, all WSN protocols should be designed with minimum control message exchange, as each wireless transmission consumes energy. Some protocols have complex algorithms that should be avoided, as CPU calculations also consume much energy. It is not surprising to see so many WSN protocols being proposed with "energy-efficient" features.

1.8 Special WSNs

There are many types of WSNs. For instance, if the sensors have video-capturing capability, the WSN is called a *VSN*. In the following two sections, we highlight two special WSNs: multimedia WSNs and underwater WSNs.

1.8.1 Wireless Multimedia Sensor Networks [Akyildiz07, Purushottam07]

A *wireless multimedia sensor network* (WMSN) is a special type of WSN technology. Such an application poses many challenges to the traditional WSN design. As the term suggests, it collects multimedia (or video/audio) data from its sensors. Multimedia data needs quite a large storage compared to the traditional data (such as floating-point value) collected by WSNs, and thus the demand for bandwidth is increased; and additional processing power is required for such networks. Despite the additional resources required for such networks, the applications are of great interest, and many are used in military and civil services.

WMSNs have many new applications, such as

- *Storage of potentially relevant activities.* For instance, a WMSN may be used to capture a criminal act in progress, such as a robbery.
- *Traffic avoidance, enforcement, and control systems.* For instance, a traffic light camera may watch for a license plate of a getaway car and report to a nearby police office of its recent location.
- *Advanced healthcare delivery.* Medical sensor networks [HU03] can be used to enable healthcare services that can receive a distress alert and locate the

distressed patients. A patient carries sensors to allow remote doctors to monitor his/her various bodily factors, including temperature, blood pressure, glucose levels, ECG, and breathing. Additionally, remote medical staff can monitor their patients via video and audio sensors, location sensors, motion or activity sensors, all of which can be embedded in wrist devices [HU03].

- *Automated assistance for the elderly.* WMSNs can be used to monitor, record, and study the behavior of the elderly so as to identify the causes of their ailments. Networks of worn video and audio sensors can detect any critical health conditions of the elderly patients.
- *Environmental monitoring.* Acoustic and video sensors can be used to perform habitat monitoring, in which information has to be conveyed in a time-critical fashion. For example, oceanographers can use video sensors to capture the process of sandbars evolving, via image-processing techniques [HOLMAN03].
- *Person locator services.* Multimedia such as live video streams and images, along with advanced multimedia signal-processing schemes, can be used to determine the location of a missing person or identify suspects and criminals.
- *Industrial process control.* Multimedia information, such as image, temperature, pressure, as well as other parameters, could be used to direct a time-critical industrial process. A VSN can be used to monitor the manufacturing process in semiconductor chips, automobiles, food products, or pharmaceutical products. Another advantage is that we may use video sensors to quickly detect manufacturing mistakes. Also, machine vision systems are able to determine the location and orientation of parts of the product to be grasped by a robotic arm. Integrating machine vision systems with WMSNs allows for the simplification of visual inspection systems and adds additional flexibility for those that require continuous, high-velocity, and high-resolution operations.

WMSNs require a design approach that accounts for several important factors, including bandwidth demand, power consumption, application-specific QoS (quality-of-service) requirements, ability to support heterogeneous applications, multimedia coverage, multimedia in-network processing, and integration into other network technologies. The ever-increasing demand for higher resolution and higher quality in the data collected by the sensor network translates into an ever-increasing demand for bandwidth.

There are several approaches to the design of a WMSN [Purushottam07]. One is to use a flat, homogeneous single-tier design with central storage. This design allows the network to expand easily, simply with the addition of another sensor. The disadvantages of such a design include single-point (central storage) failure, poor scalability (due to single-tier, centralized architecture), limited processing capability, and the limited scope of a single sensor, which disallows on-demand network utilization. For example, a surveillance network using this design would not be able to utilize multiple cameras to recognize an object and wake up on demand to perform object recognition. Another design approach is to use a *multi-tier network*, in which the higher tiers include some form of central processing. This design has better scalability and can meet different cost/performance trade-off requirements. For instance, the higher-capacity cameras may perform less frequent, but more powerful image-processing operations. Some researchers propose the single-tier clustered design approach, where each cluster contains a variety of sensors. This approach provides slightly more processing capability and visibility; yet, one cluster cannot use another cluster's collected data.

1.8.2 Underwater Acoustic Sensor Networks [Akyildiz04a]

Although the traditional terrestrial WSNs have many applications within population centers, several factors prevent such networks from being used offshore. Offshore networks (also called underwater WSNs) would require that traditional WSNs have the ability to survive underwater, need low-maintenance, and require a protocol that is tolerant of high transmission delay (due to the use of acoustic signals instead of RF ones under water) and high bit error rates. *Underwater acoustic sensor networks* present many design challenges, primarily due to the medium (water) in which they are placed, sensor corrosion by water, lack of sunlight, the propagation delay of acoustic signals (~1500 m/s) that is 10^5 times longer than the delay of radio transmission (light speed), common loss of connectivity, and high packet loss. Despite these challenges, such networks do perform well for applications such as assisted navigation, disaster prevention (namely, tsunami threats), environmental monitoring, mine reconnaissance, tactical surveillance, and the exploration of the depths of the sea.

Underwater acoustic networks primarily come in three distinct types [Akyildiz04a]:

- Static 2D underwater WSNs for ocean-bottom monitoring. Such networks consist of sensor nodes that anchor to the bottom of the seabed.
- Static 3D underwater WSNs for ocean-column monitoring. These include networks of sensors whose depth into the water can be controlled, and may be utilized for monitoring applications of several oceanic phenomena (pollution, bioactivity, chemical processes, etc.).
- Three-dimensional networks of *autonomous underwater vehicles* (*AUVs*). These networks include fixed portions composed of anchored sensors and additional sensors attached to autonomous vehicles to guide the piloting thereof. (As we discussed before, these types of networks are typically called MANETs, as the vehicles may have strong communication/data-processing capabilities.)

Three-dimensional underwater networks are used as a means of detecting, observing, and capturing underwater phenomena that cannot be effectively observed via ocean-*bottom* sensor nodes. An interesting factor in such networks is that sensor nodes suspend at different depths to observe a certain phenomenon. Of the possible solutions, one would be to attach each underwater sensor to a surface buoy via wires whose length can be adjusted to control the depth of each sensor node. Although such a solution allows for ease of deployment of the sensor network, such buoys can interfere with ships passing by, or, additionally, they are susceptible to being located and disabled by enemies in military applications. Floating buoys are also vulnerable to changes in the weather and random tampering.

With these reasons in mind, another approach would be to anchor sensor devices to the bottom rather than the top. In such an approach, each sensor device is attached to the ocean bottom and consists of a floating buoy that can be inflated by a pump. As a result of pressure, the buoy is able to push the sensor upward to the surface. The sensor's depth can be adjusted by constricting or relaxing the length of the wire that is connected to the sensor and the anchor via an electronically controlled engine that resides on the sensor device. One challenge to such an approach is that the ocean currents can sway the devices. There are various challenges with such an architecture that need to be resolved to enable 3D monitoring, including

- *Sensing coverage.* Sensors should collaboratively regulate their depth to fully cover the ocean column with their sensing ranges in consideration. Thus, the network can be capable of sampling the desired phenomenon at all depths.
- *Communication coverage.* In 3D underwater networks, it may be possible that the sink is not immediately reachable; therefore, sensors should be able to relay data to the surface station by means of multi-hop paths. As a result, network devices must coordinate their depths to ensure that the network topology is always connected, so that at least one path exists between each sensor and the sink.

AUVs can function without the need for tethers, cables, or remote control; thus, they have a plethora of applications in the fields of oceanography, environmental monitoring, as well as underwater resource study. The feasibility of inexpensive AUV submarines equipped with multiple underwater sensors that can reach any depth in the ocean has been demonstrated by prior experiments. Thus, these can be utilized to improve the abilities of underwater sensor networks in many ways. An area of research, the integration and enhancement of fixed underwater sensor networks and AUVs, calls for new network coordination algorithms such as

- *Adaptive sampling.* This includes control techniques to direct the mobile vehicles to locations where their collected data will be of most use. Such an approach is known as adaptive sampling. For instance, the density of sensor nodes can be adjusted in an area where a higher sampling rate is requested for a certain monitored phenomenon.
- *Self-configuration.* This involves control procedures to automatically detect connectivity gaps due to node failures or channel impairment. Moreover, AUVs may either be used to install and maintain the sensor network infrastructure or to install new sensors into the network. AUVs may additionally be used as temporary relay sensor nodes to restore connectivity.

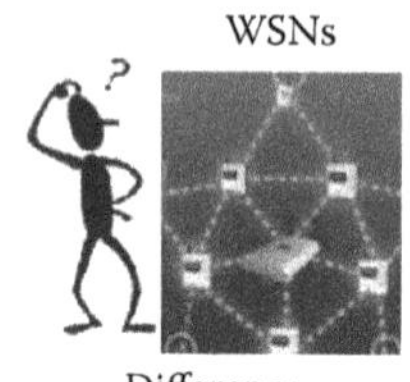

Although underwater sensor networks also use "wireless" media to transmit the data, they are different from terrestrial WSNs where RF signals are used. The typical unlicensed RF spectrum could be 433 MHz or 2.4 GHz. Underwater WSNs use "acoustic signals" as wireless media. Acoustic signals have much lower frequency than RF ones. For instance, it could be only 11 kHz. However, such acoustic signals can propagate for a much longer distance than RF ones in water environments.

1.9 WSN Applications [Hartung06, Chehri06, Manish06]

In this section, we list some typical WSN applications. Many important applications are used for environmental monitoring. Environmental monitoring applications, such as those used for the monitoring of habitats of birds, pollution detection, earthquake monitoring, planetary exploration, flood detection, forest fire detection, and pollution study, are all extremely important to protect our living environments.

Sensor networks can be strategically deployed into a forest to detect the origin of forest fires. As these sensor networks may be unattended for a long time, efficient energy-saving mechanisms and renewable energy technologies may be used. The sensors perform distributed collaboration and overcome obstacles (such as trees and rocks) that block the line-of-sight of the sensors. Researchers from the University of California (UC), Berkeley, have used WSNs in a fire environment (called FireBug) [DOOLIN05]. They could accurately measure important environmental conditions, such as relative humidity and temperature, as a flame front passed during a prescribed burn. Such a sensor network is better than the current fire detection system that typically uses high-tech airborne infrared sensors to track flame fronts and intensities over very large-scale areas [Hartung06].

The Internet can be used by remote users to control, monitor, and observe the bio-complexity of the environment. Satellite and airborne sensors are useful in observing large-sized biodiversity, but they are not fine-grained enough to observe small-sized biodiversity, which makes up most of the biodiversity in an ecosystem. So there is a need for ground-level deployment of sensor nodes to observe the bio-complexity. Figure 1.3 shows an Internet/WSN-integrated application scenario.

Sensors could be easily dropped in rugged terrain and under extremely harsh conditions. Some researchers at Harvard recently deployed a sensor network on an active volcano in South America to monitor seismic activity using vibration sensors. Though they only used a single-hop deployment strategy, they implemented

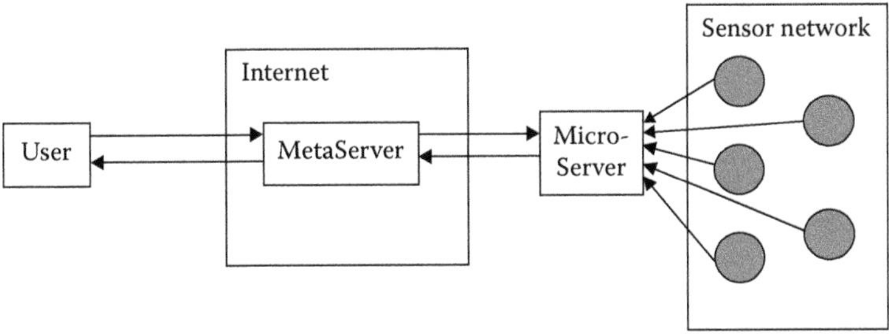

Figure 1.3 Connecting WSN to the Internet.

a fairly tight time synchronization protocol to accurately correlate their data. With this system, they hope to be able to monitor and help predict volcanic eruptions, earthquakes, and other similar volcanic activities [JOHNSON05].

On a smaller scale, a sensor network can be deployed on a single redwood tree using nodes to cover roughly 50 m. With this unique deployment, researchers were able to map the differences in the microclimate over a single tree.

Sensor networks deployed in natural parks and wildlife reserves closely monitor and aggregate data from plant and animal life. Earlier methods of field monitoring were error prone, tedious, and potentially dangerous to plant and animal life. Data gathered from sensor networks can be studied, and useful information such as nesting patterns, flowering seasons, and effects of different microenvironments can be inferred without causing harm to plant or animal life. Researchers at UC, Berkeley, have deployed a WSN at the Great Duck Island off the coast of Maine [Anderson02]. Sensors were placed in burrows to detect the modes of nesting birds, providing statistical data to biological researchers. Additionally, their work provides other researchers with good outcomes on WSN performance, routing, and topology construction. In this application, the researchers at UC, Berkeley, used Mica motes. The sensor uses an Atmega103 microcontroller running at 4 MHz and a 916 MHz radio from RF monolithics to provide bidirectional communication at 40 kbps. Thirty-two motes were placed in the area of interest. These motes transmit sensor data to a gateway, which is responsible for forwarding the data to a remote base station. Figure 1.4 represents simulation scaling for the motes versus the time delay for busy, quiet, and inactive networks [Hartung06].

Researchers from the UC, Berkeley, *Center for Embedded Networked Sensing* deployed a sensor network into the James Reserve Forest in California that could be used for a wide range of purposes, from monitoring the soil temperature to tracking wildlife [CERPA01]. They used multi-hop routing and multiple, heterogeneous nodes.

There are more habitat-monitoring applications, such as the use of a sensor system for monitoring cane toad populations [Bulusu05] and a WSN for tracking the movements of zebras [JUANG02].

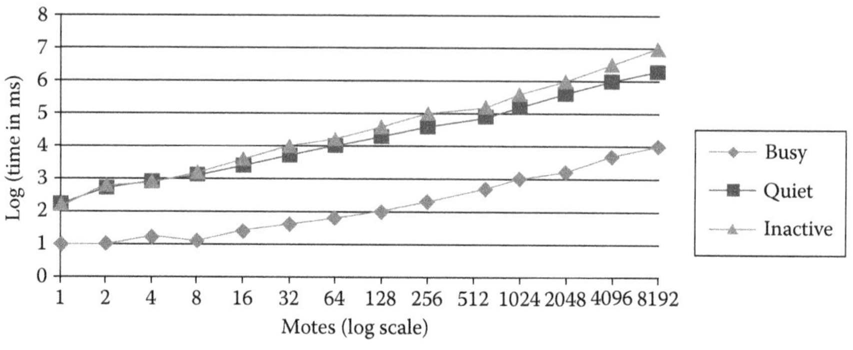

Figure 1.4 WSN delay performance. (Adapted from Anderson, J. et al., Wireless sensor networks for habitat monitoring, *Workshop on Wireless Sensor Networks and Applications (WSNA 2002)*, Atlanta, GA, September 2002.)

WSNs have proved to be very useful in both offensive and defensive military applications. Sensor networks can be used to gather data about the existing state of a military troop. The data gathered may include the amount of equipment at hand, the ammunition and troop strength, and the location of troops. These reports can be gathered and sent to higher officers in the clustering hierarchy of troop leaders, where an appropriate decision can be taken depending on the current state of affairs. Sensors used for battlefield surveillance are randomly deployed in inaccessible regions and critical areas for closely monitoring the presence of opposing forces. Moreover, these networks can also be deployed to discover new approach routes and paths in scenarios without human intervention.

Target tracking is another useful military application. Sensor networks may be used to track the path of enemy troops. The analyzed data can be fed to an intelligent ammunition system. When the target, such as a vehicle, is moving in the sensor field, target-state histories (such as a spatial trajectory) have to be estimated on the basis of sensor measurements. Each sensor node provides a local measurement useful in estimating the target state. Just before or after an attack, sensor networks can be deployed into the target area to assess the extent of the battle damage. Sensor networks can also be deployed to raise an alarm against potential nuclear, biological, or chemical attacks. The networks can be incorporated with the ability to take countermeasures against such attacks as well.

As mentioned before, a WSN can be used for patient diagnostics, drug administration in hospitals, collections of human physiological data, etc. The physiological data collected from sensors can be used for medical exploration. This data can also be stored for a long period of time. The sensor networks detect elderly people's behavior. These small sensor nodes allow the doctors to identify predefined symptoms. Each sensor node has its specific task, for example, one sensor node may detect the heart rate.

Sensor networks used for drug administration in hospitals help to minimize the chances of receiving and prescribing the wrong medication to patients.

WSNs are envisioned to be ubiquitous, integrating themselves to all homes, offices, and household appliances. Such devices can be connected to actuators, which take an action when the environment changes to a particular state. End users can communicate with these devices to make control decisions remotely. In smart homes, the sensors make intelligent decisions, such as what changes to make, and what actuations to perform based on the transforming states of the environment. The lights automatically turn on when a person enters the room at night. The temperature inside an office can vary by a few degrees. If the airflow in a room is not evenly distributed, a distributed sensor network can be used to control the airflow and temperature. Smart sensor nodes and actuators can be placed in appliances such as refrigerators, ovens, and air conditioners, so the end users can manage home via the Internet or satellite. In a warehouse, each item may have a sensor node attached. Thus, by querying the network details of an item, the type, the price, and the serial number can be collected and stored in a backend database. New items are added to the inventory by attaching a new sensor to it.

WSNs can be used for monitoring a nuclear reactor. WSNs can control the chain reaction in nuclear reactors. The sensors monitor the reaction by observing parameters like radiation and temperature. The observer uses data from sensors and maintains the nuclear reactor in a stable state. The sensor node senses information and it sends to a sink that checks abnormal conditions, like drastic changes in radiation or temperature. If an abnormal condition occurs, the sink raises an alarm.

Another sensor network application is suspicious individual detection. Consider a scenario in which a person is making frequent trips to a shop that sells chemicals and then also makes visits to a shop that sells firearms. Detecting such persons and placing them on a suspicious individuals list can help determine these persons' connections and motivations. With appropriate mining of the data collected from sensors placed in various shops, connections, links, and associations, we could detect suspicious groups or individuals.

WSNs also have many applications in *underground mines*. Sensors can measure physical phenomena in a mine, such as temperature, luminosity, and oxygen concentration. Sensors can also detect a possible anomaly, such as fire and toxic gas. With WSNs, the objects (or persons) can be localized by simply labeling them with a small node. This task is important for many applications, such as traffic management in underground mines and tracking or rescue operations [Chehri06].

In summary, WSNs have important roles for a variety of applications, such as offensive and defensive military applications, environmental monitoring, building automation, traffic management, industrial process control, civilian infrastructure protection, and target tracking. WSNs are especially useful in areas difficult to access. Wired network is not suitable in many applications due to two reasons: *cost*—wiring typically accounts for 80 percent of the cost of sensor installations; and *safety*—by using a WSN measurements can be automatically collected in locations where the wiring is very difficult or impossible to install [Legg, Chehri06].

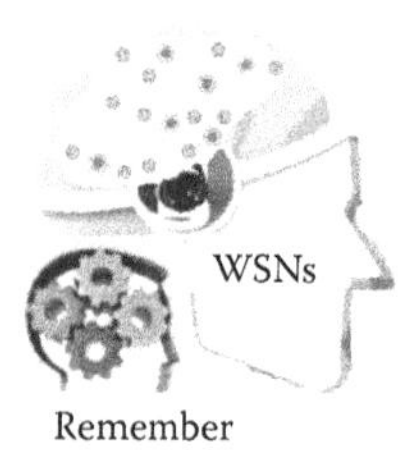

The advantages of WSNs are due not only to their self-organized nature (i.e., after deployment, numerous sensors could automatically form a connected network), but also to their "wireless" communication capabilities under harsh environments. This book does not target "wired" sensor networks, because wireless media make protocol design much more challenging than wired media.

Problems and Exercises

1.1 Multi-choice questions

1. The differences between "analog sensors" and the sensors in WSNs do not include which of the following?
   a. WSN sensors have ADC capabilities.
   b. WSN sensors have CPUs (also called microcontrollers) to do some local data processing.
   c. Traditional analog sensors typically do not need power input.
   d. Traditional analog sensors cannot self-organize themselves into a wireless network.
2. The differences between WSNs and MANETs do not include which of the following?
   a. WSNs typically have a larger scale (more nodes) than MANETs.
   b. MANETs have more mobility behaviors.
   c. MANET nodes typically have more power storage capability than WSN nodes.
   d. WSNs have a higher design/deployment cost than MANETs.
3. Underwater sensor networks are different from terrestrial sensor networks in which of the following features?
   a. Underwater sensors typically do not use RF communications. Acoustic communications are used instead.
   b. Underwater sensors are mobile while terrestrial sensors are fixed.
   c. Underwater sensors are more expensive than terrestrial ones.
   d. Underwater sensors typically use solar power.

4. WMSNs require which of the following?
   a. They need large storage due to video/audio data.
   b. They need strict QoS considerations.
   c. They need a large bandwidth.
   d. All of the above.
5. With respect to WSN localization, which of the following items is true?
   a. WSN localization typically uses a GPS.
   b. WSN localization can easily achieve <0.1 m accuracy.
   c. WSN localization can use the triangle theory to localize a node.
   d. WSN localization does not need clock synchronization.

1.2 Explain the hardware architecture of a WSN sensor node. What type of CPUs can it use? List some examples after doing some Web research.

1.3 What kinds of design and resource constraints does a WSN have?

1.4 Why cannot underwater sensor networks use RF communications?

1.5 Assume that we use WSNs for vineyard monitoring. Conduct some Web research and draw a feasible WSN system diagram (including sensors, sink, Internet server, etc.) to achieve such an application.

ENGINEERING DESIGN

II

Chapter 2

Hardware—Sensor Mote Architecture and Design

In this chapter, we study the hardware design details of sensor nodes. A WSN sensor node (also called a *mote*) consists of analog sensors, a microcontroller, memory, an RF (radio frequency) communication unit, a battery, and other components. We will use [Jason03] as the main reference, as it provides a pioneering sensor mote design.

This chapter also covers some *physical-layer* concepts in WSNs (such as modulation and wireless signal transmissions). The next few chapters will cover the details of higher layers (such as MAC layer, routing layer, and transport layer).

In this chapter, we first discuss each component of the sensor mote. Later on, we will integrate everything together into an intelligent sensor mote.

2.1 Components of a Sensor Mote [Jason03]

In the following, we explain the hardware components of a sensor mote. Each of the components should be designed from both operation performance and energy efficiency viewpoints.

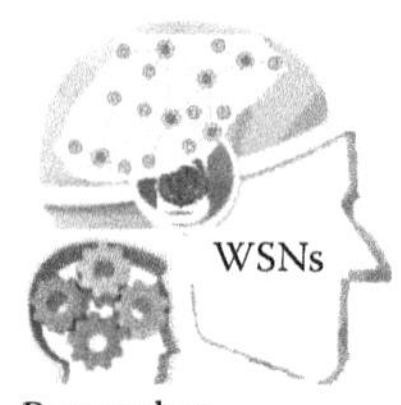

A mote (i.e., a WSN sensor node) is a typical embedded system from the computer engineering design viewpoint. As we know, any embedded system needs a microprocessor (also called a CPU or a microcontroller) to control all other chips. On the other hand, a mote needs to achieve wireless networking with other motes. Thus, its CPU needs to interface with an RF transceiver (i.e., a radio chip). How do we interface its CPU with a radio chip in a fast, low-energy way is a challenging issue.

2.1.1 Sensors

Thousands of different analog/digital sensors have been invented, and are ready to be attached to a wireless sensing platform to form a WSN node (also called a "mote"). Recent advances in MEMS and carbon nanotube technologies have enabled many different types of sensors. Some examples are chemical sensors and digital nose sensor. Table 2.1 lists some common microsensors and their main features [Jason03].

Analog and *digital* sensors have the following different characteristics:

1. *Analog sensors* generate raw analog voltage values based on the physical phenomena that they are measuring. They produce a *continual* waveform, which needs to be digitized (i.e., forming digital signals such as 0101001…) by special chips (such as an ADC, i.e., an analog-to-digital converter). These digital signals can then be easily processed by CPU and DSP (digital-signal-processing) chips.

   After receiving the raw analog data, a CPU must process this data to produce a reading in meaningful units. For example, when an accelerometer generates a raw reading of 0.815 V, it must be translated into a meaningful (i.e., human-understandable) acceleration measurement. Does 0.815 V correspond to an acceleration of 0.5 or 1.1 m/s? Such an analog data translation procedure could be a complicated process because of sensors' different timing and voltage scales.

   Because the output voltage generally has a DC offset among a time-varying signal, we typically use amplifiers and filters to match the output of the sensor to the range and fidelity of the ADC.
2. *Digital sensors* actually integrate all of the above-mentioned voltage-processing hardware into a sensor to directly provide a clean digital interface. Because they have implemented all required compensation and linearization internally, their output is already a digital reading with an appropriate scale.

Table 2.1 Power Consumption and Capabilities of Commonly Available Sensors

Discrete Sample Voltage				
Sensor Type	*Current*	*Time*	*Requirement (V)*	*Manufacturer*
Photo	1.9 mA	330 uS	2.7–5.5	Taos
Temperature	1 mA	400 mS	2.5–5.5	Dallas Semiconductor
Humidity	550 uA	300 mS	2.4–5.5	Sensirion
Pressure	1 mA	35 mS	2.2–3.6	Intersema
Magnetic field	4 mA	30 uS	Any	Honeywell
Acceleration	2 mA	10 mS	2.5–3.3	Analog Devices
Acoustic	0.5 mA	1 mS	2–10	Panasonic
Smoke	5 uA	—	6–12	Motorola
Passive IR (motion)	0 mA	1 mS	Any	Melixis
Photosynthetic light	0 mA	1 mS	Any	Li-Cor
Soil moisture	2 mA	10 mS	2–5	Ech2o

Source: Adapted from Hill, J.L., System architecture for wireless sensor networks, PhD dissertation, Department of Computer Science, University of California at Berkeley, Berkeley, CA, Spring 2003.

If you purchase a commercial microcontroller (a CPU) to interface with the above sensors, it typically has multiple interfaces with either analog or digital sensors.

Because sensors have limited power output, and the WSN sensors are typically designed to be disposable, we need to carefully control how quickly a sensor can be enabled, sampled, and disabled, because these operations have a huge impact on energy consumption. For instance, although most sensors have the capability of producing thousands of samples per second, in practice, we are interested only in a few samples per minute. Such a *low duty cycle* (percentage of active time) can greatly save energy.

Although it is important to minimize the active time of a sensor (i.e., putting the sensor to sleep for as long as possible), it is also important to minimize the "transition" time, that is, the sensor should be turned on/off *as quickly as it can be*, to save energy. For example, if a sensor takes 100 ms to turn on and read a sample, assume that the sample reading consumes just 1 mA at 3 V; it will cost 300 μJ in total to get a sample. This is the same amount of energy as a sensor that consumes 1000 mA of current at 3 V but takes only 100 us (i.e., 1000 times faster) to turn on and read a sample [Jason 03].

In some applications, the voltage requirements may not match well with the battery outputs. Hence, an extra circuit may be needed. For instance, some sensors require ±6 V. If a sensor just uses AA or lithium batteries, we need special *voltage converters* and *regulators* to use this sensor. The power consumption and turn-on times of converters and regulators' circuitry must be included in the total energy budget for the sensor.

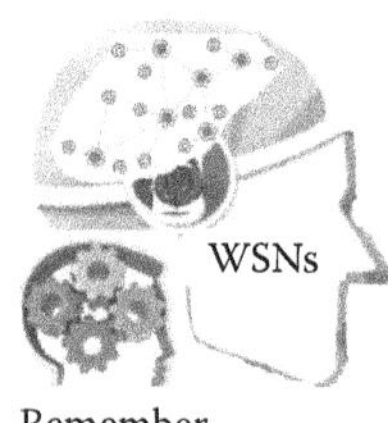

Today, almost all analog sensors convert environmental parameters into a readable low-voltage level. Interpreting these voltage levels from the event detection perspective is a difficult issue. Moreover, we need to capture such a weak current and use an ADC to receive digital signals. During the analog-to-digital conversion, the noise from hardware and environments should be eliminated.

2.1.2 Microprocessor

Another important component, called a *microcontroller* (i.e., a *tiny CPU*, also called a *microprocessor* or a *processor*), has pins (i.e., interfaces) to integrate flash storage, RAM, ADCs, and digital I/O (input/output) onto a single integrated circuit. Such tight integration makes the microcontroller ideal for use in deeply embedded systems, like WSNs.

When we select a commercial microcontroller family for a WSN application, we need to consider some of the application requirements, including power consumption, voltage requirements, cost, support for peripherals, and the number of external components required. Some of these are explained as follows:

1. *Power consumption*: Different microcontrollers have very different power consumption levels. For instance, 8 or 16 bit microcontrollers have varied power consumption between 0.25 and 2.5 mA/MHz. Such a wide difference (over ten times) between low-power and standard microcontrollers determines the WSN system performance significantly.

   Many people think that sleep can put a sensor in a completely "relaxed" state, and thus power consumption is minor in the sleep state. This is not true in reality. In the sleep mode, the CPU stops execution. However, it still maintains some basic memory control activities and time synchronization, in case, later on, it needs to timely wake up. The electric current consumption in the sleep mode varies from 1 to 50 μA across CPU families. As the CPU is expected to be idle 99.9 percent of the time, such a 50× μA difference can have a more significant impact on mote performance compared to the milliampere difference in peak power consumption.

As mentioned before, the energy consumption also depends on how much time the operation of entering into/exiting from the sleep mode takes. Such a transition time (entering sleep/wake-up time) could take 6 μs to 10 ms. The wake-up delay is used to start and stabilize system clocks. The faster a CPU can enter or leave the sleep mode, the more energy a mote can save. As a matter of fact, by a quick wake-up, we can put a mote into the sleep mode even in a very short period of inactivity. For example, when sending out a packet, the controller can even enter the sleep mode among "bits." Thus, we save a lot of energy.

2. *Voltage requirements*: CPU performance also depends on the operating voltage range. Traditional WSN microcontrollers operate between 2.7 and 3.3 V. New generations of low-power CPUs can even operate at 1.8 V. WSN applications need a wide voltage tolerance.
3. *CPU speed*: In a WSN, the CPU needs to execute the wireless communication protocols and perform local data processing. These operations do not need a high-speed CPU. That is why most of today's WSN CPUs have a speed of <4 MHz. To select a proper CPU speed, we need to know the amount of sensor data to be processed. The CPU must be able to finish the operations within delay deadlines.
4. *Dynamic CPU speed*: Some WSN CPUs can dynamically change the operating frequency (i.e., CPU speed). CMOS power consumption obeys the equation $P = CV^2F$. Therefore, higher CPU frequency brings more power consumption. But the CPU execution time is inversely proportional to frequency, that is, higher frequency makes a program run faster, which also saves energy. Therefore, we cannot say that the sensor energy consumption will change drastically by increasing or reducing CPU frequency.

Table 2.2 lists some important features to be considered when selecting a CPU, such as power, memory size, reprogrammability, A/D channels, and operating supply. It compares some suitable CPUs used in different motes in the market. Typically, Atmel AT90LS8535 offers a good performance in most WSN applications.

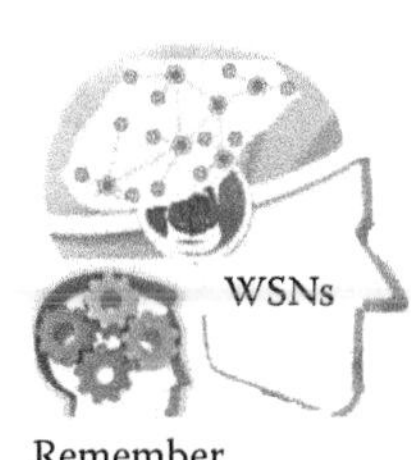

Remember

Note that the table does not intend to list all advanced CPUs used in different embedded systems. Instead, it only lists some popular microcontrollers that may be suitable to small, low-power, low-cost motes. In some products, the microcontrollers are integrated with different memories (such as flash and ROM).

- A CPU design example: SNAP/LE [Virantha04]

In [Virantha04], the author presents the design of a low-power microcontroller called *SNAP/LE* (*sensor network asynchronous processor/low energy*), optimized for data-monitoring operations in WSNs.

Table 2.2 Comparison of Microprocessors

	Atmel AVR AT90LS8535	*Microchip PIC16F877 (Preliminary)*	*MC68H(R) C 908JL3*	*Amtel AT91M404000 16/32 Bit Strong Thumb*
Flash memory	4K	8Kx14	4K	External memory
Endurance	1K	1K	10K	N/A
MIPS/mA	1.25 (minute)	1.66 (preliminary)	0.1 (typical)	0.6MIPS/mA (1.35mA static current)
A/D channels	8 (10 bit)	8 (10 bit)	12 (8 bit)	0
In-application programming (IAP)	No	Yes	Yes	Yes
Operating voltage	2.7–5.5V	2.0–5.5V	2.7–3.6V	2.7–3.6V
I/O pins	35	40	23	100

Source: Adapted from Hollar, S.E.-A., COTS dust, MS thesis, Mechanical Engineering, University of California at Berkeley, Berkeley, CA, Fall 2000.

SNAP/LE does not just simply select a conventional microprocessor for low-energy optimization. Instead, it is a self-designed brand-new microprocessor with new hardware support for commonly occurring operations in WSNs. It aims to maximize the lifetime of a network. SNAP/LE is event driven with extremely low-overhead transitions between active and idle periods.

A dominant feature of SNAP/LE is to use automatic, fine-grained power management, which can be seen from the following fact: When a circuit does not perform a particular operation, it will not have any circuit-switching activities. Such asynchronous circuits also remove glitches/switching hazards in the CPU, which avoids another cause of energy waste.

Another interesting feature of SNAP/LE is that its hardware directly supports sensor event execution, which means that we do not need an operating system (OS) such as the TinyOS! No OS reduces static and dynamic instruction counts. It also simplifies CPU design, as we do not need to worry about precise exceptions and virtual memory translation.

Most of the traditional mote CPUs adopt a commercial *off-the-shelf* (*COTS*) microcontroller, such as Berkeley motes' Atmel Mega128L [Atmel08]. SNAP/LE does not use a commercial CPU; instead, it is a processor designed *specifically* for low-energy WSNs. It not only meets the computational demands of a WSN node, but also consumes much less energy than other CPUs.

Good idea

Customized Very Large-Scale Integrated Circuit (VLSI) versus COTS design: It is hard to say which one is the winner. Typically, from time and complexity viewpoints, most researchers choose to use COTS, as so many different companies provide high-performance, low-cost chips to assemble a mote. However, from cost and performance viewpoints, customized VLSI design is the final solution, as you can minimize chip size and achieve the best speed/energy performance. Later on, we will cover Spec [Jason03]. Like SNAP/LE, it is also a customized design.

SNAP/LE aims to design a CPU having all of the following features:

1. *A simple programming model*: A good CPU design should allow easy programming. Its programming model should support the following operation mode: WSN motes sleep most of the time, periodically waking up to handle radio traffic or sensor data. Additionally, the CPU should efficiently execute the most common WSN tasks, such as scheduling internal timers or reading sensor data. SNAP/LE was designed with these features in mind.
2. *Lower-power sleep mode*: As we mentioned before, sensors remain in the sleep state during most of the time. SNAP/LE is designed for extra-low power consumption while it is in the sleep state.
3. *Low-overhead wake-up mechanism*: As a fast transition between sleep and wake-up states is needed to save energy, SNAP/LE aims to achieve around 10 ns of transition time, which is much less than a typical sensor-event-handling time (i.e., a few milliseconds).
4. *Low power consumption while awake*: Besides maintaining a low power consumption during the sleep state, SNAP/LE also minimizes the energy consumption while in an "awake" (computing) state.

SNAP/LE uses a 16 bit data path. Its instructions can be one or two 16 bit words long (two-word instructions take two CPU clock cycles to execute).

Simultaneous execution of several instructions is supported in SNAP/LE. Its potential concurrency can be seen from its microarchitecture in Figure 2.1. The

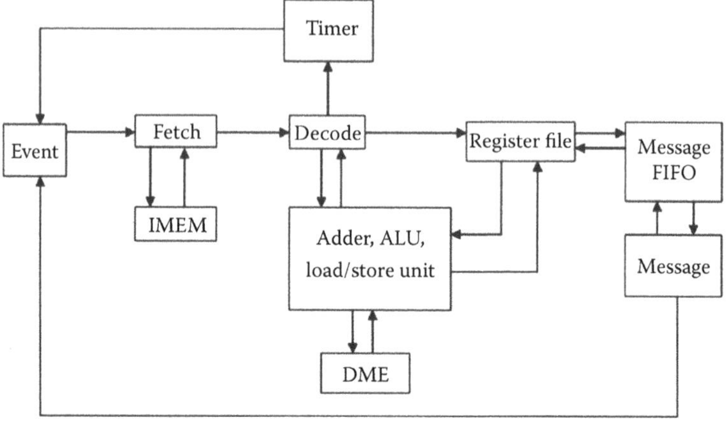

Figure 2.1 Microarchitecture of SNAP/LE showing major units. (Adapted from Ekanayake, V. et al., An ultralow-power processor for sensor networks, *ASPLOS '04*, Boston, MA, October 7–13, 2004.)

event queue stores outstanding events that are yet to be processed. These instruction tokens travel through the pipeline and are transformed by the computation blocks (adders, decoders, etc.).

SNAP/LE uses a *data-driven* switching activity to reduce the total switching capacity of the processor. It thus saves energy. The use of asynchronous (i.e., data-driven) circuits further enables energy saving. (To achieve equivalent energy saving in a clocked processor, the designer would have to clock the gate and every latch in the processor.)

The SNAP/LE CPU core includes an important component, that is, the *event queue*. It works with the *instruction fetch* unit to form a hardware implementation of a First Input, First Output (FIFO) task scheduler. The scheduler first executes the *boot* code. When the scheduler reaches the "done" instruction, which is also the last instruction in the boot code, it will stop fetching instructions and wait for an *event token* to appear at the head of the event queue.

Each *event token* tells what event has occurred. *Event tokens* are inserted into the event queue by two hardware components: (1) the *timer coprocessor* when a timeout finishes and (2) the *message coprocessor* when data arrives from the sensor node's radio or from one of its analog sensors.

SNAP/LE has only one sleep state called the "deep sleep" state. It takes only 10's ns for its CPU to wake up from this sleep state. The "deep sleep" state and the low wake-up latency both help with the energy saving. This feature is not seen in conventional WSN CPUs: Most of them have several "sleep" states. For instance, they may have a "deeper" sleep state that consumes less power, but requires more time to wake up than a "lighter" sleep state. The Atmel microcontroller, for example, has six sleep states.

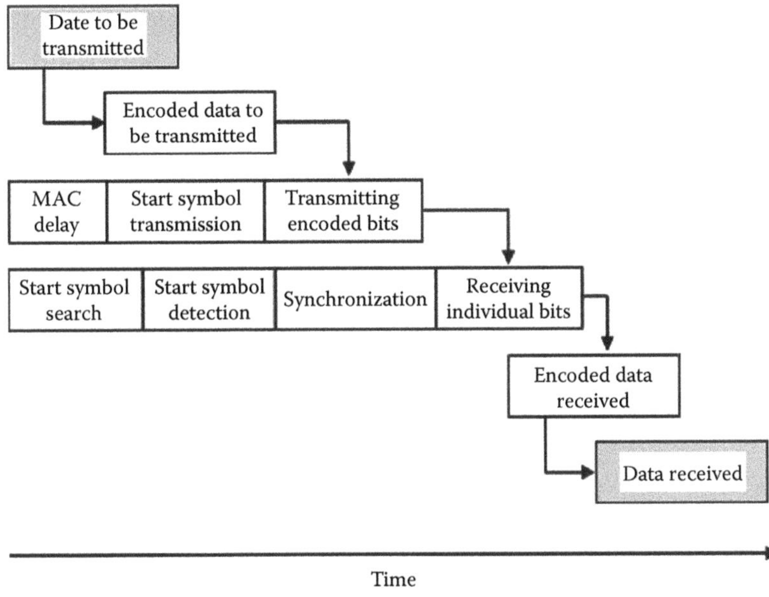

Figure 2.2 Transmission to reception wireless communication phases. (Adapted from Hill, J.L., System architecture for wireless sensor networks, PhD dissertation, Department of Computer Science, University of California at Berkeley, Berkeley, CA, Spring 2003.)

As we can see from Figure 2.2, the SNAP/LE CPU has the following hardware units: an adder, a logic unit, load-store units, a timer unit for interfacing with the timer coprocessor, a jump/branch unit, a linear-feedback shift register (for pseudorandom number generation), and a shifter. The most commonly used units (such as the adder, the logic unit, and the load-store unit) are placed on the fast busses and the rest on the slow busses. All of the function units were designed with minimal pipelining to limit SNAP/LE's power consumption while awake.

2.1.3 Memory

After discussing the CPU, we move to another important mote component—*memory*. Generally, WSN motes only require small amounts of storage and program memory. This is because the sensor data only stays in a local sensor for a short time, and then is transmitted through the network to the WSN base station.

Today, many CPUs have an *on-chip* storage (i.e., a *flash memory*) that is typically <128 K. Such an on-chip storage can be used for both program memory and temporary data storage. WSN CPUs also have a data RAM (typically 32–128 kB) that can be used for program execution.

Let us take a look at the differences between a *flash memory* and an SRAM (*static random access memory*) [Jason03]:

1. From the *storage* viewpoint, the flash technology has a higher density than SRAM. For instance, a flash memory could have a storage density of 150 kB/mm^2 in a 0.25 μm process [AMD03], while Intel's recent SRAM density record is 60 kB/mm^2 using a 90 nm process [Intel02].
2. From the *energy consumption* viewpoint, flash is a persistent storage technology that does not need energy to maintain data. However, SRAM requires more energy to retain data over time (but it does not require as much energy for the initial storage operation).
3. From the *time* viewpoint, a flash write operation requires 4 μs to complete compared to .07 μs required by SRAM—both consuming 15 mA.

Therefore, if we need to store data for long periods of time, it is more efficient to use flash instead of SRAM.

2.1.4 Radios

Now let us discuss another important hardware component in a mote: *radio transceiver*. First, let us recall a few facts on a mote's low-power, short-range transceiver:

1. It consumes around 15–300 mW of power during sending and receiving.
2. It needs approximately the same amount of energy when in the receive or the transmit mode.
3. Unlike what many people think, as long as the radio is *on*, whether or not it is receiving actual data, the energy is consumed.
4. More energy is consumed in *receiving* packets than sending packets. In a sensor, the actual power emitted out of the antenna (when sending data) only accounts for a small fraction of the transceiver's energy consumption. Therefore, the receiver power consumption dominates the overall cost of radio communication. This fact is often ignored in wireless studies.
5. If the receiver is never turned off (i.e., it is always *on*), it will be the component that consumes the largest energy. *Do not think that the reception is free when no data is received.* Therefore, try to put the transceiver into the sleep state (i.e., a complete "off" state) when no data is received.
6. If we use higher transmission power (i.e., put more energy into a radio signal to be sent), we could make the signal propagate for a longer distance. The relationship between power consumption and the distance traveled is a polynomial with an exponent between 3 and 4 (this exponent is called *path loss*, which exists due to radio interference). As an example, if we want to

transmit twice as far through an indoor environment, 8–16 times more energy must be emitted.

7. Although the data transmission distance is mainly determined by the transmitter power, other factors could also have an impact on the radio range, such as the receiving sensitivity of the RF receiver, the antenna gain and efficiency, and the channel-encoding mechanism.
8. In most WSN applications, due to *low-cost* requirements, we cannot exploit high-gain, directional antennas because they require special alignment. Therefore, most times we assume that omnidirectional antennas are used in most WSNs.

In WSNs, we typically use dBm (instead of dB) to measure both *transmission strength* and *receiver sensitivity*. (Note: The dB scale is a logarithmic scale where a 10 dB increase represents a 10× increase in power. The baseline of 0 dBm represents 1 mW, so 1 W is 30 dBm). Typical receiver sensitivities are between -85 and -110 dBm [Jason03].

Radio propagation distance can be increased by either (1) increasing the receiver's antenna sensitivity or (2) increasing a sender's transmission power level. When a sender uses a transmission power of 0 dBm, and a receiver's sensitivity is set to -85 dBm, the signal may propagate for an outdoor free space range of 25–50 m, while a sensitivity of -110 dBm (higher sensitivity than 0 dBm case) will result in a range of 100–200 m. (Note: The use of a radio with a sensitivity of -100 dBm instead of a radio with -85 dBm sensitivity will allow you to decrease the transmission power by a factor of 30 and achieve the same range [Jason03].)

A *VCO* (*voltage-controlled oscillator*)-based radio architecture is being used in most of today's RF transceivers. These transceivers have the ability to communicate at various carrier frequencies (each carrier frequency is called a *channel*). Such a multichannel communication can effectively resist interfering signals. If a channel is found making high noise, the transceiver can immediately switch to another channel.

A few important technical aspects of RF communications are explained as follows:

1. *Modulation schemes*

   When we talk about RF communications, we encounter an important subtopic called digital modulation, which puts sensor data in a high-frequency RF carrier signal. Without modulation, the data cannot be transmitted to a long distance. It also cannot resist noise signals well.

   A typical modulation example is that of a cell phone's voice signal (with low frequency, <4 kHz) that needs to be put in a high-frequency carrier signal (900 MHz) to communicate with a base-station tower that may be a few miles away. A 900 MHz signal can efficiently resist the environmental noise (also called wireless interference) from obstacles, weather, etc.

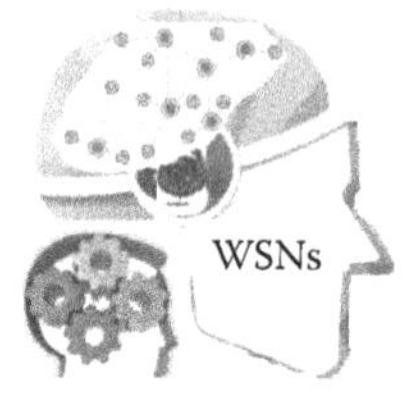

Most radio communication systems need a modem (MOdulation and DEModulation device) to put low-frequency, narrowband digital signals into high-frequency, wideband carrier signals (such as 2.4GHz signals). This is because low-frequency signals cannot resist noise well and cannot reach a long distance. Here, we will discuss a few popular modulation schemes. In fact, we have dozens of choices. It will take an entire textbook to discuss these modulation schemes. This book can only cover some basic ones.

Amplitude modulation (AM) and frequency modulation (FM) have been used for a long term. AM does not need a complex circuit. It can easily encode and decode signals. However, it is highly susceptible to noise because the data is simply encoded in the amplitude (i.e., strength) of the carrier signal. Any external noise can change such an amplitude. In contrast, FM is less susceptible to noise because all the data is transmitted at the same amplitude level.

However, FM is not the strongest alternative to resist noise. *Spread spectrum* transmission techniques can greatly increase the channel's tolerance to noise by spreading the signal over a wide range of frequencies. There are two types of spread spectrum schemes. One is called *frequency hopping* (*FH*) and the other one is called *code division multiple access* (*CDMA*).

In FH, the wideband carrier is divided into many small channels. FH changes communication channels continually based on a pseudorandom algorithm. Because an enemy does not know which channel it will switch to, it is difficult to select the right channel to add noise. Dwell times—the duration for which each channel is used—range from 100's μs to 10's ms.

But FH has some shortcomings when used in WSNs. For instance, it has a high overhead to maintain channel synchronization and to discover the current hopping sequence. (Think about this: If a sensor defines a specific channel use order, it must let other sensors know of this order for correct RF communications, because all communications must occur under the same channel at a specific time). If a sensor tries to find out what channels its neighbors use, it must attempt to search all possible channel locations. This is a high-overhead operation and not suitable to low-duty-cycle networks. It can be seen how this leads to high power consumption in Bluetooth devices.

CDMA (also called direct-sequencing spread spectrum, i.e., DSSS) does not divide the wideband signal into small channels. Instead, the signal is directly spread over a wide frequency band by multiplying the signal by a higher-rate pseudorandom sequence. During reception, the received signal is passed through a correlator that reconstructs the original input signal.

But for WSNs, CDMA also has too much overhead due to the maintenance of spreading codes and the cost of signal decorrelation. It needs high-bit-rate communications, which is not realistic in low-bit-rate WSNs.

Lester [Jason03] illustrates the power consumption of modern low-power transceivers through two commercial radios, the RF Monolithics TR1000 and the Chipcon CC1000:

a. TR1000: (1) Transmit: It consumes 21 mW of energy when it transmits at 0.75 mW. (2) Receive: TR1000 consumes 15 mW of energy when using a receiving sensitivity of −85 dBm.
b. CC1000: (1) Transmit: It consumes 50 mW to transmit at 3 mW. (2) Receive: It consumes 20 mW when using a receiving sensitivity of −105 dBm. When transmitting at the same 0.75 mW as TR1000, CC1000 consumes 31.6 mW.
c. Communication range: TR1000 provides an outdoor, line-of-sight communication range of up to 300 ft compared to 900 ft provided by CC1000.
d. Lifetime: If CC1000 does not go to sleep, it can transmit for approximately four days straight or remain in the receive mode for nine days straight. To last for one year, CC1000 must operate at a duty cycle of approximately two percent.

2. *Bit rate*

   Although the Internet prefers a high data rate (its backbone speed could be over 30 Gbps), WSN applications do not need such high-speed communication, as most times the sensors just send out some numerical values. That is why many sensors today only offer around 10–100 kbps of data rate.
3. *Turn-on time*

   We have emphasized the importance of a radio's ability to quickly enter into/exit from the sleep mode. A 5 ms response time is not acceptable. If we need to transmit data, we should minimize the time and the energy spent in configuring/powering up the radio.

If a WSN needs to detect emergency events within seconds, the radio must be powered on at least once per second. If a radio's turn-on time is 50 ms, it is difficult to achieve the required duty cycles of less than one percent.

Another interesting phenomenon is that multichannel radios based on the VCO frequency synthesizer must stabilize themselves prior to transmission or reception. A VCO locked to a high-frequency crystal should also stabilize itself. Obviously, we need to minimize the stability time. The CC1000 radio requires 2 ms for the primary crystal to stabilize. The TR1000 radio can be turned on and made ready to receive in just 300 μs. This is why TR1000 can respond to an event more than ten times faster than CC1000.

Some typical RF chips suitable to WSN communications are summarized in Table 2.3. These chips can be purchased from many semiconductor companies.

Table 2.3 Current Radios Suitable to WSNs and Their Capabilities

Radio Features	*Radio TR1000*	*Radio CC1000*	*Radio CC2400*	*Radio nRF2401*	*Radio CC2420*	*Radio MC13191/92*	*Radio ZV4002*
Max data rate (kbps)	115.2	76.8	1000	1000	250	250	723.2
RX power (mA)	3.8	9.6	24	18(25)	19.7	37(42)	65
TX power (mA/dBm)	12/1.5	6.5/10	19/0	13/0	17.4/0	34(30)/0	65/0
Power-down power (μA)	1	1	1.5	0.4	1	1	140
Turn-on time (ms)	0.02	2	1.13	3	0.58	20	[a]
Modulation	OOK/ASK	FSK	FSK, GFSK	GFSK	DSSS-O-QPSK	DSSS-O-QPSK	FHSS-GFSK
Packet detection	No	No	Programmable	Yes	Yes	Yes	Yes
Address decoding	No	No	No	Yes	Yes	Yes	Yes
Encryption support	No	No	No	No	128 bit AES	No	128 bit SC
Error detection	No	No	Yes	Yes	Yes	Yes	Yes
Error correction	No	No	No	No	Yes	Yes	Yes
Acknowledgments	No	No	No	No	Yes	Yes	Yes
Time-sync	bit	SFD/byte	SFD/packet	Packet	SFD	SFD	Bluetooth
Localization	RSSI	RSSI	RSSI	No	RSSI/LQI	RSSI/LQI	RSSI

Source: Adapted from Hill, J.L., System architecture for wireless sensor networks, PhD dissertation, Department of Computer Science, University of California at Berkeley, Berkeley, CA, Spring 2003; Hollar, S.E.-A., COTS dust, MS thesis, Mechanical Engineering, University of California at Berkeley, Berkeley, CA, Fall 2000.

[a] Manufacturer's documentation does not include additional information.

2.1.5 Power Sources

One of the most important components in a mote is the power source. If we use batteries, three common battery technologies can be used in WSNs, i.e., alkaline, lithium, and nickel metal hydride [Jason03]:

1. *Alkaline*—If you buy an AA alkaline battery, you will see that its output voltage is rated at 1.5 V. In reality, when it operates, the voltage could vary from 1.65 to 0.8 V (when it is used for a longer time, its voltage is lower). Its current is rated at 2850 mA.

   It is a cheap, high-capacity energy source. But some sensors cannot tolerate its wide voltage range. Its large physical size is also an issue. Even though no devices are driven by its power, it can self-discharge itself and becomes useless after five years (because its voltage would be too low).
2. *Lithium*—Lithium batteries have a much smaller physical size than alkaline ones (the smallest versions are just a few millimeters in diameter). Another good thing is that they have a constant voltage output. Even when the battery is almost drained, its voltage does not decay much. Another good thing is that unlike alkaline batteries, lithium batteries are able to operate at temperatures down to −40°C. CR2032 is the most common lithium battery. It is rated at 3 V and 255 mAh, and sells for just 16 cents.

   However, these batteries have a big disadvantage—they have very low nominal discharge currents. Therefore, these cannot drive most of today's motes that need more than 1000 mA of current. For instance, these may be good to drive Crossbow Mica2Dot (the smallest mote from Crossbow), but they cannot drive the Mica2 mote.
3. *Nickel metal hydride*—Nickel metal hydride batteries can be easily recharged. These have a few shortcomings: An *AA* size NiMH battery has approximately half the energy density of an alkaline battery (however, at approximately five times the cost). These only produce 1.2 V. But many WSN hardware components require 2.7 V or more.

   Table 2.4 lists the main features of the above three types of batteries [Seth00].

Table 2.4 WSN Battery Types

Battery Type	*Voltage (V)*	*Energy Density (mW-hr/g)*	*Maximum Current*
Alkaline AA P107-ND	1.5	90	130 mA at 24 g
Nickel-metal hybrid P014-ND (rechargeable)	1.2	55	>2600 mA at 26 g
Lithium	3.0	285	10 mA at 10.5 g

If a mote is designed to operate at low voltage, a battery could run for a long time. For instance, suppose that a mote consumes 250 mW and its components require 2.7 V. However, if we redesign the mote to make its components operate under a voltage down to 2.0 V, it would last approximately five times as long off of the same power source (assuming AA battery is used). Therefore, a seemingly unimportant CPU parameter (i.e., hardware voltage requirement) could result in a 5× difference in system lifetime.

Almost all batteries have a decaying voltage output as time passes by. Thus *voltage regulation* techniques have been proposed to take varying input voltages and produce a stable, constant output voltage. *Standard* voltage regulators can only generate an output voltage that is lower than the input voltage. However, if we use *boost* converters, we may get output voltages that are higher than the input voltage. But voltage regulators also have disadvantages. For instance, for a regulator, its quiescent current consumption, which is the power consumption when no current is being output, can be relatively high.

If we use alkaline batteries, as it is difficult to build a voltage regulator without quiescent power consumption, it will be highly advantageous to build motes with components that are tolerant to a wide voltage range. If the mote's components can operate over a range of 2.1–3.3 V, general alkaline batteries will be good enough.

Besides the above battery-based power sources, energy harvesting, especially solar energy harvesting, has become increasingly important as a way to improve the lifetime and the maintenance cost of WSNs. While macro-solar power systems have been well studied, micro-solar-based solar energy harvesting is more constrained in energy budget. Table 2.5 lists several micro-solar-powered designs with a specific set of requirements, such as lifetime, simplicity, cost, and so on. Heliomote [VRaghunathan05] and Trio [PDutta06] are two leading designs of micro-solar power systems. These have different designs. Heliomote [VRaghunathan05] focuses on simplicity and uses single-level energy storage and hardware-controlled battery charging. Trio is concerned more about lifetime and flexibility. It employs two-level energy storage and software-controlled battery charging.

Good idea

Energy, energy, energy.

Do you know one of the hottest R&D topics is renewable energy systems? Human beings are facing a great challenge: We cannot simply depend on gas! Look at the unlimited power source—solar! Why do we not explore it for all applications including motes? Easier said than done. We need you—smart scientists and engineers, to come up with a feasible, low-cost solution to explore solar, wind, nuclear, and other renewable energy sources.

Table 2.5 Micro-Solar Power System Examples

Micro-Solar Power System	*Goal*	*Key Features*	*Source*
Prometheus Trio	Lifetime, flexibility	Two-level storage, SW charging	[XJiang05, PDutta06]
Heliomote	Simplicity	HW charging to NiMH battery	[VRaghunathan05]
Everlast	Lifetime	MPP tracking	[FSimjee06]
RF beacon	Proof of concept	No support for power disruption	[SRoundy03]
Farm monitoring	Compactness, reliability, cost	HW charging to NiMH battery	[PSikka06]
ZebraNet	Compactness	SW charging to Li + battery	[PZhang04]

Source: Adapted from Jeong, J. et al., Design and analysis of MicroSolar power systems for wireless sensor networks, Technical Report No. UCB/EECS-2007-24, http://www.eecs.berkeley.edu/Pubs/TechRpts/2007/EECS-2007-24.html.

2.1.6 Peripheral Support

We have discussed about CPUs (i.e., microcontrollers) and their internal design principle. A CPU has some pins to specifically interact with external devices. It has the following two types of pins.

1. *Digital I/O pins*: Standard digital I/O lines are included on all CPUs as the baseline interface mechanism. It interfaces with RF transceivers, memory units, and other components that output digital signals.

   Note: In these digital I/O pins, digital communication protocols are used to read digital sensors. But some other peripheral chips connect to a CPU through serial communication protocols over a radio or an RS-232 transceiver. Overall, digital communication supports three standard communication protocols: UART (Universal Asynchronous Receiver Transmitter), I^2C (Inter-Integrated Circuit), and SPI (Serial Peripheral Interface). Both I^2C and SPI use *synchronous* protocols with explicit clock signals. However, UART uses an *asynchronous* mechanism.
2. *Analog I/O pins*: A CPU also has analog I/O pins to interface directly with analog sensors. For these pins, the CPU has internal ADCs that allow for precise control of sample timing and easy access to sample results. If an internal converter is not present in a CPU, the mote designer should include an external converter.

2.2 Put Everything Together [Jason03]

2.2.1 Typical Sensor Mote Architecture

After we have learned different hardware components in a mote, it is time to put them together. In summary, a mote mainly achieves *local* sensor data computation and *neighboring* RF communications.

This section investigates the general mote architecture that addresses the needs of computation and communications. As we target the general architecture here, we will not emphasize any particular radio or processing technology. Instead, we emphasize the general WSN hardware design principle, especially the hardware that achieves computation and communications in a low-power approach.

2.2.1.1 Wireless Communication Requirements

A mote needs to use wireless communications to talk with others. The wireless signals are actually raw electromagnetic-signaling primitives. An RF transmitter should use digital modulation to modulate the data to the RF carrier. An RF receiver then performs demodulation and data extraction.

In WSNs, a mote mainly sends out two types of data: (1) *sensor data* collected from the environment and (2) *control data*, such as wireless network protocols. These data are encapsulated into "packets" from the network protocol viewpoint. Figure 2.2 illustrates the key phases of a packet-based wireless communication protocol. Please note that many of the operations must be performed in parallel with each other. This is similar to a car-manufacturing company that assembles components in parallel. Figure 2.2 shows that distinct layers overlap in time to reflect a "parallel" nature.

As shown in Figure 2.2, encoding is the first step in the communication process. The analog sensor data is encoded into digital signals (i.e., *bits*, also called *codes*) for transmission. Note that the codes should also have some type of error detection/correction functionalities. For instance, when the wireless interference damages some bits, the error detection codes should be used to find out such errors.

To shorten the transmission delay, "encoding" is pipelined with the actual transmission process, that is, once the first byte is encoded, RF transmission immediately begins. We then keep encoding new bytes as the preceding bytes are transmitted.

Today, many coding schemes have been proposed. A simple scheme could be a DC-balancing scheme, such as Manchester encoding. A more advanced but more complex scheme could be CDMA (we covered its concept before). In all encoding schemes, data bits (either 0 or 1) are grouped into different units, called *symbols*. Each symbol is coded into a collection of radio transmission bits, called *chips*. In Manchester encoding, for 1 bit of data, we use two chips per symbol. CDMA schemes often have 15–50 chips per symbol, with each symbol containing 1–4 data bits.

When data is passed to wireless communication protocols and is ready to be sent to another mote, a MAC protocol needs to be executed first. If you could recall

MAC definitions, its main task is to make sure that neighbors can transmit data without conflict. A simple example is carrier sense media access (CSMA). A mote listens to the communication channel before it sends out data. If the channel is busy, it waits for a short, random delay, and then reinitiates the transmission.

After the MAC protocol successfully sends out data, the *routing layer* protocol will take care of the data from mote to mote. It finds out an optimal path (from the energy-saving viewpoint) to deliver the data to the destination (such as a base station).

When data continuously flows between a sender and a receiver, based on an accurate time synchronization scheme, the sender precisely controls the timing of each bit transition so that the receiver can maintain synchronization with the sender.

When a receiver receives the data, it uses decoding and demodulation to recover the original data. Noise is removed by some data-cleaning algorithms.

2.2.1.2 Key Issues

Lester [Jason03] has pointed out a few important issues during a mote design:

1. Concurrency
   To speed up data processing, it is important to provide an efficient architecture to support fine-grained concurrency. No matter whether on the sender or on the receiver side, the RF computations should occur in parallel with application-level data processing and even with network protocol processing. When an RF communication is going on, we cannot stop some necessary operations, such as sensor event detection and data calculation.
2. Flexibility
   Note that WSN applications have very different QoS (quality-of-service) requirements. Some applications need real-time data transmission, while others can tolerate some delay. Some applications need localized data compression, while others just simply send data to a sink. Some need security support, while others do not consider network attacks.

   Therefore, it is important that the mote design has a flexible architecture to support a wide range of application scenarios. Although traditional embedded devices (such as cell phones or Bluetooth devices) may use a fixed set of communication protocols that they must adhere to, WSNs should allow flexible communication protocol designs to exploit trade-offs between bandwidth, latency, and in-network processing.

   The above flexible protocol design requires a flexible hardware architecture. Different hardware architectures could lead to very different application optimizations. For instance, a video sensor network needs larger memory and a stronger CPU, while an underwater sensor needs acoustic (instead of RF) communication modems.

3. Decoupling between RF and processing speed
   A mote should not closely couple the following two operational characteristics: RF transmission rates and CPU processing speed. This is because the CPU and the RF transceiver have very different optimization requirements as follows: (1) A radio prefers to send out data at its maximum transmission rate. This is because a shorter transmitting time reduces the energy used. (2) On the other hand, modern studies in low-power CPU design and dynamic voltage scaling have disclosed a fact—CPUs prefer to spread out computation in time as much as possible so that they can run at the lowest-possible voltage.

   Therefore, from an energy-saving perspective, it would be preferred that the CPU performs all calculations as slowly as possible, and just as the computation is complete, the radio would burst out the data as quickly as possible.

   Now we know that the decoupling between the CPU and the radio is important, as it allows the above different operation patterns: the CPU slowly processes data and the radio quickly sends out data. When the speed of the microcontroller is coupled to the data transmission rate, both components of the system are forced to operate at nonoptimal points.

2.2.1.3 Traditional Wireless Design [Jason03]

Today, many embedded systems (such as cell phones, 802.11 wireless cards, and Bluetooth-enabled devices) choose to address the concurrency and decoupling issues by employing a dedicated CPU to run communication protocols. This CPU should run communication protocols that meet real-time requirements during the following operations: radio modulating and demodulating, encoding/decoding, and other operations.

As an example, in a Bluetooth device, the host channel interface (HCI) performs a high-level packet interface over a UART. Such an interface hides the intricacies of communication synchronization, signal-encoding, and MAC protocols. The speed of the CPU is then set to meet the requirements of the RF communication protocols.

Unfortunately, the above CPU operation mode is not suitable to WSN applications, because it separates radio communication and data calculation in partitioning of resources. This leads to nonoptimal resource utilization. Its chip-to-chip communication mechanisms are not efficient.

An alternative to the above approach is to use the mote design ideas in [Jason03]. Instead of using a dedicated CPU, a single execution engine is shared across application and protocol processing. The concurrency requirements of the system are met virtually (instead of physically) by fine-grained interleaving of event processing in TinyOS.

In the following few sections, we cover the main design ideas of some motes (such as Reno, Mica, and Spec) proposed in [Jason03]. Because these motes represented a pioneering WSN node design in the last decade, we could learn some basic hardware design principles on how we could make a mote work well for realistic WSN applications.

2.2.1.4 Mote Example: Reno

Reno is a mote proposed in [Jason03] with special-purpose hardware accelerators for handling the real-time, high-speed requirements of the radio.

Figure 2.3 depicts Reno's general architecture. Its CPU needs to handle multiple concurrent operations (similar to the "multi-threads" concept in MS Windows). Context switching needs to be efficiently supported. Register windows can be used to decrease the context-switching overhead. Reno's CPU includes multiple register sets, which avoid the operation of dumping data from registers to the memory. Instead, the OS simply switches to a free register set.

As shown in Figure 2.3, a shared bus interconnects memory, I/O ports, ADCs, system timers, and hardware accelerators. Because of its high-speed, low-latency interconnect, data can be moved easily between the processor, memory, and peripheral devices. Such a bus allows not only direct CPU–peripherals interactions, but also allows a peripheral device to interact with another peripheral. Note that a peripheral can use the bus to directly pull data from the memory. It can also easily push data into a UART peripheral.

Therefore, Reno can use the shared bus to enhance RF communications as follows: It allows a data-encoding peripheral to pull data directly from memory and then push it into a data transmission accelerator, such as a modulation circuit for RF communications. This is different from many computer-operating modes where the CPU has to be involved into any memory read/write. In Reno, the CPU does not get involved into communications. This frees the CPU from some heavy load, as the CPU can simply orchestrate the data transmission.

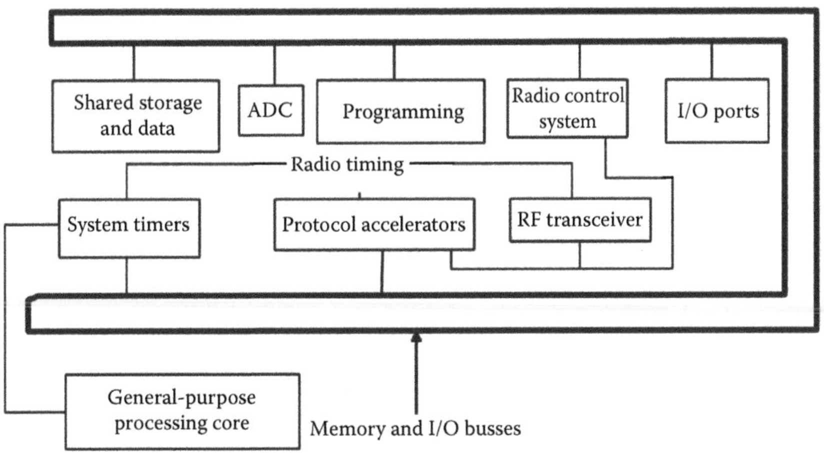

Figure 2.3 Generalized architecture for embedded wireless device. (Adapted from Hill, J.L., System architecture for wireless sensor networks, PhD dissertation, Department of Computer Science, University of California at Berkeley, Berkeley, CA, Spring 2003.)

If you could recall "Computer Architecture" or "Assembly Language" courses, we could use the same addressing schemes to name each memory location and other devices, that is, a memory address could be a real memory location or just the virtual location of a device's data buffer. The system uses a wire to link the device's data buffer to a real memory location. Reno uses such an addressing scheme. It allows components that were not originally intended to function together to be combined in new and interesting ways. Suppose that a data encoder wants to get data from a radio receiver's buffer. Because such a buffer is mapped to a memory location, the encoder can just simply read from memory, transform data, and write to memory.

Finally, remember that one of the dominant features of the Reno mote is that it consists of special-purpose hardware *accelerators*, which can implement low-level operations in a fast, energy-efficient way. By increasing the efficiency of these operations, the overall power consumption of the system can be greatly reduced.

2.3 Mica Mote Design [Jason03]

The Mica mote adds key hardware accelerators to Reno to validate the generalized architecture. Mica supplements the CPU with hardware accelerators to increase the transmission bit rates and timing accuracy.

Mica hardware components include an Atmega103 microprocessor (i.e., CPU), an RFM TR1000 radio, external storage, and communication accelerators. The hardware accelerators optionally assist to increase the performance of key phases of the wireless communication.

Figure 2.4 shows the Mica architecture. It has five major function modules: CPU, RF communication, power management, I/O expansion, and secondary storage. On the Web site http://www.tinyos.net, the readers can find a quick survey of the major modules; a general overview for the system as a whole; and a detailed bill of materials, device schematic, and datasheet for all hardware components.

The Mica mote uses Atmel ATMEGA103L or ATMEGA128 (4MHz). The Atmel CPU also connects a 128kB flash program memory; a 4kB static RAM; an internal eight-channel, 10 bit ADC; three hardware timers; 48 general-purpose I/O lines; one external UART; and one SPI port. The Mica radio module consists of an RF Monolithics TR1000 transceiver.

- *Mote ID*: To obtain a unique identification for each mote, Mica uses a Maxim DS2401 silicon serial number, which is a low-cost ROM device with a minimal electronic interface without power requirements [Dallas08].
- *Memory*: Mica uses a 4 Mbit Atmel AT45DB041B serial flash chip, which has a small footprint. The flash memory stores two types of information: (1) sensor data and (2) application programs. Typically, the flash memory should be larger than the 128kB program memory to hold a complete program. That is why Mica does not use the electronically erasable, programmable ROM-based memory, which is used by Reno and is generally smaller than 32kB.

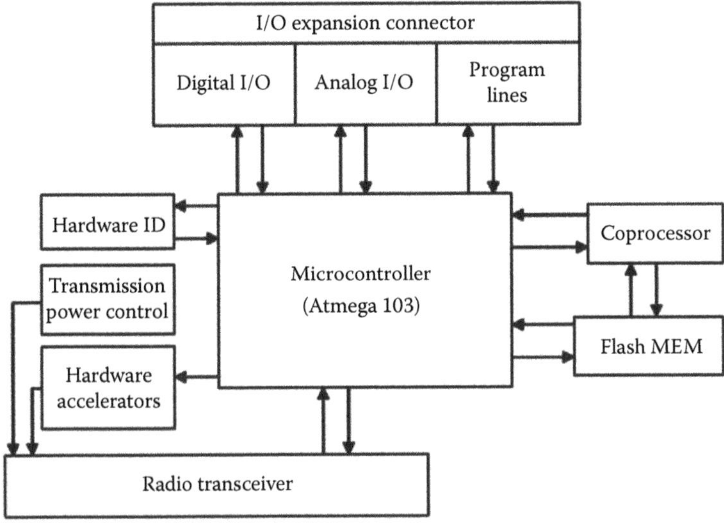

Figure 2.4 Block diagram of Mica architecture. (Adapted from Hill, J.L., System architecture for wireless sensor networks, PhD dissertation, Department of Computer Science, University of California at Berkeley, Berkeley, CA, Spring 2003.)

- *Power supply*: Mica can be driven by AA alkaline batteries and boosts their output voltage. The radio will not operate, however, without the boost converter enabled. Mica uses a Maxim1678 DC–DC converter to provide a constant 3.3 V supply. The converter accepts an input voltage as low as 1.1 V. Note that input voltages significantly affect the radio transceiver's (TR1000's) transmission strength and receiving sensitivity.

  Table 2.6 shows the power consumption levels in different Mica hardware components. When the mote is in ultra-low-power sleep mode, the power system is disabled. Then the entire system runs directly off the unregulated input voltage. This helps to reduce the power consumed by the boost converter and the CPU.
- *Peripherals*: Mica's I/O subsystem interface consists of a 51-pin expansion connector. These pins allow the mote to interface with a variety of sensing and programming boards. The 51-pin connector has the following interfaces: eight analog lines, eight power control lines, three pulse-width-modulated lines, two analog compare lines, four external interrupt lines, one serial port, a collection of lines dedicated to the programming of microcontrollers, and some bus interfaces.
- *Radio*: Mica uses a TR1000 radio to allow the CPU to directly access the signal strength of the incoming RF transmission. Such a radio interface also allows the CPU to sample the level of background noise during periods when there is no active data transmission. In multi-hop networking applications, such information (radio signal strength and noise levels) can dramatically improve the routing efficiency by selecting links with good signal-to-noise ratios.

Table 2.6 Breakdown of Active and Idle Power Consumption for Mica Hardware at 3 V

Hardware Device	*Active (mW)*	*Idle (uW)*
CPU	16.5	30
Flash drive	45	30
LED	10	0
Radio	21 (TX), 15 (RX)	0
Silicon ID	0.015	0

Source: Adapted from Hill, J.L., System architecture for wireless sensor networks, PhD dissertation, Department of Computer Science, University of California at Berkeley, Berkeley, CA, Spring 2003.

Mica allows softwares to power the radios on/off quickly and predictably. Therefore, a Mica mote can easily enter a low-duty-cycle operation without a global control.

2.4 Customized Mote—Spec [Jason03]

Although it is a quick and simple way to integrate commercial off-the-shelf (COTS) components into a mote, from manufacturing cost, energy consumption, and system performance viewpoints, it is more efficient to design a custom-integrated solution.

If using COTS chips, the chip-to-chip communications can sacrifice the system delay and power performance due to the interface overhead. Therefore, Lester [Jason03] developed a custom ASIC for the mote board, which is called *Spec*. By designing the customized silicon, it achieves orders-of-magnitude efficiency improvements on the main communication primitives.

Spec is much smaller than most commercial motes. It is just 2.5 mm on a side in a 0.25 μm CMOS process even though it integrates a microcontroller, SRAM, communication accelerators, and a 900 MHz multichannel transmitter.

Of course, although its CPU, RF transceiver, and memory are based on a single-chip design, it still needs some low-cost external components, which include a crystal, a battery, an inductor, and an antenna, to form a complete WSN mote.

Spec has a general architecture, as shown in Figure 2.5. The CPU core is a basic 8 bit RISC core with 16 bit instructions. A bank of six memory blocks (each 512 bytes) is connected to the CPU core. The reason of dividing the memory into banks is to achieve a smooth integration between instruction memory and data memory. Besides the memory controller, the CPU core is also connected to an ultra-low-power

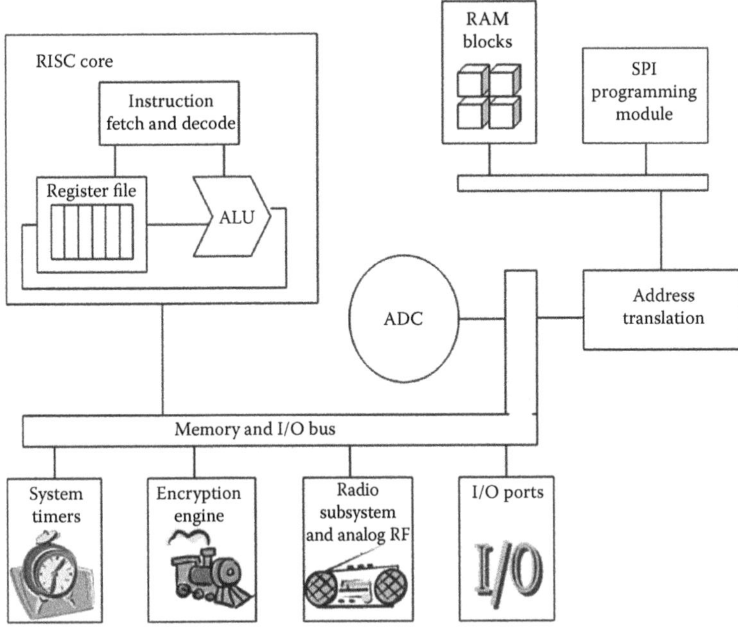

Figure 2.5 Block diagram of Spec, the single-chip wireless mote. (Adapted from Hill, J.L., System architecture for wireless sensor networks, PhD dissertation, Department of Computer Science, University of California at Berkeley, Berkeley, CA, Spring 2003.)

ADC, an encryption accelerator, general-purpose I/O ports, system timers, a chip-programming module, and an RF subsystem.

The RF subsystem performs the following tasks: it extracts and generates bits with correct sending/receiving timing control; it performs bit pattern matching to find a start symbol (thus, the receiver knows the boundaries of different data units); it forms a stream for data to be transmitted; it takes data into and out of memory; to achieve security, it can encrypt and decrypt data automatically; and it performs other tasks as well.

Lester [Jason03] first used the Very High-Speed Hardware Description Language (VHDL) digital logic tools to synthesize Spec's behavioral characteristics. After VHDL simulation, these tools map the high-level VHDL code into standard cells provided by National Semiconductor using Ambit Build Gates. The layout of this tool was performed with Silicon Ensemble—a tool from Cadence Design Systems. In addition to VHDL simulation, the functionality of the Spec core was also verified by downloading it onto a Xilinx FPGA.

Spec's data-processing speed is much higher than Mica in many applications. Spec provides significant advantages in power consumption due to its integrated design and hardware accelerators. Because Spec is a fully integrated chip, it does not offer the same interface flexibility as Mica.

2.5 COTS Dust Systems [Seth00]

In [Seth00], several interesting sensor systems were built. Their motes used the Atmel AT90LS8535 (as the microcontroller) and the RF Monolithics 916MHz RF transceiver. The mote controls seven different types of analog sensors (temperature, light, barometric pressure, two axis accelerators, and two axis magnetometers). Regarding the power source, it uses a single 3V lithium coin cell battery. It can operate for five days of continuous operation or for 1.5 years at one percent duty cycling.

It uses a slow CPU, Atmel MCU with 149.475kHz. It creates 19 instructions to send and receive raw data bits through the RF system. Each clock cycle only executes one instruction. Thus, its raw data rate is 149.475kHz/19 cycles/bit = 7.867kbps.

In wireless environments, noise/interference can damage packets of data. Hollar [Seth00] uses a cyclic redundancy check (CRC) to check packet bit errors.

Figure 2.6 shows its single-hop communication protocol, a simple procedure for sending data from one device to the next through one RF transmission–reception pair. Its communication protocol is used for two types of motes (see Figure 2.6): The *base* mote communicates to a computer (which could serve as a base station) via the serial port, and the *floating* motes communicate to the *base* mote via RF. The *floating* motes continuously send out data packets that are received by the base mote, which then displays the information on the computer screen.

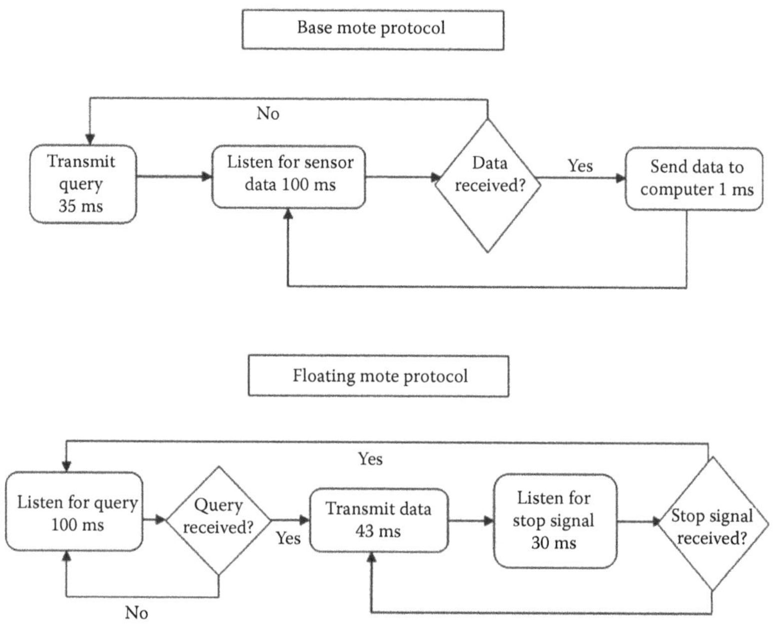

Figure 2.6 Protocols used for base and floating Motes. (Adapted from Hollar, S.E.-A., COTS dust, MS thesis, Mechanical Engineering, University of California at Berkeley, Berkeley, CA, Fall 2000.)

Hollar [Seth00] also used a simple time synchronization protocol between the two devices. To establish time synchronization, the base mote must first query one of the floating motes. As shown in Figure 2.6, after sending a query command, it listens for a response. If a response is not heard after 100 ms, it proceeds to send out another transmit query. Such a query procedure is repeated until a valid message is received. Once received, the message is sent from the base mote to the computer over the serial port. The base mote then proceeds to listen for the next data packet.

Figure 2.6 also shows that both protocols have listening periods right after transmission periods. This enables the motes to respond to queries right away. A handshaking protocol enables both motes to communicate with one another as quickly as possible.

Case study

Hollar [Seth00] only presents a very basic mote design without considering many other WSN application requirements. For instance, it does not support multi-hop communication well. Its CPU/transceiver design still has a large space to house a more energy-efficient interface design. The reason we include this example here is to show that even for a simple mote prototype design, there could be many lessons to learn. (Please see Section 2.5.1 on some lessons learned from [Seth00].)

2.5.1 Design Advice: Failures and Successes

There are some good lessons summarized in [Seth00] as follows.

1. *On the selection of CPU and RF transceiver*: In the beginning, Hollar [Seth00] used Scenix SX28AC series microprocessor that operates with clock cycles up to 50 MHz. However, when the first circuit board was populated, there was trouble getting the RF Monolithics transceiver chipset to work in the presence of the Scenix microprocessor.

   Hollar [Seth00] found out why the transceiver chipset did not work correctly. The RF transceiver was saturated with the noise generated by the CPU—because the CPU was clocked at a slow speed of 1 MHz; possibly, the fast rise and fall times of the CPU contributed to noise in the receiving band. The second possible reason is that the circuit board did not contain a ground or power plane. Ground and power planes in circuit boards help to isolate signals from one another and maintain a stable power supply voltage.
2. *On the choice of power supplies*: The Scenix CPU operated at 5 V, and the RFM chips operated at 3 V. A way is needed to generate both power supplies. A possible way is to use three alkaline batteries that provide 3 and 4.5 V. Unfortunately, the batteries could lose voltage over their lifetime. The idea

of using two voltage converters for both the MCU and the transceiver chipset was also unappealing due to the added complexity and increased component count. To solve this problem, Hollar [Seth00] aimed to use a single operating voltage, namely, that of a 3 V lithium ion battery. All components were designed to operate within the battery range of 2.75–3.25 V.

2.6 Telos Mote [Joseph05]

The Telos series mote (such as Telos-B) is a popularly used sensor platform today. Unlike *Spec*, which integrates the design into silicon, Telos uses COTS components with hardware accelerators to build a power-efficient system that does not sacrifice performance.

Table 2.7 summarizes the main features of different motes.

After comparing the CPU performance from Atmel, Motorola, and Microchip, Telos developers select the *MSP430 CPU* due to its following advantages:

1. It has the lowest power consumption in both sleep and active modes (see Table 2.7).
2. It can tolerate a low operation voltage of 1.8 V. A low-voltage operation could help extract all the energy out of a power source. If we use AA batteries, they have a cutoff voltage of 0.9 V. A Telos mote uses two batteries. Then the system cutoff voltage will be 1.8 V, which is exactly the minimum required voltage for MSP430. If we use other CPUs, say ATmega128 MCU (Mica family), it can only run down to 2.7 V, leaving almost 50 percent of the AA batteries unused.
3. We know that a faster wake-up time helps to conserve energy. Table 2.7 shows that MSP430 has the fastest wake-up time (it takes $<6\,\mu s$ to transition from the standby ($1\,\mu A$) to the active mode).
4. From the memory viewpoint, Table 2.7 shows that MSP430 has the largest on-chip RAM buffer (10 kB). It is good for on-chip signal processing. A larger RAM allows more sophisticated applications.

From the RF communication viewpoint, Telos has the following features:

1. It uses the IEEE 802.15.4 standard. Such a standardized radio allows Telos to communicate with many radio devices from other vendors.
2. It uses the Chipcon CC2420 radio. It uses a 2.4 GHz RF band, a wideband radio with Offset Quadrature Phase-Shift Keying (O-QPSK) modulation with DSSS at 250 kbps. Such a high data rate (other motes typically operate under 150 kbps) shortens the operation time (which helps to reduce energy consumption).

The Telos mote can be programmed through an on-board USB that also provides power. A USB interface is better than an RS232-based serial interface considering that many people use laptops (that have a few USB ports) to program a mote.

Table 2.7 Family of Berkeley Motes Preceding Telos and Their Capabilities

	Motes							
Mote Type	*WeC*	*Rene*	*Rene2*	*Dot*	*Mica*	*Mica2Dot*	*Mica 2*	*Telos*
Year	*1998*	*1999*	*2000*	*2000*	*2001*	*2002*	*2002*	*2004*
Mote microcontroller properties								
Type	AT90LS8535	AT90LS8535	ATmega163	ATmega163	ATmega128	ATmega128	ATmega128	TI MSP430
Program Memory (kB)	8	8	16	16	128	128	128	48
RAM (kB)	0.5	0.5	1	1	4	4	4	10
Active power (mW)	15	15	15	15	8	8	33	3
Sleep power (mW)	45	45	45	45	75	75	75	15
Wake-up time (μW)	1000	1000	36	36	180	180	180	6
Mote nonvolatile storage properties								
Chip	24LC256	24LC256	24LC256	24LC256	AT45DB041B	AT45DB041B	AT45DB041B	STM25P80
Connection type	I^2C	I^2C	I^2C	I^2C	SPI	SPI	SPI	SPI
Size (kB)	32	32	32	32	512	512	512	1024
Mote communication properties								
Radio	TR1000	TR1000	TR1000	TR1000	TR1000	CC1000	CC1000	CC2420
Date rate (kbps)	10	10	10	10	40	38.4	38.4	250

(continued)

Table 2.7 (continued) Family of Berkeley Motes Preceding Telos and Their Capabilities

	Motes							
Mote Type	*WeC*	*Rene*	*Rene2*	*Dot*	*Mica*	*Mica2Dot*	*Mica 2*	*Telos*
Year	*1998*	*1999*	*2000*	*2000*	*2001*	*2002*	*2002*	*2004*
Modulation type	OOK	OOK	OOK	OOK	ASK	FSK	FSK	O-QPSK
Receive power (mW)	9	9	9	9	12	29	29	38
Transmit power at 0 dBm (mW)	36	36	36	36	36	42	42	35
Mote power consumption properties								
Minimum Operation (V)	2.7	2.7	2.7	2.7	2.7	2.7	2.7	1.8
Total active power (mW)	24	24	24	24	27	44	89	41
Mote program and sensor interface properties								
Expansion	None	51-pin	51-pin	none	51-pin	19-pin	51-pin	16-pin
Communication	IEEE 1284 and RS232	IEEE 1284 and RS232	IEEE 1284 and RS232	IEEE 1284 and RS232	IEEE 1284 and RS232	IEEE 1284 and RS232	IEEE 1284 and RS232	USB
Integrated sensors	No	No	No	Yes	No	No	No	Yes

Source: Polastre, J. et al., Telos: Enabling ultralow power wireless research, *Fourth International Symposium on Information Processing in Sensor Networks 2005 (IPSN 2005)*, Los Angeles, CA, April 15, 2005, IEEE Press Piscataway, NJ, pp. 364–369.

The Telos mote has a user button, a reset button, and a 16-pin IDC expansion header. A programmer can re-task the reset button as a non-maskable interrupt, thus allowing it to be used as a power button instead. A developer can also export I^2C and UART over the 16-pin IDC expansion header to attach many connections found on today's legacy "Mica-style" sensor boards [JPolastre04].

In many cases, we need hardware write protection to protect the good program images in a memory. Such a write protection also prevents possible write errors when using over-the-air programming, which is used in some advanced motes. Telos is the first mote to include hardware write protection for external storage. The write protection is disabled if plugged into a USB interface. When running on batteries (without a USB), the memory is write protected.

Telos mote has some "sub-circuit" with a separate power-on/power-off switch. If any failure is detected, we could power-off a sub-circuit instead of the whole system. Such a power protection is based on the lessons learned from a real-world WSN application on the Great Duck Island (GDI) [RSzewczyk04]. In the GDI application, a small part of circuit caused the failure of a sensor. As the failure could be recognized in the software, the ability to cut power to that section of the board could have saved the system as a whole.

2.7 CargoNet [Mateusz07]

In [Mateusz07], a mote called CargoNet is designed to bridge the gap between WSNs and radio frequency identification (RFID). CargoNet originally targeted applications in environmental monitoring at the crate level for supply-chain management and asset security. It uses custom-designed circuits to minimize power consumption and cost.

The CargoNet nodes use a new concept, called *quasi-passive wake-up*, to achieve an *asynchronous*, *multimodal* wake-up, which can wake up (from the sleep mode) to perform extremely low-power operations. CargoNet can be used to monitor conditions inside a typical shipping crate while consuming <25 μW of average power.

CargoNet *uses external stimuli signals to wake up its sensor mote*. This idea is not new, as some other related systems have also explored external wake-up. But CargoNet consumes much less power than them.

As an example, a *similar wake-up strategy for vibration detection and autonomous crack monitoring* was proposed by researchers at the Northwestern University. It uses a single geophone as the input sensor and wakes up to detect aperiodic shocks to ensure structural safety of buildings. Although their analog front end consumes only 16.5 μW on an average, their processing is performed by a Mica2 mote, which adds a further 105 μW to their average power budget.

As another example, the T-mote [Tmote06] also has *comparators to generate interrupts upon acoustic or acceleration stimuli*, but it needs much energy (in the milliwatt range) due to the use of active accelerometers and microphone amplifiers.

In the following, let us first understand the general concept of RFID.

RFID is used to replace the traditional bar-code technique to improve traditional bar codes when used in the transport and distribution of goods. Traditional bar codes require a line of sight between an interrogator and a tagged object. Therefore, human operators must align a tagged object to ensure a successful read. The distance between the interrogator and bar codes is very short (typically, a few centimeters). Moreover, these bar codes have very short information.

On the other hand, RFID uses a reader to read tags that are attached to products. The distance between the reader and tags could be even a few inches to over 10 ft long (its range depends on the used RF). Moreover, the reader can read a tag's data through non-line-of-sight signals that propagate widely and permeate through most nonconductive materials, allowing identification without human involvement. Although the traditional bar code is printed onto a surface and cannot be changed, we can change the data in RFID tags because they are electronic circuits that can change state based on external stimuli.

A concept called "active RFID" has been proposed recently. It is actually a special sensor device with battery and CPUs to provide better "visibility" into their supply chains. "Active RFIDs" can accurately collect data about environmental conditions experienced by goods in transit, better manage risk, and maintain flexibility. These RFIDs can detect potentially damaged goods before they reach their destination.

The CargoNet node is such a type of "active RFID". Its *quasi-passive wake-up* is based on the following interesting fact: The external stimuli could actually be used to wake up and even provide energy to the CPU!

CargoNet can desensitize the sensors following repeating stimuli. This further reduces occurrences of redundant wake-up to save power. *Quasi-passive wake-up* allows a CargoNet RFID tag to simultaneously and continuously monitor many sensor modalities for exceptional activity without consuming much power.

Figure 2.7 is a system diagram with the CargoNet "active RFID" tags and an RFID reader. Its core hardware consists of the MSP430 microcontroller, a real-time clock (RTC), and a CC2500 2.4 GHz radio. The MSP430F135 flash-based microcontroller is manufactured by *Texas Instruments* (TI). It has a specified standby current of $<0.1\,\mu A$ when entering the sleep state.

The CargoNet tag has an internal flash memory with a capacity of 16 kB. It is a small memory. But it is big enough for most targeted applications due to two reasons:

1. Its design allows any memory not dedicated to program storage to be used for data logging. Its OS occupies a very tiny space.
2. We typically record only extraordinary events (such as extremes of temperature and significant shocks). The routine code occupies <8 kB. Suppose potentially harmful or notable events occur once per day and require 10 bytes to log, its flash memory will last over two years before it is filled!

If a developer needs to test long programs, or, in some cases, we may need to store more detailed information in the mote, CargoNet allows the attachment of an external flash memory to the tag (for instance, Atmel's AT45DB081B could be attached, which has 8 Mbits of storage capacity and current consumption of 2 μA at standby).

MSP430 has internally a fast-starting, high-frequency *clock oscillator*. A developer can certainly use external clocks, such as a low-frequency watch crystal. CargoNet suggests employing a separate Philips PCF8563 *RTC* chip, which has a low timekeeping current (only 0.35 μA). The RTC allows an "active RFID" tag to find out the exact location where the damage occurred by measuring the time from the last checkpoint. The RTC can also issue a once-per-minute polling sequence of the humidity and temperature sensors.

The "active RFID" tags use CC2500 from Chipcon to communicate wirelessly with RFID readers/interrogators (see Figure 2.7). Unlike traditional RFID systems, the CC2500 radio is fully bidirectional, such that the "active RFID" tags can also receive instructions from the RFID readers besides sending tag data to the readers. *This feature bridges the gap between "active RFIDs" and WSNs as WSN motes require bidirectional communications between nodes.* Such a bidirectional radio com-

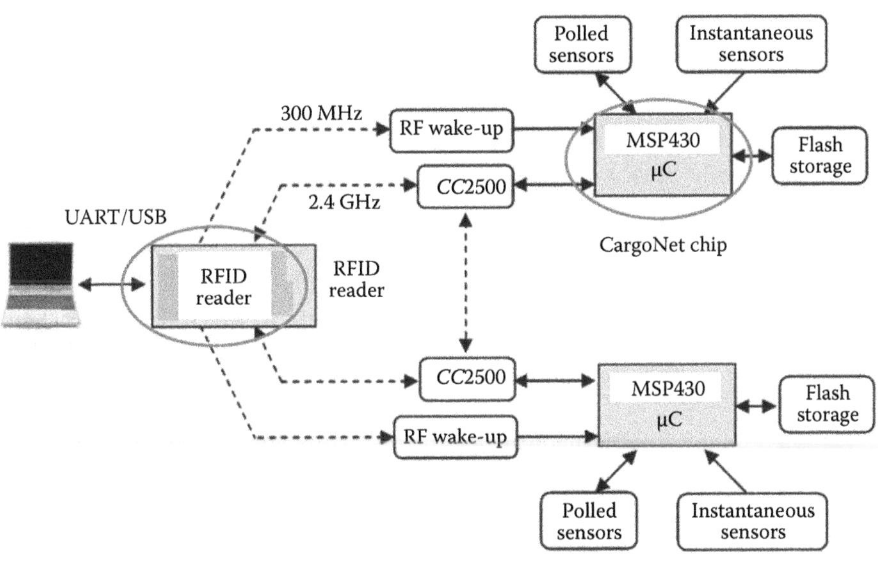

Figure 2.7 CargoNet system diagram. (Adapted from Malinowski, M. et al., CargoNet: A low-cost MicroPower sensor node exploiting quasi-passive wakeup for adaptive asychronous monitoring of exceptional events, *SenSys '07*, Sydney, Australia, November 6–9, 2007.)

munication capability enables useful applications, such as synchronizing clocks, recording the identity of neighbors, and qualifying the validity of sensor readings.

When a base station (which is actually an RFID reader/interrogator in Figure 2.7) sends out data query requests to the "active RFID" tags to check significant events (e.g., temperatures or shocks), request data dumps, or adjust tag parameters, it uses a radio burst signal. CargoNet motes are able to wake quasi-passively after receiving such an RF amplitude burst at 300 MHz over a dynamically adjustable threshold.

Note that a CargoNet "active RFID" tag does not typically poll or amplify quickly changing environmental stimuli. This is for the purpose of saving energy. Instead, it simply takes environmental stimuli and compares them against a threshold through "quasi-passive wake-up" technology. The above comparator is based on a Linear Technology product (LTC1540). Because of its nonlinear class-D operation, it typically consumes only 840 nW of quiescent power [Linear04].

But for some stimuli that do not change quickly enough, they may not be able to reach the "wake-up threshold." Such stimuli examples have temperature, humidity, etc. For this case, the "active RFID" tag will poll the stimuli.

An "active RFID" tag uses a 12 bit accuracy sequence to poll the Sensirion SHT11 temperature/humidity sensors. The polling time is around 55 ms. If it polls the sensors once a minute, this corresponds to a duty cycle of only 0.092 percent and an average power consumption of 1.5 μW. Such a low-duty-cycle polling does not dominate the power budget of the tag at all.

If an "active RFID" tag needs to wake up quickly to achieve a very fast response to a stimulus, say, a temperature event, the quasi-passive wake-up on temperature can be accommodated via a PTC thermistor or other thermal sensors, which exhibit a high impedance and a sharp characteristic response.

CargoNet systems also use the following two sensors: an *RF wake-up receiver* and a *vibration dosimeter*. They have linear amplifiers to boost or integrate weak signals.

Table 2.8 lists CargoNet sensors that assemble a suite of measurements relevant to the transport of equipment and goods.

Good idea

Normally people distinguish an RFID from a mote very clearly. But CargoNet designs a device that can serve as both an RFID tag and a WSN node. It can collect data from environment into a "tag," and then allow an RFID reader to remotely read such data. The reason we use an "active RFID" here is because CargoNet makes its RFID tag battery driven and shows good performance close to an intelligent sensor node.

Table 2.8 CargoNet Sensor Types

Sensor Type	*Measurement or Application*
Shock sensor	Potential impact damage
Vibration dosimeter	Average low-level vibrations
Tilt switch	Package orientation and shaking
Piezo microphone	Events causing loud nearby sounds
Light sensor	Container breach or box opening
Magnetic switch	Package removed or box opened
Temperature sensor	Overheating or potential spoilage
Humidity sensor	Potential moisture damage
RF wake-up	Query from reader or another tag

Source: Adapted from Malinowski, M. et al., CargoNet: A low-cost MicroPower sensor node exploiting quasi-passive wakeup for adaptive asychronous monitoring of exceptional events, *SenSys '07*, Sydney, Australia, November 6–9, 2007.

Now let us provide more details on CargoNet's "quasi-passive wake-up" strategy. Figure 2.8 shows its basic wake-up procedure. After an "active RFID" tag receives a stimulus signal, it compares the results against a threshold. If the stimulus is strong enough to warrant interest, the tag wakes up a larger system.

The above quasi-passive wake-up scheme needs to be built on the following conditions:

1. An always-enabled circuit, that is, the analog front end, should consume the order of a microwatt or less. With millivolt signal levels, a nanopower comparator (such as the LTC1540) is needed to boost the stimulus to logic levels and wake up the "active RFID" tag.
2. In terms of the wake-up time, the "active RFID" tag must wake up quickly enough to adequately process the incoming stimulus. MSP430, with a 6 μs start-up time, is therefore ideal.
3. The tag's duty cycles must be kept very low. This can help reduce the number of wake-ups and the amount of time spent in the active mode.

Besides the high-frequency, faster, longer-range radio for data communication with an RFID reader, the CargoNet "active RFID" tag also has a lower-frequency, shorter-range signaling channel for interrogation and passing of location information.

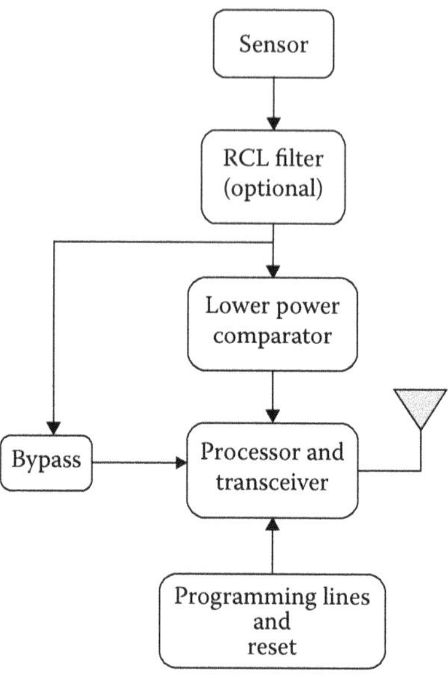

Figure 2.8 The CargoNet system quasi-passive wake-up scheme. (Adapted from Malinowski, M. et al., CargoNet: A low-cost MicroPower sensor node exploiting quasi-passive wakeup for adaptive asychronous monitoring of exceptional events, *SenSys '07,* Sydney, Australia, November 6–9, 2007.)

This is to make it compliant with other commercial RFID tags that detect the location information.

Due to the shorter range of the low-frequency radio link, high RF power can be delivered to the tag to wake it up. Then, the high-frequency radio is powered on. It consumes up to 20 mA of current when using CC2500.

Problems and Exercises

2.1 Multi-choice questions:

1. A sensor mote includes
   a. Analog/digital sensor chips
   b. RF transceiver
   c. CPU/memory
   d. All of the above
2. The differences between analog sensors and digital sensors do not include which of the following aspects?
   a. Analog sensors need standard chip-to-chip communication protocols to take with the CPU board while digital sensors do not need them.

   b. Analog sensors need compensation and linearization, but digital sensors do not need them.
   c. Digital sensors are a better choice than analog sensors from the CPU interface viewpoint.
   d. No ADC is needed in the digital sensor case.
3. In a sensor network, most of the sensor mote's energy is typically consumed in
   a. Analog-sensing part
   b. CPU local calculations on signal processing
   c. Wireless hop-to-hop communications
   d. Wake-up/sleeping transition
4. On the CPU in the sensor mote, which of the following is not correct?
   a. The CPUs used in sensor motes have much weaker capability than those used in general desktops or laptops. The sensor mote CPUs are often called microprocessors or microcontrollers.
   b. A CPU's working frequency in a sensor mote is typically below 100 MHz.
   c. When a CPU is in the idle/sleep mode, no energy consumption is involved.
   d. The main duties of a CPU are to execute communication protocols and locally process the data.
5. Which of the following is not correct on the sensor mote memory?
   a. Sensor nodes only require small amounts of storage and program memory.
   b. If data is to be stored for long periods of time, it is more efficient to use flash instead of SRAM.
   c. The program execution occurs in the flash memory instead of in SRAM.
   d. The typical SRAM size is <1 MB so far.
6. The radios on the sensor mote have which the following features?
   a. Low-power radios consume more energy in the receive mode than in the transmit mode.
   b. The sending distance of a wireless system is controlled by several key factors. The most intuitive factor is the transmission power.
   c. Most RF transceivers in the market today use a VCO-based radio architecture and have the ability to communicate at various carrier frequencies.
   d. AM is the simplest to encode and decode, and it is less susceptible to noise.
7. On the sender side, which of the following operations is not needed?
   a. Wait for the receiver's acknowledgment before sending out the next packet
   b. Encode the data by adding error detection bits
   c. Wait for collision free with the help of MAC protocols
   d. Organize sensor data to different packets

8. The reason(s) of decoupling between RF and processing speed could be which among the following?
   a. When the speed of the microcontroller is coupled to the data transmission rate, both pieces of the system are forced to operate at nonoptimal points.
   b. A radio is most efficient when data transmissions occur at its maximum transmission rate. When coupled with CPU processing, such efficiency cannot be achieved.
   c. RF and CPU are totally different chips and need to be decoupled in most cases.
   d. Both a and b.
9. Spec is better than Mica due to which of the following reasons?
   a. The Mica nodes were constrained by existing inter-chip interfaces. The development of a custom ASIC allows us to tear down the artificial constraints imposed by commercial components.
   b. It is possible to achieve orders-of-magnitude efficiency improvements on key communication primitives by using custom silicon.
   c. Both a and b.
   d. Spec can transmit signals for a longer distance than Mica.
10. Which of the following is not correct on Telos motes?
   a. Telos uses the Bluetooth communication standard (an IEEE 802.15 series), which makes it suitable for short-range radio communications.
   b. Telos uses the MSP430 microcontroller that has the lowest power consumption in sleep and active modes.
   c. Instead of integrating the design into silicon, Telos uses COTS components with hardware accelerators to build a power-efficient system that does not sacrifice performance.
   d. Telos is programmed (either with the bootstrap loader or JTAG) through an on-board USB that also provides power.

2.2 Do some Web research to find out the characteristics and the design principle of solar-based batteries.

2.3 What are the differences between "sensors" and "sensor motes"?

2.4 Read [Mateusz07] and provide more details on the integration of RFID into CargoNet sensor motes.

2.5 What advantages does the Telos mote have compared to others (such as Mica)?

NETWORK PROTOCOL STACK

Chapter 3

Medium Access Control in Wireless Sensor Networks

3.1 Introduction

A wireless sensor network (WSN) is a collection of different sensor nodes used to sense parameters such as vibration, temperature, pressure, sound, and pollutants in the environment. In WSNs, each sensor node is an autonomous device that consists of a communicating device, a computing device, a sensing device, and memory. To effectively exchange data among multiple sensor nodes, WSNs employ the medium access control (MAC) protocol to coordinate the signal transmissions over the shared wireless radio channel. Otherwise, multiple nodes may try to access the transmission medium (e.g., the wireless channel) simultaneously, which leads to signal collision, data loss, retransmission, wastage of energy, delay in data transmission, and so on.

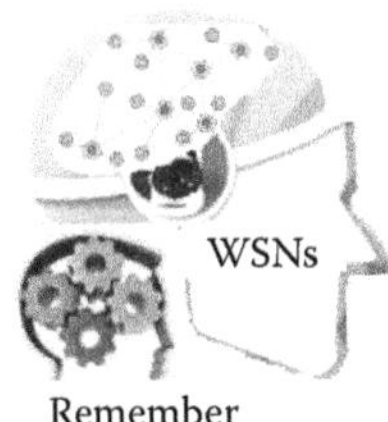

A MAC protocol determines how multiple nodes share the access to a physical medium (e.g., a wireless channel), by defining communication schedules and rules, such as (1) which nodes should occupy the channel, (2) when and how long the nodes can occupy the channel, and (3) how the nodes use the channel to talk with their neighbor nodes.

3.1.1 Medium Access Control in Wireless Networks

MAC protocols play a vital role in many network paradigms, including wired networks, mobile ad hoc networks (MANETs), and WSNs. The design of an efficient MAC protocol has to take into account the unique challenges that emerge in the respective networking paradigm. For example, unlike wired networks such as Ethernet, a wireless channel in WSNs generally experiences more data loss due to collision, signal loss, noises, and even link breakage. Signal collision occurring in a wireless link cannot be detected the same way as that in a wired link. In addition, a WSN owns very limited resources, such as energy, bandwidth, and computing capability, which constrains the applicability of the MAC protocols developed in other wireless networks, including Wi-Fi and MANET.

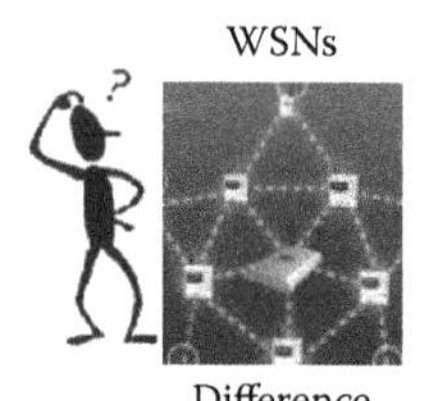

In general, the MAC schemes developed in other network paradigms cannot be directly applied in WSNs due to the unique and challenging issues posed by the wireless medium, the tiny sensor nodes, and various WSN applications.

3.1.2 MAC Design Is Challenging in WSNs

As a specific type of wireless networks, the WSN shares similar challenges faced in other wireless networking technologies. As elaborated below, such challenges and many other resource constraints in WSNs have significant impacts on how MAC is conducted in sensor nodes [AWoo01]:

1. Resource constraints
2. Signal loss in the wireless channel
3. Collision at the receiver's end
4. Hidden and exposed terminal problems

3.1.2.1 Resource Constraints

As described in Chapter 2, a wireless sensor node owns limited resources, such as power, bandwidth, computing capability, and storage space, which must be taken into account when devising MAC protocols in WSNs. Energy is a key concern in battery-powered sensor nodes. Once the battery is consumed, it is generally difficult or impractical to charge/replace exhausted batteries. That is why, the primary objective in many WSN MAC protocol designs is maximizing node/network lifetime, leaving the other performance metrics as secondary objectives. For example, the energy could be saved by turning off the devices that are not in use at the particular period of time.

Good idea

As the communication of sensor nodes is much more energy consuming than the computation, minimizing the communication while achieving the desired network operation has been one popular and effective approach in designing energy-efficient WSN MAC protocols.

The bandwidth in a WSN is pretty low when compared to that of wired networks, such as fiber optical networking. The bandwidth constraint and the dynamics of WSN topology also impose challenging issues to be considered in the MAC design. Specifically, in WSNs, the data is sensed and stored in a distributed fashion and every sensor node is an autonomous device that is independent of other nodes in the network. Sensor nodes need to communicate with one another to self-organize as a network system for data transmission, whereas redundancies should be avoided. Moreover, the sensor nodes may fail due to the fact that the tiny sensor nodes are fragile. The topology of the network changes when the nodes' failures occur in the network. Similarly, power depletion and node movement may also result in network topology change.

3.1.2.2 Signal Loss in Wireless Channel

WSNs employ wireless channels as the transmission media, which suffer signal distortion and loss due to attenuation, reflection, diffraction, scattering, and so on. Signal attenuation generally refers to the loss of energy as the transmitted signal travels from the source node to the destination node through air. The transmitted signal can get reflected when there are obstacles between the source node and the destination node. The edges of the obstacles can result in multiple signals divided from the original transmitted signal, and the rough surfaces of the obstacles can cause scattering due to multiple signal reflections. A commonly used wireless propagation mode with an omnidirectional antenna was introduced in [Rappaport96], whereby the signal power received at node j from the sender node i is given by the following equation:

$$P_j = \beta \frac{P_i}{d_{ij}^{\alpha}} \tag{3.1}$$

In Equation 3.1, P_j, P_i, and d_{ij} represent the power received at node j, the power sent out from node i, and the distance between node i and node j, respectively, while

α and β denote the energy loss constant, typically depending on the wireless transmission environment. This equation also indicates that the longer the wireless signal propagates in the air, the more the power loss can occur. In fact, a wireless node i can reach another node j (or a wireless link exists from node i to node j) if and only if node i transmits at a certain power level. Otherwise, the receiver j cannot properly decode the signal for the transmitted data information from sender i or node j cannot hear the signal from node i at all due to the power loss. In other words, each sensor node has a limited transmission range. For battery-powered WSNs, the transmission range of a node varies dynamically and the wireless links among nodes are susceptible to failures/changes, which necessitates different WSN link access control schemes.

3.1.2.3 Collisions Occurring at the Receiver's End

When two or more sensor nodes send data to other nodes simultaneously through the same channel, multiple signals might collide at the receiver side, which prevents the receiver from obtaining meaningful data information. To ensure reliable data transmission, MAC protocols have to define processes (e.g., retransmission after random delay) to recover from the collision. Collisions result in wastage of energy, lower bandwidth utilization, and larger data delivery latency. In wired networking, such as Ethernet, collisions can be easily detected by comparing the sent signal and the received signal at the sender side. Accordingly, the sender (e.g., in Ethernet) concludes that another sender is also sending data and performs some operations to recover from the collision quickly. However, in WSNs, the signal sent from the sender is not equal to that received by the receiver due to signal loss or obstacles. For example, assume that two senders are sending data to the same receiver simultaneously. If the two senders are not within each other's transmission range or there is an obstacle preventing two senders from hearing each other, the signals from the two senders collide at the receiver, which cannot be detected by either sender.

3.1.2.4 Hidden Terminal and Exposed Terminal Problems [AWoo01]

As shown in Figure 3.1, the circle around each node represents the corresponding transmission range of the node when omnidirectional antennas are employed, and we assume that all nodes have the same transmission range. Two sensor nodes are said to be in mutual range (or in the same collision domain) when the transmission ranges of the two nodes interfere with one another. For example, nodes 1 and 2 are in mutual range. Similarly, nodes 2 and 3 are in mutual range, while nodes 4 and 3 share the same collision domain. Obviously, when a node is receiving data from a neighbor, there can be only one valid transmission (or sender) within the node's mutual range. Otherwise, multiple signals will collide at the receiver, which results in data loss and energy wastage. To minimize collisions, carrier sense is widely used in the design of MAC protocols. With carrier sense, the transmitter listens to the

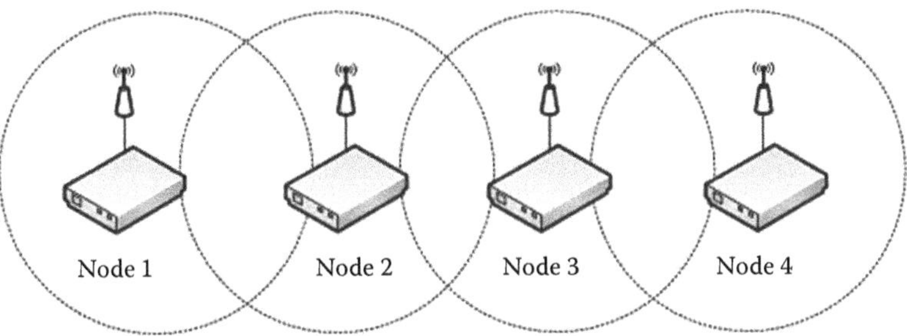

Figure 3.1 Hidden and exposed terminal problems.

transmission channel for a carrier signal to detect if there is an ongoing transmission from another node before attempting to send data. If a carrier is sensed or there is an ongoing transmission in the medium, the node can wait for the transmission in progress to finish before initiating its own transmission.

The carrier sense scheme serves well in the Ethernet MAC protocol. However, the significant difference between the signal sent from the sender and the signal received by the receiver makes carrier sense much ineffective in a wireless environment. For example, assume that there is no ongoing communications among the nodes in Figure 3.1. At one time, both sensor node 1 and sensor node 3 sense some events and decide to notify node 2. Before sending data, both node 1 and node 3 sense the channel and learn that the channel is free. Hence, if node 1 and node 3 start to send data to node 2 simultaneously, this leads to data collision at node 2. Even when there is an ongoing transmission from node 3 to node 2, node 1 cannot sense the signal from node 3 as it does not share the mutual range. Subsequently, node 1 may assume that the channel is unoccupied and is not aware that node 2 is already engaged in a transmission. The signal from node 1 may disrupt the transmission from node 3 to node 2. This is because node 3 is invisible to node 1 even though both can reach node 2. This is well known in wireless networking as the hidden terminal problem.

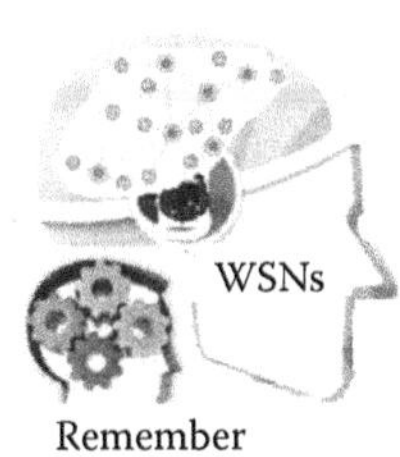

The hidden terminal problem indicates that carrier sense in wireless networking may fail to avoid collisions. In addition, carrier sense can result in channel underutilization in wireless networking.

Consider that node 2 is transmitting data to node 1 and node 3 also intends to send data to node 4. Node 3 performs carrier sense and finds that the transmission channel is occupied and has to wait for the finish of the transmission from node 2 to node 1. However, only the signal interference or collision at the receiver side leads to data loss, and energy and bandwidth wastage. In fact, node 3 and node 2 can simultaneously send data to node 4 and node 1, respectively. This is because the interference between the two transmissions (i.e., node 2 to node 1 and node 3 to node 4) does not occur at the receiver side. Hence, node 3 is prevented from sending data to node 4 even though both node 4 and node 1 should be able to receive the respective data properly. This is called the exposed terminal problem in wireless networking.

There has been extensive research in the MAC protocol design to resolve the above challenges and problems in WSNs (e.g., [Ftobagi75, Pkarn90, Bharghavan93]). Several earlier MAC research results on carrier sense and the hidden terminal problem in wireless networking are collectively adopted by the IEEE 802.11 standard [IEEE07], which also serves as the basis for many MAC protocols proposed for WSNs. Hence, in the rest of this chapter, we first briefly go through the IEEE 802.11 project. Then, we present the classification of MAC protocols, followed by discussions on several typical sensor MAC protocols of each category, which include Sensor Medium Access Control (S-MAC) [Wye02], Timeout Medium Access Control (T-MAC) [Tvdam03], Traffic Adaptive Medium Access (TRAMA) [Vrajendran06], Sift Medium Access Control [Kjamieson03], Zebra MAC (Z-MAC) [Irhee08], and Berkeley MAC (B-MAC) [Jpolastre04].

3.2 Overview of Project IEEE 802.11

Figure 3.2 shows the layered architecture of the IEEE 802.11 project. The physical layer in IEEE 802.11 contains direct-sequencing spread spectrum (DSSS),

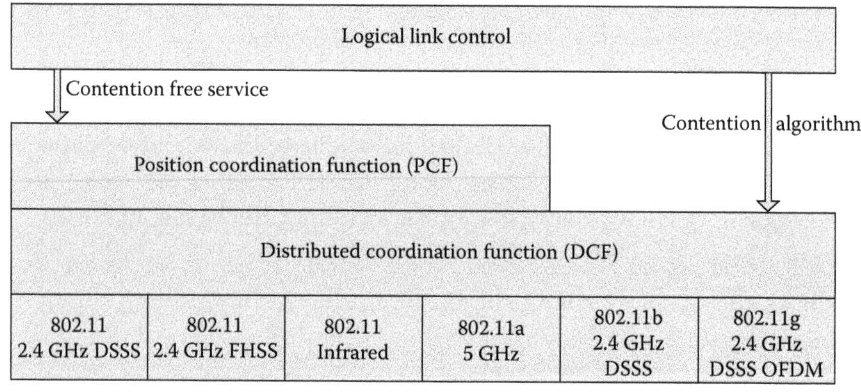

Figure 3.2 IEEE 802.11 protocol architecture. (Adapted from Stallings, W., *IT Prof.*, 6, 32, September–October, 2004.)

frequency-hopping spread spectrum (FHSS), infrared, 802.11a, 802.11b, and 802.11g. The DSSS defines the physical medium in the frequency of 2.4 GHz or 5 GHz ISM band, at the data rates of 1–54 Mbps. The FHSS employs the physical media of the same frequency and the same data rates as those of the DSSS. But the basic difference between them is the number of channels. The number of channels depends on the network regulatory agencies in every country. For the DSSS, it varies between 13 in European nations and 1 in Japan. While for the FHSS, it varies between 70 in the United States and 23 in Japan. Similarly, the infrared has the same data rates as that of the FHSS and the DSSS. But the infrared uses the wavelengths in the range of 850–950 nm.

The data link layer includes logical link control and MAC, which defines two access methods: the distributed coordination function (DCF) and the point coordination function (PCF).

3.2.1 Point Coordination Function

The 802.11 MAC defines the point coordination access scheme called PCF to provide contention-free service, which is available only in the infrastructure mode, as shown in Figure 3.3. In the infrastructure mode, stations are connected to the

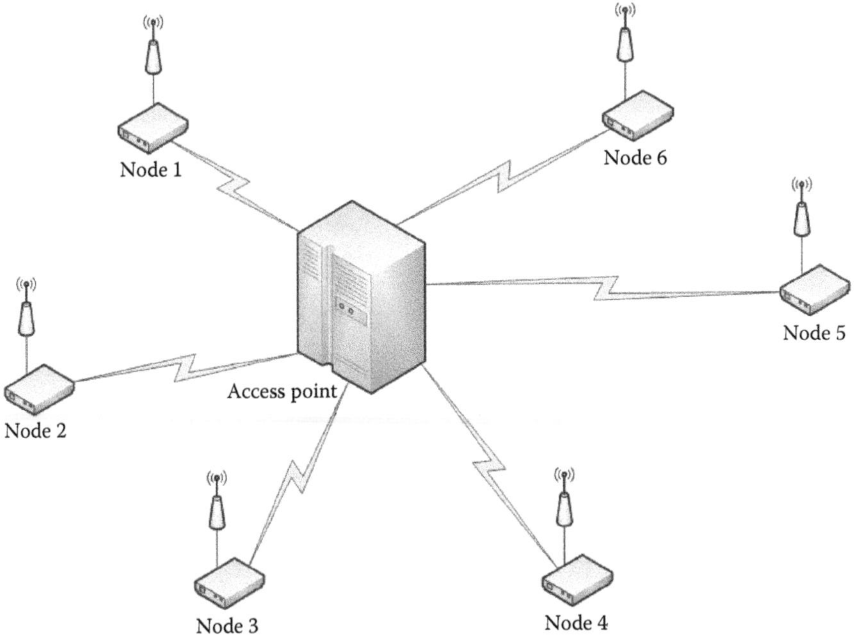

Figure 3.3 IEEE 802.11 infrastructure mode. (Adapted from Stallings, W., *IT Prof.*, 6, 32, September–October, 2004.)

network through an access point (AP), which employs a centralized MAC algorithm. This mode can conveniently support high traffic priority.

3.2.2 Distributed Coordination Function

The DCF is defined to share the medium between multiple stations in an ad hoc mode, as shown in Figure 3.4. The DCF enables the stations to exchange data asynchronously by applying Carrier Sense Multiple Access with Collision Avoidance (CSMA/CA) and the IEEE 802.11 RTS/CTS to share the medium between stations [Pkarn90, Bharghavan93]. In CSMA/CA, a station has to first listen to the channel for a predetermined amount of time so as to check whether the channel is free before sending data. If the channel is sensed busy before transmission, then the transmission is deferred for a "random" interval to avoid collisions. Note that, as mentioned earlier, collision detection is not feasible due to the nature of the wireless channel and the hidden terminal problem. Hence, the scheme of exchanging the request-to-send (RTS) packet and the clear-to-send (CTS) packet is introduced in IEEE 802.11 to alert all nodes within the range of the sender, the receiver, or both, to keep quiet during the main data transmission.

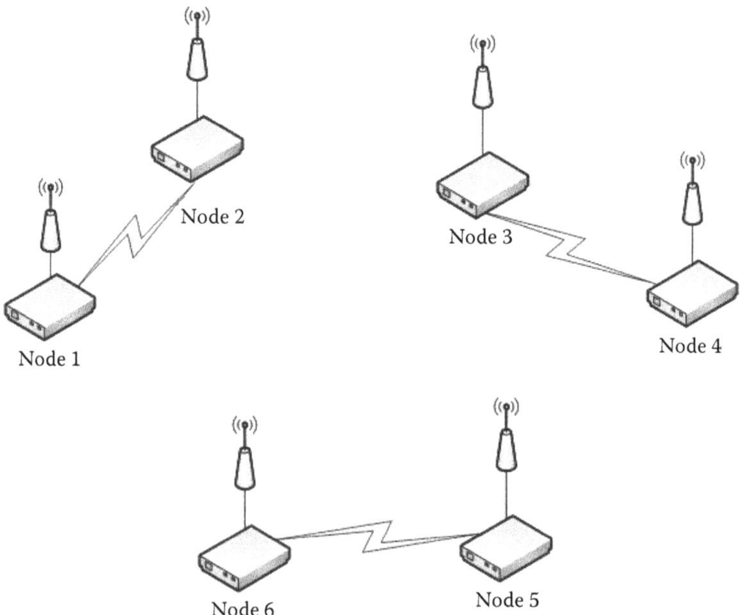

Figure 3.4 IEEE 802.11 ad hoc mode. (Adapted from Stallings, W., *IT Prof.*, 6, 32, September–October, 2004.)

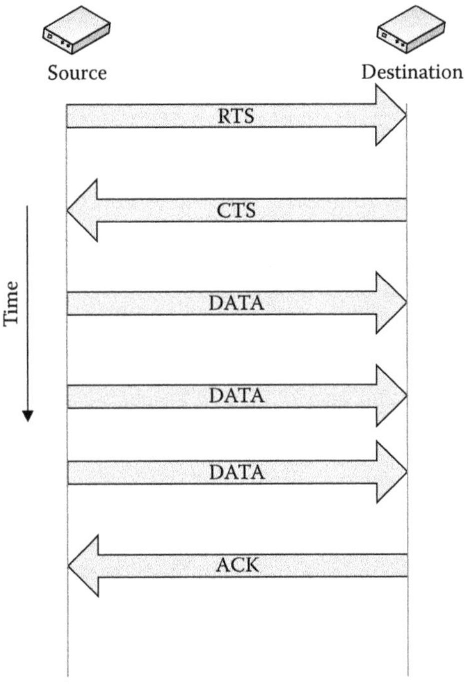

Figure 3.5 Data transfer using RTS/CTS mechanism.

As shown in Figure 3.5, the source node (or sender) sends an RTS packet to the destination (or the receiver) if the sender wants to send data to the receiver. The destination node replies with a CTS packet. Any other node receiving the RTS or the CTS packet should refrain from sending data for a given time to avoid collisions (or solve the hidden node problem). The amount of time for which the node should wait before trying to get access to the wireless medium is included in both the RTS and the CTS packet. The primary fields of the RTS and the CTS packet are shown in Figures 3.6 and 3.7, respectively.

There are five fields in the RTS packet format, which are

1. Frame control (2 bytes): This field contains the information about the version of the protocol used, power management, whether there are more fragments of the data, and whether or not the packet is protected.
2. Duration (2 bytes): It is the time remaining for transmitting the data or management information plus one CTS frame and one ACK (acknowledgment) frame.
3. Receiver Address (RA) (6 bytes): It is the address of the intended destination.
4. Transmitter Address (TA) (6 bytes): It is the address of the source that initiated the data transfer.

Bytes	2	2	6	6	4
	Frame control	Duration	RA	TA	FCS

Figure 3.6 RTS packet format. (Adapted from IEEE Standard for Information technology—Telecommunications and information exchange between systems—Local and Metropolitan area networks—Specific requirements, *Part 11: Wireless LAN Medium Access Control (MAC) and Physical Layer (PHY) Specifications,* pp. 120–121, July 2007.)

Bytes	2	2	6	4
	Frame control	Duration	RA	FCS

Figure 3.7 CTS packet format. (Adapted from IEEE Standard for Information technology—Telecommunications and information exchange between systems—Local and Metropolitan area networks—Specific requirements, *Part 11: Wireless LAN Medium Access Control (MAC) and Physical Layer (PHY) Specifications,* pp. 120–121, July 2007.)

5. Frame Check Sequence (FCS) (4 bytes): It is used to check for errors in data transmission. It is a cyclic redundancy code (CRC) of a length of 32 bits. It is calculated for all the fields, including the header, using a 32 degree polynomial.

There are four fields in the CTS packet format, which are

1. Frame control (2 bytes): This field contains the information about the version of the protocol used, power management, whether there are more fragments of the data, and whether or not the packet is protected.
2. Duration (2 bytes): It is the difference between the duration field received from the source and the time in the CTS frame.
3. RA (6 bytes): It is the address of the intended destination, which is copied from the TA field in the RTS packet format. If the CTS packet is the first packet that the destination is transmitting, then RA is the transmitter's MAC address.
4. FCS (4 bytes): It is used to check for errors in data transmission. It is a CRC of a length of 32 bits. It is calculated for all the fields, including the header, using a 32 degree polynomial.

Upon receiving the CTS packet, the sender can initiate data transmission to the receiver. If the data is successfully received by the receiver, then the receiver sends an ACK to the sender, as shown in Figure 3.5.

3.3 Classification of MAC Protocols

Traditionally, there are four different channel access schemes: time division multiple access (TDMA), frequency division multiple access (FDMA), code division multiple access (CDMA), and space division multiple access (SDMA) [Keoliver05]. In TDMA, all the nodes use the same frequency channel and each node is assigned with a designated time slot(s) for data transmission. The nodes transmit in rapid succession, one after the other, each using its own time slot. Time synchronization among the nodes accessing the shared medium is required for the success of the TDMA scheme. The technique that FDMA uses is similar to that of TDMA. The only difference is that instead of dividing the time, FDMA allocates different frequencies to each node. CDMA employs the spread spectrum technology and a special coding scheme to allow multiple users to share the same physical channel where each node is assigned a unique code. SDMA, on the other hand, uses the spatial separation of the nodes for multiple channel access through spatial multiplexing or diversity. In general, different networking technologies may share access via different methods or a combination of multiple methods, such as TDMA, FDMA, CDMA, or SDMA.

In wireless networks, the medium access scheme can be distributed and centralized [Achandra00]. Based on the mode of operation, wireless MAC protocols can also be broadly classified as random access protocol, guaranteed access protocol, and hybrid access protocol. In the random access MAC protocol, each node tries to access the transmission medium in a random manner, while in the guaranteed access MAC protocol, nodes access the transmission medium in a systematic manner by employing a master–slave procedure or sharing the token to take their turn. Hybrid protocols use a blend of guaranteed access and random access for accessing the transmission medium. Similarly, for resolving the challenges such as the hidden terminal problem, resource constraints, and application requirements, researchers in the literature have investigated a number of MAC protocols specifically for WSNs through either extension of existing MAC protocols or proposing new medium access concepts. Based on the method used for contention avoidance, MAC protocols in WSNs can be roughly classified into three categories as follows:

1. Contention-based MAC protocols
2. Schedule-based MAC protocols
3. Hybrid and event-based MAC protocols

3.3.1 Contention-Based MAC Protocols

Contention-based MAC protocols allow multiple nodes to access the medium at the same time. Collisions may then occur, but are handled with different contention resolutions, such as random backoff, RTS/CTS exchange, and collision avoidance techniques. A classic example is Carrier Sense Multiple Access (CSMA), in which a node senses the medium for ongoing communication before attempting a message transmission. If the node finds that the medium is busy, it will back off and retry later. When the medium is sensed to be clear, the node waits for a random period, the contention period, before transmitting. The contention period decreases the probability of two nodes beginning to transmit at the same moment and, therefore, reducing data collision. The scheme of RTS/CTS message exchange in IEEE 802.11 DCF and time-out are often combined with the unique features of WSN applications in contention-based MAC protocols to optimize the network performance in terms of energy consumption, lifetime, latency, or throughput. Examples of contention-based MAC protocols in WSNs are S-MAC [Wye02], T-MAC [Tvdam03], WiseMAC [Aelhoiydi04], DMAC [Glu04], DSMAC [Plin04], AC-MAC [Fli06, Jai04], and so on. In the following text, we introduce the basic protocol designs of S-MAC and T-MAC protocols.

3.3.1.1 Sensor Medium Access Control [Wye02]

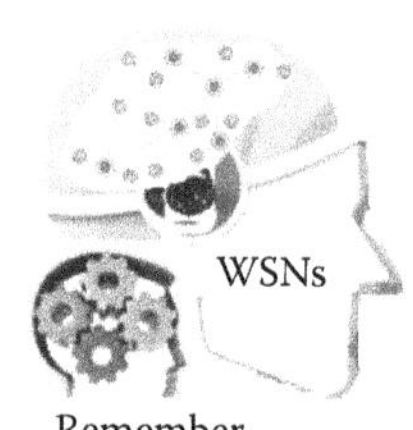

It is observed that energy in WSNs is wasted in multiple processes, including *idle listening*, *data collisions*, *overhearing*, and *control overhead*.

Idle listening is a state where the sensor node waits for another node to possibly transmit the data to it. In many sensor network applications, if nothing is sensed, nodes are in the idle mode for most of the time. However, traditional MAC protocols, such as IEEE 802.11 and CDMA, require nodes to listen to the channel for possible transmission. Studies show that idle listening consumes 50 percent to 100 percent of the energy required for receiving [Stemm97]. In many WSN applications, the nodes stay in the idle state far longer than in the communication state, which in fact consumes a significant portion of the nodes' energy. Data collisions lead to the corruption of data in the transmitted packet that has to be discarded, and the follow-on retransmissions increase energy consumption as well as network latency. Similarly, overhearing transmissions among other nodes and the control overhead can contribute to the energy wastage within WSN nodes.

The S-MAC protocol tends to reduce the aforementioned energy wastage using periodic sleep and a listen cycle, while introducing some penalty on the per-hop latency of data transmission. The S-MAC protocol assumes that all the nodes are used for one application or a set of applications. As the sensor nodes have one common application goal and there might be a situation where one node holds more information than the other nodes, S-MAC applies the concept of message passing to allow the node holding more data to access the channel longer, which in fact preserves the more important application-level fairness rather than per-hop fairness.

3.3.1.1.1 Periodic Listen and Sleep

As sensor nodes stay in the idle state quite often in many WSN applications, the S-MAC protocol introduces a set of sleep and wake-up states to reduce the energy wasted in idle listening. In the sleep state, the nodes turn off their communication devices (which contribute the most to the energy consumption) and keep other components on. By following a schedule, the nodes move from the sleep state to the wake-up state, after a certain time interval. This time interval depends on the application for which the nodes are being used. In the wake-up state, the nodes turn on their communication devices and participate in the necessary communication with other nodes.

To follow the schedule for sleep/wake-up and communicate with neighbors in time, the nodes in the S-MAC protocol require periodic synchronization among their neighbors. To avoid time synchronization errors, S-MAC uses two techniques. First, all the time stamps used for synchronization are not absolute but relative. Second, the listen period is longer than the clock drift. In the S-MAC protocol, the nodes are free to choose the listen/sleep schedules, but it is preferred that the neighboring nodes should synchronize with each other to reduce the control overhead because a node can communicate with another node only if both are in the wake-up state. In other words, it is ideal for the neighborhood nodes to listen at the same time and go to sleep at the same time. However, in a multi-hop network, as shown in Figure 3.8, not all neighboring nodes can synchronize together to follow the same listen/sleep schedule. For example, sensor nodes A and B follow the same listen/sleep schedule. Similarly, sensor nodes C and D follow the same listen/sleep schedule, which might be different from the schedule followed by sensor nodes A and B. A node exchanges its schedule by broadcasting it to all its immediate neighbors. This ensures that all neighboring nodes can talk to each other even if they have different schedules.

Figure 3.8 An example of a four-node network. (Adapted from Ye, W. et al., An energy-efficient MAC protocol for wireless sensor networks, *Proceedings of IEEE INFOCOM*, New York, June 2002, Vol. 3, 1567–1576.

In the S-MAC protocol, if a node wants to talk to a neighbor, the node must wait until the neighbor listens (or is in the wake-up state). If more than one neighbor wants to talk with a node, the neighbors have to contend to access the medium when the node is in the wake-up state. For this contention, the scheme of RTS/CTS exchange is adopted. The node which first sends out the RTS packet owns the right to access the medium, and the receiver replies with a

CTS packet. Upon receiving the CTS packet, the node can finish the data transmission and follow the sleep or the listen schedule.

3.3.1.1.2 Choosing and Maintaining Schedules

Each node should choose a schedule, before the periodic listen and sleep, and exchange the schedule with its neighbors. The schedule is stored in a table that contains the schedules of the neighboring nodes. The selection of the schedule and the insertion of schedules of its neighbors are performed as follows.

1. Every node listens to the transmission channel for a certain period of time. If the node does not receive a schedule advertisement, the node randomly selects its own listen/sleep schedule and broadcasts the schedule in a SYNC packet to its neighbors specifying that it moves into the sleep state after t seconds. In the absence of a neighbor's schedule to follow, this node chooses its schedule independently, and is called a synchronizer.
2. During the listen period, if a node receives a SYNC packet from its neighbor prior to randomly selecting its own schedule, the node will follow the schedule specified in the SYNC packet it received from the neighbor node. Such a node is called a follower. Assume that the follower recognizes that the sender of the SYNC packet will move into the sleep state in t seconds. After waiting for a random delay of t_d seconds to avoid potential collision from other followers, the follower rebroadcasts the schedule and specifies that it moves into the sleep state after $t - t_d$ seconds.
3. If a node selects a schedule and then receives a different schedule from its neighbor, then the node stays in the wake-up state by following both the received schedule from its neighbor and its original schedule.

3.3.1.1.3 Maintaining Synchronization

Synchronization among the nodes in WSNs is maintained by sending SYNC packets. SYNC packets contain the address of the source node and the time of its next sleep. To remove clock synchronization errors, the time of the next sleep is not absolute but is relative to the time of transmission of the SYNC packet, which is approximately equal to the time of reception of the packet by the destination. The destination node will start the timer immediately after receiving the SYNC packet. When the timer expires, the node moves into the sleep state. To send data packets and SYNC packets, the wake-up period is divided into two parts. In the first part, the nodes receive the SYNC packets, and the second part is for receiving the RTS packets. Each part is further divided into slots for carrier sense prior to the channel access of SYNC or data packet transmission.

Each node periodically broadcasts its schedule in SYNC packets to its neighbors such that the newly joined nodes can follow the same schedule. For the newly joined nodes, the schedule selection process is the same as that described above. Before identifying itself as a synchronizer, the newly joined node will set the initial listen period long enough to increase the probability of picking up a neighbor's schedule.

3.3.1.1.4 Collision and Overhearing Avoidance

To avoid multiple neighbors sending data to a node simultaneously, S-MAC adopts the RTS/CTS exchange as well as the virtual and physical carrier sense mechanisms, which is proven to be an effective approach to address the hidden terminal problem [Pkarn90, Bharghavan93, IEEE07]. All the nodes initially should sense the carrier before initiating a data transmission. If the source node senses the channel and concludes that the channel is busy, it moves into the sleep state. The source node wakes up again when the destination node is in the wake-up state. S-MAC sends the broadcast packets, such as SYNC packets, directly without employing RTS/CTS exchange. For unicast packets, the source and destination nodes follow the RTS/CTS/data/ACK sequence during the data transmission process. In addition, every data packet contains a field that indicates the remaining transmission time it needs. This is similar to the concept of network allocation vector (NAV) in IEEE 802.11. Hence, a node knows how long it has to keep silent or move back to the sleep state (prior to accessing the channel) after receiving a packet destined to another node.

The S-MAC protocol reduces the energy that is wasted in overhearing. S-MAC moves any node into the sleep state whenever the node hears an RTS or a CTS packet. This is due to the fact that subsequent data and ACK transmission will normally take much longer time. For example, in Figure 3.9, node C is sending the data to node D. It is clear that node D and node C should not be in the sleep state. As the collisions occur on the receiver's end, node E cannot send data and should be in the sleep state to avoid collisions at node D. Node B theoretically can send data to node A while it is in the wake-up state, because node D is not in node B's transmission range. However, node B cannot receive any reply from node A, and node B's transmission could cause collisions at node C when node C tries to receive the ACK. Similarly, node E cannot participate in the data transmission when node C is talking with node D. Hence, all immediate neighbors of both the sender and the receiver should sleep after they hear the RTS or the CTS packet. In other words, based on the NAV information carried in the RTS/CTS packet, the node can sleep to avoid overhearing until the current transmission is over.

Figure 3.9 An example of overhearing avoidance. (Adapted from Ye, W. et al., An energy-efficient MAC protocol for wireless sensor network, *Proceedings of IEEE INFOCOM*, New York, June 2002, Vol. 3, 1567–1576.)

3.3.1.1.5 Message Passing

A message is a collection of meaningful data that can be one large packet or a series of small packets. On the one hand, when a long message embedded in one packet is a corrupted message, the retransmission is costly in terms of energy consumption, latency, and bandwidth utilization. On the other hand, the transmission of a long message by using multiple short and independent packets results in significant control overhead, such as the RTS/CTS exchange. Hence, the S-MAC protocol breaks up the long message into many small fragments and transmits them in burst. Only one RTS/CTS exchange is employed for the whole burst to reserve the medium for transmitting all the fragments. The transmission of a data fragment is assumed to be successful only if the sender receives the ACK from the receiver. If the sender does not receive the ACK packet, it will extend the reserved transmission time for one more fragment and retransmit the current fragment immediately. The ACK packet is used after receiving each data fragment to overcome the hidden terminal problem. The NAV information of the current transmission is also present in the ACK and data packets. In this way, a node in the path could know about the remaining duration for the ongoing data transmission even when there are corrupted packets or the node wakes up in the middle of the data transmission.

3.3.1.1.6 Energy Saving versus Increased Latency

To analyze the delay penalty introduced by S-MAC, let us first take a look at the delays that are inherent to contention-based MAC protocols (e.g., IEEE 802.11 DCF) in a multi-hop network. The delays include carrier sense delay, backoff delay, transmission delay, propagation delay, processing delay, and queuing delay. However, S-MAC introduces an extra delay, called *sleep delay*, which is experienced by the source when it finds the intended destination in the sleep state. In this case, the source node needs to wait until the destination node moves into the wake-up state. Assume that a frame is defined as a complete cycle of listen and sleep. Then, the average sleep delay will be as given by Equation 3.2 when the data packet arrives with equal probability during a frame:

$$D_s = \frac{T_{\text{frame}}}{2} \tag{3.2}$$

where

D_s denotes the sleep delay

T_{frame} denotes the time frame and is a sum of T_{listen}, denoting the time period of the listen state, and T_{sleep}, denoting the time period of the sleep state, as given by Equation 3.3

$$T_{\text{frame}} = T_{\text{sleep}} + T_{\text{listen}} \tag{3.3}$$

The relative energy saving from S-MAC is given by Equation 3.4, where the last item is the duty cycle of the node. It can be seen that the smaller the listen period, the shorter is the average sleep delay.

$$E_s = \frac{T_{\text{sleep}}}{T_{\text{frame}}} = \frac{T_{\text{frame}} - T_{\text{listen}}}{T_{\text{frame}}} = 1 - \frac{T_{\text{listen}}}{T_{\text{frame}}} \tag{3.4}$$

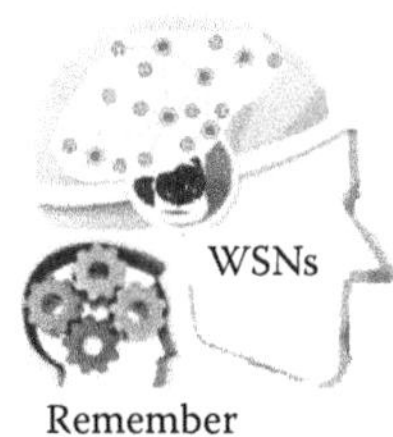

The S-MAC protocol reduces the energy consumption of the nodes, thereby increasing the lifetime of the entire network. But S-MAC introduces some delay to lessen the energy consumption. Thus, it may not be a good idea to apply S-MAC for MAC in a WSN that is used for time-critical applications.

3.3.1.1.7 Evaluating Performance of S-MAC Protocol

Experiments in [Wye02] show that S-MAC has very good energy-conserving properties as compared to those of IEEE 802.11 DCF. On a source node, an IEEE 802.11-like MAC consumes two to six times more energy than S-MAC for traffic load with messages sent every 1–10 s. To reduce the latency in S-MAC, a new technique called *adaptive listen* is introduced in [Wye04]. The basic idea is to switch the nodes from the low-duty-cycle mode to a more active mode. Specifically, adaptive listen lets the node that overhears its neighbor's transmissions (ideally, only RTS or CTS) wake up for a short period of time at the end of the ongoing transmission. Rather than waiting for the scheduled listen time, the wake-up node can immediately receive data from the neighbor if it is the next-hop node. Otherwise, the node will go back to sleep until its next scheduled listen time.

3.3.1.2 Timeout MAC [Tvdam03]

To further resolve the problem of idle listening in a WSN, T-MAC is proposed as another contention-based MAC protocol to reduce energy consumption by turning off the radio components of the nodes when they are not needed [Tvdam03]. The basic idea of T-MAC is to turn on the radio components of the node at a synchronized time and turn them off after a certain time-out when no communication occurs for some time. Unlike its predecessor, S-MAC, which turns on the radio according to a predefined schedule, T-MAC dynamically adapts a listen/sleep duty cycle in a different way through fine-grained time-outs. As a result, the T-MAC protocol can save more energy than S-MAC in a network where message rates vary.

3.3.1.2.1 Protocol Design

Similar to the S-MAC protocol, the T-MAC protocol also uses the periodic sleep and wake-up states to save energy in WSNs. In the sleep state, the node has the sensing devices turned on and the data sensed is put into the queues. The node in the sleep state also accepts new messages from the neighboring nodes, and these messages are queued. In the active or the wake-up state, the nodes keep listening and transmitting data as needed. For data transmission in the active state, T-MAC adopts the RTS/CTS/data/ACK scheme, to provide collision avoidance and reliable transmission. A node transits from the active state to the sleep state when no *activation event* occurs within a time span of TA. An activation event is defined as one of the following:

1. The expiration of a periodic frame timer
2. The reception of any data on the radio
3. The sensing of communication on the radio
4. The end of transmission of a node's own data or ACK packet
5. The knowledge (obtained through overhearing prior to RTS and CTS packets) that a data exchange of a neighbor has ended

The minimal amount of idle listening per frame is determined by the value of TA. As messages received in the sleep state must be buffered, the maximum frame time is bounded by the buffer capacity.

3.3.1.2.1.1 Clustering and Synchronization — Synchronization in the T-MAC protocol is done using a technique called virtual clustering [Wye02]. In virtual clustering, nodes with the same schedule form clusters, without enforcing the same schedule to all nodes in the network. Virtual clustering allows a node to broadcast the schedule and anticipates that the node maintains the schedules of its neighboring nodes. Initially, every node starts its operation by listening and waiting. If a node receives nothing after listening and waiting for a certain period of time, it chooses a frame schedule and broadcasts its SYNC packet to the neighbors. On the other hand, if the node receives a SYNC packet from any one of the neighbors, then it follows the same schedule in the SYNC packet it received. Furthermore, if the node receives a SYNC packet after broadcasting its own SYNC packet, then it follows both schedules and notifies the sender of the SYNC packet that there exists more than one schedule. Nodes broadcast their schedules once in a while. At irregular intervals, the nodes listen for complete time frames, so that the nodes could detect different schedules that exist in the same cluster. The nodes should transmit the data at the start of the active state, as the neighbor nodes within the virtual cluster (with the same schedule) and the neighbors that have adopted the schedule as extra are awaken in the active state.

3.3.1.2.1.2 Contention Resolution — In T-MAC, a frame consists of an active state and a sleep state. In the sleep state, the sensed data is queued to be transmitted. Hence, at the beginning of the active state in a frame, each node may have buffered a large amount of data in the form of data burst (to be sent out). This results in higher contention for the medium access at the beginning of the active state. RTS/CTS exchange is employed in T-MAC for the channel contention. A node keeps sensing the medium for a random time with a fixed contention interval before sending an RTS packet. After sending an RTS packet, the sender may not receive a CTS reply if the receiver is in the sleep state. Even in the active state, the receiver may not be able to send a CTS reply if the RTS packet is lost due to collision or the receiver is prohibited from replying due to an overheard RTS or CTS. Because the receiver could be in the active state, it makes sense for the sender to retry the RTS transmission. The sender will go to sleep if there is still no CTS reply after two retires.

As mentioned earlier, the active state ends when no activation has occurred for a period of TA, which means the sender will automatically transit to the sleep state if it does not receive the CTS packet in time. Therefore, the value of TA must be selected such that the sender is able to receive the CTS reply [Tvdam03]. For a third neighbor node that overhears the RTS or the CTS packet, unlike S-MAC, which requires the node to go to the sleep state, T-MAC keeps overhearing as an option. The argument is that the overhearing avoidance could dramatically decrease the throughput of WSNs, as it is very possible that the node that overhears RTS/CTS is the receiver of a subsequent message.

3.3.1.2.1.3 Early Sleeping in T-MAC — The research in [Tvdam03] found that T-MAC does not perform well when all the nodes send the data to a data sink. For example, assume that there are four sensor nodes, A, B, C, and, D, and the messages flow only in one direction: A→ B→ C→ D, as shown in Figure 3.10.

To communicate with node D, node C has to contend for the transmission channel. Node C may lose the contention of the transmission channel to node A or node B. If node C loses the contention due to an RTS packet from node B, node C shall send a CTS reply to node B, which will be overheard by node D. Accordingly, node D can anticipate itself as the subsequent receiver and wake up when the communication between node C and node B is over. However, node C must remain silent if it loses the contention due to overhearing the CTS packet from node B to node A. In this case, node D, which is totally blind to the communication between node A and node B, will go to sleep after the expiration of the TA timer. Hence, in the next contention round, although node C wins the

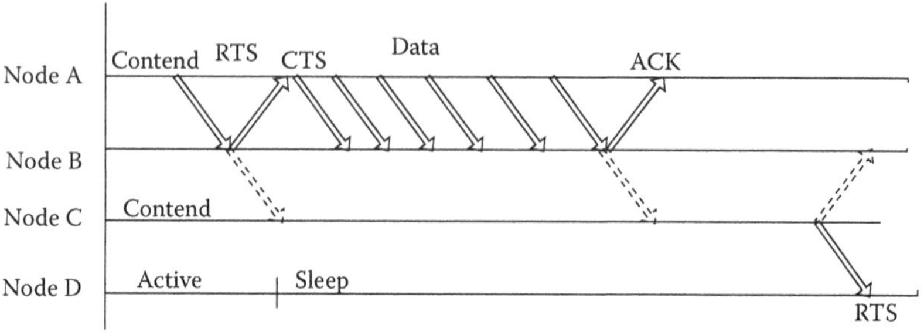

Figure 3.10 Early-sleeping problem. (Adapted from van Dam, T. and Langendoen, K., An adaptive energy-efficient MAC protocol for wireless sensor networks, *Proceedings of the First International Conference on Embedded Networked Sensor Systems,* Los Angeles, CA, November 2003, ACM, New York, 171–180.)

contention, it cannot talk with node D who is in the sleep state. This observed behavior is called the *early-sleeping problem*, as a node moves to the sleep state even though a neighbor intends to communicate with it. There are two possible solutions for the *early-sleeping problem*: future request to send (FRTS) and taking priority on full buffers.

3.3.1.2.1.4 Future Request to Send — The basic idea of *FRTS* is to inform another node that there will be a message for it even though the transmission medium is not available at the current time. The operation of FRTS is shown in Figure 3.11. Once node C overhears the CTS packet from node B to node A, node C can immediately transmit a special packet called the FRTS packet to node D if node C has data for node D. The FRTS packet contains its own destination as well

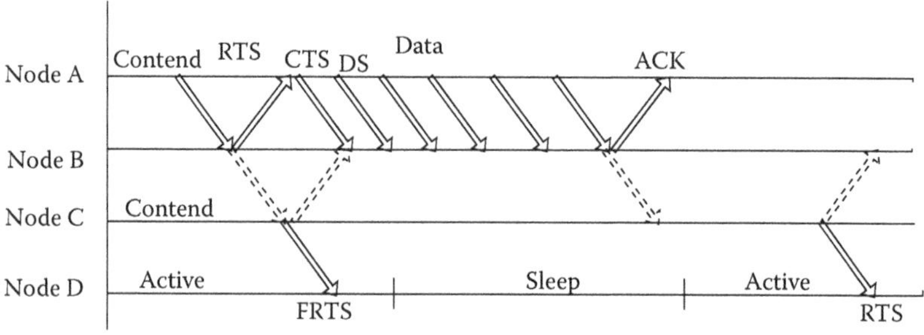

Figure 3.11 FRTS. (Adapted from van Dam, T. and Langendoen, K., An adaptive energy-efficient MAC protocol for wireless sensor networks, *Proceedings of the First International Conference on Embedded Networked Sensor Systems,* Los Angeles, CA, November 2003, ACM, New York, 171–180.)

as the information of the length of the ongoing data transmission, which prevents node C sending data to node D. A node should not send the FRTS packet if it is prohibited from data transmission.

The destination node of the FRTS packet must be in the active or the wake-up state to receive data from the sender of the FRTS packet when the ongoing communication is performed. The destination node gets this information from the FRTS packet. To prevent another node from occupying the transmission medium, the winner of the previous contention (i.e., node A) sends a small dummy data-send (DS) packet prior to sending a burst of data. The DS packet contains no useful information. Hence, the collision between the DS packet and the FRTS packet does not affect the following data transmission.

3.3.1.2.1.5 Taking Priority on Full Buffers — The second solution is based on the observation that a node may prefer sending to receiving when its transmit/routing buffers are almost full. As shown in Figure 3.12, assume that node B sends the RTS packet to node C, whose buffers are almost full. Instead of sending a CTS reply to node B, node C initiates a data transfer with node D by sending an RTS packet to node D.

However, with this scheme, the node that is the intended receiver of the prior contention winner has higher probability of controlling the transmission medium. In this example, node C loses the contention with node B. But luckily, node C is the winner's receiver and node C (not node B) actually owns the transmission medium now. Obviously, node D will not have the *early-sleeping problem* in this case. In addition, *the full-buffer priority scheme* introduces a limited form of flow control into the network, which actually is useful for many nodes-to-sink communication scenarios in WSNs.

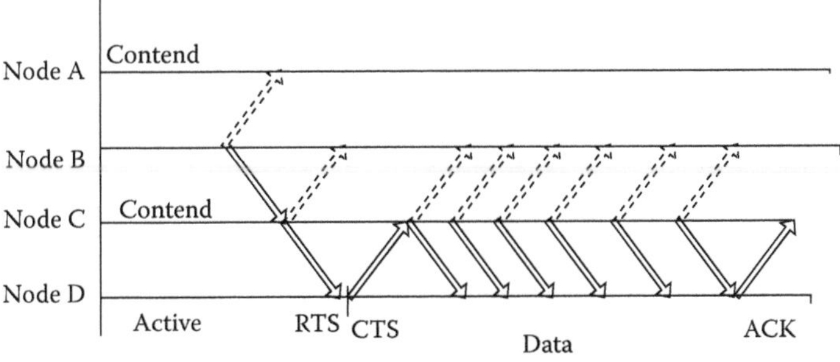

Figure 3.12 Taking priority on full buffers. (Adapted from van Dam, T. and Langendoen, K., An adaptive energy-efficient MAC protocol for wireless sensor networks, *Proceedings of the First International Conference on Embedded Networked Sensor Systems*, Los Angeles, CA, November 2003, ACM, New York, 171–180.)

However, when the high-load traffic is not flowed in a nodes-to-sink communication pattern, the data flow must be applied carefully. The probability of collisions increases rapidly when the nodes in a random communication pattern start taking priority. These collisions reduce the overall performance of the WSN. Therefore, T-MAC uses a threshold to limit nodes taking priority on full buffers.

3.3.1.2.2 Evaluating T-MAC Protocol

T-MAC introduces the concept of turning off the radio when a certain time-out occurs, which presents an effective way to address the idle-listening problem and decreases the energy consumption in a volatile environment where the message rate fluctuates, either in time or in location [Tvdam03]. Simulations show that the T-MAC protocol can save as much as 96 percent of the energy compared to a traditional CSMA-based protocol, by using the radios for as little as 2.5 percent under a very low traffic load. With a high traffic load, the T-MAC protocol does not increase the latency and ensures a high throughput by not entering the sleep state. Under homogeneous traffic load, T-MAC and S-MAC achieve similar reductions in energy consumption (up to 98 percent) compared to the CSMA protocol. However, in a network where message rates vary, the T-MAC protocol saves more energy than its predecessor, S-MAC, which only turns on the node radio for a fixed period.

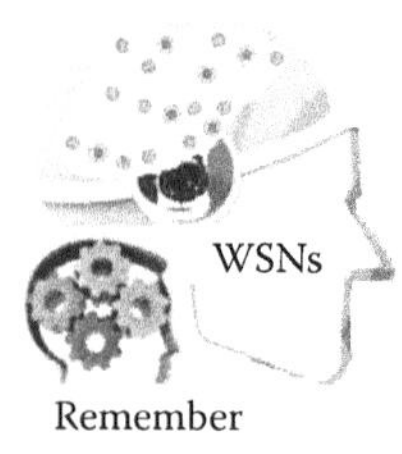

The T-MAC protocol reduces the energy consumption of the nodes, thereby increasing the lifetime of the network without introducing any latency. It reduces the time required for the transmission of data from the source node to the destination node. It also solves the early-sleeping problem by introducing FRTS and taking priority on the buffers.

3.3.2 Schedule-Based MAC Protocols

In schedule-based medium access, each node uses the shared transmission media based on a schedule. Similar to TDMA-based protocols, time normally is divided into so-called time slots of fixed length. The schedule determines the assignment of the time slots in a way such that conflicts do not exist and each node gets an opportunity to use the medium. Often, these schedules are repeated after a certain period and the nodes form a cluster. As each node can access the shared medium only in the dedicated time slot, schedule-based MAC protocols generally can avoid contentions, collisions, and idle listening. Without additional overhead, the schedule can also easily transit a node into the sleep state for energy saving. In addition,

QoS and priority support can be conveniently achieved with schedule-based MAC protocols. However, a number of challenging issues arise when designing schedule-based medium access schemes for resource-constrained WSNs, as follows:

1. High-quality clock synchronization among the nodes is not easy to achieve.
2. The dynamics of WSNs, including nodes addition, nodes failure, and mobility, make effective slot assignment difficult.
3. Slot assignment in multi-hop WSNs is challenging.
4. Poor scalability and complexity in the schedule maintenance may significantly degrade the network performance.

In the literature, a number of studies have been conducted to design efficient schedule-based medium access schemes while resolving the aforementioned challenges. Examples of schedule-based MAC protocols are TRAMA [Vrajendran06], LEACH (low-Energy Adaptive Clustering Hierarchy) [Heinzelman02], Self-Organizing Medium Access Control for Sensor Networks (SMACS) [Ksohrabi00], FLAMA (Flow-Aware Medium Access) [Vrajendran05], SPARE MAC (Slot Periodic Assignment of Reception) [Lcampelli07], μ-MAC [Abarroso05], VTS-MAC (Virtual Time Division Medium Access) [Eelopez06], ER-MAC [Rkannan03], and BMA MAC (Bitmap-Assisted MAC) [Jli04]. The LEACH protocol introduces the concept of hierarchy into WSNs for transferring data from the sensor nodes to the base station, while the FLAMA protocol uses distributed election, by using the information of the flow, the two-hop neighborhood, and the simple traffic adaptive scheme for energy-efficient channel access. In the SPARE MAC protocol, the nodes, which are receivers at a particular instance of time, receive the reception schedule and propagate the information of the reception schedule to all the neighbors. The μ-MAC protocol divides the transmission channel into contention and contention-free periods and relies on the information provided by upper layers. The VTS-MAC protocol divides the nodes into clusters. In VTS-MAC, the time line is divided into time slots such that the number of nodes in the network is equal to the number of time slots. On the other hand, the BMA MAC protocol proposes an intra-cluster MAC protocol, which divides the nodes in the network into clusters. The nodes in the cluster can communicate to the cluster only when there is an occurrence of significant events.

In the rest of this section, we particularly introduce the basic idea of the TRAMA protocol [Vrajendran06].

3.3.2.1 Traffic Adaptive Medium Access Protocol [Vrajendran06]

The TRAMA protocol is a schedule-based MAC protocol for WSNs, which saves energy by making sure that there will be no collisions in the data transmission and by making the nodes enter a low-power state whenever the nodes are not the intended receivers or transmitters. This protocol uses an adaptive selection for electing the nodes that transmit at a particular period of time and allows nodes to

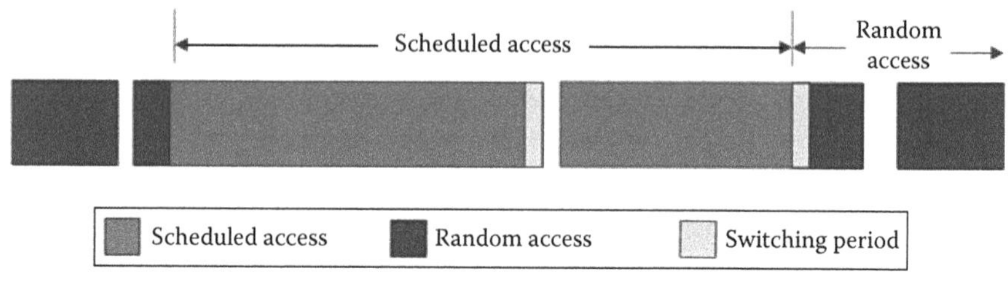

Figure 3.13 Time line of the TRAMA protocol. (Adapted from Rajendran, V. et al., Energy-efficient, collision-free medium access control for wireless sensor networks, *Proceedings of the First International Conference on Embedded Sensor Systems (SenSys '03)*, Los Angeles, CA, February 2006, ACM, New York, Vol. 12, No. 1, 63–78.)

determine when they can transit into the sleep mode. With traffic information, TRAMA can avoid assigning time slots to nodes having no traffic to send.

The time line of the TRAMA protocol is shown in Figure 3.13, which includes *random access* and *scheduled access* slots. The random access slots are called the signaling period, and the scheduled access slots are called the transmission period. During the signaling period, the nodes broadcast the one-hop neighborhood information among neighboring nodes such that each node can obtain a two-hop topology information around itself. During the transmission access slots, the nodes transmit the data and propagate the schedule for contention-free data exchange. The schedule information consists of a set of receivers for the traffic originating at the node, and TRAMA assumes that clock synchronization was done previously. In the contention-free period, time is divided into small time slots and the schedule is fixed. When the contention-free period is over, the nodes fall back to the random access period. The lengths of the signaling time slots and the transmission time slots depend on the type of application. The signaling time slot occurs more often for the dynamic scenarios, where the nodes move from one location in the network to other locations. On the other hand, for the static scenarios, where the nodes do not have mobility, the signaling time slot is shorter. Because, in WSNs, the sensor node does not move very often from one location to another, the signaling time slot is shorter.

TRAMA consists of three components, which are

1. Neighbor protocol
2. Adaptive election algorithm
3. Schedule exchange protocol

3.3.2.1.1 Neighbor Protocol

The TRAMA protocol starts its operation in the signaling period. In the signaling period, every node chooses a random time slot and broadcasts the one-hop neighborhood

Type	Source address	Destination address	Delete number	Add number	Deleted node ID's	Added node ID's

Figure 3.14 Signal header. (Adapted from Rajendran, V. et al., Energy-efficient, collision-free medium access control for wireless sensor networks, *Proceedings of the First International Conference on Embedded Sensor Systems (SenSys '03)*, Los Angeles, CA, February 2006, ACM, New York, Vol. 12, No. 1, 63–78.)

information among neighboring nodes. At the end of the signaling period, it is expected that all the nodes are able to discover their neighbors. Hence, the main purpose of the signaling period is to permit node additions and deletions such that the changes in the topology can be discovered. The connectivity information in the network is found by these signaling packets. Figure 3.14 shows the format of the header of the signaling packets. Signaling packets carry incremental neighborhood updates, and if there are no updates, signaling packets are sent as "keep-alive" beacons. Otherwise, if a node is not heard for a certain period of time, the node is assumed to have been disconnected from the network. The incremental updates from a node include the one-hop neighborhood information of this node in terms of added and deleted neighbors.

Once all the one-hop neighbors of a node, say node B, send the corresponding one-hop information to node B, it can earn all its neighbors' neighbors. In other words, node B eventually will have all the information of its two-hop neighbor nodes and can construct a two-hop local topology around itself.

Note that during the random access period, signaling packets may be lost due to collisions, which can result in inconsistent neighborhood information across the network. To ensure consistent neighborhood information, the length of the random access period and the number of retransmissions of the signaling packets should be set according to the real network or the application scenario.

3.3.2.1.2 Adaptive Election Algorithm

After discovering the neighbors, the TRAMA protocol employs the *adaptive election algorithm* to establish a schedule. Nodes locally compute which one is the absolute winner among the two-hop neighbors in a certain time slot by calculating the priority function as in Equation 3.5:

$$\text{prio}(u,t) = \text{hash}(u \oplus t) \tag{3.5}$$

where

u is the node identification

t is the slot number

hash $(u \oplus t)$ is a networkwide known hash function

Based upon the results of the priority function, time slots are reserved to the winner (i.e., the node with the highest priority). For energy efficiency, TRAMA switches nodes to the sleep state whenever possible and reuses slots that are not used by the winner. For example, the winner may give up its transmission slot if it does not have any data to send, and the slot could be used by another node.

At any given time slot, *t*, during the transmission period, the state of a node, *u*, is determined according to the two-hop neighborhood information and the schedules announced by *u*'s one-hop neighbors. Each node has three possible states:

1. Sleep state
2. Receive state
3. Transmit state

A node is in the *transmit state* if it has data to send and is the winner, i.e., has the highest priority based on the result calculated from Equation 3.5. When a node is the intended receiver of the current sender, the node is in the *receive state*. Otherwise, the communication system of the node can be switched off, and it moves into the *sleep state*, as it does not participate in any data exchange.

3.3.2.1.3 Schedule Exchange Protocol

The traffic-based schedule information is established and maintained by the *schedule exchange protocol*, which is further broadcasted among the neighboring nodes periodically during the transmission slots. The schedule is generated as follows.

Step 1: Each node computes the number of time slots required to transmit the data through the transmission channel, SCHEDULE_INTERVAL, based on the rate at which packets are generated at this node.

Step 2: The node then precomputes the number of slots in the interval [*t*, *t* + SCHEDULE_INTERVAL], for which the node will be selected as the transmitter. In other words, during this interval, the node has the highest priority among its two-hop neighbors and is assumed to be the winner of this interval.

Step 3: The node informs the intended receivers of these slots to avoid the collisions, because all the neighboring nodes of the present node will have the information about the transmission schedule of the present node.

However, if the node does not have data to send, it marks the slots as VACANT and sends the information to the neighboring nodes so that other nodes can make use of the vacant slots. The last time slot in the winning interval is used for broadcasting the node's schedule for the next interval.

The nodes announce the schedule information by using schedule packets, as shown in Figure 3.15. A schedule packet includes fields such as *source address*, *time-out*, *width*, *number of slots*, and *bitmap*. The *source address* identifies which node is announcing

Bits	32	8	8	8		
	Source address	Timeout	Width	Number of slots	Bitmap	...

Figure 3.15 Schedule packet format. (Adapted from Rajendran, V. et al., Energy-efficient, collision-free medium access control for wireless sensor networks, *Proceedings of the First International Conference on Embedded Sensor Systems (SenSys '03)*, Los Angeles, CA, February 2006, ACM, New York, Vol. 12, No. 1, 63–78.)

the schedule, the *time-out* indicates how long this schedule is valid, the *width* is the number of bits in the *bitmap*, the *number of slots* is the total number of winning slots, and the *bitmap* identifies the intended receivers. As the data from the MAC layer is targeting only one-hop neighbors of the sender and the neighboring information is already provided by the neighboring protocol, there is no need to specify the receiver's address in the schedule packet. Instead, TRAMA adopts a bitmap scheme to identify the intended receivers. The length of the bitmap is equal to the number of one-hop neighbors. Each bit in the bitmap represents a particular one-hop neighbor, and the order is based on the IDs of the neighbor nodes. If the sender wants to send the data to a particular neighboring node, the sender will set the corresponding bit in the bitmap to 1. Otherwise, the corresponding bit is set 0 if the node is not the intended receiver. Hence, when all the bits in the bitmap are set to l, the schedule packet is a broadcasting packet, because all the one-hop neighbors are the intended receivers. Similarly, multicast can be easily supported only by setting the multicast group of bits to 1.

The summary of a node's schedule is also sent with every data packet to minimize the impact of loss in the schedule dissemination. Nodes maintain the schedule information for all the one-hop neighbors. The information is consulted when a node needs to decide where to transmit or giving up the slot. The updated schedule based on this decision will be carried by the summary within the data packet.

3.3.2.1.4 Performance Evaluation of TRAMA

TRAMA assumes that time is slotted and uses a distributed election scheme based on the information about traffic at each node to determine which node can access the channel for transmission at any particular time slot. With the traffic information, TRAMA avoids assigning time slots to nodes having no traffic to send and also allows nodes to determine when they can switch to the sleep mode. The TRAMA protocol ensures that a distance of three hops or more can concurrently transmit data. The performance of TRAMA depends mainly on the traffic pattern, while the performance of S-MAC depends on the duty cycle. Simulations in [Vrajendran06] show that TRAMA outperforms contention-based protocols (CSMA, 802.11, and S-MAC) in terms of energy consumption and throughput. However, TRAMA experiences a higher delay than the static scheduled access protocols (e.g., [Bao01]) due to the scheduling overhead. Similar to TDMA-based protocols, TRAMA is well suited for sensor applications like periodic data collection and monitoring, which are not delay sensitive but require high delivery guarantees and energy efficiency.

3.3.3 Hybrid and Event-Based MAC Protocols

There are also a number of MAC protocols that are neither based solely on schedule nor on contention (e.g., [Ksarvakar08, Ngajaweera08, Kjamieson03, Szhou07, Jpolastre04, Irhee08]), which are developed for WSNs in the literature. Some MAC protocols use a hybrid of contention-based and schedule-based concepts and some are event-based. Examples of hybrid and event-based MAC protocols are Zebra MAC [Irhee08], Sift MAC [Kjamieson03], FAMA/TDMA (Floor Acquisition Multiple Access/Time Division Multiple Access) Hybrid MAC [Ngajaweera08], EZ-MAC (Utilized ZigBee MAC) [Ksarvakar08], and A^2-MAC (Application Adaptive Medium Access Control) [Szhou07]. Particularly, the FAMA/TDMA.

Hybrid protocol combines both FAMA and TDMA for providing medium access to all the nodes in a network. Initially, the nodes in a network contend for gaining access to the transmission by sending RTS packets to the base station. The node with the first successful RTS packet is given absolute access to the transmission channel to transmit the sensed data. In the EZ-MAC protocol, the data is sent with a low service access delay, keeping the access blocking ratio low by an optimized structural sequence. It also uses the scheduling scheme for WSNs. A^2-MAC is a data collection protocol. It is a hybrid, slotted CSMA/TDMA protocol. In the following, we introduce examples of event-based protocols, followed by a hybrid MAC protocol developed for WSNs.

3.3.3.1 Sift Medium Access Control [Kjamieson03]

In many WSN applications, the purpose of the sensor nodes is to detect events and report to a specific node called the base station. Whenever an event occurs, all the nodes that sense the event will start transmitting the details of the event to the base station. As multiple nodes that detect an event are quite possibly within a short distance, they share the same transmission medium. When all the nodes report at the same time, there will be contention in the transmission channel. Such a situation is known as spatially *correlated contention*. However, as multiple nodes detect the same event and may report similar sensed data to the base station, it is not necessary that all the sensor nodes report the event that has been detected. The event would be reported to the base station even if only a subset of the sensor nodes in the event's neighborhood actually reports the event. On the other hand, sensor nodes in WSNs may fail or die due to battery or other causes, and the density of the sensor nodes in a particular geographical area varies. Thus, it is desirable that the MAC protocols for such WSNs effectively handle the *correlated contention* along with the time-varying *density*, which is the goal of the Sift MAC protocol [Kjamieson03].

3.3.3.1.1 Protocol Design

Similar to the traditional CSMA protocols, Sift MAC uses a contention window of fixed size of a length of 32 slots. The difference between the CSMA protocols and Sift MAC is that the probability of picking a slot in Sift MAC in a given interval is not uniform. In Sift MAC, nodes compete to transmit the data in slot $r \in [1, \mathrm{CW}]$, where CW is the length of the contention window. The nodes compete for a particular slot based on the shared belief on the size of the current living population, N, which changes after every slot in which no transmission occurs. The believed population starts off at some large value, indicating a correspondingly small per-node probability of winning the channel access. If no node transmits in the first slot, then each node sensing the medium reduces the believed number of competing nodes by multiplicatively increasing its transmission probability for the next slot. This process is repeated to enable the winner to be chosen rapidly across a wide range of potential population sizes without incurring long latency due to collisions.

For example, if only one node competes for the transmission medium, it then gains the access in a particular slot of the contention window to transmit the data. After the completion of data transmission, all the nodes compete for the new slots to transmit the data and estimate the values of N.

3.3.3.1.2 Backoff Probability Distribution in Sift MAC

Assume that every node picks up a slot $r \in [1, \mathrm{CW}]$ using a nonuniform probability function, p_r. A slot $r \in [1, \mathrm{CW}]$ in the contention window is said to be silent if no node chooses to transmit the data in that slot. Similarly, it is said that a slot $r \in [1, \mathrm{CW}]$ has a collision if more than one node chooses the same slot. A sensor node can win a slot in the contention window only if one node chooses slot r for data transmission, which means slot r is the first non-silent slot in the contention window. We call it a success if some sensor nodes win some slots. Sift MAC uses an increasing and truncated geometric distribution, as in Equation 3.6, for the nonuniform probability process, p_r:

$$p_r = \frac{(1-\alpha)\alpha^{\mathrm{CW}}\alpha^{-r}}{1-\alpha^{\mathrm{CW}}} \quad \text{for } r \in [1, \mathrm{CW}] \tag{3.6}$$

In Equation 3.6, α is a distribution parameter in the range of (0,1), which results in an exponential increase of p_r. This means the later slots in the contention windows have higher probabilities.

Each station's choice of which slot to pick can be viewed as a decision procedure having CW stages. A sensor node starts at stage 1 by estimating the value of the current living population, N, as N_1, and then chooses slot 1 in the contention window with the same probability. If no node chooses slot 1, then each node assumes

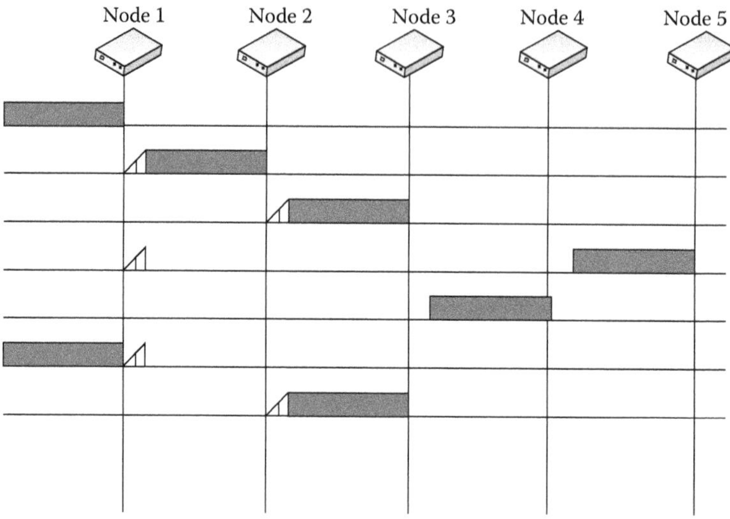

Figure 3.16 Time line of five nodes running the Sift MAC protocol. Shaded bars indicate packet transmission times. When the medium becomes idle, stations backoff at random according to the Sift distribution before transmitting. (Adapted from Jamieson, K. et al., Sift: A MAC protocol for event-driven wireless sensor networks, *Proceedings of the Third European Workshop on Wireless Sensor Networks*, Zurich, Switzerland, Lecture Notes in Computer Science, Vol. 3868, 260–275, Springer, New York, May 2003.)

that the estimation was not correct and modifies the estimated value by decreasing it from N_1 to N_2. And then the node chooses slot 2 with certain probability. If slot 2 turns out to be another silent slot, then the estimated value is further reduced to N_3 and the above process is continued. Let N_r be one of the estimated values of N, which is the updated belief after having $r - 1$ number of silent slots in the contention window (Figure 3.16).

As the current living population $N \in [1, N_1]$, the probability of success should be kept constantly high during the decision process. Thus, the following two properties should be held [Kjamieson03]:

1. The probability of success should be high when $N = N_1$
2. The probability of success should be constant

Assume that there are r silent slots in the contention window. Let p_r^1 be the probability of a node choosing a slot r, while there are $r - 1$ silent slots. Then, the probability that slot $r + 1$ is a success slot is given by Equation 3.7:

$$N_r p_r^1 (1 - p_r^1)^{N_r - 1} \approx N_r p_r^1 e^{-N_r p_r^1} \quad \text{(for large } N_r \text{ and small } p_r^1\text{)} \tag{3.7}$$

Thus, property (2) holds good only when $N_r p_r^1$ remains almost constant, so that the probability of success, $N_r p_r^1 e^{-N_r p_r^1}$, does not vary significantly along with time.

To identify a distribution that yields a constant $N_r p_r^1$, an exponential scheme is chosen, as in Equation 3.8, for covering a large N, while only a small number of slots in the contention window exist, by which the belief of the population reduces:

$$\beta = \frac{N_{r+1}}{N_r} \tag{3.8}$$

In Equation 3.8, β is constant and is given by $0 < \beta < 1$. Assume that there will be no collisions or no two sensor nodes will choose the same slot in the contention window; then, for sensor node S,

$$\begin{aligned} p_r^1 &= P_r(S \text{ chooses } r \mid \text{silence in earlier slots}) \\ &= P_r(S \text{ chooses } r \mid S \text{ did not choose earlier slots}) \\ &= \frac{P_r(S \text{ chooses } r)}{P_r(S \text{ did not choose earlier slots})} \\ &= \frac{p_r}{1-(p_1 + p_2 + \cdots + p_{r-1})} \end{aligned} \tag{3.9}$$

$$= \frac{(1-\alpha)\alpha^{\mathrm{CW}-r}}{1-\alpha^{\mathrm{CW}-r+1}} \tag{3.10}$$

$$\frac{p_r^1}{p_{r+1}^1} = \frac{(1-\alpha)\alpha^{\mathrm{CW}-r}}{1-\alpha^{\mathrm{CW}-r+1}}\alpha \approx \alpha \quad (\text{for small } \alpha^{\mathrm{CW}-r}) \tag{3.11}$$

If the values of α and β are equated, then Equations 3.8 and 3.11 can be equated, which results in

$$N_{r+1} p_{r+1}^1 \approx N_r p_r^1$$

This proves that the probability of success should be constant even when the value of N changes from N_1 to 1. As in property (1), the probability of success is high if $N = N_1$. Equation 3.10 also implies that $p_{\mathrm{CW}}^1 = 1$; so, if all the slots in the contention window are silent, the last slot must be chosen by a node. Therefore, α should be chosen such that a node in stage CW believes that it is the only active node. This can be further illustrated by setting the value of the current living population as 1, which implies that $N = 1$.

From Equation 3.9, if $\alpha = \beta$ and $1 = N_{CW} = \alpha^{CW-1}N_1$, the value of α will thus be given by $\alpha = N_1^{\frac{-1}{CW-1}}$.

3.3.3.1.3 Protocol Specification

In Sift MAC, every node has four states as follows:

1. Idle state: in which a node waits for the data that is sent from other nodes
2. Contend state: in which a node contends for the transmission channel and tries to gain access to the transmission medium
3. Receive state: in which a node receives the data that is sent from another node
4. AckWait state: in which a node waits for the ACK from another node after sending data to the node

The pseudocode for the transitions between the states is given in Figure 3.17. In this figure, the function *pickslot()* is used for picking up a slot for transmitting the data using the Sift distribution specified in Equation 3.6. The directive *moveto(state)* changes the state of a particular node from the present state to the given state. The directive *wait(time)* waits for the period of time that is specified in the parameter.

t_{slot} is the minimal time separation such that if two nodes transmit more than t_{slot} seconds apart, the two nodes will hear the onset of each others' transmission. t_{sifs} is the period of time delayed at the beginning of a data ACK packet for turning around from transmitting the packet to receiving the ACK. t_{difs} is the quantity of time delay added at the beginning of a new data transmission. Thus, $t_{difs} + \text{slot} * t_{slot}$ is the time taken for one complete data transmission and the subsequent transmission of the ACK. $t_{ACKtime\text{-}out}$ is the time for which a node waits to receive the ACK.

3.3.3.1.4 Request-to-Send and Clear-to-Send Mechanisms

For avoiding collisions, all the nodes in the sensor networks that implement the Sift MAC protocol employ the RTS/CTS exchange scheme. Similar to the way the Sift backoff distribution is used to compete on data packets, it can also be used to compete on sending the RTS packet. Hence, one can just replace "frame" with "RTS" and "ACK" with "CTS" in the pseudocode to achieve the RTS competition.

3.3.3.1.5 Performance Evaluation of Sift MAC

The basic idea of Sift MAC is to use an increasing, nonuniform probability distribution within a fixed-size contention window, instead of using a time-varying contention window, from which a node randomly picks a transmission slot, as in traditional contention-based MAC protocols. The Sift MAC protocol is tuned for sensor networks, where every node does not have to report every detected event.

```
Idle State
wait (channel idle)
if (recv frame for self)
moveto Receive
end if
if (xmit queue not empty)
moveto Contend
end if
Contend state
slot _ pickslot ()
wait t_difs+ slot*t_slot
if (channel busy)
moveto Idle
end if
Transmit frame
moveto AckWait
Receive state
Check frame CRC
wait t_sifs
Send ACK
moveto Idle
AckWait state
Wait t_ACK timeout
if (recv an ACK for self)
discard frame
moveto Idle
end if
Retransmit frame
moveto AckWait
```

Figure 3.17 Pseudocode for state transition in Sift MAC. (From Jamieson, K. et al., Sift: A MAC protocol for event-driven wireless sensor networks, *Proceedings of the Third European Workshop on Wireless Sensor Networks*, Zurich, Switzerland, Lecture Notes in Computer Science, Vol. 3868, 260–275, Springer, New York, May 2003.)

Simulation studies show that the Sift MAC protocol performs well when a spatially correlated contention occurs and adapts well to changes in the active-population size. In specific, results show that the Sift MAC protocol improves over 802.11 in terms of report latency up to a factor of 7 as the number of nodes reporting an event scales up to 512.

3.3.3.2 Berkeley Medium Access Control [Jpolastre04]

To meet the requirements of WSN deployments and monitoring applications, the B-MAC protocol is designed to achieve the following goals:

1. Low-power listening (LPL)
2. Effective collision avoidance
3. Simple implementation, and small code and RAM sizes
4. Effective channel utilization at low and high data rates
5. Reconfigurable by the network protocols
6. Tolerant to changes in radio frequencies and network topology
7. Scalable to large number of nodes

The B-MAC protocol provides certain interfaces for achieving these goals. These interfaces are listed in Figure 3.18. For sensing the transmission channel, the B-MAC protocol uses clear channel assessment (CCA) and packet backoffs.

3.3.3.2.1 Protocol Design

In B-MAC, signal strength is sampled when it is assumed that the transmission channel is free. For example, the transmission channel is free when the ongoing

```
interface MacControl {
command result _ t EnableCCA();
command result _ t EnableCCA();
command result _ t DisableCCA();
command result _ t EnableAck();
command result _ t DisableAck();
command void* HaltTx();
}
interface MacBackoff {
event uint16 _ t initialBackoff(void* msg);
event uint16 _ t congestionBackoff(void* msg);
}
interface LowPowerListening {
command result _ t SetListeningMode(uint8 _ t mode);
command uint8 _ t GetListeningMode();
command result _ t SetTransmitMode(uint8 _ t mode);
command uint8 _ t GetTransmitMode();
command result _ t SetPreambleLength(uint16 _ t bytes);
command uint16 _ t GetPreambleLength();
command result _ t SetCheckInterval(uint16 _ t ms);
command uint16 _ t GetCheckInterval();
}
```

Figure 3.18 Interfaces of B-MAC protocol. (Adapted from Polastre, J., Interfacing Telos to 51-pin sensorboards, http://www.tinyos.net/hardware/telos/telos-legacy-adapter.pdf, October 2004.)

transmission is completed or when the communication device is not receiving any data. The sampled data is entered into a queue. The median of the sampled data is found and is added to an exponentially weighted moving average with decay, α. The median is used for adding robustness to the noise floor estimate. After the noise floor is estimated, the request for monitoring the received signal strength starts monitoring the transmission channel. The B-MAC protocol searches for outliers in the received signal strength. For instance, if a node senses an outlier, it declares that the channel is unoccupied, as a valid packet could never have an outlier below the noise floor. If there is no outlier found in the samples, then it is concluded that the channel is busy.

Using the MacControl interface in Figure 3.18, nodes in the B-MAC protocol can turn the CCA on or off. If the CCA is disabled, the scheduling protocol is implemented in the B-MAC protocol. This protocol uses packet backoff when the CCA is enabled. For packet backoff, instead of setting a backoff time, B-MAC uses an event-driven approach, which may return the backoff time or ignore the event. If the event is ignored, a small backoff time is set. After the initial backoff time, the CCA outlier algorithm is run. If the channel is not clear, the service for the congestion of the backoff time is signaled by the event.

The B-MAC protocol also provides a link-layer ACK support. If the application requires an ACK, then the ACK is sent to the source node from the receiver node. After receiving the ACK, the source node sets a bit in the sender's transmission message buffer. B-MAC uses LPL for periodic transmission channel sampling. Every node in the B-MAC protocol senses for activity in the transmission channel. If it senses an ongoing data transmission, then it waits for the completion of the transmission. After the data transmission, the node moves into the sleep state. If no packet is received, then a timer pushes the node into the sleep state. The interval between two LPL samples is maximized to minimize the time spent in sampling the transmission channel.

3.3.3.2.2 Performance Evaluation of B-MAC

The B-MAC protocol performs better as compared to S-MAC and T-MAC in terms of throughput and energy consumption. The performance of S-MAC and T-MAC protocols is dependent on the length of the duty cycle. B-MAC provides a flexible interface to obtain ultralow power operation, effective collision avoidance, and high channel utilization in WSNs. B-MAC effectively performs clear channel estimation. While supporting on-the-fly reconfiguration and providing bidirectional interfaces for system services, B-MAC employs an adaptive preamble sampling scheme to reduce the duty cycle, minimize idle listening, and achieve low-power operation. B-MAC may be configured to run at extremely low duty cycles and does not force applications to incur overheads of synchronization and state maintenance like other MAC protocols. Experimental studies show that B-MAC's flexibility results in better packet-delivery rates, throughput, latency, and energy consumption than those of S-MAC [Jpolastre04].

3.3.3.3 *Zebra Medium Access Control [Irhee08]*

The Z-MAC protocol is a hybrid protocol that combines the merits of TDMA and CSMA while offsetting the demerits of both the schemes. Z-MAC uses CSMA at the base, but follows TDMA depending on the contention level. The overhead of the Z-MAC protocol is incurred at the setup phase, which occurs at the beginning. In the setup phase, the nodes are assigned the time slots for data transmission. The nodes use the assigned time slots for the transmission of the sensed data in a particular period of time known as a frame. A node is called the owner of a time slot if it wins the access to the transmission medium; otherwise, the node is known as a nonowner. The nonowners of the time slot have lower priority to transmit the data when compared to that of the owners of the time slot. The priority is set using the contention window size. If, at a particular point of time, the owners do not transmit the data, then the nonowners of the time slot may transmit the data by using the time slot that is left unused by the owners of the time slot. The Z-MAC protocol performs similar to TDMA when the level of contention is low (or the traffic load is low), and it performs similar to CSMA when the level of contention is high (or the traffic load is high).

3.3.3.3.1 Z-MAC Setup Phase

Initially, the Z-MAC protocol runs the setup phase, which consists of the following steps:

1. Neighbor discovery
2. Slot assignment
3. Local frame exchange
4. Global time synchronization

3.3.3.3.1.1 Neighbor Discovery — In the *neighbor discovery* step, every node in the network finds the one-hop neighborhood by sending *ping* messages, containing the current list of one-hop neighbors, to its one-hop neighbors. The two-hop neighborhood information can then be found by combining all the received one-hop neighborhood information of its neighbors.

3.3.3.3.1.2 Slot Assignment — In the slot assignment step, Z-MAC uses the distributed RAND (DRAND) [Irhee06] algorithm to assign the time slots for data transmission. This algorithm is the distributed implementation of the RAND algorithm [Rramanathan97]. The RAND algorithm is a centralized algorithm for assignment of time slots. The DRAND algorithm runs in rounds. There are four states in DRAND, which are IDLE state, REQUEST state, RELEASE state, and GRANT state. The state diagram of DRAND is shown in Figure 3.19. Initially,

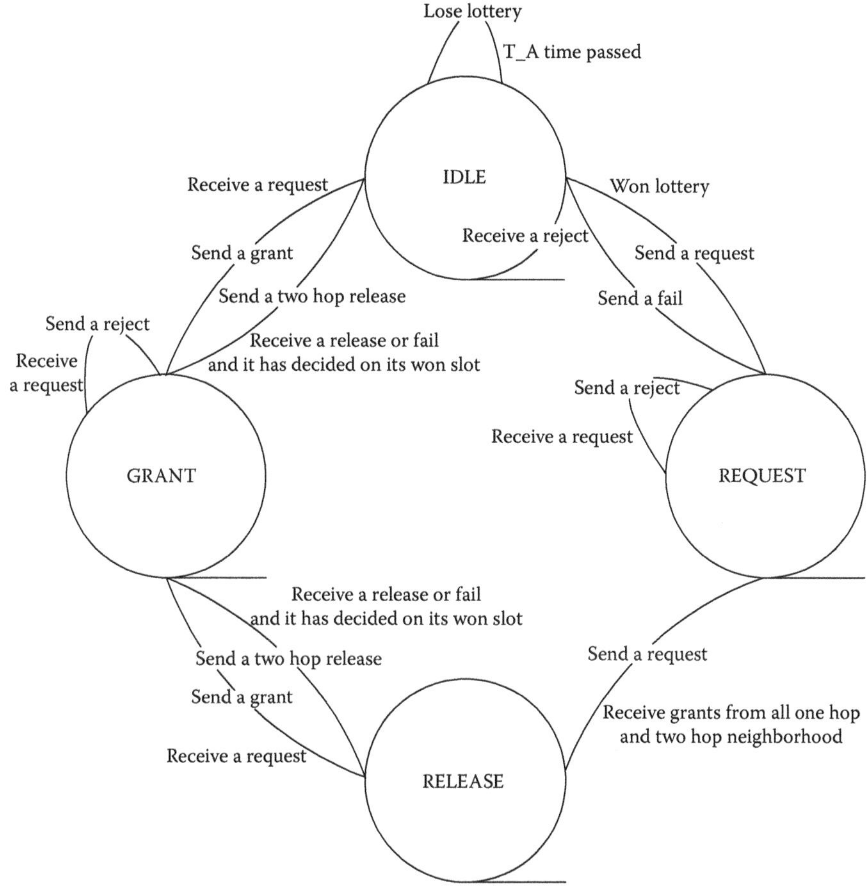

Figure 3.19 DRAND state diagram. (Adapted from Rhee, I. et al., DRAND: Distributed randomized TDMA scheduling for wireless ad-hoc networks, *Proceedings of the IEEE MobiHoc*, Florence, Italy, May 2006, 190–201.)

every node is in the IDLE state. During the IDLE state, the node tosses the coin for which the probability of getting a head or a tail is 1/2. If the result is a head, then it runs a lottery. If it loses the lottery, then it remains in the same state. If it wins the lottery, it moves into the REQUEST state, where the node broadcasts a request message to all its one-hop neighbors.

Consider node B, which is a one-hop neighbor to node A. If node B receives a REQUEST message from node A when node B is in the IDLE state or the RELEASE state, then it responds with a *grant* message and moves to the GRANT state. If node B is in the REQUEST state or the GRANT state, then it responds with a *reject* message to node A. If node A does not receive a *grant* message or a *reject* message within the specified period of time, it sends the same request message again.

3.3.3.3.1.3 Local Framing — After slot assignment, every node in the network receives a time slot for transmitting its data to its intended destination. Then, the node needs to decide on the period in which it can use the time slot for transmitting the data to the intended destination. This period is called *the time frame* of the node. After the node decides the period for transmitting its data, the node should propagate the *maximum slot number* (MSN) to the entire network and adapt to local time slot changes. If there is any addition of new nodes in the network, then DRAND will assign the new time slots for the newly added nodes. The change in the MSN should also be propagated to all nodes in the network.

Under high-contention conditions, Z-MAC requires clock synchronization. The Z-MAC protocol uses the Real-Time Transport Protocol (RTP/RCTP) [Hschulzrinne96] for clock synchronization under high-contention conditions. In RTP/RCTP, every node in the network sends the control message at a rate that is limited to a small fraction of session bandwidth and every node in the session adjusts its bandwidth according to the allocated session bandwidth. In the Z-MAC protocol, every node limits the data-sending rate to a predetermined data-sending rate, which is determined based on the energy and the bandwidth.

3.3.3.3.1.4 Global Time Synchronization — Local framing assumes that all the nodes are synchronized initially at time slot 0. This could be achieved by fixing a predetermined time to synchronize the time slot 0. All the nodes are synchronized at slot 0 using the Timing-sync protocol for WSNs (TPSN) [SGaneriwal03]. TPSN assumes that every node has a 16 bit register that acts as a clock that is triggered by the crystal oscillator. TPSN runs in two steps. In the first step, the nodes in the network construct a hierarchical structure. Every node k belongs to a level i. The nodes in level i can communicate with the nodes in level $i - 1$. Only one node will be at level zero. This is called the root level. In the second step, which is the *synchronization* step, all the nodes at level i synchronize with the nodes at level $i - 1$. Thus, every node in the network synchronizes with the root level. Thus, all the nodes in the network get synchronized at slot 0. After global time synchronization, each node in the network performs local time synchronization.

3.3.3.3.2 Transmission Control of Z-MAC Protocol

Every node in the Z-MAC protocol can be in any one of the following two modes:

1. Low-contention level (LCL)
2. High-contention level (HCL)

Every node will be in an LCL until it receives an *explicit contention notification* (ECN) message from a two-hop neighbor node. A node sends an ECN message if it experiences high contention. Once an ECN message is received, the node transits to the HCL mode.

Explicit Contention Notification Message

ECN messages intend to notify the two-hop neighbors of the current owner of the time slot not to act as hidden terminals when the contention level is high. In the Z-MAC protocol, every node needs to decide the contention level based on the estimate of the contention level. The nodes can estimate the contention level using the following two methods.

1. To measure the packet loss in the ACKs
   As two-hop contention may result in collision and hence data loss, the source node could measure the contention level by measuring the packet loss in the transmission. Based on the received ACK packets, a node could calculate the packet loss percentage and decide the contention level. However, this technique requires the receiver to send the ACK back to the sender, and incurs extra overhead and decreases channel utilization.
2. To measure the noise level of the channel
   Whenever the contention level is high, the noise level in the transmission channel increases. Measuring the noise level in the transmission channel does not require any extra overhead. For measuring the noise in the transmission channel, the nodes calculate the number of noise backoffs. A noise backoff is a backoff transmitted by the source node when it senses the transmission channel using clear channel assessment (CCA). With CCA, a node in the network can transmit only when the node senses the channel to be clear. When the node experiences the contention, it takes the backoff message. When more than one destination take the backoff message, the node sends the one-hop ECN message to the node indicating an HCL. If node j receives the ECN message sent by node i, node j first checks whether it is the destination of the ECN message. If it is not the destination node, it simply discards the message. If it is the destination node, node j broadcasts the ECN message to its one-hop neighbors. Once a node receives the ECN message from a node in a two-hop neighborhood, it sets the HCL flag.

3.3.3.3.3 Transmission Rule

When a node in a network needs to transmit data, it first checks whether it is the owner of the time slot. If the node finds that it is the owner of the time slot, then it checks whether the transmission channel is unoccupied. If the node finds that the transmission channel is unoccupied, then it transmits the data to the intended destination. Otherwise, it sets a timer and waits for a period of T_0. When the timer expires, it runs the CCA, and if the transmission channel is clear, it transmits the data. Otherwise, it waits for random time and repeats the same process. If a node is in an HCL and is the nonowner of the time slot, it postpones the transmission for time T_0, and then performs random backoff within the contention window, $[T_0, T_{n0}]$. When the backoff timer expires, the node senses the transmission channel, and if the channel is unoccupied, then the node transmits the data. Otherwise, the node waits until the channel is clear and repeats the above process.

3.3.3.3.4 Receiving Schedule of Z-MAC Protocol

The Z-MAC protocol relies on the B-MAC [Jpolastre04] protocol for receiving a schedule. Z-MAC uses the LPL mode, wherein each node maintains a listening duty cycle separated by a check period and each transmission is preceded by a preamble as large as the check period. Therefore, under low duty cycles, the energy consumption of Z-MAC in idle listening is comparable to that of B-MAC. The check period is one of the important factors in the receiving schedule because the check period must allow one complete transmission of the data packet. Thus, the size of the slot must be larger than the sum of check period, T_0, T_{n0}, CCA period, and time required for the propagation of one data packet.

3.3.3.3.5 Performance of Z-MAC Protocol

The Z-MAC protocol can dynamically adjust the behavior of medium access between CSMA and TDMA depending on the level of contention in the network. The protocol takes advantage of the two-hop neighbor topology information and loosely synchronized clocks to improve MAC performance under high contention. Like TDMA, Z-MAC achieves high channel utilization under high contention and reduces collision among two-hop neighbors at low cost. Under low contention, the protocol behaves like CSMA and achieves high channel utilization and low latency. A unique feature of Z-MAC is that its performance is robust to synchronization errors, slot assignment failures, and time-varying channel conditions. In the worst case, its performance always falls back to that of CSMA. Compared to B-MAC [Jpolastre04], Z-MAC has an advantage under medium to high contention and is competitive under low contention (especially in terms of energy efficiency).

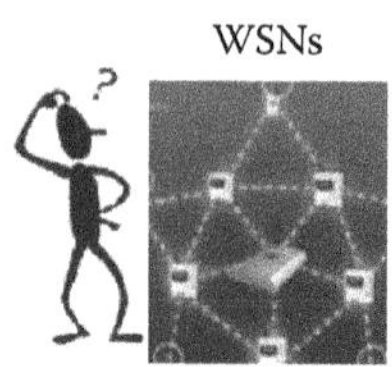

Sift MAC [Kjamieson03] shows high performance under one-hop contention, but under two-hop contention, it needs to rely on RTS/CTS and incurs high overhead. Z-MAC can be favorably adopted in applications where expected data rates and two-hop contention are medium to high.

3.4 Conclusion

In this chapter, we have gone through the challenges of MAC design in WSNs. To resolve these challenges, much research in the literature has been conducted on the design of effective MAC protocols suitable for different WSN applications. Hence, this chapter also briefly introduced several classical MAC protocols, including contention-based S-MAC and T-MAC, schedule-based TRAMA, as well as hybrid and event-based MAC like Sift MAC, Z-MAC, and B-MAC protocols.

Problems and Exercises

3.1 Multi-choice questions:

1. Which of the following is not a state in the TRAMA protocol?
   a. Sleep state
   b. Receive state
   c. Transmit state
   d. Wake-up state
2. The size of the contention window in Sift MAC protocol is
   a. 16
   b. 32
   c. 512
   d. None of the above
3. The Z-MAC protocol combines two traditional protocols for MAC. Which two?
   a. CDMA and TDMA
   b. FDMA and CSMA
   c. CDMA and SDMA
   d. CSMA and TDMA

3.2 Why is energy an important concern in the design of medium access protocols in WSNs?

3.3 Does the performance of the Sift MAC protocol depend on the number of the nodes in a WSN? What is the reason for the variation in performance when the number of nodes in the network increases?

3.4 What are the major differences between the S-MAC and the T-MAC protocol?

3.5 What are the different states for the nodes in WSNs using the TRAMA protocol? Explain the operation of different states for the nodes in WSNs using TRAMA.

3.6 Explain the operation of each state for nodes using the Z-MAC protocol.

3.7 Explain the importance of LPL and clear channel assessment in the B-MAC protocol.

3.8 Explain the early-sleeping problem in the T-MAC protocol and also how to resolve it.

3.9 Explain how the nodes in WSNs using the S-MAC protocol choose and exchange their schedules.

3.10 Explain the hidden and the exposed terminal problems in WSNs. Give examples on how to handle these problems in WSNs.

Chapter 4

Routing in Wireless Sensor Networks

4.1 Introduction

A wireless sensor network (WSN) is a distributed wireless ad hoc network comprising of a number of sensor nodes that are used for sensing the environment to track climatic changes, seismic activities, movement of enemy troops in a war zone, industrial monitoring and control, etc. In WSNs, the transmission range of a tiny sensor node is limited. However, the sensed information from such a sensor node normally has to be transmitted to and processed at the base station (BS) or at a control center (also called sink), which could be far away from the sensor node and out of the transmission range of the sensor node. In other words, the data may have to travel multiple hops before reaching the sink. Similarly, the query commands issued by users or the sink may have to travel multiple hops through the network to obtain some particular information, collected by different sensor nodes at different locations. Therefore, it is essential to deploy an efficient scheme in the WSN to select paths going through multiple hops and forward data from the source to the destination, which is a major function of the routing process.

Routing plays an important role in wired networks, wireless networks, and mobile ad hoc networks (MANETs), which attracted a large number of studies in the past. However, due to the unique constraints and application requirements in WSNs, the routing schemes developed for the Internet and the MANET are often not feasible or cannot deliver promising performance as needed in WSNs. For example, most Internet routing protocols assume highly reliable wired links with very low bit error rates, while MANET routing solutions are normally optimized for highly mobile nodes with symmetric links between neighbors. However, these assumptions are not true for WSNs. While facing the challenges arisen from the wireless environment and links such as MANETs or wireless LANs, routing schemes in WSNs also have to consider unique issues, including limited resources (such as energy, bandwidth, and computing), lossy wireless links and fault tolerance, data aggregation and data reporting, node deployment, scalability, coverage, network dynamics, and node/link heterogeneity [Njamal04].

4.1.1 Limited Resources in WSNs

As the sensor nodes are normally powered by batteries, which are not easy to replace or recharge, the energy of each network node in WSNs is limited. Energy-efficient routing is one of the critical design criteria for WSNs, because power failure of a sensor node not only affects the node itself but also its ability to forward packets on behalf of others and, thus, the overall network lifetime. In addition, if the data is lost due to a failure in the routing scheme or suboptimal route selection, the retransmission consumes extra energy and results in additional delay to the network, which also wastes the limited available bandwidth in the sensor network.

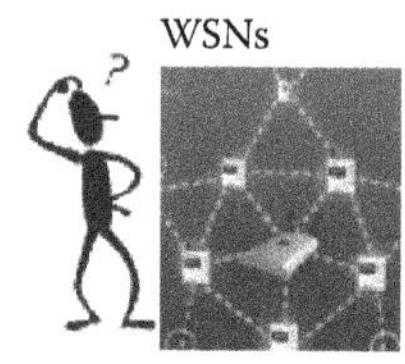

Energy efficiency is one of the most common concerns to be effectively addressed in the design of routing schemes for WSNs. Many research efforts have been devoted to developing energy-aware routing protocols for WSNs. However, in general, energy consumption is not a concern at all in traditional Internet routing protocols.

Similarly, the limited resources, such as bandwidth, memory, and computing, deserve careful consideration in the design of WSN routing protocols. For example, with limited memory space, a sensor node cannot store the whole topology information of a large-scale network for a routing decision. Neither is a large routing table feasible in the sensor node.

4.1.2 Fault Tolerance

Unlike the traditional wired networks, the nodes and the links in sensor networks are more prone to errors or failure. The sensor node may not work due to lack of power or physical damage. The wireless link can be broken by the failure of sensor nodes, or by environmental interference or obstacles. Thus, the routing procedures in WSNs should function effectively even when there are node or link failures in the network. To achieve this, routing schemes in WSNs may have to find alternative paths dynamically or take advantage of the redundancy in the network to tolerate the unpredictable failures in the sensor networks.

4.1.3 Data Reporting and Aggregation

The data sensed by the sensor nodes in WSNs needs to be reported to the users of the system by transmitting the data to the BS. Data reporting can be performed through different approaches, such as time driven, event driven, query driven, and the hybrid of all these three methods. In the time-driven approach, the nodes periodically report the data after a certain interval of time. In the event-driven approach, whenever an event occurs in the environment, the nodes report the information associated with that event to the BS. In the query-driven approach, the BS issues a query to some nodes in the network and expects the corresponding nodes to collect the necessary information delivered to the BS. These different data-reporting approaches have different advantages and disadvantages, which requires different routing schemes to address the needs.

In addition, sensor nodes in WSNs may produce a significant amount of redundant data. For example, multiple sensors may report the same information or different aspects of the same events occurring in the vicinity of a particular location. To reduce the number of transmissions and related resources' consumption, aggregating similar packets from multiple nodes based on certain criteria should be considered. Aggregation techniques include duplicate suppression, signal processing, data fusion, and so on.

4.1.4 Node Deployment

Node deployment in WSNs can be done in two ways: randomly and manually. When node deployment is done in a random fashion, the nodes in the network form a wireless ad hoc structure. Thus, the routing protocols deployed in the network have to self-learn the topology information and dynamically forward data through energy-efficient operation. When node deployment is done manually, the routes for transmitting data can be calculated optimally to achieve some goals using an off-line algorithm. In other words, the routes can be predefined. However, in case of the topology changes due to node/link failures, dynamical routing schemes are still necessary in manually deployed WSNs.

4.1.5 Scalability and Coverage

In many WSN applications, the number of deployed nodes in the physical area to be monitored can be significantly large. The routing scheme employed in WSNs should be scalable and working toward a similar efficiency even though the network size is large. The number of events or the data information to be delivered in WSNs can be enormous at a particular instance of time, which also requires high scalability from the routing process in the network.

The transmission and sense range of sensor nodes in WSNs are constrained by the physical size and capability of the nodes, which are normally small compared with the physical area covered by the network. Efficient routing schemes have to consider the particular requirements of the specific applications to maintain necessary network connectivity and coverage.

4.1.6 Network Dynamics and Heterogeneity

Some nodes in WSNs may contain the devices that make the nodes move from one position to another. Due to the movement of the nodes in the network, the network topology and connectivity change. The number of the nodes in the network and network connectivity may also change from time to time due to node/link failures. Such network dynamics have to be considered together with the dynamic events in the design of routing protocols for WSNs.

In many WSN applications, the nodes and the link between any two nodes are assumed to be homogenous. But, in reality, these are not always homogenous. The amounts of energy, transmission range (i.e., the maximum distance to which the node can directly transmit the data), memory, and processing capability may vary among the nodes and during the lifetime of the nodes. For example, symmetric link is the default feature for wired networks, such as Ethernet or optical networking. However, this symmetric link feature does not hold true for the wireless links in sensor networks, which necessitates the discovery of different routing processes for sensor networks.

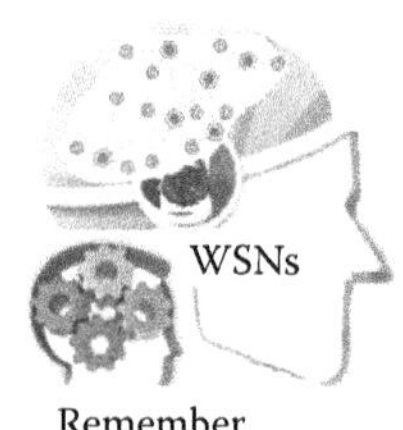

Sensor nodes have constraints due to limited resources such as energy, bandwidth, memory, and computing capability. Such constraints combined with the aforementioned challenging issues necessitate the invention and development of new routing solutions for WSNs.

4.2 Layout for the Chapter

With the challenging issues in mind, we first introduce some general concepts adopted in the design of routing protocols for sensor networks. These concepts include flooding, gossiping, and ideal dissemination. Next, the classification of routing schemes developed in the literature is described. Then, this chapter highlights several typical routing protocols, including Sensor Protocols for Information via Negotiation (SPIN), Directed Diffusion, Low-Energy Adaptive Clustering Hierarchy routing protocol (LEACH), Threshold-Sensitive Energy-Efficient Sensor Network (TEEN), Geographical and Energy-Aware Routing (GEAR), and multi-path routing, in sensor networks.

4.3 Classification of Routing Protocols in WSNs [Njamal04]

Routing in sensor networks is very challenging and different from contemporary wired/wireless networks, such as Ethernet and MANETs [Rwheinzelman99, Jkulik02]. The popular IP-based protocols cannot be applied to sensor networks because it is infeasible to build a global addressing scheme for the deployment and maintenance of thousands of tiny sensor nodes having limited resources. Many new algorithms have hence been developed for routing and forwarding data in sensor networks.

4.3.1 Proactive and Reactive Routing

Routing protocols can be *proactive*, *reactive*, or *hybrid* depending on how the route is found. Proactive protocols attempt to continuously evaluate the routes within the network so that all routes are computed before they are needed. In other words, when a packet needs to be forwarded, the route is already available and can be immediately adopted. Reactive protocols, on the other hand, invoke a route determination procedure only on demand. Hence, some sort of search procedure has to be employed to identify a route prior to data forwarding. Hybrid routing protocols attempt to integrate the above two ideas to take the advantages of both.

The advantage of proactive schemes is that there is little or no delay in determining a route whenever a route is needed. On the other hand, reactive protocols have to start a route discovery process to identify proper path information when a route is needed, which means that the time for determining a route can be quite significant. This leads to increased latency for packet delivery, and may not be applicable to real-time communication. However, proactive schemes are likewise not appropriate for the ad hoc networking environment whereas network topology changes fast and constantly. Such network dynamics may result in continuous route evaluation and maintenance, which use a large portion of the network resources. Particularly, when the changes are more frequent than the route requests, the routing information from the continuous evaluation process may not be necessary and never be used.

4.3.2 Flat and Hierarchical Routing

Based on the network structure, routing protocols in WSNs can also be broadly divided into *flat* routing and *hierarchical* routing. In *flat* routing schemes, equal roles and functionality are typically assigned to each node. *Flat* routing protocols distribute information as needed to any node that can be reached, or receive information. *Hierarchical* routing protocols often group nodes together by function into a hierarchy or cluster. By assigning different roles to different type of nodes or performing traffic aggregation to reduce redundancy, a hierarchical protocol allows WSNs to make best use of the heterogeneous nodes' capability. In many hierarchical routing protocols, each cluster designates a cluster-head (CH) node to aggregate and relay intercluster traffic. These CH nodes may become the bottleneck, potentially resulting in network congestion and single point of failure. In addition, maintaining the hierarchy or cluster can be costly in terms of energy or bandwidth consumption for small- to moderate-sized WSNs, which indicates that flat schemes are favorable in this case. On the other hand, *hierarchical* routing protocols are often better suited to large WSNs due to their scalability.

In fact, there are many other ways to classify routing protocols based on different criteria, such as protocol operation, network flow, energy, and QoS awareness. In the remaining part of this chapter, we will focus on four typical categories: data-centric protocols, hierarchical routing protocols, location-based routing protocols, and multipath routing in WSNs.

4.4 Data-Centric Routing Protocols in WSNs

In many applications of WSNs, the physical area covered by the sensors and the number of deployed sensor nodes can be enormous. Typically, the meaningful data traffic is generated due to the sensors' response to a query from the users (e.g., sink or BS) or actively reporting a detected event. In either case, multiple sensors having the data of interest will initiate the data transmission, which may result in significant redundancy and resource wastage. Certainly, if sensor nodes are as reliable as the IP routers and globally addressable, the redundancy issue will be trivial to resolve. However, it is infeasible (if not impossible) to assign a unique identifier to each sensor node and make each sensor node globally addressable like the IP router in the Internet. Accordingly, *data-centric routing protocols* are proposed for WSNs. In the data-centric routing scheme, the sink sends queries to specific regions and waits for answers from these regions. These queries are described in a high-level language. As data is being requested through queries, attribute-based naming is necessary to specify the properties of the data of interest.

SPIN and Directed Diffusion are among the earliest data-centric protocols [Jkulik02, Rwheinzelman99, CIntanagonwiwat00], which consider data negotiation between nodes to eliminate redundant data and save energy. These two protocols motivated the design of many other protocols that followed similar concepts. Examples of such data-centric routing protocols are Rumor Routing [Bdavid02], Minimum Cost Forwarding Algorithm (MCFA) [Fye01], Gradient-Based Routing (GBR) [Cschurgers01], COUGAR [Yyao02], Energy-Aware

Routing [Rcshah02], etc. Rumor Routing is mainly intended for contexts in which geographic routing criteria is not applicable. The Rumor Routing protocol uses a set of long-lived agents to create paths that are directed toward the events they encounter. MCFA uses the information about the direction of routing. It gets rid of the unique ID and routing table; instead each node in MCFA maintains the least-cost estimate from the node itself to the BS. COUGAR considers the whole network as a distributed database system and uses declarative queries for query processing from network layer functions. It also utilizes in-network data aggregation for more energy saving. Energy-Aware Routing uses a set of suboptimal paths occasionally to increase the lifetime of the network. These paths are chosen by means of a probability function, which depends on the energy consumption of each path.

In the rest of this section, we describe three typical data dissemination schemes—flooding/gossiping, SPIN, and Directed Diffusion—in detail with a focus on their key ideas and performance issues.

4.4.1 Flooding and Gossiping

Flooding is a classical and straightforward mechanism to disseminate data in WSNs, which takes advantage of the broadcasting nature of the wireless medium. To deliver a particular packet from the source to the destination node with flooding, the source node broadcasts the data to all the neighbors. Upon receiving the packet, each neighbor will broadcast a copy of the packet to its neighbors. This process continues until the packet arrives at the destination or the packet is dropped. Flooding is very easy to implement, but it has a major drawback of increasing the network load with redundant traffic. In classical flooding, a node may blindly broadcast whatever it receives, regardless of whether or not the neighbor has already received a copy from another source. This leads to the *implosion* problem [Rwheinzelman99, Jkulik02]. Figure 4.1 shows the implosion problem where the same message goes to node C, from nodes A and B, thereby creating redundancy [Rwheinzelman99]. In Figure 4.1, a WSN with four nodes, A, B, C, and D, is shown. Assume that the data needs to be sent from node D to node C using flooding. Node D broadcasts the data (a) to its neighbors, which are nodes A and B. The nodes A and B forward the same data (a) to node C. Here the issue is that node C receives the same data twice. This *implosion* results in multiple copies of the same data packet floating around the network, and a node may receive multiple copies of the data information.

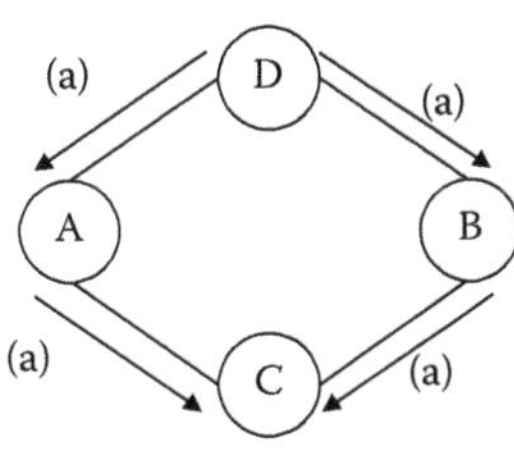

Figure 4.1 Implosion problem. (From Heidemann, R.W. et al., Adaptive protocols for information dissemination in wireless sensor networks, *ACM Mobicom '99*, Seattle, WA, August 1999, 174–185; Joanna, K. et al., *Wireless Netw., ACM*, 8(2/3), 169, March–May 2002.)

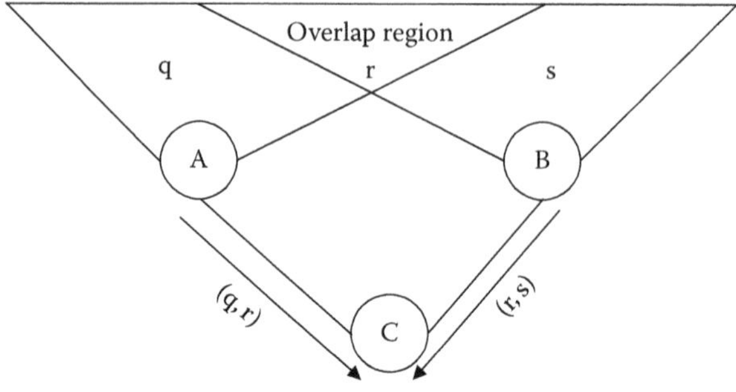

Figure 4.2 Example of overlap region. (From Heidemann, R.W. et al., Adaptive protocols for information dissemination in wireless sensor networks, *ACM Mobicom '99*, Seattle, WA, August 1999, 174–185; Joanna, K. et al., *Wireless Netw., ACM,* 8(2/3), 169, March–May 2002.)

Sensor nodes often cover overlapping geographic areas and gather overlapping pieces of event data. The sensed data received by the neighbors of the nodes would contain some part of the data that is redundant, which is known as *overlap* [Rwheinzelman99, Jkulik02]. Figure 4.2 shows an example of the overlapping issue. Node A in Figure 4.2 senses the data in the regions q and r. Similarly, node B senses the data in the regions r and s. Assume that the data sensed in the regions q and r is (q, r) and the data sensed in the regions r and s is (r, s). After sensing the data, the nodes A and B send the data (q, r) and (r, s) to node C. Obviously, the redundant copy of data (r) received at the destination node C is unnecessary.

The *implosion* and *overlap* issues lead to additional traffic in the network, which is unnecessary. The limited resources, such as energy and bandwidth, in WSNs will be wasted by this naïve flooding process. Hence, many studies have brought up techniques such as probability and packet ID to control the redundancies generated from the flooding process. For example, after assigning a unique ID for the packet, a sensor node can remember the IDs for the packets it broadcasted earlier. Then, the node can ignore the broadcast requests when it sees the same packet ID again. Similarly, a node may ignore a broadcast request according to a certain probability distribution. However, such techniques still cannot totally eliminate the flooding redundancies and may have considerable negative impacts on the network performance. To avoid the problem of flooding redundancy, gossiping takes a step further by just selecting one random node to forward the packet rather than broadcasting. In other words, in gossiping, the receiving node sends the packet to a randomly selected neighbor. The received packet is forwarded to another next-hop neighbor, which is also picked randomly to forward the packet and so on. However, the random selection of next-hop neighbors can cause delays in the propagation of data through the network.

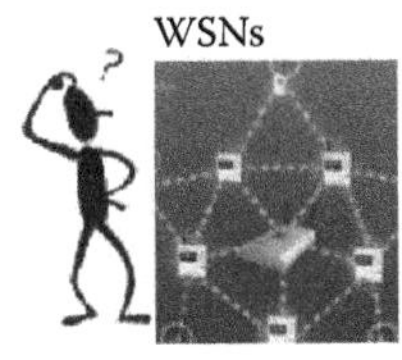

Gossiping can suppress the implosion issues in the flooding scheme. However, both flooding and gossiping routing schemes actually do nothing in reducing the redundant reports and packets in the *overlap* scenario, where an event is detected by all the sensors in the region or multiple sensors in the region reply with similar information on a particular query.

4.4.1.1 Ideal Dissemination

To disseminate data in WSNs, ideally, sensor nodes send observed data along the optimal routing path (taking into consideration the number of hops, time to transmit the data, energy consumption) and the intended nodes receive each piece of distinct data only once. This phenomenon is called *ideal dissemination* in [Rwheinzelman99, Jkulik02]. For example, assume that node A initially possesses data (a,c) and node B only possesses data (c), as shown in Figure 4.3. To efficiently disseminate the data throughout the network, node D, in Figure 4.3, uses the *ideal dissemination* scheme and will only transmit the data in the order shown in the boxed number. First, node D delivers data (a,c) and (a) to node A and B, respectively, while node B sends data (c) to node D. Then, either node B or node C sends data (a) to node D. *Ideal dissemination* does not waste energy on transmitting and receiving useless data. Of course, in a real distributed ad hoc sensor network, it is extremely challenging (if not impossible) to achieve ideal dissemination.

4.4.2 Sensor Protocols for Information via Negotiation [Jkulik02, Rwheinzelman99]

To overcome the aforementioned implosion and overlap issues, a family of adaptive protocols, called SPIN, was proposed to use negotiations for diffusing data in

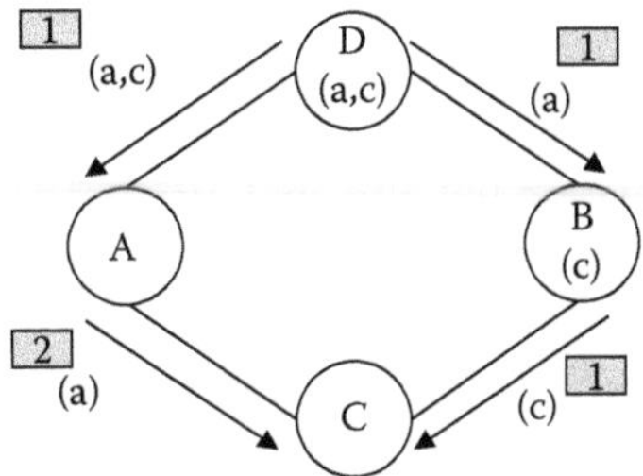

Figure 4.3 Example of ideal dissemination. (From Heidemann, R.W. et al., Adaptive protocols for information dissemination in wireless sensor networks, *ACM Mobicom '99*, Seattle, WA, August 1999, 174–185; Joanna, K. et al., *Wireless Netw., ACM*, 8(2/3), 169, March–May 2002.)

WSNs [Jkulik02, Rwheinzelman99]. SPIN uses metadata for describing the sensed data. The metadata is exchanged among sensors via data advertisement. Upon obtaining new data, each node advertises the availability of new data information to its neighbors. The interested neighbor node, who also would like to possess the new data, can send a request message to the advertiser. Then the advertiser will reply with the data to the requested nodes. Unlike the classical flooding and gossiping protocols, which are blind to the resources' consumption in the network, SPIN uses a resource manager to become resource aware and resource adaptive in the process of data dissemination. The major goal of SPIN's metadata negotiation is to resolve the classical flooding problems, such as redundant information passing, overlapping of sensing areas, and resource blindness, and, thus, achieve better energy efficiency.

4.4.2.1 Design of SPIN

The design of SPIN is motivated by application-level framing (ALF) [Ddclark90]. Using ALF, the network protocols choose transmission units that are meaningful to applications. In other words, the packetization is best done in terms of application data units. Hence, SPIN designs metadata to ensure common naming data in both the transmission protocol and the application. Instead of sending the actual data, sensor nodes send metadata to interested neighbors in the form of an advertisement. The metadata must be smaller than the actual data for SPIN to be energy efficient. If the actual data is distinguishable, then the corresponding metadata should also be distinguishable. Similarly, two pieces of data that happen to be indistinguishable should have the same metadata. Generally, the format of metadata depends on the particular application [Rwheinzelman99].

Another important aspect of SPIN is that it uses a resource manager to monitor the available resources in the node and make the corresponding decision whether or not to participate in a particular data dissemination. Applications probe the resource manager before transmitting or processing data. The nodes using SPIN calculate the energy and resources available by means of polling the resource system. Hence, the routing decisions in SPIN are made by combining the knowledge of not only topology information but also application data layout and the status of resources available at each node.

There are three different types of messages in SPIN, which are new data advertisement (ADV), data request (REQ), and DATA message. When a node has a DATA message to share, the node can advertise this fact by transmitting an ADV packet containing the metadata of the message. A node that is interested in the details of the message based on the received metadata packet can send an REQ packet to the advertiser. Then, the requesting node will receive the DATA message containing actual details of the message with a metadata header from the advertiser.

4.4.2.2 Different Types of SPIN

The above SPIN philosophy is tuned to accommodate different WSN application and network scenarios. Four SPIN protocols are proposed in [Rwheinzelman99]: SPIN-PP, SPIN-BC, SPIN-EC, and SPIN-RL:

- SPIN-PP—for point-to-point transmission media. Assume that there is plentiful energy and packets are never lost in the network.
- SPIN-BC—for broadcast transmission media. Assume that there is plentiful energy and packets are never lost in the network.
- SPIN-EC—an energy-conserving version of SPIN-PP.
- SPIN-RL—a reliable version of SPIN-BC.

4.4.2.2.1 SPIN-PP

SPIN-PP employs three stages of message exchange for networks using point-to-point transmission media, which allow nodes A and B to communicate exclusively with each other without interfering with other nodes. The three stages correspond to the three messages described above. The protocol starts with a node sending an ADV message to its neighbors to advertise the data it intends to disseminate. In the next stage, the neighbors check whether they are interested in the advertised data after receiving the ADV packet. If a node determines to possess a copy of the data, the node sends back an REQ message to the node that sent the ADV message. Then, in the final stage, the actual data in the form of DATA message is delivered from the advertiser to the requester. Based on the received new DATA message and its own data in the memory, a node could perform some aggregation or redundancy-reducing processes prior to re-advertising the aggregated metadata to the neighbors.

4.4.2.2.2 SPIN-EC

SPIN-EC adds an energy-conserving scheme to the SPIN-PP protocol. When a node receives a new data, it will consult the resource manager before initiating the SPIN protocol and advertising the new metadata. The SPIN protocol will be started if and only if it turns out that the node has enough energy to complete all the stages of the protocol. Otherwise, it simply refrains from participating in the protocol. Similarly, upon receiving an advertisement, a node does not send out a request if it does not have enough energy to transmit the request and receive the corresponding data.

4.4.2.2.3 SPIN-BC

SPIN-BC is developed for broadcast transmission media. In SPIN-BC, the nodes use a single channel to broadcast the data to all the nodes in the receiving range. SPIN-BC employs the one-to-many communication scheme for delivering the same data to multiple sensor nodes in one transmission. Similar to SPIN-PP, SPIN-BC also operates in three stages. There are three primary aspects in which SPIN-BC is different from SPIN-PP.

1. In SPIN-PP, one transmission can only target one specific node. Hence, a node has to send the advertised metadata to every neighbor in a separate transmission. However, taking advantage of the broadcast transmission media, every node within the transmission range could receive the same data in SPIN-BC.
2. Unlike SPIN-PP, SPIN-BC does not allow nodes to respond to the ADV packets immediately. In SPIN-BC, upon receiving the ADV packet, the nodes check whether they already possess the data advertised. If a node does not possess the data, the node sets a random timer. When the timer expires, the node broadcasts an REQ message to the original advertiser if the node does not receive the advertised data yet. Then the node advertising the metadata will respond to the REQ with the DATA message. When nodes other than the original advertiser receive the REQ, they cancel their own request timers to avoid redundant copies of the same request.
3. SPIN-BC will broadcast the DATA message only once and will not respond to multiple requests for the same data.

4.4.2.2.4 SPIN-RL

To handle the lossy link in WSNs, the SPIN-RL protocol makes two adjustments on SPIN-BC for reliable transmission. First, nodes employing the SPIN-RL protocol keep track of all the advertisements that are received. If a node does not receive the data within a particular period of time after sending out the request, the node consults the track of all advertisements received and sends another request to a randomly selected advertiser with the same piece of data. Second, nodes in SPIN-RL limit the frequency with which they will resend data to the neighbors. After a node sends the requested data, say (a), to other nodes, the node waits for some period of time before responding to any further requests demanding the same piece of data (a).

4.4.2.3 Evaluating SPIN Protocols [Rwheinzelman99, Jkulik02]

Using metadata names, nodes in SPIN negotiate with each other about the necessary data exchange. These negotiations ensure that nodes only transmit data when necessary and energy is not wasted on useless or redundant transmissions. With the resource manager, each node is aware of the available resources and is able to cut back on the activities to expand the lifetime of the network.

Table 4.1 shows the related parameters in the simulation with a randomly generated 25-node network [Jkulik02]. Each node in the network is initialized with 3 data items, randomly chosen from a set of 25 possible data items. No network loss and queuing delay is considered.

Table 4.1 Simulation Test Bed for SPIN

Nodes	25
Edges	59
Average degree	4.7 neighbors
Diameter	8 hops
Average shortest path	3.2 hops
Antenna reach	10 m
Radio propagation speed	3 × 8 m/s
Processing delay	5–10 ms
Radio speed	1 Mbps
Transmit cost	600 mW
Receive cost	200 mW
Data size	500 bytes
Metadata size	16 bytes

Source: Adapted from Joanna, K. et al., *Wireless Netw., ACM,* 8(2/3), 169, March–May 2002.

Table 4.2 shows the simulation results from SPIN-PP in which the *ideal dissemination* scheme is used as the baseline. Comparing the *flooding* and *gossiping* schemes, SPIN-PP consumes much less energy; it uses energy less by approximately a factor of 3.5 than flooding. This is partially due to the fact that flooding and gossiping schemes introduce much redundant data. As shown in Table 4.2, simulation shows that 77 percent of the transmitted DATA messages are redundant in the flooding scheme and 96 percent of them are redundant in the gossiping scheme. Note that SPIN-PP also introduces limited overhead traffic, such as the ADV and REQ packets, to the network.

The convergence time is defined as the time it takes to ensure that all the nodes in the network receive the intended data. SPIN-PP takes 80 ms longer to converge than flooding, whereas flooding takes only 10 ms longer to converge than the *ideal dissemination* scheme. Although it appears that SPIN-PP performs much worse than the flooding scheme in terms of the convergence time, this increase is actually a constant amount, regardless of the length of the simulation. Thus, for longer simulations, the increase in the convergence time for the SPIN-PP protocol will be negligible [Rwheinzelman99].

Other simulation and analysis results also indicate that SPIN-EC distributes 60 percent more data per unit energy than the flooding scheme. SPIN-PP and SPIN-EC outperform the gossiping scheme and come close to the *ideal dissemination*

Table 4.2 Results for Simulations of the SPIN-PP Protocol

Performance Relative to Ideal	*SPIN*	*Flooding*	*Gossiping*
Increase in energy dissipation	0.45 J	6.3 J	44.1 J
Increase in convergence time	90 ms	10 ms	3025 ms
Slope of energy Dissipation versus node degree correlation line	1.25*x*	5*x*	25*x*
Percent of total data messages that are redundant	0	77 percent	96 percent

Source: Adapted from Heidemann, R.W. et al., Adaptive protocols for information dissemination in wireless sensor networks, *ACM Mobicom '99*, Seattle, WA, August 1999, 174–185.

protocol. In addition, SPIN-BC and SPIN-RL are able to use one-to-many communications exclusively, while still acquiring data faster and using less energy than the flooding scheme. SPIN-RL can efficiently handle packet loss and dissipate twice the amount of data per unit energy as the flooding scheme.

4.4.3 Directed Diffusion [CIntanagonwiwat00]

Directed Diffusion differs from SPIN in terms of the way data transmission is initiated. The basic idea of *Directed Diffusion* is to diffuse data through sensor nodes by using a naming scheme for the data. With the naming scheme, the sink can issue a query to the sensor nodes regarding the data the sink is interested in. Then the corresponding sensor nodes reply with the necessary information to the sink. To achieve this, *Direct Diffusion* assigns attribute-value pairs to the data and queries on an on-demand basis. To issue a query indicating the type of data the sink is looking for, an *interest* is defined using the attribute-value pairs, such as name of objects, geographical area, duration, interval, etc. The sink disseminates the *interest* through its neighbors. The *interest* is cached in the sensor nodes. Whenever a node receives data, the node can compare the received data with the values of the *interest.* If there is a match, the node will establish paths to the sink from which the node receives the interest. These paths are known as events. Then, the sink can choose paths to resend the *interest* and expect the sensor node to reply with the data back to the sink.

Directed diffusion consists of several elements:

- *Data* is named using attribute-value pairs.
- *Interest* is a sensing task for named data.

- *Gradient* is a reply link to a neighbor from which the interest is received.
- *Events* start flowing toward the originators of interests along multiple paths.
- *Reinforcement* is a mechanism for the sink to select paths receiving the sensed data.

4.4.3.1 Naming

The task descriptions in Directed Diffusion are named by several attribute-value pairs that identify the specific task. For example, a task of tracing animals may be described as follows:

```
Type = animal                     //detect animals
Interval = 0.5 s                  //send back events every
                                  0.5 s
Timestamp = 02:02:19              //interest-generated time
ExpiresAt = 02:12:19              //not interested in this
                                  afterward
RECT = [-100, 100, 200, 400]      //sensors within the spec-
                                  ified region perform the
                                  task
```

A task description is called an *interest*, which specifies an interest in a particular kind of data, which matches the attributes in the task description; similarly, the data sent in response to the interest is also named using attribute-value pairs. For example, a sensor that detects animals in the specified region may create a return data message (or reply) as follows:

```
Type = animal             //detect animals
Instance = cow            //instance type
Location = [122, 210]     //node location
Confidence = 0.90         //confidence in the match
Timestamp = 02:02:20      //event-occurring time
```

How to choose the naming schemes, such as attributes and value ranges, heavily depends on the applications the sensor networks are deployed for. The choice of the naming scheme has to be done carefully, as it can affect the expressivity of tasks and may impact performance of the network.

4.4.3.2 Interest Propagation and Gradient Establishment

An interest message is a query that specifies what the sink wants the sensors to report. Assume that there are five fields in an *interest*, as in the previous example: Type, Interval, Timestamp, ExpiresAt, and RECT. The sink periodically broadcasts the interest message to its neighbors. Initially, this *interest* message

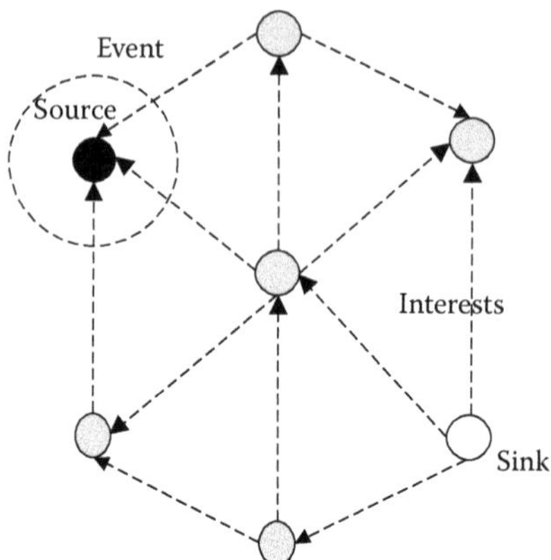

Figure 4.4 Interest propagation in the network. (Adapted from Intanagonwiwat, C. et al., Directed diffusion: A scalable and robust communication paradigm for sensor networks, *Proceedings of the Sixth Annual International Conference on Mobile Computing and Networking (MobiCOM '00),* Boston, MA, August 2000, ACM Press, New York, 56–67.)

serves as an exploratory method to see if there are sensors that detect animals in the specified region. Hence, the initial interest just specifies the node to return the event information at a lower rate. In our example, the sink expects an event report every half second. Later on, when the sink can revise the interest and adopt the reinforcement scheme (described later) to ask for event reports at a higher data rate.

Figure 4.4 shows an example of how the interest is propagated in WSNs. Basically, upon receiving the sensing task from the end user (e.g., an application residing in the sink), the sink broadcasts the *interest* message to all its neighboring nodes. The neighboring nodes of the sink then further forward the message to the network. There are several possible options for the node to re-send the interest to some subset of its neighbors, as shown in Table 4.3.

Each node maintains a cache for the interests where each entry is unique. When a node receives a particular interest, it is compared with all the entries in the cache. If the particular entry is not present in the cache, the entry is entered into the cache and a gradient is set up toward the neighbor node, which sends the *interest* packet. Figure 4.5 shows that the gradients are established from the source node to the sink node in the opposite direction of the interests' propagation, as in Figure 4.4. Each gradient contains data rate and expiration time, which are derived from the *interest* packet. On the other hand, if there is an interest entry in the cache, aggregation on the interests and updation of the gradient fields can be performed by the node. The *interest* entry is removed from the cache when all the gradients associated with the interest entry have expired.

Table 4.3 Design Options for Diffusion

Diffusion Element	*Potential Choices*
Interest propagation	Flooding Constrained or directional flooding based on location Directional propagation based on previously cached data
Data propagation	Reinforcement to single-path delivery Multipath delivery with selective quality along different paths Multipath delivery with probabilistic forwarding
Data caching and aggregation	For robust data delivery in the face of node failure For coordinated sensing and data reduction For directing interests
Reinforcement	Rules for how many neighbors to reinforce Rules for deciding when to reinforce Negative reinforcement mechanisms and rules

Source: Adapted from Intanagonwiwat, C. et al., Directed diffusion: A scalable and robust communication paradigm for sensor networks, *Proceedings of the Sixth Annual International Conference on Mobile Computing and Networking (MobiCOM '00),* Boston, MA, August 2000, ACM Press, New York, 56–67.

4.4.3.3 Data Propagation

When a sensor node detects a target corresponding to one interest entry in its *interest* cache, the node computes the highest requested rate among all its outgoing gradients. After generating the corresponding reply (or return data message), the source node sends the reply to each neighbor for whom it has a gradient. Upon receiving the data message, the intermediate node checks its cache to see whether there is a match. If a match is found, the node checks the data cache associated with the matched interest entry to decide whether to drop the redundant message, or update the cache and resend the new data message to its neighbors according to the gradients' settings. On the other hand, if no interest match is identified, the data message is simply dropped. As shown in Table 4.3, in the process of data propagation, each node can prevent a loop by using the data cache and can down-convert to the appropriate gradient with lower data rates when necessary.

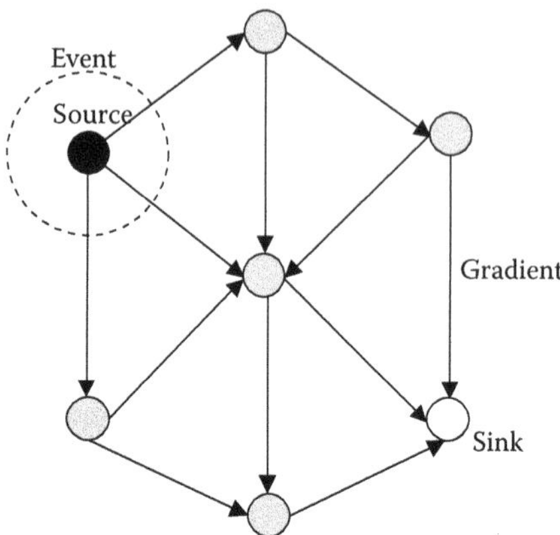

Figure 4.5 Gradient establishment in the network. (Adapted from Intanagonwiwat, C. et al., Directed diffusion: A scalable and robust communication paradigm for sensor networks, *Proceedings of the Sixth Annual International Conference on Mobile Computing and Networking (MobiCOM '00)*, Boston, MA, August 2000, ACM Press, New York, 56–67.)

4.4.3.4 Reinforcement

Initially, the sink node disseminates the *interest* message at a low rate. The source sensors that have data matching the *interest* also reply with data at a low rate, possibly along multiple paths toward the sink. When the sink receives the low-rate data, it chooses one particular neighbor node and sends a reinforcement packet to request for a higher-rate data report. The reinforcement packet is similar to the original *interest* message, but with a smaller interval or a higher data rate, as follows:

```
Type = animal                   //detect animals
Interval = 10ms                 //send back events every 10ms
Timestamp = 03:02:19            //interest-generated time
ExpiresAt = 03:12:19            //not interested in this
                                afterward
RECT = [-100, 100, 200, 400]    //sensors within the speci-
                                fied region perform the task
```

When the specific node receives this reinforcement, the node notices that a gradient toward the neighbor already exists and a higher rate is requested. If the new data rate is higher than that of any existing gradients, the node must also forward the reinforcement to at least one neighbor node. As shown in Table 4.3, different techniques can be applied to choose the neighbor node to which the reinforcement will be sent. In general, whenever one path delivers an event faster than other paths, the sink tends to use this path to request high-quality data. Through this sequence of reinforcement, paths with corresponding data rates are established from the source node to the sink node.

However, the above process may result in more than one path being reinforced. Assume that two paths, P_1 and P_2, are used for delivering data as a result of the above reinforcement. Now, if the sink prefers the path P_2 over P_1, the sink can use the above scheme to continuously send reinforcement through path P_2, but it also needs the scheme of negative reinforcement to degrade the necessary rate through path P_1. Two approaches can be adopted for negative reinforcement. One approach is to time out the high-data-rate gradients in the network unless they are explicitly reinforced. Hence, the sink can just keep reinforcing path P_2, and eventually path P_1 will degrade to a lower data rate. Another approach is to explicitly degrade the path P_1 by resending the interest with a lower data rate.

In fact, in networks with multiple sources or multiple sinks, the schemes we described above can also be effectively applied. In addition, in Directed Diffusion, the intermediate nodes on a previously reinforced path can also initiate the reinforcement process when local repairs of failed or degraded paths are identified. Causes of failure or degradation include node-energy depletion, environmental interference, raid fade, and so on.

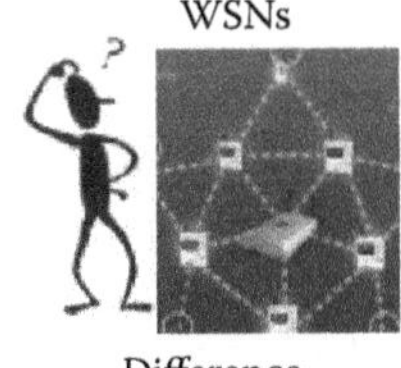

The data dissemination with Directed Diffusion is different from traditional networking in several aspects:

All communication in Directed Diffusion is neighbor to neighbor, not end to end.

Directed Diffusion in WSNs is suitable for task-oriented applications.

No globally unique identifier or address is needed.

4.4.3.5 Evaluating Directed Diffusion

One of the motivations of the Directed Diffusion scheme is the physical systems (e.g., ant colonies) that build up transmission paths strictly using local communication well and are extraordinarily robust. Another reason behind using such a scheme is to get rid of unnecessary and complicated operations of network layer routing for energy saving. Hence, *interest* and *gradients* are defined and utilized in Directed Diffusion to establish paths between the sink and sources for data transmission.

Studies have shown that Directed Diffusion has the potential for significant energy saving and outperforms the idealized traditional data dissemination scheme: omniscient multicast [CIntanagonwiwat00]. In addition, *Directed Diffusion* owns features such as data-centric dissemination, reinforcement-based adaptation to the empirically best path, and in-network data aggregation and caching. Because it is data centric, all communication is neighbor to neighbor with no need either for a global addressing mechanism or for maintaining global network topology. Aggregation and caching offers better network performance in terms of energy efficiency and delay. However, Directed Diffusion cannot be applied to all sorts of sensor network applications, because it is based on a query-driven data-delivery model. The applications that require continuous data delivery to the sink may not work efficiently with a query-driven on-demand model, such as Directed Diffusion [Njamal04].

4.5 Hierarchical Routing Protocols in WSNs

Scalability is one of the major design concerns of sensor networks, particularly for many applications with a large number of sensor nodes deployed to cover a pretty large physical area. A single-tier or flat network operation in such large-scale sensor networks can cause

- Large convergence time for many algorithms and protocols
- Overload with increase in sensor density
- Large memory space required for storing the network information
- Increased latency, complexity, and instability in communication
- Inadequate tracking of events

Because the tiny sensors with limited resources are typically not capable of performing long-haul communication, the concept of clustering or hierarchical network routing has been pursued in many routing approaches to allow the system to cover a large area of interest without degrading the service. A cluster is generally a collection of nodes with similar missions, within similar vicinity, or having similar functionalities/resources. A hierarchical routing protocol can be viewed as a set of flat routing protocols, each operating at different levels of granularity. For example, in a two-tier hierarchical routing protocol, the intercluster component is essentially a flat routing protocol that computes the routes between clusters. Likewise, the intra-cluster component is a flat routing protocol, which generates routes between nodes in each cluster. Hierarchical routing protocols provide global routes to the network clusters, rather than individual nodes, which can simplify many aforementioned scalability issues in the network and are often better suited to very large networks compared to flat routing protocols. In addition, data aggregation and fusion can be performed within the cluster to decrease the number of messages transmitted to the sink, which can enhance the network performance in terms of energy efficiency.

Examples of hierarchical network routing protocols include LEACH [Heinzelman02], TEEN [AManjeshwar01], Adaptive Periodic Threshold-Sensitive Energy-Efficient Sensor Network protocol (APTEEN) [Marati02], Power-Efficient Gathering in Sensor Information Systems (PEGASIS) [Slindsay02], Hierarchical-PEGASIS [ASavvides01], Minimum Energy Consumption Network (MECN) [Vrodoplu99], Small Minimum Energy Consumption Network (SMECN) [Lli01], Self-Organizing Protocol (SOP) [Lsubramanian00], Sensor Aggregate Routing

[Qfang03], Virtual Grid Architecture Routing [JNal-karaki04], Hierarchical Power-Aware Routing [Qli01], and Two-Tier Data Dimension (TTDD) [FYe02].

LEACH forms clusters of the sensor nodes based on the received signal strength and uses the local CHs as the gateway to the BS. TEEN is a hierarchical protocol designed to be responsive to sudden and drastic changes in the sensed attributes, such as temperature, pressure, and rainfall. APTEEN aims at both capturing periodic data collections and reacting to time-critical events. PEGASIS forms chains of sensor nodes so that each node transmits and receives from a neighbor and only one node is selected from that chain to transmit to the BS rather than forming multiple clusters. Hierarchical-PEGASIS, an extension of PEGASIS, aims at decreasing the delay incurred for packets during transmission to the BS. MECN finds a subnetwork of the WSN with less number of nodes and also finds the minimum global energy required for data transfer. SMECN, an extension of MECN, considers the obstacles in data transmission while relaxing the assumption in MECN that every node in the network can transmit to each other node. Sensor Aggregate Routing comprises of the sensor nodes with a grouping predicate for a collaborative, cooperative processing task. The parameters of the predicate depend on the task and the resource requirements. Virtual Grid Architecture Routing uses data processing and in-network processing to maximize the network lifetime. The network is divided into zones based on the global positioning system (GPS) information. Data aggregation is performed at two levels: local aggregation and global aggregation. Each zone has a local aggregator and master aggregator. In Hierarchical Power-Aware Routing, the network is divided into groups based on geographical proximity and each group is allowed to decide how to route the data such that the energy consumed for routing will be minimum. In TTDD, each source node builds a grid structure for disseminating the data to the mobile sinks. The nodes that sense an event process the signal, and one of the nodes in the group that sensed the event becomes the source of the sensed data. The source node then builds a grid structure to route the data to the other nodes in the network.

4.5.1 Low-Energy Adaptive Clustering Hierarchy Protocol [Heinzelman02]

The LEACH protocol is a self-organizing, adaptive clustering protocol that uses randomization to distribute the energy consumption evenly among the sensor nodes in the network. The LEACH protocol aims at increasing the system lifetime and reducing the latency for transferring the data. The LEACH protocol uses the following techniques to achieve its goals [Heinzelman02]:

1. Localized control for data transfers
2. Low-energy medium access control
3. Self-configuring, randomized, and adaptive cluster formation
4. Application-specific data processing, like data compression and data aggregation

The LEACH routing protocol divides the sensor nodes in the network into groups called clusters. The clusters have special types of nodes, called CH nodes. These nodes are used for transmission of the data to the BS. These nodes are also responsible for the medium access among the nodes in the cluster.

The CH nodes in the cluster consume more energy as compared to the non-CH nodes. To make the energy consumption uniform among the nodes in the network, the LEACH protocol uses randomized rotation for the selection of CHs among the other nodes in the network. LEACH uses the CH nodes for the transmission of data from the non-CH nodes to the BS. The data sensed by the nodes is sent to the CH nodes initially, and then the CH nodes transmit the data to the BS.

4.5.1.1 Protocol Design

As described in Chapters 1 and 2, in sensor networks deployed for environment monitoring or surveillance applications, the overlap in the sensing range and application-specific requirements makes the sensed data in a specific region redundant and strongly correlated. The basic idea of LEACH is to form sensor nodes into clusters and locally process the correlated data such that the useless or redundant transmissions in the network are reduced.

In LEACH, the sensor nodes in the network organize themselves into groups, also called clusters. Then LEACH randomly selects a few nodes as CHs and rotates this role to evenly distribute the energy load among all the sensors in the network. All non-CH nodes will collect sensed data and send the data to the CH. The CH aggregates and compresses the data arriving from nodes that belong to the respective cluster before it sends the aggregated packet to the sink (or BS).

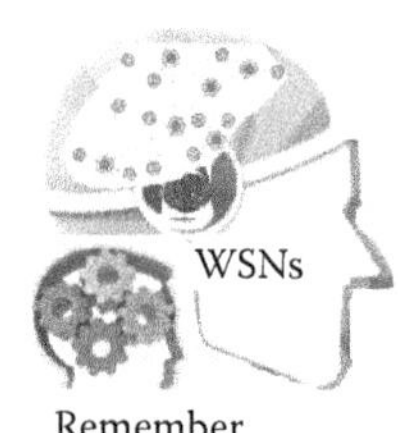

Remember

The operation of LEACH is divided into *rounds*. Each round consists of two phases: the *setup* phase and the *steady state* phase. Each round starts with a *setup* phase when the clusters are organized and CHs are selected. What is followed is the *steady state* phase when the CHs collect and process the data from the nodes within their clusters before the aggregated data is transferred to the sink.

4.5.1.2 Setup Phase: Cluster Formation and Cluster-Head Selection

In LEACH, sensor nodes are organized into clusters by using a distributed algorithm where nodes make autonomous decisions without any centralized control. The goal is to maintain k clusters during each round and evenly distribute the load among all the nodes such that no node is overloaded or runs out of energy before the others. The LEACH protocol assumes that every node is initialized with equal power and can apply power control to vary transmission power. LEACH assumes that every node in the network can reach the sink with enough power. Intuitively, the CH will suffer more energy consumption than other nodes in the cluster due to its responsibility to process data aggregation and deliver data to the remote sink. To avoid quick energy depletion in the CH, LEACH incorporates randomized rotation of the CH role among the high-energy sensors [Hwendi00].

At the beginning of the *setup* phase, each node decides whether or not to act as a CH of the current round. This decision is based on the predetermined percentage of CHs in the network and the number of times the node has been a CH so far. More specifically, a sensor node n chooses a random number between 0 and 1. If this number is less than a threshold value, $T(n)$, the node becomes a CH for the current round. The threshold value is calculated as in Equation 4.1:

$$T(n)=\begin{cases}\dfrac{p}{1-p*(r \bmod (1/p))} & \text{if } n\in G\\ 0 & \text{otherwise}\end{cases} \tag{4.1}$$

where

p is the desired percentage of CHs (e.g., 5 percent)
r is the current round
G is the set of nodes that have not been CHs in the last $1/p$ rounds

As we can see from Equation 4.1, each node has a probability p of becoming a CH in round 0 (i.e., $r = 0$). After that, the CH nodes in round 0 cannot be elected as CHs for the next $1/p$ rounds. Hence, the probability that another node in G is elected as the CH increases, as there are fewer nodes that are eligible to become CHs. Eventually, the threshold in Equation 4.1 will ensure a node to be a CH at some point within $1/p$ rounds.

After electing itself as the CH, the node broadcasts an advertisement message to the network indicating that a cluster is created and the advertiser is the CH. All the non-CH nodes, after receiving this advertisement, decide on the cluster to which they want to join. This decision is based on the received signal strength of the advertisement from the CHs. A non-CH node then sends a join-request message to the appropriate CH. After receiving all the join-request messages, the CH node sets up a TDMA schedule and assigns each node a time slot when it can transmit. This schedule is broadcast to all the nodes in the cluster. As shown in the flowchart of the distributed cluster formation scheme in Figure 4.6, the *setup* phase ends with the reception of the TDMA schedule by all nodes in the cluster.

4.5.1.3 Steady State Phase

After the non-CH nodes receive the TDMA schedule created by their CH nodes, the nodes start transmitting data depending on the TDMA schedule. To synchronize and start the *steady state* phase at the same time, the sink can issue the corresponding

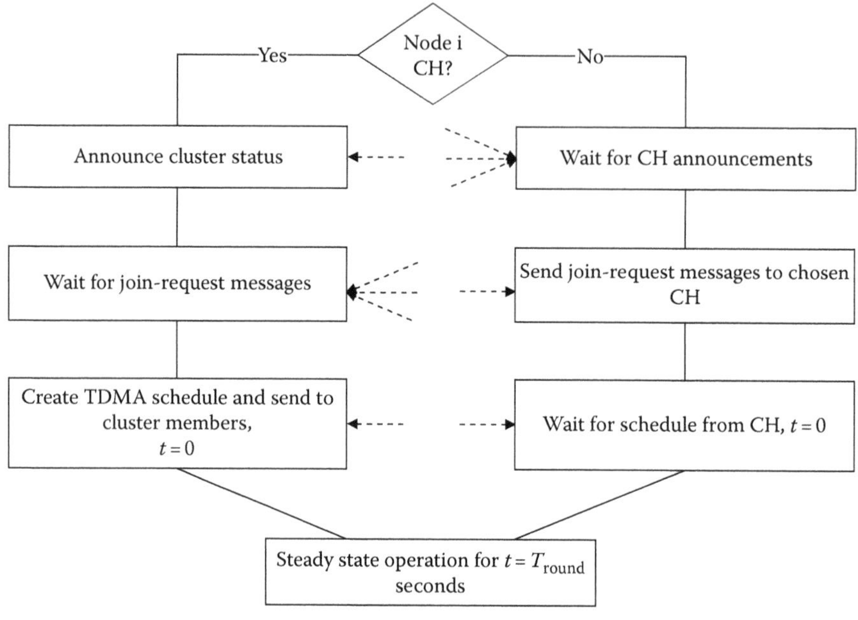

Figure 4.6 Flowchart of the formation of clusters in the LEACH protocol. (Adapted from Heinzelman, W. et al., *IEEE Trans. Wireless Commun.*, 1(4), 660, October 2002.)

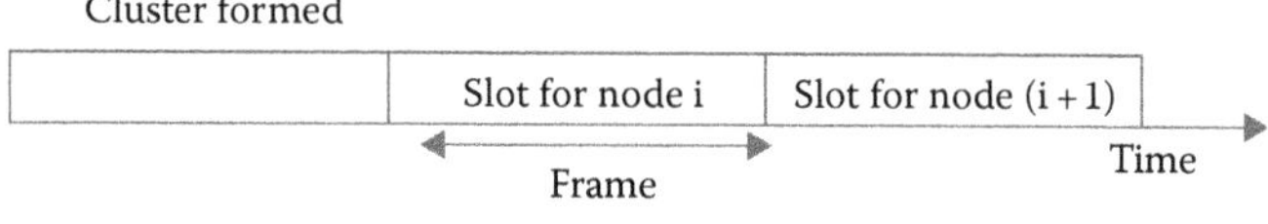

Figure 4.7 Time line for the LEACH protocol. (From Heinzelman, W. et al., *IEEE Trans. Wireless Commun.*, 1(4), 660, October 2002.)

synchronization pulses to all the nodes. As shown in Figure 4.7, the *steady state* phase is further divided into frames. During the assigned frame, the sensor node can transmit the data to the CH node. The duration of each frame slot is constant and depends on the number of nodes in the cluster.

As the CH is normally nearby, each non-CH node can apply a power control scheme to set the minimal amount of energy (based on the received signal strength of the CH advertisement) required for data transmission to the CH. To further improve the energy efficiency in the network, the radio of each non-CH node is turned off (sleep) until its allocated transmission time. However, the CH node must keep its receiver on to receive all the data from the nodes in the cluster. When the data from all the nodes in the cluster has been received, the CH node performs aggregation and signal-processing functions to compress the data. Then the CH node sets the necessary power level and sends the aggregated data to the sink. After a certain time, which is determined a priori, the network goes into the next round to start the *setup* and *steady state* phases again. The duration of the steady state phase is longer than the duration of the setup phase to minimize the overhead.

4.5.1.4 LEACH-Centralized

The previous approach provides certain advantages in forming a cluster using the preceding algorithm. But the previous algorithm has many disadvantages, such as the LEACH protocol offers no guarantee about the number of clusters in a particular area and the placement of the clusters. LEACH-C uses a centralized algorithm for clusters' formation and produces better results as compared to those of the LEACH protocol. In the LEACH-C protocol, each node scans the current location using GPS during the *setup* phase and transmits its current location as well as energy level to the sink. Based on the energy level and location information of all the nodes in the network, the sink can select the CHs and form the clusters optimally in terms of minimizing the amount of energy consumed for data transmission. Then the sink broadcasts the information of the cluster formation to the network. If a node is not assigned as a CH in this round, the node can go to sleep based on its TDMA transmission schedule. However, the CH node has to receive, aggregate, and forward data to the sink.

4.5.1.5 Evaluating LEACH Protocol

In hierarchical routing protocols, each cluster designates a single CH node to relay intercluster traffic. To prevent the CH node from becoming the traffic/energy "hot spot," potentially resulting in network congestion and single point of failure, LEACH adopts a distributed scheme to rotate CH roles to evenly distribute the load among all the nodes in the network. In addition, LEACH employs dynamic clustering, in-network data processing, power-controlled transmission, and collision avoidance schemes to increase the network lifetime. Studies in [Heinzelman02] show that LEACH can achieve over a factor of 7 reductions in energy dissipation compared to direct communication and a factor of 4–8 compared to the minimum transmission energy routing protocol. In addition, the LEACH-C protocol can further improve the network performance by forming better clusters using the global knowledge of the location and energy levels of each node in the network.

However, restricting nodes accessing through CHs can lead to suboptimal routes and data transmission, as potential neighbors in different clusters are prohibited from communicating directly. The idea of dynamic clustering incurs extra overhead for the cluster formation/maintenance, which may diminish the gain in energy consumption. Moreover, LEACH assumes that each node can transmit directly to the CH and the sink, which may be not applicable for networks deployed in large regions. Hence, LEACH has been extended to account for heterogeneous sensor nodes, better scalability, and energy efficiency in the literature.

4.5.2 Threshold-Sensitive Energy-Efficient Sensor Network Protocol [AManjeshwar01]

SPIN, LEACH, and Directed Diffusion protocols have been developed for applications requiring periodic environment monitoring or querying a snapshot of the relevant parameters at certain intervals. On the other hand, two hierarchical routing protocols called TEEN and APTEEN are proposed in [AManjeshwar01, Marati02] for time-critical applications, where responsiveness to changes in the sensed attributes is important. TEEN pursues a hierarchical approach along with the use of a data-centric mechanism to provide the end user with the ability to control the trade-off between energy efficiency, accuracy, and response time dynamically.

4.5.2.1 Sensor Network Model in TEEN

In TEEN, the sink or BS can transmit data to all the nodes in the network at any point of time. However, the sensor node cannot always reach the sink directly due to the constraints of power and transmission range. Unlike LEACH with only one-tier hierarchy, the network architecture in TEEN is based on multilevel hierarchical grouping, as shown in Figure 4.8, where closer nodes form clusters and this process takes place for the multiple levels (or tiers). The CH in each cluster collects data from its cluster members, aggregates the data, and sends the data to an upper-level CH or the BS. Figure 4.8 shows an example of multi-tier clustering. Nodes 1.1.1, 1.1.2, 1.1.3, 1.1.4, and 1.1.5 form a low-level cluster with node 1.1 as the CH. Similarly, nodes 1.2 and 1 serve as the CHs for the respective low-level clusters. The CHs 1.1, 1.2, and 1 from the low-level clusters, in turn, form a cluster with node 1 as the CH. Hence, node 1 also becomes the CH of the second-level cluster. This hierarchy pattern is repeated through the network to form multilevel hierarchies. The uppermost-level cluster nodes will be able to send data directly to the BS, which acts as the root of the uppermost hierarchy and supervises the entire network.

With this network architecture, TEEN allows the nodes to communicate with their immediate CH. Hence, a node does not have to reach the BS directly (as required in LEACH). The data from low-level clusters may travel through multiple CHs before reaching the BS. The CHs at each level will perform necessary data processing, such as aggregation and compression, to conserve energy for the transmission. To evenly distribute the energy consumption, the nodes take turns to serve as CHs, which is similar to LEACH.

4.5.2.2 Operation of TEEN Protocol

Figure 4.9 shows the time line of the TEEN operation. After the clusters are formed, the CH broadcasts two thresholds to the nodes: *hard threshold* and *soft threshold*.

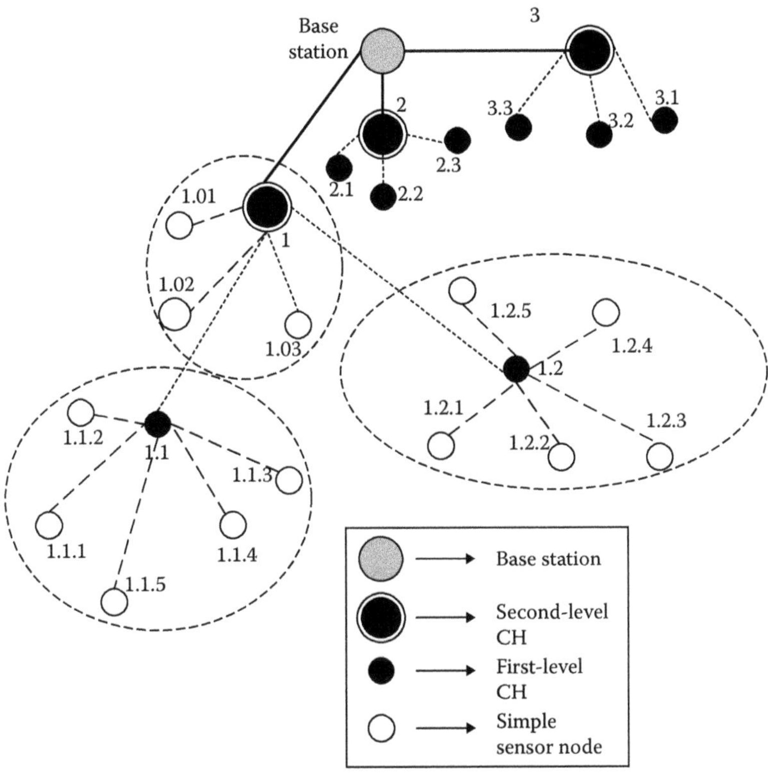

Figure 4.8 An example of network hierarchies in TEEN. (From Manjeshwar, A. and Agarwal, D.P., TEEN: A routing protocol for enhanced efficiency in wireless sensor networks, *Proceedings of 15th IEEE International Parallel and Distributed Processing Symposium,* San Francisco, CA, April 2001, 2009–2015.)

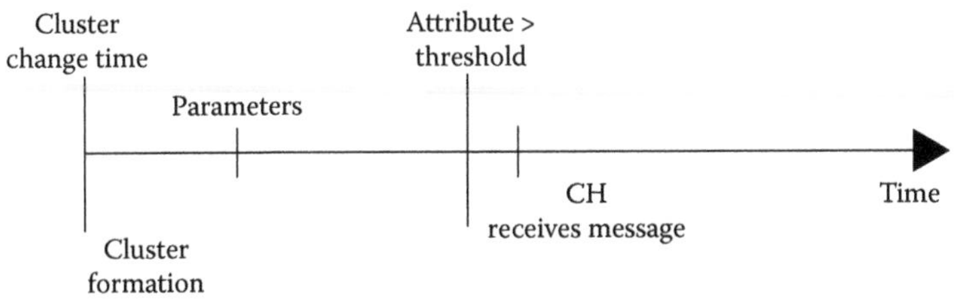

Figure 4.9 Operation of the TEEN protocol. (From Manjeshwar, A. and Agarwal, D.P., TEEN: A routing protocol for enhanced efficiency in wireless sensor networks, *Proceedings of 15th IEEE International Parallel and Distributed Processing Symposium,* San Francisco, CA, April 2001, 2009–2015.)

Hard threshold
Hard threshold is the threshold *value* of the attribute beyond which the sensing node must switch on its transmitter and report the value to its CH. Therefore, the hard threshold allows the nodes to transmit only when the sensed attribute is in the range of interest, which may result in significant reduction in the number of transmissions.

Soft threshold
Soft threshold is the small *change* in the value of the sensed attribute, which triggers the node to switch on its transmitter and to transmit the sensed data to the BS. In other words, once a node senses a value at or beyond the hard threshold, it transmits data only when the value of that attribute changes by an amount equal to or greater than the *soft threshold*. As a consequence, the soft threshold will further reduce the number of transmissions that might otherwise occur when there is little or no change in the sensed attribute.

One can adjust both hard and soft threshold values to control the number of data transmissions. A smaller value of the *soft threshold* gives a more accurate picture of the network, at the expense of increased data transmission and, thus, energy consumption. This indicates that the end user can control the trade-off between energy efficiency and data accuracy by adjusting the values of the threshold. In fact, TEEN allows the user to assign new threshold values and broadcast them to the network when CHs are to change (as shown in Figure 4.9).

As shown in Figure 4.9, the TEEN protocol initially forms the clusters and the parameters are sent to the nodes in the network. The nodes continuously monitor their environment. The first time the value of an attribute reaches its hard threshold value, the node switches on its transmitter and transmits the sensed data to the CH. The sensed data is also stored in an internal variable of the node, called the *sensed value* (SV), which is also updated whenever a node transmits data. The nodes will transmit data in any cluster period, only when both the following conditions are true [AManjeshwar01]:

1. The current value of the sensed attribute has to be greater than the hard threshold
2. The current value of the sensed attribute differs from the SV by an amount equal to or greater than the soft threshold

4.5.2.3 Evaluating TEEN Protocol

The important features of the TEEN protocol include its suitability for time-critical sensing applications. A sudden or drastic change in the value of a sensed attribute in these applications will reach the sink or user almost instantaneously. Also, as message transmission consumes much more energy than data sensing, TEEN can reduce unnecessary transmission, and hence the energy consumption in this scheme can potentially be much less when compared to that in the proactive network. By adjusting the threshold values according to the criticality of the sensed attribute and the target application, TEEN can quickly adapt to the network's real condition and the user's specific requirements.

The simulation has been performed on a network of 100 nodes with a fixed BS in [AManjeshwar01]. The nodes are placed in a random fashion with an initial energy of 2 J in each node. Cluster formation is done as in the LEACH protocol. The energy consumption of the node is modeled as idle-time power dissipation (equal to the radio electronics energy) and sensing power dissipation (equal to 10 percent of the radio electronics energy). Two performance metrics are used to analyze and evaluate the protocols: *average energy dissipated* and *total number of nodes alive*. The *average energy dissipated* is defined as the average dissipation of energy per node over time in the network (as it performs various functions, such as transmitting, receiving, sensing, and aggregation of data). The *total number of nodes alive* indicates the overall lifetime of the network. Simulation results show that TEEN performs better than LEACH-C and LEACH.

However, TEEN is not suitable for applications where periodic reports are needed, because the user may not get any data at all whether or not the thresholds are reached. Thus, the user may not get any data and will never be able to know whether there are any nodes in the network that are alive.

4.5.2.4 Adaptive Periodic Threshold-Sensitive Energy-Efficient Network Protocol [Marati02]

As an extension to TEEN, the APTEEN protocol, on the other hand, is a hybrid protocol that changes the periodicity or threshold values used in the TEEN protocol according to user needs and the application type. APTEEN aims at proactively capturing periodic data collections and reactively responding to time-critical events. Its network-clustering architecture is the same as in TEEN. When the BS forms the clusters, the CH nodes broadcast the attributes, the threshold values, and the transmission schedule to all the nodes.

- *Attributes* are a set of physical parameters that need to be sensed in the network.
- *Thresholds* include soft and hard thresholds, which are the same as the thresholds in the TEEN protocol and serve the same purposes as in the TEEN protocol.
- *Count time* (*CT*) is the period of time after which the sensed data needs to be sent to the CHs.
- *Schedule* refers to the time division multiple access schedule, which is used for sharing the transmission medium among the sensor nodes in the network.

Similar to TEEN, the node in APTEEN senses the environment continuously, and only those nodes that sense a data value at or beyond the thresholds report the data to CHs. If a node does not send data for a time period equal to CT, APTEEN forces the node to sense and transmit the data. APTEEN supports three different query types:

- *Historical:* To analyze past data
- *One time:* To take a snapshot view of the network
- *Persistent:* To monitor an event for a period of time

A TDMA schedule is used, and each node in the cluster is assigned a transmission slot. APTEEN also allows the user to set the CT interval and the threshold values for energy efficiency. Simulations show that APTEEN's performance is somewhere between LEACH and TEEN in terms of energy dissipation and network lifetime. TEEN gives the best performance because it decreases the number of transmissions more significantly than APTEEN does. The drawbacks of TEEN and APTEEN, are the overhead and complexity associated with forming clusters at multiple levels, threshold-based functions, managing counter time and schedule, as well as dealing with attribute-based naming of queries.

4.6 Location-Based Routing Protocols in WSNs

With advances in sensor technologies, many applications densely deploy a large number of sensor nodes carrying a global positioning system (GPS) or a ranging device to facilitate the monitoring, tracing, or surveillance tasks. In the absence of a GPS unit, the location of nodes can be estimated through intelligent localization methods based on techniques such as coarse-grained connectivity, trilateration principle, robust quadrilaterals, and acoustic and multimodal sensing [Bulusu00, Ward97, Moore04, Girod01]. The location information of the sensors can be used to calculate the distance between the source and the destinations so that the energy consumption can be estimated or the transmission power level can be properly adjusted. In addition, recall the routing scheme called Directed Diffusion described earlier in this chapter; the location information can facilitate the sink to issue the query specifying the region in the *interest* message. Accordingly, location-based protocols are proposed to utilize position information to relay the data to the desired regions. Instead of diffusing the data to the whole network, nodes can target the data on a particular region or direction with the help of the geographical information, which potentially reduces the number of transmissions significantly, hence improving the network performance. Examples of location-based routing protocols are Geographic Adaptive Fidelity (GAF) [Yxu01], GEAR [Yyan01], Greedy Other Adaptive Face Routing (GOAFR) [Fkuhn03], and SPAN [Bchen02].

More specifically, GAF is an energy-aware location-based routing algorithm, designed primarily for MANETs, but may be applicable to sensor networks as well. GAF conducts routing based on the location of the node, which is associated with a point in the virtual grid formed for the covered area. GEAR uses energy-aware and geographically informed neighbor selection heuristics to route a packet toward the target region. The protocol suggests the use of geographical information while disseminating queries to appropriate regions, because data queries often include geographic attributes. GOAFR routes the data by picking up the nearest neighbor to the node to be the next hop in the routing process. SPAN identifies some nodes as coordinators based on their positions to form a backbone network for data transmission.

4.6.1 Geographical and Energy-Aware Routing Protocol [Yyan01]

Unlike unicast communication, the GEAR protocol attempts to deliver data to all the nodes inside a target region, which is a common primitive in data-centric WSN applications. GEAR uses energy-aware and geographically informed neighbor selection heuristics to route data toward the specified region. Each node keeps an estimated cost and a learned cost of reaching the destination region through each neighbor. The estimated cost is a combination of residual energy and distance to the destination region, while the learned cost is a refinement of the estimated cost that accounts for routing around holes in the network. Based on the cost information, GEAR picks the next-hop neighbors intelligently to route the data to the destination region in an energy-efficient way. Once the data reaches the region, GEAR employs a recursive geographic forwarding technique to disseminate the packet within the region.

In fact, GEAR complements Directed Diffusion by restricting the number of interests' dissemination to a certain region rather than sending the interests to the whole network, thus conserving more energy.

4.6.1.1 Phases of GEAR

GEAR employs two phases in the process of forwarding data to all the nodes in the target region:

1. Forwarding the packet toward the target region
2. Disseminating the packet within the region

In the first phase, GEAR routes the data toward the target region. To forward the data toward the target region in an energy-efficient way, GEAR takes advantage of the geographical and energy information of sensor nodes to make routing decisions.

In the second phase, GEAR disseminates the data in the target region by using either recursive geographical forwarding or restricted flooding schemes. When the density in the target region is high, the region is further divided into four subregions. Four copies of the data are created and delivered to the subregions. This splitting and forwarding process continues until all the nodes in the target region are covered. On the other hand, when the density in the target region is low, restricted flooding is a better fit to save energy.

4.6.1.2 Energy-Aware Neighbor Computation

Assume that node N is forwarding the packet P to the target region R, where D is the centroid. When receiving the packet P, node N progressively routes the packet P to the target region while trying to balance the energy consumption among all N's neighbors. To achieve this, GEAR introduces the concepts of *estimated* cost and *learned* cost to facilitate the routing decision.

Each node, say N, maintains a state $h(N_i, R)$ called the *learned* cost of region R. If a node does not maintain the *learned* cost of region R, $h(N_i, R)$, then an estimated cost, $c(N_i, R)$, is computed as the default value of $h(N_i, R)$, which is defined in Equation 4.2:

$$c(N_i,R) = \alpha d(N_i,R) + (1-\alpha)e(N_i) \tag{4.2}$$

where

α is the tunable coefficient

$d(N_i, R)$ is the distance from N_i to the centroid D of region R, normalized by the largest such distance among all neighbors of N

$e(N_i)$ is the energy consumed at node N_i, normalized by the largest-consumed energy among neighbors of N

When a node picks a next-hop neighbor, N_{min}, to forward the packet, the *learned* cost of region R is updated as in Equation 4.3:

$$h(N,R) = h(N_{min},R) + C(N,N_{min}) \tag{4.3}$$

where $C(N, N_{min})$ is the cost of transmitting a packet from N to N_{min}, and can also be a combination function of both the remaining energy levels of N, N_{min} and the distance between these two nodes

Once node N has a *learned* cost or an *estimated* cost for each neighbor, node N has to determine which neighbor should be the next-hop node for the following two scenarios:

1. There is at least one neighbor of node N who is closer to D than N
2. All N's neighbors are further away from D than N

1. Closer neighbor exists
   When there are neighbor nodes closer to the destination, GEAR uses a greedy technique to forward the data to the destination. In specific, node N picks the next-hop node among the neighbors that are closer to the destination, minimizing the learned-cost value, $h(N_i, R)$, at the same time. According to Equations 4.2 and 4.3, we make three observations:

   a. If all N's neighbors are equal in terms of energy consumption, node N will choose the neighbor who has the shortest distance to D
   b. If all N's neighbors have the same distance to D, node N will split the load among neighbors
   c. Otherwise, node N selects the next-hop node based on the trade-off between routing toward the neighbor nearest to the destination and balancing energy consumption.
2. All the nodes are further away from the node N
   When there are no neighbor nodes closer to the destination, that is, all neighbors are farther away from the destination, we say a *hole* is identified. In other words, a *hole* occurs when a node does not have any neighbor closer to the target region than itself. In this scenario, the *learned* cost will be combined with an update rule to forward the packets circumventing the holes.

For example, assume that nodes G, H, and I have their energy depleted completely, as shown in Figure 4.10, which is a grid topology with a distance of 1 between two neighbors in the same row or column. Thus, these nodes cannot not forward the data. For simplicity purposes, we set the coefficient α in Equation

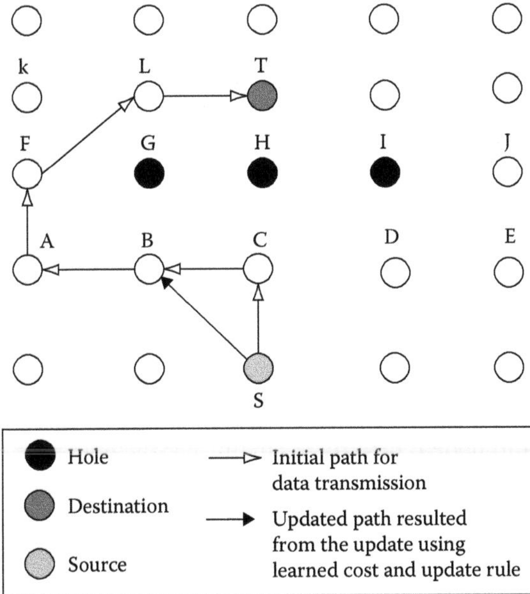

Figure 4.10 Example of learning routes when holes are present. (Adapted from Yan, Y. et al., Geographical and energy aware routing: A recursive data dissemination protocol for wireless sensor networks, Technical Report UCLA-CSD TR-010023, August 2001.)

4.2 as 1 and use the distance instead of the *normalized distance* mentioned earlier. Initially, node S assumes that the neighbor nodes B, C, and D are closer to T based on the *learned costs* in the following equations:

$$h(\mathrm{B,T}) = c(\mathrm{B,T}) = \sqrt{5}$$
$$h(\mathrm{C,T}) = c(\mathrm{C,T}) = 2 \qquad (4.4)$$
$$h(\mathrm{D,T}) = c(\mathrm{D,T}) = \sqrt{5}$$

Hence, to route a packet to T, node S will choose C (which has the lowest *learned cost*) as the next-hop node and forward the packet to C. However, node C will find itself in a hole, because all C's neighbors are further away from T than itself. Then, node C will perform two operations:

- Node C will forward the packet to a neighbor with minimal *learned cost*. Ties are broken based on some predefined ordering (e.g., node ID). In this case, node B will receive the packet forwarded by node C.
- Node C updates its own *learned cost* as $h(\mathrm{C, T}) = h(\mathrm{B, T}) + C(\mathrm{C, G})$, where $h(\mathrm{B,T}) = \sqrt{5}$ and $C(\mathrm{C, B}) = 1$ (assume that one-hop transmission cost is 1), and sends the *learned cost* back to node S.

Next time, upon receiving a packet destined to T, the *learned-cost* values of its neighbors are given by the following equations:

$$h(\mathrm{B,T}) = \sqrt{5}$$
$$h(\mathrm{C,T}) = \sqrt{5} + 1 \qquad (4.5)$$
$$h(\mathrm{D,T}) = \sqrt{5}$$

At node S, instead of delivering the packet to node C (which will forward the packet to node B, causing two transmissions from node S to node B), node S will forward the packet to node B directly to circumvent the hole.

Hence, the *learned cost* is propagated one hop back every time a packet reaches the destination, so that the route setup for the next packet will be adjusted. By propagating the *learned-cost* values upstream through the update rule, GEAR will enable the packet to have an earlier chance to avoid holes (i.e., more effectively circumnavigate holes) and, at the same time, avoid depleting the nodes surrounding the holes. In addition, as the cost is a combination of the normalized distance and energy consumption, the coefficient α in Equation 4.2 can be tuned to emphasize minimizing the path length to the destination or balancing energy consumption.

4.6.1.3 Recursive Geographic Forwarding

When the query packet destined to all nodes in region R reaches the target region, a simple flooding scheme with duplicate suppression (or restricted flooding scheme [Finn87]) can be adopted to disseminate the packet inside the region, particularly, in low-density scenarios. Restricted flooding exploits the broadcast medium of the wireless channel; it only sends one broadcast message to all its neighbors, but every node in its transmission range receives this broadcast message.

However, flooding is expensive in terms of energy consumption, due to the fact that a significant number of redundant and useless transmissions may be introduced by the flooding. The redundant transmission can be especially expensive in high-density networks, which is the case for some WSN applications, where nodes are densely and redundantly deployed for robustness. Hence, recursive geographic forwarding is proposed to disseminate the packet inside the target region when the node density is high. As shown in Figure 4.11, assume that the big rectangle is the target region, R, and a particular node, N_i, receives a data packet, P, for this region, R. The node N_i finds that the packet, P, is sent to the region where it resides. Then, node N_i creates four new copies of the packet, P, and forwards it to four subregions

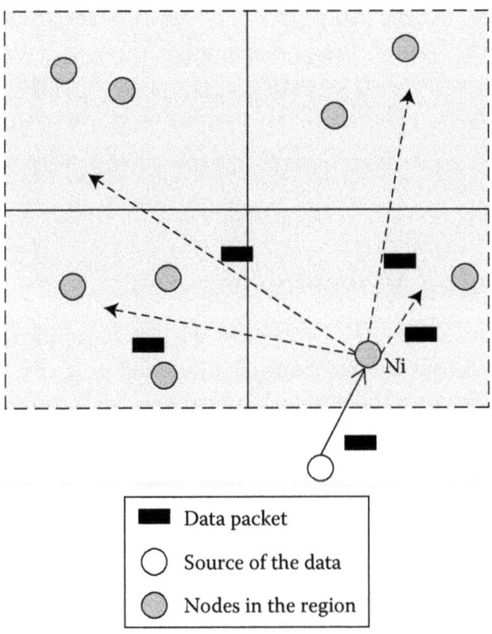

Figure 4.11 Recursive forwarding and splitting process. (Adapted from Yan, Y. et al., Geographical and energy aware routing: A recursive data dissemination protocol for wireless sensor networks, Technical Report UCLA-CSD TR-010023, August 2001.)

of region R. This recursive splitting and forwarding procedure continues until the current node finds itself as the only member in the subregion.

In the case of low node density, the *recursive geographic forwarding* is subject to nonterminated and useless packet transmission. In *recursive geographic forwarding*, packet forwarding and splitting terminates if the subregion is found empty. However, the transmission range of a sensor node is small compared to the subregion size. Hence, the node that is close to the subregion cannot reach the other end of the subregion and has no idea whether the region is empty or not. As a result, *recursive geographic forwarding* still searches for routes to get into the empty subregion. This search will not terminate until the packet is dropped because the number of hops it traversed exceeds the limit (e.g., time to live, or TTL). This kind of daunted search process can heavily drain the node around the subregion, particularly, in networks with low density, in which the probability that the target region is empty is high. In addition, unicast communication in *recursive geographic forwarding* cannot take advantage of the broadcast nature of the wireless medium and requires multiple transmissions in this scenario, which could result in suboptimal energy usage. For these reasons, in the case of low node density, *restricted flooding* is employed by GEAR, replacing *recursive geographic forwarding* [Yyan01].

GEAR proposes to use the degree of a node for differentiating low density with respect to high density. When the packet reaches the first node, N, in a region, whether to use *restricted flooding* or *recursive geographic forwarding* depends on the number of neighbors of node N. If the number is below a threshold, then the packet is flooded inside the region; otherwise, *recursive geographic forwarding* will be triggered.

4.6.1.4 Evaluating GEAR Protocol

GEAR use energy-aware metrics together with geographical information to make energy-efficient routing decisions. While balancing the energy consumption and thereby increasing the network lifetime, GEAR progressively forwards data to the target region based on the proposed cost function and update rule. Within a region, it uses a restricted flooding or a recursive geographic forwarding technique to disseminate the data. GEAR is compared to a similar non-energy-aware routing protocol, greedy perimeter stateless routing (GPSR) [Bkrap00], in which the packets follow the perimeter of the planar graph to find their route. GEAR not only reduces the energy consumption for the route setup, but also outperforms GPSR in terms of packet delivery. The simulation results show that for an uneven traffic distribution, GEAR delivers 70 percent to 80 percent more packets than GPSR. For uniform traffic pairs, GEAR delivers 25 percent to 35 percent more packets than GPSR. Moreover, in both cases, GEAR achieves better connectivity after the initial partition [Yyan01].

4.7 Multipath and QoS-Based Routing

To maintain network reliability, enhance the throughput, or balance the traffic, the techniques using multipath routing are often employed. Multipath routing can provide route resilience through redundant packets delivering over multiple paths or fast route recovery from network disruption. As the bandwidth may be limited in a sensor network, routing along a single path may not provide enough bandwidth for some applications, such as camera or video capture. If multiple paths are employed simultaneously to route the data, a larger aggregated bandwidth and a smaller end-to-end delay may be achieved. Similarly, load balancing can be achieved by spreading the traffic along multiple routes, which can alleviate congestion and bottlenecks in the network.

Therefore, we can see that different strategies to use multiple paths can result in enhancements in different network performance metrics, which actually occurs in many QoS-based routing protocols. In the remaining part of this chapter, we introduce some basic principles of multipath routing, followed by QoS-based routing schemes in sensor networks.

4.7.1 Multipath Routing

In multipath routing, there are multiple, say k, paths between the source and destination nodes. The k paths are *link-disjoint* if they have no common links. The k paths are *node-disjoint* if they have no common intermediate nodes. We call two or more paths non-disjoint (or braided) if they share some links or intermediate nodes (i.e., the node/link disjointedness constraint is relaxed). Multipath routing has been explored for several important reasons. The first is to increase the likelihood of *reliable data delivery*. Sending multiple copies of data along different paths simultaneously offers resilience to failure of a certain number of paths [Ganesan01]. Duplicate data transmission along multipaths can result in more accurate delivery and better data quality for WSNs, at the possible expense of increased traffic redundancy and energy consumption. The second is to enhance the *throughput* from a source to a destination. In these approaches, data for the same source–destination pair is sent out through multiple paths, which create multiple data flows and, hence, potentially increase the throughput from the source to the destination. These multiple flows are better considered together with the wireless interferences among the nodes in the MAC layer to achieve optimized performance. Another major benefit of multipath routing is *load balancing*. In this case, the source and the destination use only one path for routing the data, which is called the primary path. Multiple-path candidates alternatively serve as the primary path for routing data from the same source–destination pair, which can spread energy consumption across nodes on multipaths in the network. This approach can avoid depleting the energy resources of some nodes through constant usage of the same route, potentially resulting in longer network lifetime. Moreover, if there are node failures in the primary path, multipath routing can immediately employ the alternate paths, which are constructed along with the primary path to continuously deliver data from the source node to the destination node.

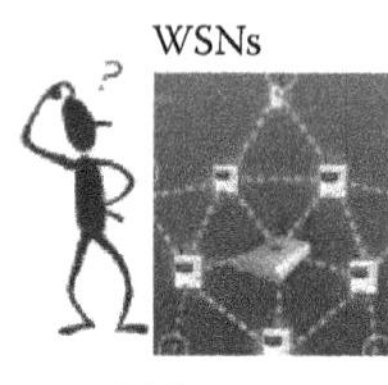

Generally, in routing, when there are node failures in the primary path, then the nodes in this path use flooding for routing the data in the network, to reconstruct the path and recover from the failure. However, the multipath routing scheme can quickly recover from the failure by selecting the alternate paths that are already constructed along with the primary path (without any cost for searching for another one).

For example, the authors in [Chang04] assume that the transmitter power level can be adjusted to use the minimum energy required to reach the intended next-hop receiver. Hence, the energy consumption rate per unit information transmission heavily depends on the choice of the next-hop node, i.e., the routing decision. The routing problem is formulated as a linear programming problem with the objective of maximizing the network lifetime. A routing algorithm is also proposed to route data through shortest-cost-path routing whose link cost is a combination of transmission and reception energy consumption and the residual energy levels at the two end nodes. An alternative path is employed whenever a better path is discovered. Two different models are considered for the information-generation processes: constant rate and arbitrary. Simulation results with both information-generation process models show that the proposed routing algorithm can achieve network lifetime that is very close to the optimal performance obtained by solving the linear programming formulations [Chang04]. Another example of multipath routing is demonstrated in [Dulman03], whereas the techniques of multipath routing are used to enhance the reliability of WSNs. As mentioned earlier, network reliability can be increased by providing several paths from the source to the destination and sending the same packet on multiple paths, which may result in significant traffic redundancy. Hence, there is a trade-off between the amount of traffic redundancy and the reliability of the network. This trade-off is investigated in [Dulman03] using a redundancy function that is dependent on the multipath degree and failing probabilities of the available paths. The proposed idea is to split the original data packet into subpackets, which are sent through the multiple paths. As a result, even if some of these subpackets are lost, the original message can still be reconstructed due to the redundancy added into the data transmission process. In addition, *Directed Diffusion* [CIntanagonwiwat00] is also a good example employing robust multipath routing and delivery. Based on the Directed Diffusion paradigm, the authors in [Ganesan01] investigate how to construct a small number of multipaths in WSNs such that failures on the primary path can be recovered without invoking networkwide flooding for path discovery (thus enhancing network energy performance). Two typical multipath designs—node-disjoint multipath and braided multipath (which consists of partially disjoint/overlapped alternate paths) schemes—are evaluated in terms of the energy/resilience trade-offs under independent and geographically correlated failures. The study has found that for a disjoint multipath configuration whose patterned failure resilience is comparable to that of braided multipaths, the braided multipaths have about 50 percent higher resilience to isolated failures and a third of the overhead for alternate path maintenance [Ganesan01]. Therefore, the braided multipaths are a viable alternative for energy-efficient recovery from isolated and patterned failures with lower maintenance cost in WSNs.

4.7.2 QoS-Based Routing Protocols in WSNs [Ksohrabi00]

For different WSN applications, the constraints and the QoS metrics (delay, energy, priority, bandwidth, fairness, robustness, etc.) to be optimized can be different. Many principles for QoS in the traditional networks can find a counterpart in sensor networks. Some of the QoS metrics are still important and challenging under the constraints of WSNs, and others are not as significant as they are in the Internet. For example, delay, robustness, bandwidth, and energy are paramount and optimized goals in many WSN applications. However, in many cases, the sensors are designed to collectively and cooperatively carry out a task, which diminishes the importance of the QoS metrics, such as fairness. Examples of QoS-based and multipath routing include SPEED [The03] and Sequential Assignment Routing (SAR) [Ksohrabi00]. SPEED is a routing protocol for sensor networks that provides soft real-time end-to-end guarantees, requires each node to maintain information about its neighbors, and uses geographic forwarding to find the paths.

Demand-driven routing protocols are those that find the route between the source and destination systems after a request or demand is issued. These protocols eliminate the overhead associated with table or neighbor update in high-mobility scenarios. However, demand-driven protocols may take a longer time and energy to find the route in a reactive way. SAR, on the other hand, is a table-driven multipath approach, striving to achieve energy efficiency and fault tolerance. Based on the observation that the possibility of protection on the failures in WSNs is tightly related to the degree of disjointedness (i.e., the number of paths with no common branches) in the network, the SAR protocol creates trees rooted at one-hop neighbors of the sink by taking the QoS metrics, the energy resources on each path, and the priority level of each packet into consideration. By using the trees, multiple paths from the sink to sensors are formed. One of these paths is selected according to the energy resources, the QoS on each path, and the priority level of a packet. Failure recovery is performed by enforcing routing table consistency between upstream and downstream nodes on each path, which is done using a handshaking procedure. Any local failure is taken care of, by an automatic path restoration procedure, which is done locally. The simulation studies in [Ksohrabi00] show that SAR can achieve energy efficiency while taking packet priorities into account. The multiple paths maintained by SAR ensure that the system is fault tolerant and easily recoverable. However, the overhead of table maintenance at each sensor node makes SAR infeasible for very-large-scale WSNs.

4.8 Conclusion

In this chapter, we have gone through the challenges and concerns in designing routing protocols for WSNs. The classification of the routing protocols proposed in the literature and several typical routing schemes, such as SPIN, Directed Diffusion, LEACH, TEEN, GEAR, SAR, and multipath routing schemes, are elaborated.

Problems and Exercises

4.1 Multi-choice questions:

1. Out of the following protocols, which is an example of a data-centric protocol?
   a. Rumor Routing
   b. MCFA
   c. SPEED
   d. GBR
2. Which of the following is not a choice for data diffusion?
   a. Flooding
   b. Gossiping
   c. Directional propagation
   d. None of the above
3. SPIN does not have which one of these data packets?
   a. ADV
   b. REQ
   c. ACK
   d. Data

4.2 Explain three challenges for data routing in sensor networks?

4.3 Explain the differences between SPIN-PP and SPIN-BC. Explain the use of the resource manager in the SPIN protocol.

4.4 Explain how the gradient in Directed Diffusion is created.

4.5 Explain how the clusters are formed in the LEACH protocol. How are the CHs determined in LEACH and LEACH-C?

4.6 Explain the purposes of hard threshold and soft threshold in TEEN.

4.7 Explain the different phases in GEAR. Explain why restricted flooding and recursive geographic forwarding are used in GEAR.

4.8 Explain the differences between disjoint multipath routing and braided multipath routing schemes.

Chapter 5

Transport Layer in Wireless Sensor Networks

As we recall from the general network layers concept, the major tasks of the transport layer are (1) to guarantee reliable transmission of network packets through end-to-end retransmissions or other strategies and (2) to reduce or avoid network congestion due to too heavy traffic flow in the routers or other relay points. TCP is used in the Internet. However, we cannot use TCP in the WSN transport layer design. This chapter explains WSN transport layer design requirements and some good protocol examples.

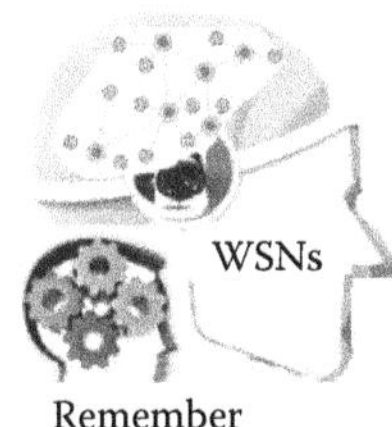

When you design a transport layer protocol for any network, it typically consists of two tasks: (1) It is responsible for an *end-to-end* reliable transmission (i.e., no packet loss) instead of a *hop-to-hop* reliable transmission (which is a MAC [Medium Access Control] layer task). However, you could use hop-to-hop strategies to achieve end-to-end reliability. For instance, later on, we will discuss some WSN transport schemes that use hop-to-hop packet loss recovery to achieve end-to-end reliability. (2) A transport layer protocol should also take care of network congestion issues, such as how to detect the congestion places and how to avoid those congestion events. Although the above two tasks are supposed to be implemented in the same transport protocol, some transport schemes only focus on one of them (either reliability or congestion issues). This is acceptable. However, we point out that it is not a complete transport protocol if only one of them is achieved.

5.1 Introduction

We can summarize the requirements of a transport layer protocol for sensor networks as follows [YIyer05]:

1. *Generic design*: The WSN transport layer protocol should be independent of the application, network, and MAC layer protocols. If a transport layer heavily depends on network topology assumptions (such as a tree-based architecture), it may not be suitable to some applications that use a flat topology.
2. *Heterogeneous data flow support*: A transport protocol should support both continuous and event-driven flows in the same network. Continuous (i.e., streaming) data needs to use fast response rate control algorithms to limit the stream-flow speed to reduce congestion. An event-driven flow has lesser requirements on the rate control sensibility. But it requires a highly reliable event capture (i.e., no data loss).
3. *Controlled variable reliability*: The reliability could be complete (i.e., no packet loss) or incomplete (i.e., some packet loss may be tolerated). In some WSN applications, we can conserve energy at the nodes by achieving incomplete reliability. For instance, if the system does not need a 100 percent packet arrival rate, we may not invoke a packet retransmission scheme.
4. *Congestion detection and avoidance*: This is perhaps the first task in a transport protocol. Congestion detection is not so easy in WSNs because congestion exists only in some specific "hot spots," where the amount of traffic is significantly higher than in other places. But how do we quickly detect these "hot spots"?
5. *Localized or centralized congestion control*: Although we should distribute computation-intensive tasks at the base station; however, if we could distribute some congestion detection and avoidance tasks in sensors, we could obtain a better congestion avoidance effect, because it is the sensors that need to reduce their sending rates to reduce the traffic.
6. *Scalability*: A WSN may have thousands of nodes. Hence, the protocol should be scalable. Unfortunately, it is not easy to find all sensors with buffer overflow.
7. *Leaving space for extension*: The protocol design should leave space for future optimizations to improve network performance and support new applications.

5.2 Pump Slowly, Fetch Quickly [Chieh-Yih05]

5.2.1 Why Does TCP Not Work Well in WSNs?

Why do we need a transport protocol in WSNs? This is because WSNs also have the following two requirements as does the Internet:

1. *Reliable end-to-end data transmission*: The data should be transmitted with no or very few losses between the two ends (a sensor and a base station).

   Typically, the sensor data is transmitted *from a sensor to a base station*. The new detected event is important. We may need a 100 percent reliability for it, that is, no transmission errors or loss at all. If it is general sensor data without urgent processing requirements, we may tolerate certain loss, that is, the reliability could be less than 100 percent. As an example, considering temperature monitoring or animal-location tracking, the system could tolerate the occasional loss of sensor readings. Therefore, we do not need the complex protocol that would ensure reliable delivery of data.

   On the other hand, from a base station to a sensor, typically, the transmitted data includes important data query or sensor control commands. Such data needs a 100 percent reliability (i.e., no error or loss). In [Chieh-Yih05], the authors proposed an application that needs base station-to-sensor transport layer control, which requires the reprogramming of groups of sensors over the air. Today, WSNs are typically hard-wired to perform a specific task efficiently at low cost. We need to build more powerful hardware and software capable of reprogramming sensors to do different things. When we disseminate a program image to sensor nodes, we cannot tolerate the loss of a single message associated with the code segment or the script, as a loss would render the image useless and the reprogramming operation a failure.
2. *Congestion detection and avoidance*: In a WSN, when many sensors send out data simultaneously, some sensors that help to relay data get congested. It is important to identify these congested sensors and to use efficient ways to avoid congestion events.

   The most popular transport protocol, TCP, has been successfully used in the Internet for a few decades. The TCP protocol stack uses a three-way handshake protocol to establish a communication pipe first. Then, a window-based streaming protocol keeps running to control the sending rate. When it detects timer-out or three duplicate acknowledgment (ACK) packets, it assumes packet loss and retransmits the data. It aims to achieve a 100 percent reliability.

   TCP uses a 20 byte header to hold some congestion control and other information. The overhead from headers can consume a lot of resources, especially with small packets. In WSNs, the sensor data is composed of typically some numerical values. It only needs a few bytes to represent such data. Then the TCP overhead is relatively large.

   TCP is designed to make the base station (most times it is the receiver side) as simple as possible. The base station simply acknowledges the sender's packet (if the data is correct, it sends out an ACK; otherwise, it sends nothing back). The sender needs to perform a series of complex rate control operations. However, in WSNs, the sender (sensors) have very constrained resources and the base station has unlimited energy. It is better to put more load on the base-station side.

Moreover, TCP provides 100 percent reliability, that is, it does not allow any packet loss. As mentioned before, complete reliability is not required in many WSN applications.

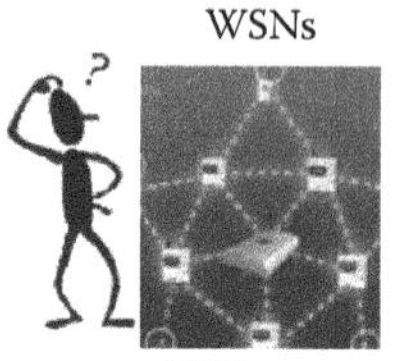

In the Internet, TCP always achieves a 100 percent reliability, that is, no packet is lost. (By the way, we see packet errors as packet loss, because a receiver will not accept any packets with bit errors.) In a WSN, we allow less than 100 percent reliability in the upstream direction (sensors → sink) due to the existence of some redundant sensor data. But the downstream direction (sink → sensors) should have a 100 percent reliability, because a sink always sends out important data (such as a sensor query or sensor control commands).

In this section, we focus on the first function of the transport protocol—reliability. We will defer congestion issues to future discussions. We answer a question as follows: How do we design a WSN transport protocol to achieve reliable data transmission? Such a transport protocol should have low complexity and energy efficiency to be realized on low-end sensor nodes (such as the Berkeley mote series of sensors), and can isolate applications from the unreliable nature of wireless sensor networks in an efficient and robust manner.

A WSN transport protocol, called pump slowly, fetch quickly (PSFQ), is proposed in [Chieh-Yih05]. It aims to make the WSN transport layer less complex, robust, scalable, and customizable to the needs of different applications.

PSFQ has minimum requirements on the routing infrastructure (as opposed to IP multicast routing requirements). It also uses minimum signaling (signaling means protocol message exchanges among sensors), which helps to reduce the communication cost for data reliability. PSFQ is responsive to high error rates in wireless communications, which allows successful operations even under highly error-prone conditions.

5.2.2 Key Ideas

How do we achieve minimum packet loss/errors? PSFQ uses the following interesting, straightforward idea: *when sending data to a sensor, it should be done at a relatively slow speed* (*i.e., "pump slowly"*). This is because pumping data too fast increases wireless loss rate. On the other hand, *if a sensor experiences data loss, that sensor should fetch* (*i.e., recover*) *any missing segments from its upstream neighbor very aggressively to perform local recovery. This is called "fetch quickly."* Note that it is important to use such a quick, *local* data recovery to minimize the lost recovery cost. If not *local*, we need to resort the sender to retransmit the data, which is painful when considering multi-hop, unreliable wireless links.

1. *Using hop-to-hop (i.e., local) error recovery*: Let us take a look at traditional end-to-end error recovery mechanisms, in which only the final destination node is responsible for detecting loss and requesting retransmission. Why does *end-to-end* error recovery not work well in WSNs? In many applications, we drop lots of inexpensive sensors (from a plane) in a large area with irregular terrain and harsh radio environments. Due to the long distance between an event area and the base station, a WSN needs to rely on multi-hop forwarding techniques to exchange messages.

   Based on the *probability theory*, if one hop has an error rate of $0 < p < 1$, each hop keeps dropping packets (all erroneous packets will be dropped by a relay sensor), and errors accumulate exponentially over multiple hops. After we pass many hops, the final destination will have little chance of receiving a high percentage of good packets.

   Using a simple math model, assume that the packet error rate of a wireless channel is p; then, the chances of exchanging a message successfully across n hops decreases quickly to $(1 - p)^n$.

   Figure 5.1 [Chieh-Yih05] numerically shows such a phenomenon. Its y-axis plots packet success arrival rate. The x-axis plots the network size in

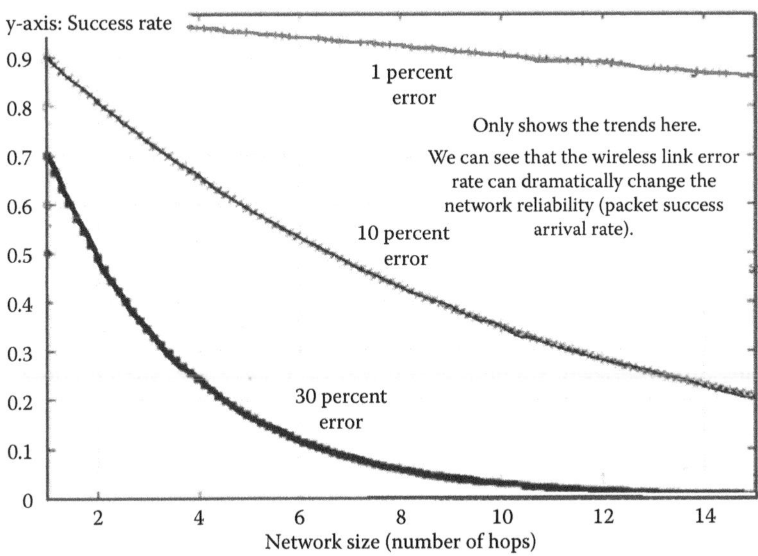

Figure 5.1 Probability of successful delivery of a message using an end-to-end model across a multi-hop network. (Adapted from Wan, C.-Y. et al., *IEEE J. Sel. Areas Commun.*, 23(4), 862, April 2005.)

number of hops. Based on this figure, we can see that in larger WSNs (where hops >14) it is very difficult to deliver a single message using an end-to-end error recovery approach when the error rate is larger than 10 percent. This is because so many packets get lost after passing so many hops, and it becomes very inefficient to recover more than 80 percent of lost packets.

Let us use an analogy: If a student failed one course, he or she may retake it and catch the four-year graduation time. But if he or she failed ten courses, there is no way for him or her to participate in the graduation ceremony, because he or she may need five years to finish all courses (including the retaking of all courses which he or she failed).

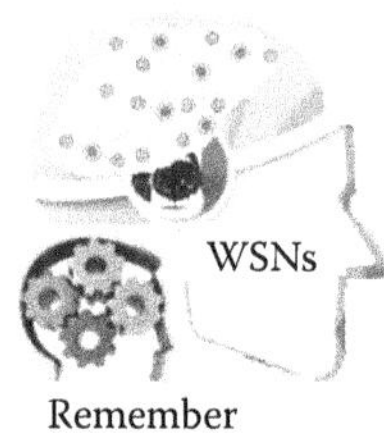

Remember

Always remember this "snowball" effect: If the loss cannot be overcome in one wireless link, the next link will make the situation worse. In the traditional Internet, we normally do not have this loss accumulation issue, as the Internet backbone is built on highly reliable fiber optics. But WSNs use radio links among low-cost, energy-constrained sensors. High bit error rate is unavoidable.

Another bad news is that [JZhao03] shows that the error rates of a WSN are typically around 10 percent or above. We can imagine that the error rate could be even higher in certain harsh environments, such as military applications, industrial process monitoring, and disaster recovery activities.

All of the above observations tell us that we should not wait till the end to recover the erroneous data, that is, *end-to-end error recovery is not a good candidate for reliable transport in WSNs*. Therefore, PSFQ proposes to use hop-to-hop error recovery, in which intermediate sensors also take the responsibility of loss detection and recovery. In other words, reliable data exchange is achieved on a hop-to-hop basis, rather than on an end-to-end basis.

Such a hop-to-hop error recovery approach efficiently eliminates wireless error accumulation, because it divides multi-hop forwarding operations into a series of single-hop transmission processes. Such a hop-to-hop approach uses local data processing to scale better and become more tolerable to wireless errors, while reducing the likelihood of packet reordering in comparison to end-to-end approaches.

2. *Multiple retransmissions for the same lost packet*: In WSNs, to handle an erroneous packet, retransmission should occur. Sometimes multiple packet retransmissions can occur in each hop. Therefore, the data delivery latency would be dependent on the expected number of retransmissions for successful delivery.

The receiver uses a queue (i.e., a memory buffer) to hold all failed packets. It will not clear the queue until these packets are retransmitted and successfully received. To reduce the latency, it is essential to maximize the probability of successful delivery of a packet within a "controllable time frame."

We may use *multiple retransmissions* of the same packet i (thus increasing the chances of successful delivery) before the next packet $i + 1$ arrives. This is called "fetch quickly"; in other words, we use multiple retransmissions to quickly recover a lost packet, which quickly clears the queue at a receiver (e.g., an intermediate sensor) before new packets arrive to keep the queue length small, and, hence, reduce the entire communication delay.

Wan and Campbell [Chieh-Yih05] have analyzed the optimal number of retransmissions that trade off the success rate (i.e., the probability of successful delivery of a packet within a time limit) against energy consumption on retransmissions. Using strict math models, the authors found out the relationship between packet success arrival rate and packet loss rate under different retransmission scenarios. As shown in Figure 5.2, substantial improvements in the success rate can be gained when the channel error rate is less than 60 percent. However, the additional benefit of allowing more retransmissions diminishes quickly and becomes negligible when the number of retransmissions (for the same packet) is larger than five. *This is why PSFQ sets up the ratio between the timers associated with the pump and fetch operations to five.*

3. *Recover data in the earliest time*: If a packet is not timely recovered, will we get incomplete data in a downstream sensor? But how does a downstream sensor know that a packet is lost? It knows this using sequence numbers! Each packet has a sequence ID in its header. If a downstream sensor receives packets 3 and 5, it knows that packet 4 is missing (i.e., lost).

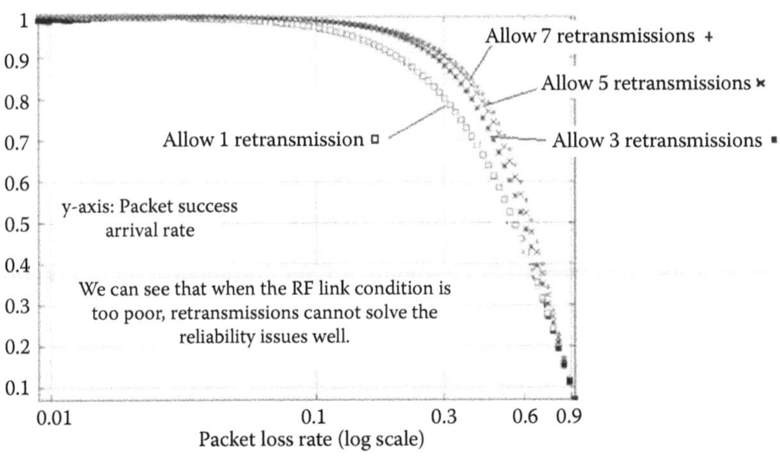

Figure 5.2 Probability of successful delivery of a message over one hop when the mechanism allows multiple retransmissions before the next packet arrival. (Adapted from Wan, C.-Y. et al., *IEEE J. Sel. Areas Commun.*, 23(4), 862, April 2005.)

Now, we face a choice: Suppose that a packet (ID = 99) is lost between sensors 1 and 2. But sensor 1 is a little "lazy" and does not want to timely recover such a packet using retransmissions. It may expect that one of its downstream sensors will recover the data. Is this a good idea? No, we cannot do this. Why not? This is because only sensor 1 has packet #99, and its downstream sensors do not have packet #99 in their buffer for retransmission even when they want to recover such a packet. Therefore, eventually, a downstream sensor, say sensor 12, still needs sensor 1's help to retransmit packet #99. If this is the story, why does a sensor not recover a lost packet at the first time? This is because sensor 2 will feedback to sensor 1 (through a negative acknowledgment [NACK] packet) to tell it to retransmit packet #99.

If any missing packet is immediately recovered in that corresponding hop, any future (downstream) sensors would not see any broken packet sequence IDs. Therefore, we could add a rule to each sensor: *All intermediate nodes only relay messages with continuous sequence numbers.* To ensure in-sequence data forwarding and the complete recovery for any fetch operations from downstream nodes, we need a data cache (i.e., a buffer) in each sensor. Note that the cache size should be determined.

Good idea

Transmission using in-order packet sequence numbers is an important idea in many networks. For example, the Internet TCP protocol uses a window-based packet-sending scheme. All packets have the in-order sequence IDs. A window of packets with higher IDs will not be flushed out if the previous window (with lower IDs) has unrecovered data. If you use out-of-order packets, you could make the transport protocol much more complex, as you need to remember all ID "gaps" (i.e., broken ID chains due to packet loss).

5.2.3 Protocol Description

From the network implementation viewpoint, a PSFQ protocol actually comprises three sub-protocol functions:

- *Message relaying (pump operation)*: A source node (could be a sensor in an event area or a base station) injects messages into the network, and intermediate nodes buffer and relay messages with the proper schedule to achieve loose delay bounds.
- *Relay-initiated error recovery (fetch operation)*: A relay sensor maintains a data cache and uses cached information to detect data loss (by checking sequence number gaps). It also initiates error recovery operations by sending ACK or NACK back to its upstream sensor.

- *Selective status reporting (report operation)*: The source (i.e., the sender) needs to obtain the statistics (such as error rate) of the dissemination status in the network, and it uses such statistical data as a basis for subsequent decision making, such as adjusting the pump rate. Therefore, a feedback and reporting mechanism is needed; such a reporting protocol should be flexible (i.e., adaptive to the environment) and scalable (i.e., minimize the overhead).

The following text will provide more details on the above three protocols (i.e., pump, fetch, and report).

Good idea

Pump slowly, fetch quickly: This idea is not difficult to understand. In WSNs with high bit error rates, we really should not insert data into the network too quickly, as sensors need time to "digest" previous packets—Just think that you could not put too many cars in a slow, single-lane road. On the other hand, if packet loss really happens, can you wait to recover the loss slowly? No way! Packet loss can result in the "snowball" effect (mentioned before). Just like in the above car example, we should quickly clear a slow, single-lane road if a car accident occurs, as all following cars are waiting for the jam to be cleared!

5.2.3.1 Pump Operation

Although PSFQ uses error recovery in individual hop, it is not a routing solution, but a transport scheme. PSFQ operates on top of the existing routing schemes to support reliable data transport. It will not search a routing path. To enable local loss recovery and in-sequence data delivery, a data cache is created and maintained at intermediate nodes.

This section focuses on pump operation. The pump operation *slowly* "pumps" data to the network (from a sender). Slow pumping helps to avoid congestion, which is one of the concerns in the transport layer.

The pump operation uses a simple packet-sending scheduling scheme. The scheduling is based on the concept of pump timers (T_{min} and T_{max}). The basic pump procedure is as follows:

A sender sends a packet to its downstream sensor every T_{min}. A sensor that receives this packet will check against its local data cache. If the packet sequence number is the same as an existing packet, it will discard such a duplicate. If this is a new message, PSFQ will buffer the packet.

For any received packet, the receiver tries to detect a gap in the sequence numbers. If a gap really exists, it will move to the "fetch" operation to perform error recovery (see the next section). Otherwise, it will continue the pump operation (see the next step).

The receiver intentionally delays the packet for a random period between T_{min} and T_{max}, and then relays to its downstream neighbor. Such a random delay before forwarding a packet is necessary to avoid potential transmission collisions.

Now we explain the roles of pump timers (T_{min} and T_{max}).

T_{min} is an important parameter. There is need to provide a *time buffer* for local packet recovery. PSFQ requires the recovery of lost packets *quickly* within a controllable time frame. T_{min} could be used for such a purpose. This is because a node has an opportunity to recover any missing segment before the next segment comes from its upstream neighbors, as a node must wait at least T_{min} anyway before forwarding a packet as part of the pump operation.

T_{max} is used to provide a loose statistical *delay bound* for the *last hop* to successfully receive the *last segment* of a complete file (e.g., a program image or script). Assuming that any missing data is recovered within one interval using the aggressive fetch operation (described in the next section), the relationship between the delay bound, $D(n)$, and T_{max} is as follows:

$$D(n) = T_{max} \times n \times \text{number of hops}$$

where n is the number of fragments of a file.

5.2.3.2 Fetch Operation

As mentioned before, a sensor enters the "fetch" mode once a sequence number gap among received packets is detected. A fetch operation invokes a retransmission from an upstream sensor once loss is detected at a receiving node.

Interestingly, PSFQ uses the concept of "loss aggregation" whenever loss is detected; that is, it can batch up all message losses in a single fetch operation whenever possible.

1. *Loss aggregation*: Researchers have found out that data loss in a wireless environment often occurs in a "bursty" way due to the strong correlation of radio fading models; that is, if a wireless link does not work well, such a poor communication condition can last for a little while and damage a batch of data. The radio noise is not an even distribution. It may work well for a long time and then work poorly for a short period. As a result, packet loss usually occurs in batches (called bursty loss). PSFQ aggregates loss such that the fetch operation deals with a "window" of lost packets instead of a single-packet loss.
   Because of bursty loss, it is not unusual to have multiple gaps in the sequence number of packets received by a sensor. Aggregating multiple loss windows in the fetch operation increases the likelihood of successful recovery.
2. *Fetch timer*: We have mentioned "pump timers" in the last section. In the *fetch* mode, we also need to define a timer. Typically, when a sensor finds out

packet loss (by looking at the sequence number gap), it aggressively sends out NACK messages to its upstream sensor to request for missing segments.

If no retransmission occurs or only a partial set of missing segments in a loss aggregation window are recovered within a fetch timer, T_r ($T_r < T_{max}$) then the receiver will resend the NACK every T_r interval (note: here, we can add a little randomization to this interval to avoid absolute synchronization between neighbors) until all the missing segments are recovered or the number of retries exceed a preset threshold, thereby ending the fetch operation.

PSFQ schedules the first NACK to be sent out within a short delay chosen between 0 and Δ. (Note: $\Delta \ll T_r$.) It cancels the first NACK to keep the number of duplicates low when a NACK for the same missing segments is overheard by another node before the NACK is sent. As Δ is small, the chance of this happening is relatively small. In general, retransmissions in response to a NACK coming from other nodes are not guaranteed to be overheard by the node that canceled its first NACK.

NACK messages do not propagate to avoid network congestion. In other words, an upstream sensor that receives a NACK (from a downstream sensor) will not relay the NACK message back to one more level toward the upstream direction.

Of course, there is an exception. For instance, if the number of times it receives the same NACK exceeds a predefined threshold, and the missing packets requested by the NACK message are no longer retained in a node's data cache, then the NACK could be relayed once, which, in effect, broadens the NACK scope to one more hop to increase the chances of error recovery.

3. *Proactive fetch*: We could notice a "blind spot" in the above fetch operation: The fetch operation is a reactive loss recovery scheme, that is, a loss is detected only when a packet with a higher sequence number is received.

   However, how do we deal with the case where the last segment of a file is lost? We cannot ask the receiving node to detect this loss, because no packet with a higher sequence number will be sent. In addition, if the file has a small size (e.g., a script instead of a binary code), a bursty loss could cause the loss of all subsequent segments up to the last segment. In this case, the loss is also undetectable, and, thus, not recoverable with such a *reactive* loss detection scheme.

   To solve the "last loss" problem, PSFQ proposes a timer-based "proactive fetch" (different from the *reactive* fetch) operation as follows: If the last segment has not been received and no new packet is delivered after a period of time, T_{Pro}, a sensor can also enter the *fetch* mode proactively and send a NACK message for the next segment or the remaining segments.

   How do we determine the value of a proactive fetch timer, T_{Pro}? Obviously, if the proactive fetch is triggered too early, then extra control messaging might be wasted, as upstream nodes may still be relaying the last message.

In contrast, if the fetch mode is triggered too late, then the target node might wait too long for the last segment of a file, significantly increasing the overall delivery latency of a file transfer.

PSFQ makes a good choice of T_{Pro}: It makes T_{Pro} proportional to the gap between the last-highest sequence number (S_{last}) among the received packets and the largest sequence number (S_{max}) of the file (the difference is equal to the number of the remaining segments associated with the file), that is, $T_{Pro} = \alpha(S_{max} - S_{last})T_{max}$ ($\alpha \geq 1$), where α is a scaling factor to adjust the delay in triggering the proactive fetch and should be set to 1 for most operational cases. Therefore, T_{Pro} ensures that a sensor starts the proactive fetch earlier when it is closer to the end of a file, and waits longer when it is farther from completion.

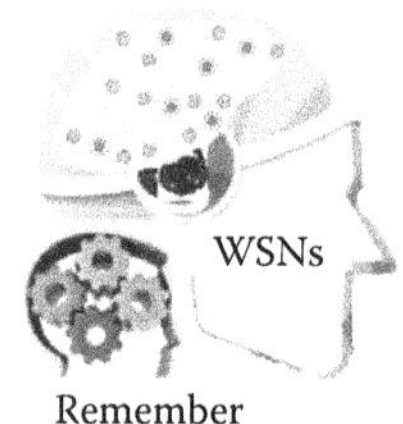

It is not an easy task to design a network protocol. It is not like just writing some C codes. We need to consider many, many details. For example, the above "timer" concept is a difficult issue to handle. This is because we cannot set the timer expiration too early or too late.

4. *Signal-strength-based fetch*: When a sensor detects a gap in the sequence number upon receiving a packet, it only responds and sends out a NACK if this packet comes from an upstream sensor with the strongest average signal quality measurement. This effectively suppresses unnecessary NACK messages triggered by the reception of the packets that come from the upstream sensors that are multiple hops away. Similarly, when a node transmits a NACK message, it includes the preferred parent with the strongest average signal in the message.

5.2.3.3 Report Operation

Report operation is designed to feedback the data delivery status to the sender in a simple and scalable manner. A node enters the report mode when it receives a data message with the "report bit" set in the message header.

Each node along the routing path toward the source node will piggyback its report message by adding its status information into the report, and then propagate the aggregated report toward the user node. Each node will ignore the report if it finds its own ID in the report, to avoid looping.

Sometimes, we have many hops between the source and the destination, and a long report is needed. A node that receives a report message may have no space to append its own state information. To solve this problem, a node will create a new report message and send it prior to relaying the previously received report. This makes other nodes in the route to report messages rather than create new reports.

5.3 Another WSN Transport Protocol—ESRT [Akan05]

ESRT (event-to-sink reliable transport) [Akan05] has been designed for WSN applications that need imperfect reliability. But it is *not for guaranteed* end-to-end data delivery services.

ESRT has considered the fact that the sink (i.e., the base station) typically is only interested in a reliable detection of event features from the *collective* information provided by numerous sensor nodes and not in their *individual* reports. Therefore, it is called "event-to-sink" reliability. This makes ESRT different from other existing transport layer models that focus on end-to-end reliability. For instance, the above PSFQ is more suitable to a *sink-to-event* reliability control, which is actually a downstream (i.e., from the base station to sensors) communication issue.

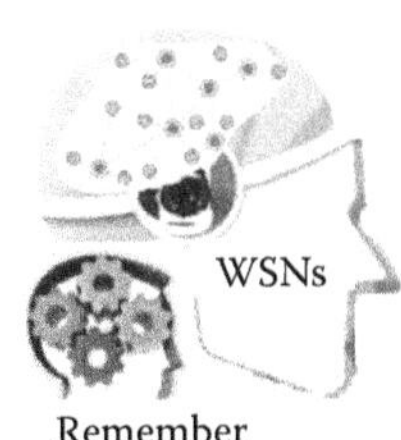

We have mentioned the different directions in a WSN (upstream: from sensors to sink; downstream: from sink to sensors). These two directions have different reliability requirements and communication characteristics. Therefore, ESRT only focuses on one direction—upstream. Later on, we will discuss the downstream reliability scheme (called GARUDA, in Section 5.7).

5.3.1 Reliable Transport Problem

Akan and Akyildiz [Akan05] have formally defined the reliable transport problem in WSNs. Many WSN applications require the reliable detection and estimation of event features based on the collective reports of sensors in the event area. Let us assume that for reliable temporal tracking, the sink must decide on the event features every τ time units. Here, τ represents the duration of a decision interval, and its setup depends on different application requirements. A WSN sink derives an *event reliability indicator* at the end of the decision interval. It should be noted that it must be calculated only using parameters available at the sink. Hence, notions of high throughput, which are based on the number of source packets sent out, are inappropriate in the event reliability calculation here.

ESRT uses a simple way to measure the reliable transport of event features from source nodes to the sink: the number of received data packets. It then defines *observed* and *desired* event reliabilities as follows.

Definition 5.1: The *observed* (i.e., actual) event reliability, r_i, is the number of received data packets in decision interval i at the sink.

Definition 5.2: The *desired* (i.e., targeted) event reliability, R, is the number of data packets required for reliable event detection. This value depends on different applications.

We require that the observed event reliability, r_i, is greater than the desired event reliability, R. In this case, the event is deemed to be reliably detected. Otherwise, we need to use the ESRT scheme to achieve the desired event reliability, R.

A WSN can assign different IDs to different types of events detected by the sensors that keep sending event information to a sink. Then, a sink can compute the observed reliability, r_i, based on data packets with an event ID. It increments the received packet count at the sink each time the ID in a packet is detected. The sink does not care which sensor sends the data.

A sensor can report event information more frequently to make the sink calculate the reliability more accurately from a statistical viewpoint. ESRT thus defines the reporting rate, f, of sensor nodes, as follows:

Definition 5.3: The reporting frequency rate, f, of a sensor node is the number of packets sent out per unit time by that node.

Definition 5.4: The *transport layer* problem (from the *reliability* viewpoint, not from the *congestion control* viewpoint) in a WSN is to configure the reporting rate, f, of source nodes so as to achieve the *required event detection reliability*, R, at the sink with minimum resource utilization.

A source sensor can adjust the reporting frequency, f, by adjusting the sampling rate, the number of quantization levels, the number of sensing modalities, etc. The reporting frequency rate, f, actually controls the amount of traffic injected into the sensor field.

5.3.2 Relationship between Normalized Event Reliability and Report Frequency

To find out how the *observed event reliability* (r) at the sink changes with the *reporting frequency rate* (f) of sensor nodes, Akan and Akyildiz [Akan05] used simulations based on ns-2 tools to construct a WSN using 200 sensor nodes that were randomly positioned in a 100 × 100 sensor field. Assume that the randomly created topology does not vary.

The *desired event reliability*, R, varies with different applications. Akan and Akyildiz [Akan05] use a better parameter to measure event reliability, that is, $\eta = r/R$. Here, η denotes the normalized event reliability at the end of each decision interval i.

Such a *normalized* reliability, η, is better than the *observed reliability*, r, because the former reflects the weight (importance) of r in the desired reliability R. *Our aim is to reach a system status with* $\eta = 1$. *Note*: η could be larger than 1, that is, the actual reliability is larger than the desired reliability. This case looks "attractive." However, it is not what we want, as higher reliability consumes more energy and accumulates more data in the network (which can cause *congestion*).

Interestingly, the simulation results in [Akan05] show that the relationship between η and f can be seen from some characteristic regions, that is, in different f ranges, we have different η trends.

Our aim is to operate as close to $\eta = 1$ as possible no matter whether $\eta > 1$ or $\eta < 1$. Suppose that when $f = f^*$, we have $\eta = 1$. We call f^* as the optimal operating point (OOP), marked as P_1 in Figure 5.3.

From this figure, we can see that the $\eta = 1$ line intersects the event reliability curve at two distinct points, P_1 and P_2. It looks like both P_1 and P_2 are OOPs. Although the event can be reliably detected at P_2, the network is somewhat congested because the reporting frequency, f, goes beyond the peak point, f_{max} (see Figure 5.3), and some source data packets are lost. Therefore, we do not call P_2 as an OOP.

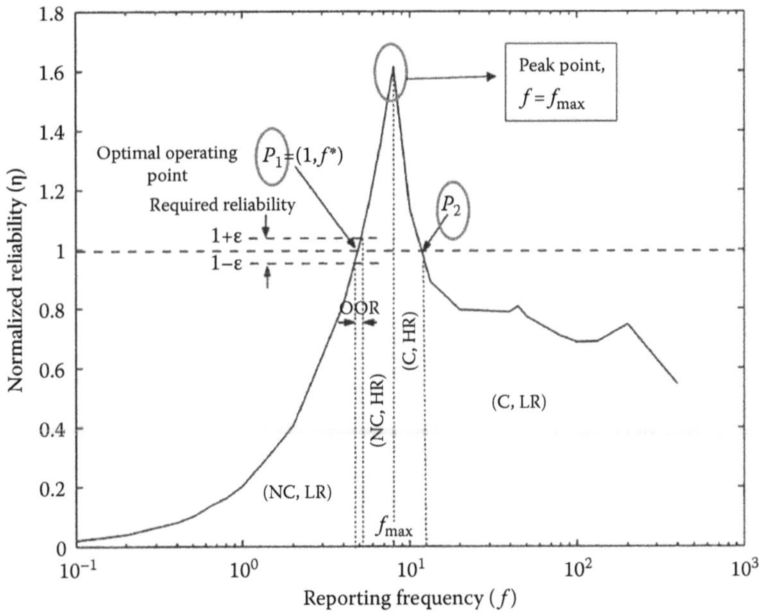

Figure 5.3 Five characteristic regions in the normalized event reliability, η, versus reporting frequency, f, behavior. (Adapted from Akan, Ö.B. and Akyildiz, I.F., *IEEE/ACM Trans. Netw.*, 13(5), 1003, October 2005.)

Good idea

What a good research methodology! Normally, researchers conduct research in this way: First, they define some challenging unsolved issues. Then, they try to use theoretical models to get some quantitative results. These math analysis results are important, because all practical engineering design is based on certain theories. Next, they use software simulations or practical hardware experiments to verify the correctness of their math analysis. However, here, ESRT uses a different research strategy: It uses simulations to find out an interesting, five-region reliability–frequency relationship! Then, the researchers move to theory models and algorithm designs.

We define a tolerance zone with a width 2ε around P_1, as shown in Figure 5.3. Here, ε is a protocol parameter. In the figure, we can then see five characteristic regions (bounded by dotted lines) with the following decision boundaries (η: *normalized reliability indicator*):

Region 1: (NC, LR), which means no congestion, low reliability

$$f < f_{\max}, \quad \eta < 1 - \varepsilon$$

This region is not good enough because it has low reliability.

Region 2: (NC, HR), which means no congestion, high reliability

$$f \leq f_{\max}, \quad \eta > 1 + \varepsilon$$

This region is good because it has high reliability and does not cause network congestion (because its event-reporting frequency is not so high, i.e., $f < f_{\max}$).

Region 3: (OOR), which means optimal operating region

$$f < f_{\max}, \quad 1 - \varepsilon \leq \eta \leq 1 + \varepsilon$$

This is the best region. All other regions should get closer to this region by changing f.

Region 4: (C, HR), which means congestion, high reliability

$$f > f_{\max}, \quad \eta > 1$$

This region is not so good, as it has network congestion issues (because $f > f_{max}$). The good thing is that it still has satisfactory reliability.

Region 5: (C, LR), which means congestion, low reliability

$$f > f_{max}, \quad \eta \leq 1$$

This is the worst region, because it has both low reliability and network congestion issues.

As analyzed above, we need to know two time-varying parameters (reporting frequency, f, and *normalized reliability*, η) and two fixed parameters (peak point frequency, f_{max}, and tolerance zone parameter, ε) before we tell in which of the five regions the system is now.

Let S_i denote the network state variable at the end of decision interval i. Then,

$$S_i \in \{(NC, LR), (NC, HR), (C, HR), (C, LR), OOR\}$$

We can see that the above five states are determined by two things: What is the current event reliability? Does it cause network congestion? Therefore, in practical network implementations, ESRT identifies the current state, S_i, from two aspects: (1) the reliability indicator, η_i, computed by the sink in each decision interval, i; and (2) a congestion detection mechanism.

Note that a sink gets to know the actual values of f and η in each decision period, say, every 5s is a decision period. Suppose that a sink knows f_i and η_i in decision period i. *Now its task is to calculate a new value of reporting frequency, f_{i+1}, in decision period $i+1$ based on a certain state transition algorithm*. Such an algorithm makes sure that all states get to the *OOR* state. We will discuss the algorithm later. Figure 5.4 shows the basic state transition principle.

Good idea

Finite state machine (FSM)—This is a basic research approach to solve some system control problems. Although we could use any advanced, complex control models or math algorithms to control a system, eventually, we need to use an FSM to define all system "states" and corresponding "actions" to transit from one state to another. As a matter of fact, all network "protocols" are written based on FSM models. Think about an interesting problem: How do you define humans as an FSM model? Possibly you could say that a human is in the "sleep" state, "eat," "study," "love," "sick," and many other states. And you can define the state transition conditions/actions. For instance, to get into the "eat" state, we need at least one "condition," called "hungry." Then, the "action" is "open your mouth and grab the food."

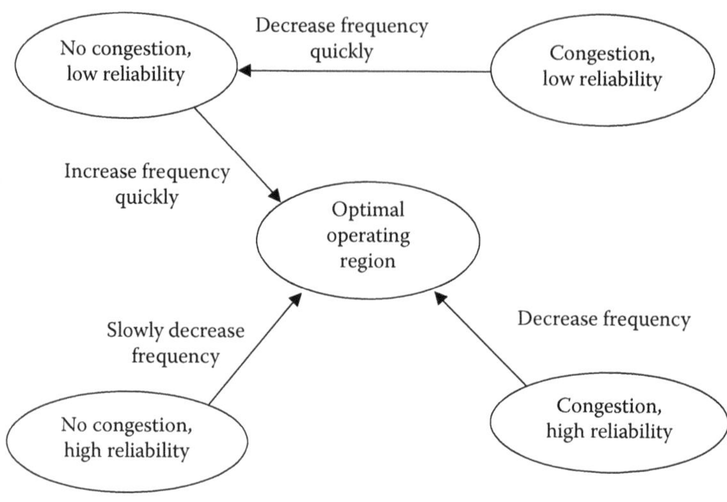

Figure 5.4 ESRT protocol state model and transitions. (Adapted from Akan, Ö.B. and Akyildiz, I.F., *IEEE/ACM Trans. Netw.*, 13(5), 1003, October 2005.)

The state transition algorithm includes the following five aspects.

1. (NC, LR): In this state, we do not have network congestion. But we do not even achieve the desired reliability. In Figure 5.3, we can see that $\eta < 1 - \varepsilon$ and $f < f_{max}$. The reason of getting into this state could be due to failure/power-down of intermediate routing nodes, packet errors due to strong wireless interference, etc. The following text explains these two reasons in more detail.
   If the reason is the failure/power-down of intermediate nodes, the packets that need to be routed through these nodes are dropped. It causes a decrease in reliability even if enough source information is sent out. However, fault-tolerant routing/rerouting in WSNs is provided by several existing algorithms [Cintanagonwiwat00]. ESRT can work with any of these schemes.
   If the congestion is because of packet loss due to RF interference, the number of lost packets is expected to scale proportionally with the reporting frequency rate, f. In most cases, we could assume that the net effect of RF channel conditions on packet losses does not deviate considerably in successive decision intervals. This is a reasonable assumption with static sensor nodes, and slowly time-varying [EShih01] and spatially separated channels for communication from the event to the sink in WSN applications. Hence, even in the presence of packet losses due to link errors, the initial reliability increase is expected to be linear.
   Anyway, when the system gets to the (NC, LR) state, the sink needs to tell the source node to aggressively increase the reporting frequency rate, f,

to attain the required reliability as soon as possible. We can achieve such an aggressive increase by invoking the fact that the r–f relationship in the absence of congestion, that is, for the range of $f < f_{max}$ (see Figure 5.3), is linear. This prompts the use of the following multiplicative increase strategy to calculate the reporting frequency rate in a new decision space, f_{i+1}, as follows:

$$f_{i+1} = \frac{f_i}{\eta_i}$$

where η_i is the reliability observed at the sink in the decision interval i.

2. (NC, HR): In this state, the required reliability level is exceeded and there is no congestion in the network, that is,

$$\eta > 1 - \varepsilon \quad \text{and} \quad f \leq f_{max}$$

This is not a bad state, as no congestion occurs and reliability is achieved. But because source nodes report more frequently than required, it wastes excessive energy in sensor nodes. Therefore, the reporting frequency should be reduced to conserve energy.

But we should not reduce the frequency aggressively (as in the last case), as it is very close to the OOP. Hence, the sink reduces the reporting frequency rate, f, in a controlled manner with half the slope. The updated reporting frequency rate can be expressed as

$$f_{i+1} = \frac{f_i}{2}\left(1 + \frac{1}{\eta_i}\right)$$

3. (C, HR): In this state, the reliability is higher than required and congestion is experienced, that is,

$$\eta > 1 \quad \text{and} \quad f > f_{max}$$

This is not a good state. First, we do not want to see congestion happening. And higher reliability (which makes η even higher than 1) is not necessary (we just need to keep the normalized reliability $\eta = 1$).

But, as no congestion occurs, it means that the frequency is not so high. We should decrease the frequency carefully (i.e., not so aggressively) such that the event-to-sink reliability is always maintained. However, the network operating in the state (C, HR) is farther from the OOP than the network operating in the state (NC,HR). Therefore, we should relieve congestion in an aggressive approach and enter the state (NC,HR) as soon as possible. ESRT uses a multiplicative decrease as follows:

$$f_{i+1} = \frac{f_i}{\eta_i}$$

4. (C, LR): Here, the reliability is inadequate and congestion also exists, that is, $\eta \le 1$ and $f > f_{\max}$. This is perhaps the worst state, as we have both reliability and congestion issues. Therefore, ESRT reduces the reporting frequency *aggressively* to bring the network to the OOR state as soon as possible.

   An aggressive way to reduce the frequency is to exponentially decrease it, as follows:

$$f_{i+1} = f_i^{\left(\frac{\eta_i}{k}\right)}$$

   where k denotes the number of *successive decision periods* for which the network has remained in the (C, LR) state, including the current decision interval, that is, $k \ge 1$. The aim is to decrease the reporting frequency with greater aggression if a state transition is not detected. Such a policy also ensures convergence to $\eta = 1$ in the (C, LR) state.
5. OOR: This is the best state. The network is operating within the tolerance of the optimal point, where the required reliability is attained with minimum energy expenditure. Hence, the reporting frequency rate is left unchanged for the next decision interval:

$$f_{i+1} = f_i$$

Good idea

If you want to slowly approach a point, you could use "log" or "linear" speed. But, for a fast approach, "multiplicative" speed could be a good idea. Of course, "exponential" speed typically gives a fast-enough approach.

5.3.3 Congestion Detection

Although ESRT's main purpose is to guarantee an optimized reliability, it also has certain impacts on network congestion. This can be seen from the above five states. On the other hand, to determine the current network state in ESRT, the sink must be able to detect congestion in the network. Now, the question is "how does a sink know that congestion occurs?"

Because TCP is not used here, we cannot use the traditional approach to determine congestion levels. Hence, ESRT uses a local buffer-level monitoring scheme

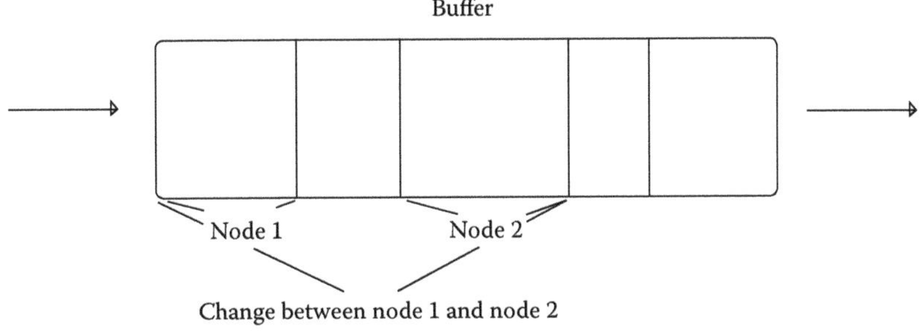

Figure 5.5 Illustration of buffer-level monitoring in sensor nodes. (Adapted from Akan, Ö.B. and Akyildiz, I.F., *IEEE/ACM Trans. Netw.*, 13(5), 1003, October 2005.)

in individual sensor nodes to find out a congestion event. Basically, a sensor will inform the sink of the congestion as long as its *routing buffer overflows due to excessive incoming packets*. The details of this mechanism are as follows.

We denote b_k and b_{k-1} as the buffer fullness levels at the end of kth and $(k - 1)$th decision intervals, respectively, and b as the buffer size, as in Figure 5.5. For a given sensor node, let Δb be the buffer length increment observed at the end of the last reporting period, that is,

$$\Delta b = b_k - b_{k-1}$$

Thus, if the sum of the current buffer level at the end of the ith reporting interval and the last experienced buffer length increment exceeds the buffer size, that is, $b_k + \Delta b > B$, the sensor node infers that it is going to experience congestion in the next reporting interval.

Checking a node's local buffer size is a typical way to find out the congestion level. TCP is based on this principle. But it does this in the source node only.

Good idea

5.4 E²SRT: Enhanced ESRT Performance [Sunil08]

Although the above algorithms could make different states go to the OOR state, in [Sunil08], their simulation results, shown in Figure 5.6, have revealed that when the *desired* reliability (R) is set up *beyond the capability of current network settings*

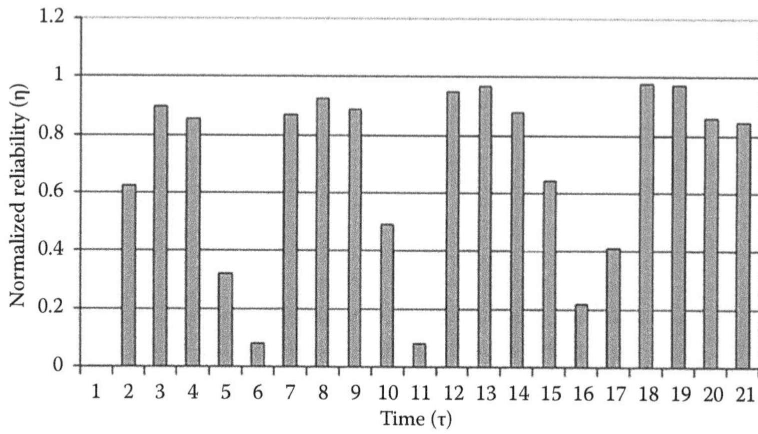

Figure 5.6 Normalized reliability fluctuates in the ESRT scheme in case of over-demanding desired reliability requirements. (Adapted from Feng, Z. et al., *Wireless Commun. Mob. Comput.*, November 2008. Accessible online.)

(*such as the network's sensor deployment strategy, sensor resources, and network scale*), the network will never be able to converge to the OOR state.

Their simulation results also show that the original ESRT scheme (such as the above-described buffer-level monitoring scheme) cannot detect this situation by itself. When we use the original ESRT algorithm to generate a new reporting frequency (for the next decision period) according to this desired reliability value, these values either lead to tremendous network congestion or make the network operate at a very low frequency rate, thus wasting most of the bandwidth. As a result, the network oscillates between the (C, LR) state and the (NC, LR) state.

The actual reliability (r) reached with this oscillation is far below the desired reliability (R). Apparently, it is also not the maximum reliability we could have obtained with current network settings. This generally means that the system was running in a very expensive and inefficient mode: The network is always trying to touch reliability far beyond its capability, which leads to more congestion, more collision, and a longer delay. Subsequently, the network throughput and overall reliability are significantly compromised.

The extensive simulations of Feng et al. [Sunil08] show that there is a threshold for this reliability demand that is decided by the current network settings, such as network size, radio type, underlying infrastructures, and protocol choices. When the desired reliability is lower than the threshold, the ESRT algorithm can always converge to the OOR mode in several control loops. However, when this requirement is above the threshold, the network soon falls into oscillation.

When the network cannot support the desired event reliability, only two network states, (NC, LR) and (C, LR), exist (see Figure 5.7).

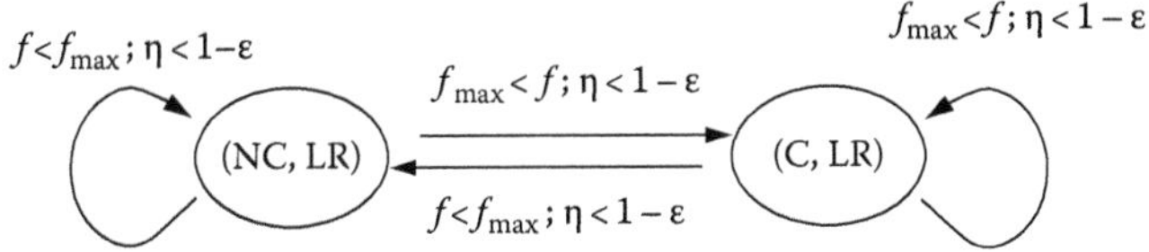

Figure 5.7 ESRT protocol state model and transitions when desired reliability is over demanding. (Adapted from Feng, Z. et al., *Wireless Commun. Mob. Comput.*, November 2008. Accessible online.)

As an example, suppose that the desired reliability is to successfully receive 4000 packets at a sink in each ten second interval. However, the network can only handle around 3500 packets per ten second interval in our simulation settings. Obviously, the reliability requirement is beyond the network capability; no OOR state exists. ESRT does not take this situation into account, and the network would fluctuate between (NC, LR) and (C, LR) states.

5.4.1 The Proposed Scheme—E^{2}SRT

Before discussing the solution proposed in [Sunil08], which is called the *enhanced event-to-sink reliability transport* (*E^{2}SRT*), we formally define the over-demanding desired reliability problem in ESRT in this section.

The over-demanding desired event reliability problem in E^{2}SRT represents a situation where the desired reliability, R, is sufficiently larger than R_{max}, so that $(R_{max}/R) < 1 - \varepsilon$. When the desired event reliability is over demanding, we call the network is in the OR (over-demanding reliability) state. We shall represent this desired reliability situation as R_{od}.

We use the following mathematical analysis to demonstrate that when the desired event reliability is over demanding, ESRT does not converge to the OOR state, and fluctuates between two low-reliability states, (NC, LR) and (C, LR).

Lemma 5.1: In the OR state, the normalized reliability, $\eta = r/R$, will never fall into the region of $[1 - \varepsilon, \infty)$.

Proof: As R_{max} is the maximum reliability that the network can reach with the current network settings, it follows that the observed event reliability $r_i \le R_{max}$. Then,

$$\eta_i = r_i/R \le R_{max}/R < 1-\varepsilon$$

We conclude that $\eta_i \in (0,1-\varepsilon)$. ■

Lemma 5.2: In the OR state, the network only has two possible working states, namely, (NC, LR) and (C, LR).

Lemma 5.2 is a straightforward extension of Lemma 5.1. However, it reveals the most distinct characteristic of the OR state, which is the base for the operations of E^{2}SRT.

Note that these results are obtained for the situation where the desired reliability is beyond the capability of the sensor network, which implies the following assumptions:

$$\eta_{max} < 1 - \varepsilon, \quad R_{max} < R$$

Only two states, (NC, LR) and (C, LR), are available.

Lemma 5.3: In and only in the OR state, starting from S_i = (NC, LR), and with a linear reliability behavior when the network is not congested, the network state will transit to S_{i+1} = (C, LR).

Proof: From S_i = (NC, LR), ESRT aggressively increments f_i as follows:

$$f_{i+1} = \frac{f_i}{\eta_i}$$

Hence,

$$f_{i+1} = \frac{f_i}{\eta_i} = \frac{f_i}{\dfrac{\eta_i}{\eta_{max}} \cdot \eta_{max}}$$

As

$$f_{max} = f_i \cdot \frac{R_{max}}{r_i} \text{ and } R_{max}/R < 1 - \varepsilon$$

it follows that

$$f_{i+1} = \frac{f_i}{\eta_i} = \frac{f_i}{\dfrac{r_i}{R} \cdot \dfrac{R_{max}}{R_{max}}} = f_{max} \cdot \frac{R}{R_{max}} > f_{max} \cdot \frac{1}{1-\varepsilon}$$

■

To address this issue, Feng et al. [Sunil08] have divided the problem into the following two subproblems:

a. How to detect the over-demanding desired event reliability situation?
b. If the above situation exists, how to quickly converge to the maximum reliability the network can reach without requiring the full knowledge of the network conditions?

The major design consideration is how to push the network to approach the maximum reliability point (MRP) (f_{max}, η_{max}) for a given network setting. Similar to the ESRT scheme, we also allow a tolerance zone of width ε around the MRP. If, at the end of a decision interval *i*, the normalized reliability, η_i, is within [$\eta_{max} - \varepsilon$, η_{max}] and no congestion is detected in the network, the network is in the maximum operating region (MOR).

Here, we follow the definition of tolerance zone of ESRT. It is a protocol parameter decided by the user based on the requirement. A smaller ε will generally provide greater proximity to the MRP, while it may take longer convergence time.

If the MRP is known, the sink can reduce the desired reliability such that the network can converge to the OOR as in ESRT. However, it is difficult to calculate the exact value of the MRP (f_{max}, η_{max}) due to the following reasons:

- Initial deployment
- Nodes moving or dying, or other reasons that cause the network topology to change
- Relocation of events
- Radio interference

Consequently, algorithms that assume *a priori* of a constant MOR are not feasible. More advanced algorithms should be adaptable to the changing network environment. They should be able to read feedback from the sensor network and predict the MRP in a recursive manner.

The proposed new algorithm in E²SRT inherits all the major features of ESRT, such as communication model and network modes definitions. It is sink based, energy efficient, and has fast convergence time. As an enhanced version, E²SRT is more resilient to abrupt network changes and resource constraints due to its operations in OR states.

In the following section, we will describe how E²SRT can approach the MOR and how E²SRT operates in each of the three OR states, in detail.

In each decision interval, the sink calculates the normalized reliability, η_i. In conjunction with congestion reports, the current network state, S_i, will be determined. Using the decision boundaries defined in ESRT, with the knowledge of state S_i, and the values of f_i and η_i, E²SRT will request the sink to update the event-reporting frequency to f_{i+1}, and the sink will broadcast the new frequency value to the sensor nodes. On receiving this updated frequency, the relevant sensor nodes will report to the sink according to the new frequency in the next decision interval. This process will repeat until the MOR state is reached. The state transition graph is shown in Figure 5.8.

E²SRT introduces a recursive algorithm that converges to the MOR in a few rounds of estimation of the MRP. As observed from Figure 5.9, the network shows some linear and symmetry properties around the MOR in the curve of normalized reliability as a function of the reporting frequency (in a logarithm format). And as we

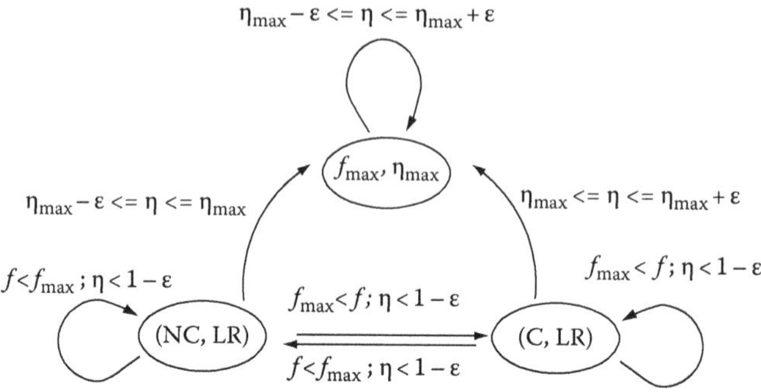

Figure 5.8 E²SRT protocol state model and transitions when desired reliability is over demanding.

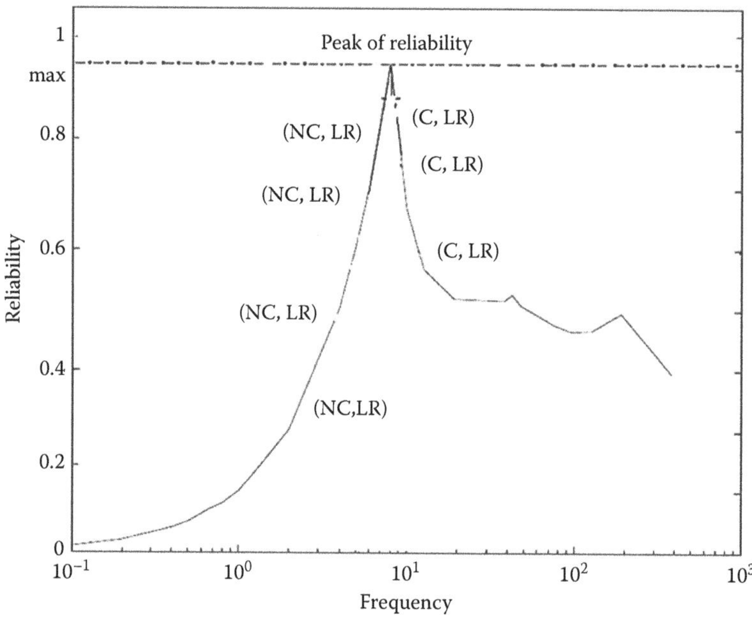

Figure 5.9 Recursive convergence of E²SRT. Starting from (NC, LR)1, the network bounces in the cone area of the curve and, finally, falls into the MOR.

previously discussed, the network fluctuates between only two states, (NC, LR) and (C, LR).

Obviously, (NC, LR) is always on the left of the MRP, while (C, LR) is always on the right of the MRP. Thus, the MRP is always somewhere between a (NC, LR) state and a (C, LR) state. We will record the reporting frequency of the last (C, LR) state as $f_{(C,LR)}$ and the frequency of the last (NC, LR) state as $f_{(NC,LR)}$. The x-axis of the graph is based on logarithm.

We estimate the frequency of the MRP as

$$f_{i+1} = 10^{\frac{\log f_{(NC,LR)} + \log f_{(C,LR)}}{2}}$$

With the above formula, starting from any of the two states, the network may stay in either (NC, LR) state or (C, LR) state for more than one consecutive decision periods. This is because the last (NC, LR)/(C, LR) state point is too far apart from the MRP compared with the last (C, LR)/(NC, LR) state. In the case of (C, LR), which means that the last (C, LR) operating point is too far away from the MRP, we can add a multiplying factor to give more importance to the last (NC, LR) operating point as

$$f_{i+1} = 10^{\frac{k}{k+1}\log f_{(NC,LR)} + \frac{1}{k+1}\log f_{(C,LR)}}$$

In the case of (NC, LR), we have the following formula:

$$f_{i+1} = 10^{\frac{1}{k+1}\log f_{(NC,LR)} + \frac{k}{k+1}\log f_{(C,LR)}}$$

A detailed description of the E^{2}SRT operation in each of the three available states is presented below.

1. *(NC, LR)*: Because the OOR state is not feasible, the goal of the updating policy is to drive the network to the MOR instead of the OOR. As pointed out by lemma 3, using the ESRT algorithm, the network would inevitably jump into the most undesirable (C, LR) state. Here, we already know that the network is in the OR state, as it has jumped at least once into the (C, LR) state and then fell back into the (NC, R) state.

   We record the frequency of the last (C, LR) state as $f_{(C,LR)}$ and the frequency of the last (NC, LR) state as $f_{(NC,LR)}$. As observed in the basic ESRT scheme, the network would show some linear and symmetry properties around the MOR region in the curve of normalized reliability as a function of the reporting frequency (in a logarithm format). This prompts us to update the reporting frequency as follows:

$$f_{i+1} = 10^{\frac{1}{k+1}\log f_{(NC,LR)} + \frac{k}{k+1}\log f_{(C,LR)}}$$

2. *(C, LR)*: In this state, we either detect a transition from the (NC, LR) state (so that we know the network is now in the OR state), or we transit from (C, LR) state itself (which means the frequency has to be further reduced). We use

a parameter k to count the time intervals for which the network has successively remained in (C, LR). As k increases, it generally means $f_{(NC,LR)}$ is closer to the MOR than $f_{(C,LR)}$. We, therefore, assign a higher frequency than $f_{(C,LR)}$. Putting together all these considerations, we update the reporting frequency based on the following formula:

$$f_{i+1} = 10^{\frac{k}{k+1}\log f_{(NC,LR)} + \frac{1}{k+1}\log f_{(C,LR)}}$$

3. *MOR*: In this state, the network is operating within ε tolerance of the maximum operating point, where the network is making its best effort to fulfill the reliability requirement with minimum energy consumption. The reporting frequency remains unchanged for the next decision interval as

$$f_{i+1} = f_i$$

The entire E²SRT protocol algorithm is summarized in the pseudocode in Figure 5.10.

Good idea

Many students keep asking a question: "How do I conduct some research?" Take a look at this E²SRT example. It starts from an existing scheme (ESRT), tries to find the "hidden" drawbacks or any unsolved issues, and finally thinks of a good way to overcome those issues. "Improving" is a good way to start your research. But eventually, you need to reach a high-level research—define an interesting, important research issue by yourself; then use a brand-new way (which other people did not find) to solve it! Look at those professors. They are trying to do the same thing—"finding a new problem; thinking of a new solution."

5.5 CODA: Congestion Detection and Avoidance in Sensor Networks [Wan03]

The above-discussed transport schemes have achieved the first goal of a WSN transport layer—reliability. In this section, we discuss a solution to achieve the second goal, that is, congestion control.

To illustrate the congestion problem, Wan et al. [Wan03] have used simulation results (see Figure 5.11) to show the impact of congestion on data dissemination in a WSN for a moderate number of active sources with varying reporting rates.

Figure 5.11 shows an interesting conclusion: There exists a water boiling point, that is, *when the source rate increases beyond a certain network capacity threshold*

```
k = 1,
ESRT=1;
/* ESRT=1 indicates that the network is in normal ESRT operation*/
E^2SRT()
/* Probe the network state*/
If S_{i-1}=(NC, LR) and S_i=(C, LR)
ESRT=0 /* OR state is detected*/
End;
If (ESRT)
/* ESRT operations takes action*/
...
end;
else if (ESRT = 0)
        if Si=(NC, LR) and \n_{i-1}-n_i\<= ε/2
        /*network is in MOR states*/
        /*keep f toward frequency used in last state */
        f_{i+1} = f_i
        end;
        If (C, LR) /*state=(C, LR)*/
        /* decrease f toward frequency used in last (NC, LR) state */
```

$$f_{i+1} = 10^{\frac{k}{k+1}\log f_{(NC,LR)} + \frac{1}{k+1}\log f_{(C,LR)}}$$

```
        K = k+1;
        end;
        else if(NC, LR) and \n_{i-1}-n_i\ > ε/2
        /* state=(C, LR)*/
        /* increase f toward frequency used in last (C, LR) state */
```

$$f_{i+1} = 10^{\frac{\log f_{(NC,LR)} + \log f_{(C,LR)}}{2}} \quad k=1$$

```
            end;
end;
```

Figure 5.10 Algorithm of the E²SRT protocol operation.

(ten events/s in this network), congestion occurs more frequently and the total number of packets dropped at the sink increases rapidly. It also shows that congestion could occur even with low to moderate source event rates. Dropped packets can include MAC signaling, data event packets themselves, and diffusion messaging packets.

The drop rates shown in Figure 5.11 represent not only significant packet losses in the sensor network, they also indicate the existence of network congestion. More importantly, a lot of energy is wasted by the failed packet transmissions! In WSNs, we care about energy resources so much!

Different WSN applications can bring either occasional or more frequent data-rate "bursts" (i.e., suddenly generate a large amount of event data). Some applications (such as lighting monitoring) may only generate light traffic from small regions of the network, while other applications (such as image sensor networks) may generate large waves of impulses potentially across the whole sensing area, which causes high loss, as shown in Figure 5.11.

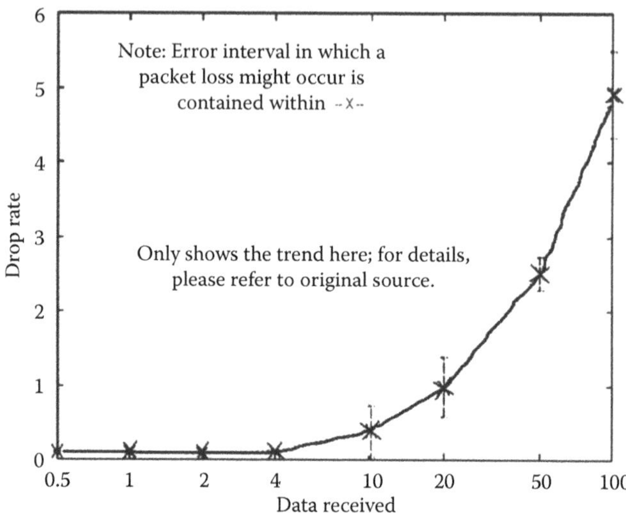

Figure 5.11 Total number of packets dropped by the WSN at the sink (drop rate) as a function of the source rate. The x-axis is plotted on a log scale to highlight data points with low reporting rates. (Adapted from Wan, C. et al., CODA: Congestion detection and avoidance in sensor networks, *Proceedings of the First International Conference on Embedded Networked Sensor Systems* (SenSys '03), Los Angeles, CA, November 5–7, 2003, ACM, New York, pp. 266–279.)

WSN congestion control mechanisms must be capable of maintaining acceptable fidelity (i.e., rate of events) of the delivered signal at the sink during periods of transient and more persistent congestion. Here, we focus on three distinct congestion scenarios.

Densely deployed sensors: Persistent hot spots proportional to the impulse rate of source sensors could occur within the first few hops from the source. In this scenario, the congestion control should be localized (around the source), fast, and capable of providing backpressure from the points of congestion back to the sources, to be effective.

Sparsely deployed sensors with low data rates: Transient hot spots could occur anywhere in the sensor field but likely farther from the sources, toward the sink. In this case, a fast scheme that combines *localized backpressure* (between nodes identified in a hot-spot region) and *packet-dropping* techniques would be more effective. Because of the transient nature of congestion, source nodes may not be involved in the backpressure.

Sparsely deployed sensors generating high-data-rate events: In this scenario, both transient and persistent hot spots are distributed throughout the sensor field. To control congestion, we need a fast scheme to resolve localized transient hot spots, and to perform a closed-loop rate regulation of all source nodes that contribute toward creating persistent hot spots.

Wan et al. [Wan03] proposed an energy-efficient congestion control scheme for sensor networks, called *CODA* (*congestion detection and avoidance*), that comprises three mechanisms.

- *Congestion detection.* The first step toward congestion control is to accurately and efficiently *detect* congestion; that is, we need to find out whether or not congestion occurs in the network. If it does, where does it occur? Congestion detection is based on the observations by each sensor: What are the present and past communication channel traffic conditions in the current sensor? What is the current buffer occupancy in the sensor? We must know the state of the communication channel, because neighboring sensors may simultaneously use such a channel to transmit data. However, we cannot *persistently* listen to the channel to measure local loading, as it could cause high energy costs. Therefore, CODA uses a sampling scheme that only activates local channel monitoring at a certain time. Once congestion is detected, nodes signal their upstream neighbors via a backpressure mechanism.
- *Open-loop, hop-to-hop backpressure.* If a node detects congestion, it propagates backpressure signals one-hop upstream toward the source. If a node receives backpressure signals, it throttles its sending rates, or it may drop packets based on the local congestion policy (e.g., packet drop). When an upstream node (toward the source) receives a backpressure message, it checks its own local network conditions. If it also detects congestion, it will further propagate the backpressure upstream.
- *Closed-loop, multisource regulation.* Closed-loop rate regulation operates over a slower time scale than the above open-loop control. But it is capable of asserting congestion control over *multiple source nodes from a single sink* in the event of persistent congestion. Each source node compares its data rate to some fraction of the maximum theoretical throughput of the channel (refer to [Wan03] for more details). If its data rate is less than this fractional throughput, it simply regulates its rate. However, when its rate is higher than the throughput, it could make a contribution to network congestion. Under this circumstance, the closed-loop congestion control is triggered. And the source enters *sink regulation*, that is, it uses feedback (e.g., ACK) from the sink to maintain its rate. The reception of ACKs in a source node serves as a self-clocking mechanism to help the source maintain its current event rate. However, if a source fails to receive ACKs, it will force itself to reduce its own rate.

The relationship between open-loop and closed-loop control is as follows: Because hot spots (i.e., congestion locations) can occur in different regions of a sensor field due to the above different scenarios, CODA needs both open-loop, hop-to-hop backpressure and closed-loop, multisource regulation mechanisms. These two control mechanisms can be used separately. But it is more efficient to use them together, as they complement each other well.

From the above description, we can also see that the rate control scheme performs different operations in source nodes, the sink, or intermediate nodes. Sources know the properties of the sending traffic, while intermediate nodes do not. A sink has the best understanding of the fidelity rate for the received signal, and, in some applications, sinks are powerful nodes that are capable of performing complicated heuristics. The goal of CODA is to do nothing during no-congestion conditions, but be responsive enough to quickly mitigate congestion around hot spots once congestion is detected.

Good idea

Open-loop and closed-loop control: These have been used in many system control applications. Open-loop control is simpler and easier to implement. But closed-loop control uses output feedback to adjust the input, which typically brings more accurate, stable system control.

5.5.1 Open-Loop, Hop-to-Hop Backpressure

The above discussions have briefly described fast/slow time-scale congestion control. *Backpressure* belongs to the fast time-scale control mechanism. If a sensor detects congestion, it broadcasts a suppression message to its one-hop upstream neighbors. It knows where the upstream nodes are located, by checking the routing protocol, which is located below the transport layer protocol in the WSN protocol stack.

When an upstream node (toward the source) receives a backpressure message, a node may keep propagating backpressure signals if it finds serious congestion. But it may not send back backpressure signals, and just simply drops its incoming data packets upon receiving a backpressure message to prevent its queue from building up.

The above discussion is concerned with open-loop control. For closed-loop congestion control, it is required to deal with any persistent congestion locally, instead of propagating the backpressure signal.

CODA defines *depth of congestion* as the number of hops that the backpressure message has traversed before a non-congested node is encountered. The *depth of congestion* can be used by the routing protocol as follows.

Select a better route path: If the depth of congestion is too high, a routing protocol may give up the current path and find a new one. This can reduce traffic over the paths suffering deep congestion.

Intentionally drop command messages to reduce congestion: The nodes can silently suppress or drop important signaling (i.e., command) messages associated with routing or data dissemination protocols. Such actions would help to push data flows out of congested regions and away from hot spots in a more transparent way.

5.5.2 Congestion Detection

To detect congestion, there are some easy ways, such as checking whether or not a queue in the sensor is full, or measuring the current communication channel traffic load—if the load is approaching the upper bound, it is an indication of congestion.

The first detection approach, monitoring queue size, has low execution overhead. But it may not provide accurate congestion detection, as the queue can overflow due to many local conditions. The second approach, listening to the communication channel shared among neighbors, can tell us the channel loading or even give us protocol-signaling information on the collision detection effect. Therefore, we prefer the second approach. However, because listening to channels continuously can incur high energy cost, we should use it only at an appropriate time to minimize system cost.

So, what is the preferred time to activate channel monitoring? Let us utilize a trick in MAC protocols. As we know, typically a sensor listens to the channels before sending packets. Such a channel-listening procedure is called "carrier sense" in MAC protocols. If the channel is clear during this period, then the radio switches into the transmission mode and sends out a packet.

Therefore, the best time to perform channel monitoring is when "carrier sense" occurs. This is because there will be no extra cost to listen and measure channel loading when a node wants to transmit a packet, as carrier sense is required anyway before a packet transmission.

In Figure 5.12, we can see a typical scenario with hot spots or congestion areas. In this example, node 1 sends data to node 3 and node 4 sends data to node 5. Both data flows pass through node 2.

As we can see from the "channel load" of Figure 5.12, node 2 has high buffer occupancy. Then node 2 activates the channel-loading measurement. The channel-loading measurement will stop naturally when the buffer is cleared, which indicates with high probability that any congestion is mitigated and data flows smoothly around the neighborhood.

5.5.3 Listening to Channel Based on Sampling

Let us define *epoch time* as a time period of transmitting multiple packets. When a node listens to the channel, we require it to listen for at least 1 epoch time to measure the channel load. During an epoch period, if a node continuously listens to the channel, it would incur high energy cost. Therefore, CODA only performs periodic sampling (i.e., listening to the channel once for a while), so that the radio can be turned off if sampling is not being performed.

We use a simple sampling scheme as follows: We measure the channel load for N consecutive epoch times of length E. In each epoch time, a predefined sampling

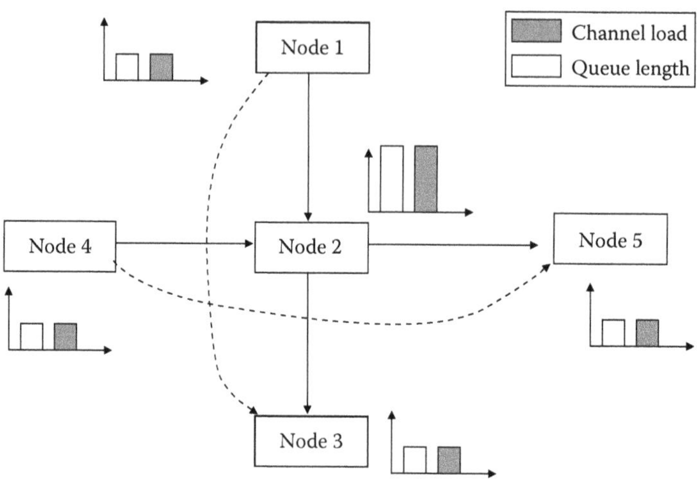

Figure 5.12 Wireless network illustrating receiver-based congestion detection. (Adapted from Wan, C. et al., CODA: Congestion detection and avoidance in sensor networks, in *Proceedings of the First International Conference on Embedded Networked Sensor Systems* (*SenSys '03*), Los Angeles, CA, November 5–7, 2003, ACM, New York, pp. 266–279.)

rate is used to obtain channel state information, that is, we count the number of times that the channel state is busy or idle within a single sensing epoch.

We then calculate the sensed *channel load*, Φ, as the *exponential average* of Φ_n (the measured channel load during epoch n) with parameter α ($0 < \alpha < 1$) over the previous N consecutive sensing epochs, as shown in the equation below.

$$\bar{\Phi}_{n+1} = \alpha\bar{\Phi}_n + (1-\alpha)\bar{\Phi}_n, \quad (n \in \{1,2,\ldots,N\}, \bar{\Phi}_1 = \Phi_1)$$

If the send buffer is cleared before n counts to N, then the average value is ignored and n is reset to 1. *Note: The tuple (N, E, α) can be used to tune the sampling scheme to accurately measure the channel load* for specific radio and system architectures.

Based on the above equation, we obtain the time-varying sensed channel load. When such a load exceeds a threshold, it means network congestion. In this case, a node broadcasts a *suppression* message as a backpressure signal and, at the same time, exercises the local congestion policy. A node will continue broadcasting this message up to a certain maximum number of times with minimum separation as long as congestion persists.

The suppression message provides the basis for the *open-loop* backpressure mechanism.

5.6 STCP: A Generic Transport Layer Protocol for WSNs [YIyer05]

STCP [YIyer05] provides a generic, scalable, and reliable transport layer paradigm for sensor networks. The WSN base station implements the majority of STCP functionalities, as it has unlimited resources compared to sensors.

5.6.1 Data Transmission Sequence in STCP

Similar to the principle of the TCP three-way handshake protocol that aims to establish an end-to-end TCP connection, before transmitting packets, a sensor node establishes an *association (similar to TCP's connection concept)* with the base station via a session initiation packet.

The session initiation packet conveys the base station the following information: the number of flows originating from the node, the type of data flow, transmission rate, and required reliability. When the session initiation packet arrives at the base station, it stores all the information, sets the timers and other parameters for each flow, and acknowledges (ACK) this packet. In the reverse path, the base station transmits an ACK or NACK depending on the type of data flow.

5.6.2 STCP Packet Formats

Figure 5.13 shows the format of a session initiation packet. A source node transmits packets associated with each data flow independently, as the transmission characteristics may be different in different flows. In Figure 5.13, the first field is the sequence number (16 bit long). It is set to zero for the session initiation packet. The

Sequence number (16)		Flows (8)	Options (8)
Clock (32)			
Flow ID #1 (8)	Flow bit (8)	Trans. rate (8)	Reliability (8)
Flow ID #2 (8)	Flow bit (8)	Trans. rate (8)	Reliability (8)
Flow ID #*N* (8)	Flow bit (8)	Trans. rate (8)	Reliability (8)

Figure 5.13 Session initiation packet. (Adapted from Iyer, Y. et al., STCP: A generic transport layer protocol for sensor networks, *Proceedings of the 14th IEEE International Conference on Computer Communications and Networks*, San Diego, CA, October 2005.)

second field (Flows, 8 bit long) indicates the number of flows originating at the node. The "clock" field indicates the local clock value at the time of transmission. The Flow ID is used to differentiate packets from different flows. The Flow Bit field specifies whether the flow is continuous (i.e., the data flow does not stop) or event driven (i.e., only sends out packets when an event is detected). For continuous flows, the Transmission Rate field indicates the rate at which a packet will be transmitted by the source node.

The *Reliability* field directly relates to WSN transport layer tasks. Again, it means the packet arrival success rate. Here, this field gives the expected reliability required by the flow.

An STCP data packet header is shown in Figure 5.14. It is similar to the session initiation packet header. The Sequence number for a data packet is a *nonzero* positive integer (for a session initiation packet, it is zero). The Flow ID indicates the flow type, which helps the base station identify the characteristics of the packet for that node.

The packet header includes an important field that is related to congestion control, called Congestion Notification (CN). As it is a 1 bit field, when it is 1, it means that congestion occurs. The Clock field gives the local time at which the packet was transmitted. The base station uses the clock value to calculate the *estimated trip time* (*ETT*) for that node and the Flow ID.

The ACK packet format is shown in Figure 5.15. All fields are as explained before. The ACK/NACK field tells that it is a positive or negative acknowledgment. STCP uses the 32 bit clock field in conjunction with the sequence number field to avoid issues related to wraparound. The Options field is for future extension purposes.

Sequence number (16)	Flow ID (8)	CN (1)	Options (7)	Clock (32)

Figure 5.14 STCP data packet header. (Adapted from Iyer, Y. et al., STCP: A generic transport layer protocol for sensor networks, *Proceedings of the 14th IEEE International Conference on Computer Communications and Networks*, San Diego, CA, October 2005.)

Sequence number (16)	Flow ID (8)	CN (1)	ACK/NACK (1)	Options (6)

Figure 5.15 STCP acknowledgment packets. (Adapted from Iyer, Y. et al., STCP: A generic transport layer protocol for sensor networks, *Proceedings of the 14th IEEE International Conference on Computer Communications and Networks*, San Diego, CA, October 2005.)

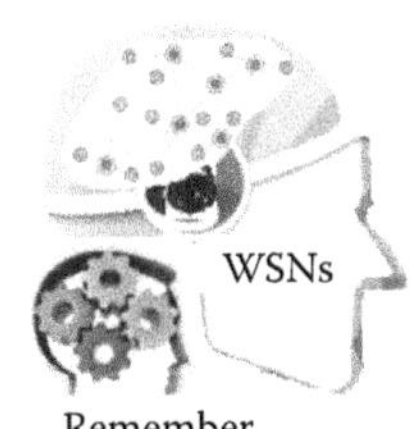

On packet format: When you design a network protocol, you should know the packet format first. This is because protocol operations are different when the field content in a packet header is different. Sometimes we do not have a standardized packet format to use. In this case, you need to define a packet format by yourself. Try to minimize the field length—if you could use 3 bits to cover five cases, why should you use 4 bits in that field?

5.6.3 Continuous Flows

This section focuses on the "continuous flow" case. The next section describes the "event-based flow" case. Note that the base station can use a session initiation packet to get to know the sending rate of the source. Thus, it can estimate the expected arrival time for the next packet. The base station maintains a timer and sends a NACK if it does not receive a packet within the expected time.

When the base station receives a packet from a sensor node, it calculates the *ETT* for the next packet to reach the base station by one of the following methods.

1. The time-out value is determined by ($T + \alpha \times$ ETT), where T is the time between two successive transmissions and alpha (α) is a positive integer that varies with ETT. The base station constantly checks to see if it has received a packet within ($T + \alpha \times$ ETT) time units for each sensor node. If a packet has been received within time, it decreases alpha (α) by 0.5. If a packet is lost (i.e., time-out occurs), or if the base station receives a packet after transmitting a NACK for it, it increases alpha (α) by 0.5.
2. The second approach is to use the Jacobson/Karels algorithm [VJacobson88], which considers the variance of the round trip time (RTT). Here, we use the ETT instead of the RTT. In this approach, we can modify the Jacobson/Karels algorithm by considering the ETT. The base station dynamically varies the values of delta (δ), mu (μ), and phi (ϕ) in the following expressions:

$$\textit{Sample ETT} = \textit{base station clock} - \textit{packet clock value}$$
$$\textit{Difference} = \textit{Sample ETT} - \textit{EstimatedETT}$$
$$\textit{Estimated ETT} = \textit{Estimated ETT} + (\delta \times \textit{Difference})$$
$$\textit{Deviation} = \textit{Deviation} + \delta\,(|\textit{Difference}| - \textit{Deviation})$$
$$\textit{Time-Out} = \mu \times \textit{ETT} + \phi \times \textit{Deviation}$$

A source node retransmits packets after it receives a NACK. Otherwise, the packet must have reached the base station. But, sometimes, maybe the NACK is lost. Therefore, the base station maintains a record of all packets for which it has sent a NACK. If a packet that has been NACKed successfully arrives, the base station clears the corresponding entry from the record. The base station periodically checks this record, and, if it finds an entry, it retransmits a NACK.

5.6.4 Event-Driven Flows

The previous case used NACK, as it is preferred for "continuous" flow. We assume that not many packets are lost; thus, NACK is sent back occasionally. If we used ACK (positive acknowledgment) in that case, we would have too many ACKs, as continuous data flows have heavy traffic.

In this section, we move to "event-driven" flows. In this case, the flow data is much less than in the former case, because the data transmission is triggered only when a new event occurs. The positive acknowledgments are used to let a source node know if a packet has reached the base station. Because the data is received occasionally, there could be big gaps between two packet arrivals. Thus, the base station cannot estimate the arrival times of the next data packet.

Similar to the TCP principle, the source node buffers each transmitted packet and also invokes a timer. When an ACK is received, the corresponding packet is deleted from the buffer. When the timer fires before an ACK is received, packets in the buffer are assumed to be lost and are retransmitted.

5.6.5 Reliability

We mentioned before that a sensor node can specify the required reliability for each flow in the session initiation packet. For *continuous* flows, the base station calculates a running average of the reliability. We know that reliability can be measured by the percentage of packets successfully received.

However, the base station will not send a NACK back if the current reliability satisfies the required reliability. The base station transmits NACKs only when the reliability goes below the required level.

5.6.6 Congestion Detection and Avoidance

How does STCP achieve the final goal—congestion detection and avoidance? We may refer to some of the traditional schemes. The random early detection (RED) mechanism designed by Floyd and Jacobson [SFloyd93] simply asks an intermediate node to drop a packet when it sees congestion. Such a packet drop can cause time-out or NACK in the source side. Because dropping of packets is detrimental to sensor networks, STCP does not adopt this approach.

In the scheme proposed in [KRamakrishnan90], intermediate nodes monitor the traffic load and *explicitly* notify the end nodes by setting a binary congestion bit in the packets. STCP adopts this method of explicit CN with some modification.

Each STCP data packet has a *CN bit* in its header. Every sensor node maintains two thresholds in its buffer: th_{lower} and th_{higher}. When the buffer reaches th_{lower}, the congestion bit is set *with a certain probability*. The value of this probability can be determined by an approach similar to that employed in RED. When the buffer reaches th_{higher}, it means the congestion is serious; then the node will set the CN bit *in every packet it forwards*.

After receiving this packet with the CN field, the base station informs the source of the congested path by setting the congestion bit in the ACK packet. When receiving such a special CN, the source will either route successive packets along a different path or slow down the transmission rate. Note that the nodes rely on the routing layer algorithm to find alternate routes.

5.6.7 Data-Centric Applications

In data-centric applications, we typically are only interested in *collective* network-wide information, instead of an individual sensor node's data. A few examples are monitoring of seismic activity and finding the maximum temperature in the network. In such applications, a sensor could aggregate the correlated data, which is called data aggregation. Due to data aggregation from a large number of source nodes, we should not ask a base station to acknowledge all the source nodes by an ACK or a NACK, because this can deplete network resources and energy.

Hence, for data-centric applications, STCP does not provide any acknowledgment scheme. This is similar to the UDP case in the Internet. STCP assumes that data from different sensors are correlated and loss tolerant to the extent that events are very likely sent to the base station in a collective and reliable way. This view is supported by the authors in ESRT.

5.7 GARUDA: Achieving Effective Reliability for Downstream Communication [Seung-Jong08]

ESRT takes care of event-to-sink (upstream) reliability issues. In this section, we consider the problem of reliable *downstream* point-to-multipoint data delivery, *from a sink to multiple sensors*. Especially, we will discuss GARUDA (a mythological bird that reliably transported gods) proposed in [Seung-Jong08], which can efficiently achieve such a downstream reliability.

Because a sink typically sends out important data (such as data query commands) to sensors, we require that any message from the sink has to reach the sensors reliably. Consider an image sensor network application. The sink may send one of the following three classes of messages, all of which have to be delivered reliably to the sensors: (1) *Over-the-air programming codes*: Suppose that the WSN has reconfigurable sensors that can be reprogrammed. A sink may want to send an upgraded image detection/image-processing software to the sensors. (2) *Data query data*: A sink may send data query commands to the sensors. (3) *Data collection commands*: Finally, the sink requests data results from sensors.

5.7.1 Challenges to the Downstream Reliability of WSNs

5.7.1.1 Environment Constraints

To implement downstream reliability, we need to overcome some challenges. One of them is to consider the limited network bandwidth and energy sources in a WSN. We need to minimize the number of retransmission overheads to ensure reliability, because this can reduce both bandwidth and energy consumption of the message overheads.

We should also realize that node failures (due to power draining) lead to dynamic network topology. The downstream reliability should be adaptive to such a dynamic topology, that is, it should not use a statically constructed mechanism (say, a broadcast tree) that does not account for the dynamics of the network.

Another challenge occurs due to the scale of the sensor network. A WSN has thousands of nodes, and the diameter of the network could be large. Therefore, there could be a tremendous amount of spatial reuse possible in the network that could be utilized to reduce delay. However, the specific *loss recovery* mechanism used may severely limit such spatial reuse, as we will elaborate later.

5.7.1.2 Acknowledgment (ACK)/NACK Paradox

Should a receiver use an ACK or a NACK to notify the sender of the packet arrival? This depends on different conditions. For instance, if the packet loss rate is very low, a NACK-based approach can save more bandwidth, as there will be few NACKs sent back to the sender. But for a high-packet-loss environment, an ACK-based approach can save more message overhead.

In addition, if we use a NACK-based approach, we need to handle the last-packet-loss issue. This issue was discussed before. The NACK-based loss recovery scheme will inherently require in-sequence forwarding of data by nodes in the network to prevent a NACK implosion [CYWan02]. This will clearly limit the spatial reuse achieved in the network.

5.7.1.3 Reliability Semantics

In WSNs, we need to consider sensor data *location dependency and redundancy.*

- *Location dependency*: In many cases, we need to find where the event is exactly located. A data query command (sent from a base station) can be *location dependent*, such as "Send temperature readings from rooms X, Y, and Z."
- *Location redundancy*: Due to large sensor density in most WSN applications, it is not necessary for all sensors in the same event area to *reliably* deliver their locally sensed data to the sink. Such upstream (event-to-sink) "partial reliability" can save network bandwidth. GARUDA is a downstream (sink-to-event) reliability scheme, which also uses "partial reliability," that is, the sink only guarantees reliable communications with part of the sensors in a neighborhood area.

GARUDA defines the "reliability semantics" that are required in WSNs based on the above characteristics. It classifies the reliability semantics into four categories:

- Delivery to the entire field (i.e., the whole WSN), which is the default semantics
- Delivery to sensors in a subregion of the field (which is called location-based delivery)
- Delivery to sensors such that the entire sensing field is covered (which is called *redundancy-aware* delivery)
- Delivery to a *probabilistic* subset of sensors (this strategy is used in WSN *resolution scoping*)

5.7.2 GARUDA Design Basics

Let us first obtain an overview of GARUDA's design. The centerpiece of GARUDA's design is an instantaneously constructible *loss recovery infrastructure* called the *core*. The *core* can be seen as an approximation of the *minimum dominating set* (*MDS*) of the network topology. The *dominating set* is a set of nodes through which we could reach all other nodes easily (such as using at most one-hop communication from one of the dominating-set nodes).

MDS is not a new concept for solving networking problems [RSivakumar99]. But GARUDA makes a new contribution to establishing an optimal core for the loss recovery process. It constructs the core during the course of a single packet flood and uses a two-phase loss recovery strategy. Its loss recovery uses out-of-sequence forwarding and is tailored to satisfy the goal of minimizing the retransmission overheads and the delay. It also uses a candidacy-based approach for the core construction to support multiple reliability semantics (Figure 5.16).

Figure 5.16a through d illustrates categories 1 through 4, respectively.

GARUDA is a pulsing-based approach, which means that it can deliver a single packet reliably to all network nodes. It can ensure the reliable delivery of the first packet of messages of any size. It has the advantages of NACK-based schemes, but, at the same time, avoids any pitfalls that consequently arise.

In the following GARUDA overview, we discuss its core infrastructure based on the assumption that the first packet is reliably delivered. Then, we see how it can achieve reliable delivery of the first packet.

5.7.2.1 Loss Recovery Servers: Core

GARUDA calls its core a set of local designated loss recovery servers (here, servers are not machines; they simply refer to nodes providing loss recovery services). We need to solve two problems when using an algorithm to construct such a core: (1) How does the algorithm choose the core nodes for the purpose of minimizing the retransmission overheads? (2) How does the core construction algorithm adapt to the dynamic network topology change due to node failures (or other reasons)?

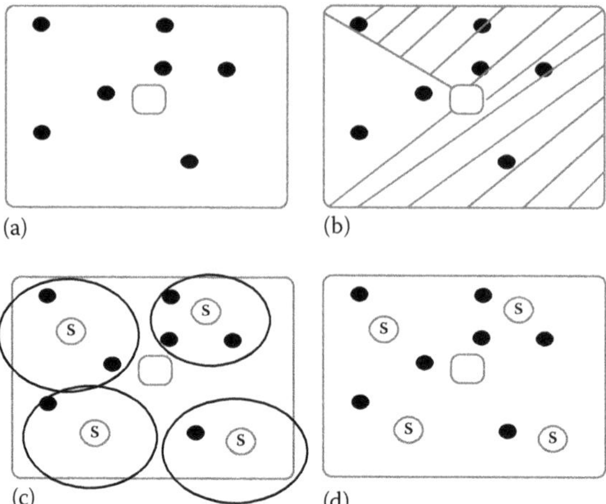

Figure 5.16 Types of reliability semantics. (a) Reliable delivery to all sensors. (b) Reliable delivery to a subregion. (c) Reliable delivery to minimal sensors to cover the sensing field. (d) Probabilistic reliable delivery to 80 percent of the sensors. (Adapted from Park, S.-J. et al., *IEEE Trans. Mobile Comput.*, 7(2), 214, February 2008.)

Believe it or not, GARUDA finishes the core construction during the first packet delivery. As long as the first packet is reliably delivered, we could determine the *hop count* of each node, which is the distance of a node from the sink. Any node with *a hop count* that is a multiple of three (such as 3, 6, and 9) will elect itself as a *core node* if it has not heard from any other core nodes. The reason we select a node at a $3i$ hop distance as a core node is because it can cover the other nodes at $3i + 1$ or $3i - 1$ hop distances, so that it can behave like one of the MDS in the direction from a sink to sensors.

In summary, the instantaneous construction of the core nodes during the first packet delivery of every new message efficiently addresses any vulnerability in the network in terms of node failures.

5.7.2.2 Loss Recovery Process

5.7.2.2.1 Out-of-Sequence Packet Forwarding

In a traditional transport protocol, such as TCP in the Internet, we deliver all packets with in-order sequence IDs; that is, a sender will not move to higher sequence IDs if lower ones are not ACKed by the receiver side. Sometimes the network can lose a packet. Then we need to retransmit these lost packets before we send the packets with higher sequence IDs. The main drawback of the in-sequence forwarding strategy is that precious downstream (sink-to-event) network resources can be left underutilized when the forwarding of higher-sequence-number packets is suppressed in the event of a loss.

Therefore, GARUDA uses an out-of-sequence packet-forwarding strategy that can overcome the above drawback, as nodes that have lost a packet can continue to forward any higher (or lower)-sequence-number packets.

5.7.2.2.2 Two-Stage Loss Recovery

Once the core is constructed, a two-stage loss recovery is used: (1) The core nodes recover all lost packets; (2) then the non-core nodes recover the lost packets.

Because we only select nodes with a *hop count* of $3i$ as core nodes, the number of non-core nodes will be a substantial portion of the total number of nodes in the network. Therefore, we ask core nodes to recover the lost packets first, which can preclude any contention from lots of non-core nodes.

The second phase of the loss recovery will not start until a non-core node overhears a message from the core node indicating that it has received all the packets. Hence, the second phase does not overlap with the first phase in each local area, preventing any contention with the first phase recovery.

5.7.3 GARUDA Framework

To observe more details on the GARUDA scheme, let us assume a network topology as shown in Figure 5.17. As mentioned before, the first-packet delivery procedure can find core nodes with a hop count of $3i$. We call all nodes with the same hop count from the sink a "band." The band ID (bID) is the same as the hop count.

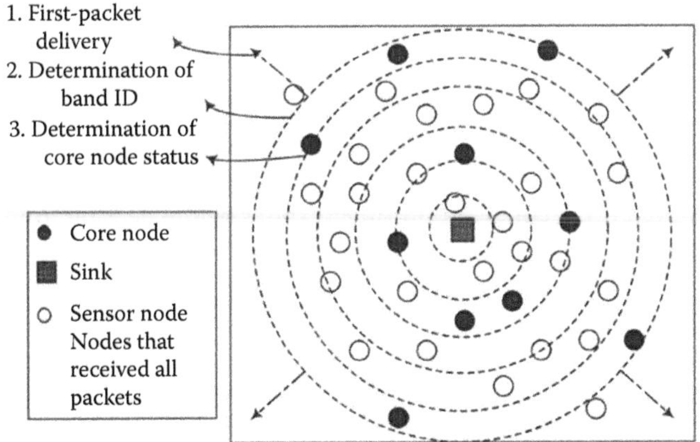

Figure 5.17 Instantaneous core construction in GARUDA. (Adapted from Park, S.-J. et al., *IEEE Trans. Mobile Comput.*, 7(2), 214, February 2008.)

We consider all nodes with the same bID (i.e., in the same "band"). Obviously, the bands can be viewed as concentric circles around the sink. Moreover, every core node should have a (bID) of 3, 6, 9, etc.

5.7.3.1 Core Construction Procedure

- In the sink (i.e., the base station):
  When the sink sends the first packet, it stamps the packet with a bID of 0. When a node receives the first packet, it increments its bID by 1 and sets the resulting value as its own bID.
- In the nodes in $3i$ bands:
  All nodes in $3i$ bands can possibly become core nodes. When a node with a bID of $3i$ forwards the packet (after a random waiting delay from the time it received the packet), it will first see if it has heard from any other core node in the same band. If it has not heard of any other nodes that claim themselves to be core nodes, it will claim itself to be a core node. The reason of doing this is to reduce the communication conflict between any two core nodes (and thus minimize the number of core nodes).

  If any node in the core band ($3i$) has not selected itself to be a core yet, when it receives a *core solicitation* message explicitly, it chooses itself as a core node at that stage.

  To maintain band-to-band communications, every core node in the $3(i + 1)$ band should also know of at least one core node in the ($3i$) band. If it receives the first packet through a core node in the ($3i$) band, it can determine this information implicitly, as every packet carries the previously visited core node's identifier bID.
- Nodes in $3i + 1$ bands:
  When a node A with a bID of ($3i + 1$) receives the first packet, it first checks to see if the packet arrived from a core node or from a non-core node. If the source S_0 was a core node, node A sets its core node as S_0. Otherwise, it sets S_0 as a candidate core node and starts a core election timer that is set to a value larger than that of the retransmission timer for the first-packet delivery. If S_1 hears from a core node S_0' before the core election timer expires, it sets its core node to S_0'.

  However, if the core election timer expires before hearing from any other core node, it sets S_0 as its core node and sends a one-to-one (unicast) message to S_0 informing it of the decision.
- Nodes in $3i + 2$ bands:
  When a node A with a bID of the form ($3i + 2$) receives the first packet, at that point, it does not know any $3(i + 1)$ nodes. Hence, it invokes its core election timer. If it hears back from a *core node* in the $3(i + 1)$ band before the timer expires, it chooses that core node as its core node. If it does not hear from the $3(i + 1)$ band, it sends an anycast *core solicitation message* with the target bID set to $3(i + 1)$. Any $3(i + 1)$ band nodes that receive the anycast message are

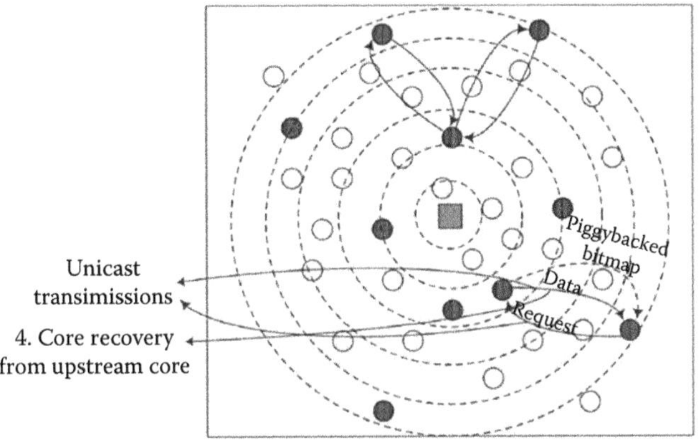

Figure 5.18 Loss recovery for core nodes in GARUDA. (Adapted from Park, S.-J. et al., *IEEE Trans. Mobile Comput.*, 7(2), 214, February 2008.)

allowed to respond after a random waiting delay. The delay is set to a smaller value for core nodes to facilitate the reuse of an already elected core node.

5.7.3.2 Two-Phase Loss Recovery

a. *Loss detection.* When a core node receives an out-of-sequence packet, the core node infers a loss, and it sends a request to an upstream (closer to the sink) core node only if it is notified that the missing packet is available at the upstream core node.
b. *Loss recovery.* When an upstream core node receives a retransmission request from a downstream core node, it performs a retransmission for the lost packet. Figure 5.18 shows the loss detection and the loss recovery principles between core nodes at the $(3i)$ band and core nodes at the $3(i + 1)$ band. If any of the non-core nodes overhears the requested packet, it retransmits the requested packet.

Good idea

GARUDA uses "bands" to define WSN core and non-core nodes. This is an interesting idea. A researcher has used the "throwing a stone in water" phenomenon to find an interesting way to define network topology—"ripples." That is, he tries to generate "ripples" when a sender broadcasts a message. The "ripple" concept is similar to the "band" concept here. But defining the band/ripple generation procedure is not an easy task, as we need to consider many details, such as broadcasting time, hop count, and neighbors' communication conflicts.

Problems and Exercises

5.1 Multi-choice questions

1. Which of the following is not one of the tasks of the transport layer?
   a. Reliable source-to-destination transmission
   b. Network congestion detection
   c. Network congestion avoidance
   d. Buffer management
2. Why does TCP not work in WSNs?
   a. TCP incurs too much overhead when used in sensors.
   b. The errors accumulate in each wireless hop.
   c. TCP leads to large power consumption.
   d. Both A and B.
3. Which of the following belong(s) to PSFQ's features?
   a. Send out data slowly
   b. Recover data quickly
   c. Hop-to-hop error recovery
   d. All of the above
4. If a single hop incurs a wireless loss rate of 10 percent, five-hop links will cause a loss rate of
   a. 40 percent
   b. 5 percent
   c. 10–5
   d. 0.2
5. The PSFQ protocol does not have which of the following functions?
   a. Data pump
   b. Error recovery (fetch)
   c. End-to-end retransmission and timer setup
   d. Status reporting
6. ESRT does not have which of the following features?
   a. It can achieve sink-to-sensors reliability.
   b. It adjusts the sensor's reporting frequency based on the reliability requirement.
   c. It aims to reach the OOR state.
   d. If in (C, LR), it needs to quickly decrease the reporting frequency.
7. E^2SRT improves ESRT in which of the following aspect(s)?
   a. When the desired reliability is beyond the capability of current network settings, the network will never be able to converge to the OOR state where the normalized reliability equals to 1.
   b. The network oscillates between the (C, LR) state with a fairly high reporting rate and the (NC, LR) state with a very low reporting rate.
   c. It greatly saves the sending power consumption during approaching to OOR.
   d. Both a and b.

8. CODA has which of the following features?
   a. CODA is to achieve reliability.
   b. CODA achieves congestion reduction.
   c. CODA achieves both reliability and congestion avoidance.
   d. None of the above.
9. STCP does not have which of the following features?
   a. STCP protocols are implemented in the sensor nodes.
   b. STCP is a generic, scalable, and reliable transport layer paradigm for WSNs.
   c. STCP provides both reliability and congestion control.
   d. STCP protocols mostly run in the base station.
10. GARUDA follows which of the following procedure(s)?
   a. It achieves reliable sink-to-sensors transmission.
   b. It uses the concept of dominant set to build the core.
   c. It recovers data in different ways for core and non-core nodes.
   d. All of the above.

5.2 Explain why TCP does not work well in WSNs.

5.3 Explain how PSFQ sets up a retransmission timer in each node for the packet loss case.

5.4 Why does ESRT propose the concept of states and use OOR as the aim?

5.5 Besides the formula that ESRT uses when approaching to OOR, can you think of other good functions that can also achieve a similar approaching speed?

5.6 How does ESRT detect congestion?

5.7 How does E^2SRT improve ESRT?

5.8 Explain how GARUDA forms the core nodes.

COMPUTER SCIENCE PRINCIPLES

Chapter 6

Operating System in Sensors

Although operating system (OS) is a typical computer science (CS) topic, WSN engineers who design sensor hardware should also understand WSN OS characteristics, as a successful WSN system needs a tight integration of hardware and software. For instance, if a WSN OS has a set of interrupt commands, how do we design these interrupt wires between a microcontroller and analog sensors? If an OS has a wake-up command, how do we design a wake-up circuit to trigger a radio transceiver if there is data to send? This chapter introduces some most popular WSN OSs, such as TinyOS.

6.1 TinyOS [Levis06]

Due to serious resource constraints in sensors, TinyOS [Levis06] is designed to be a tiny (smaller than 400 bytes), flexible OS with a set of reusable components. These components could be programmed and assembled into application-specific systems. TinyOS is an event-driven OS, that is, it defines a set of functions to be triggered by asynchronous sensor network events, such as fire event. TinyOS is implemented in the NesC language [TinyOS07], which has a similar syntax as that of the regular C language.

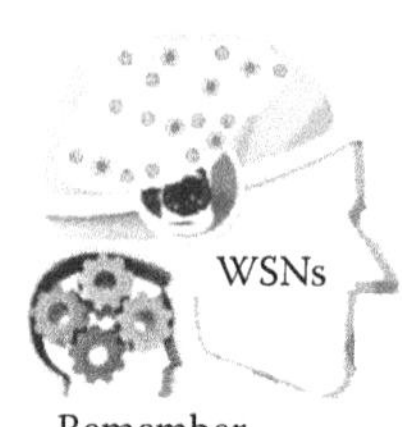

As TinyOS is still an OS, it needs to have the common functionalities of an OS. For instance, it needs to manage files, allocate memory for applications, and recycle unused CPU resources. However, different from other OSs, TinyOS should fit in a WSN's tiny memory and slow CPU features. It also needs to minimize energy consumption to elongate sensors' battery lifetime.

6.1.1 Overview

Any TinyOS program can be represented as a graph of software components. Each component is an independent computational entity. There are interfaces among different components to ensure that they can refer to each other.

Components have three computational abstractions: *commands*, *events*, and *tasks*. *Commands* and *events* are mechanisms for inter-component (i.e., between components) communication, while *tasks* are used to express intra-component (inside one component) concurrency.

A *command* is sent out by one component to request another component to execute operations (i.e., services). For instance, a software entity may request the sensor to report current readings.

An *event* is a special software entity that is generated from three sources: (1) When a *command* is executed, an *event* message is generated to signal the completion of that service. (2) When the sensor hardware has some special event (such as the wake-up of a radio transceiver), a hardware interrupt may be generated to signal such a new *event*. (3) *Events* may also be signaled asynchronously due to the network message arrival in a sensor or a base station.

From a traditional OS's viewpoint, *commands* are analogous to "downcalls" and *events* are like "upcalls." Commands and events cannot block each other. They may be carried out in different time phases. For instance, TinyOS uses a phase to issue the *request for service* (i.e., sending out the *command*) and uses another phase to send out the completion signal (i.e., generating the corresponding *event*). These two phases are decoupled. The *command* returns immediately; however, the *event* signals could be completed at a later time.

Why do we use *tasks*? In many cases, we cannot finish all operations in a command/event handler immediately, especially if these operations need to use multiple sensor hardware resources (such as radio transceiver, analog sensor, and flash memory). Thus, commands and event handlers may post a *task*, a function to be executed by the TinyOS scheduler at a later time. By executing the *tasks* at a later time, we make commands and events "look" very responsive, that is, they return results immediately. However, internally we defer any extensive computation to the *tasks*.

Although we could use *tasks* to perform significant computation, a *task* cannot run indefinitely, that is, run to completion is its basic execution model. *Tasks* represent internal concurrency within a component and may only access the state within that component (i.e., a *task* cannot access two components during its execution). The standard TinyOS *task* scheduler uses a non-preemptive, FIFO (first-in, first-out) scheduling policy.

On *components*: In TinyOS, all hardware resources are represented as *components*. For example, after a *component* receives the `getData()` *command*, later on it will signal a `dataReady()` *event* as long as a hardware interrupt fires.

TinyOS has defined many *components* for WSN programmers. An application developer writes components to compose an application. Then, these components are wired to TinyOS components to provide implementations of the required services. In the following text, we further look into the *component* model.

6.1.2 Component Model

Components encapsulate a specific set of services that are specified by *interfaces*. Not only does each WSN program consist of a series of *components*, TinyOS itself consists of a set of reusable system *components* along with a *task* scheduler.

A wiring specification can be used to connect an application to a series of components. And the wiring specification defines the complete set of components that the application uses. The concrete component implementations are independent of the wiring specification.

A TinyOS compiler can eliminate some unnecessary components after an analysis and *inlining* of the entire program. The procedure of *inlining* can operate across different component boundaries to improve both program size and efficiency.

On the concept of "interfaces": As shown in Figure 6.1, any component can have two types of interfaces: (1) the interfaces it *provides* and (2) the interfaces it *uses*.

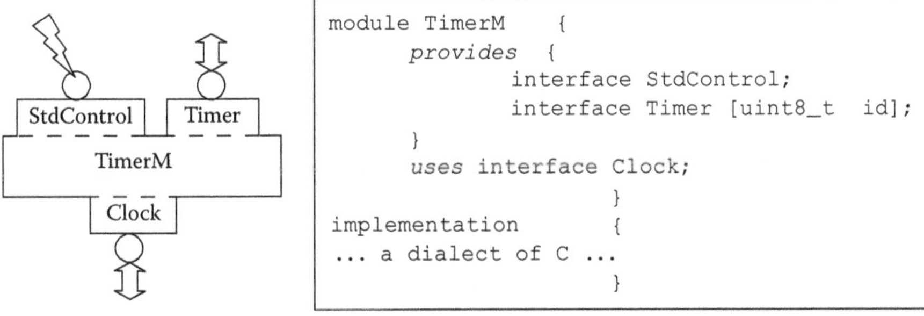
StdControl
Timer
TimerM
Clock
module TimerM {
provides {
interface StdControl;
interface Timer [uint8_t id];
}
uses interface Clock;
}
implementation {
... a dialect of C ...
}

Figure 6.1 Specification and graphical depiction of the TimerM component. (Adapted from Levis, P. et al., TinyOS: An operating system for sensor network, *Ambient Intelligence*, Weber, W. et al., Eds., Springer-Verlag, New York, 2004.)

Through these interfaces, a component can directly interact with other components. A component can *provide* or *use* the same interface type several times as long as it gives each instance a separate name.

A component uses an interface to represent a specific service (e.g., sending a message). In Figure 6.1, a component called TimerM has in total three interfaces: (1) It *provides* the `StdControl` and `Timer` interfaces and (2) *uses* a `Clock` interface. In Figure 6.1, the *provided* interfaces are shown above the TimerM component and the *used* interfaces are shown below. Bidirectional arrows depict commands and events. The lightning depicts commands.

All interface details are shown in Figure 6.2.

As shown in Figure 6.2, interfaces are *bidirectional* and contain both *commands* and *events*. The *providers* of an interface implement the function of a *command*, while the *users* of an interface implement the function of an *event*. For instance, the `Timer` interface (Figure 6.2) defines two commands that are "start" and "stop," and an event is called "fire."

Note: In this example, we do not use two separate interfaces (one for its *commands* and another for its *events*) to represent the interaction between the timer and its client. This is because combining them in the same interface makes the specification much simpler and helps to reduce bugs when wiring components together.

TinyOS is implemented in the NesC language. Components written in NesC consist of two types: *modules* and *configurations*.

```
interface StdControl {
command result_t init();
command result_t start();
command result_t stop();
 }
interface Timer {
command result_t start(char type, uint32_t interval);
command result_t stop();
event result_t fired();
 }
interface Clock {
command result_t setRate(char interval, char scale);
event result_t fire();
 }
interface SendMsg {
command result_t send(uint16_t address, uint8_t length, TOS_MsgPtr
 msg);
event result_t sendDone(TOS_MsgPtr msg, result_t success);
        }
```

Figure 6.2 Sample TinyOS interface types. (Adapted from Levis, P. et al., TinyOS: An operating system for sensor networks, *Ambient Intelligence*, Weber, W. et al., Eds., Springer-Verlag, New York, 2004.)

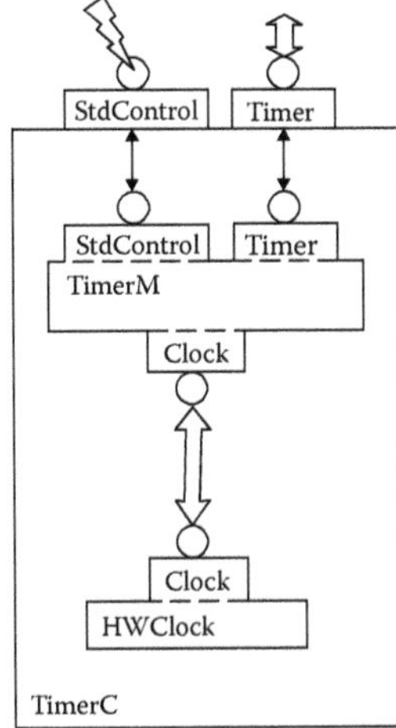


```
configuration TimerC {
 provides {
    interface StdControl;
    interface Timer[uint8_tid];
  }
}
implementation {
  components TimerM, HWClock;
    StdControl = TimerM.StdControl;
 Timer = TimerM.Timer;
 TimerM.Clk -> HWClock.Clock;
}
```

Figure 6.3 TinyOS's timer service: the TimerC configuration. (Adapted from Levis, P. et al., TinyOS: An operating system for sensor networks, *Ambient Intelligence*, Weber, W. et al., Eds., Springer-Verlag, New York, 2004.)

Modules can be used to call or execute *commands* and *events*. A module can declare private state variables and data buffers.

Configurations use interfaces to wire other components together. Figure 6.3 defines the TinyOS timer service. Its implementation is based on a *configuration* (called `TimerC`), which wires the timer *module* (`TimerM`) to the hardware clock *component* (`HWClock`). *Configurations* allow multiple *components* to be aggregated together into a single *macro component* that exposes a single set of *interfaces*.

A *component* uses its *interfaces* (called *interface namespace*) to refer to the *commands* and *events* that it uses. A configuration wires interfaces together by connecting the local names of different interfaces together, that is, a component invokes an interface without referring explicitly to its implementation. This makes it easy to introduce a new component in the component graph that uses the same interface.

An interface can be wired to other interfaces multiple times. Figure 6.4 illustrates an example. The `StdControl` interface of `Main` is wired to `Photo`, `TimerC`, and `Multihop`.

Parameterized interface: In a component, *parameterized interface* can be used to export many instances of the same interface, parameterized by an identifier (typically a small integer). For example, in Figure 6.1, the `Timer` interface is a *parameterized interface* that uses an 8 bit id, which is an extra parameter. Such a parameterized interface allows the single `Timer` component to implement multiple, separate timer interfaces, one for each client component. Because the selection

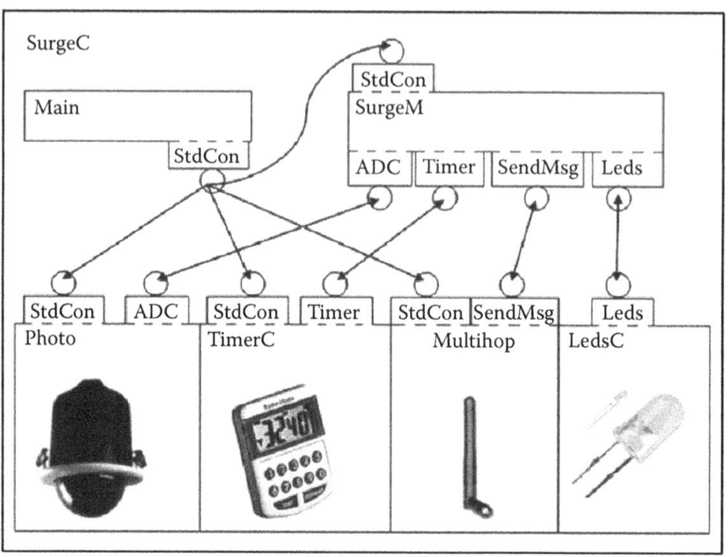

Figure 6.4 The top-level configuration for the Surge application. (Adapted from Levis, P. et al., TinyOS: An operating system for sensor networks, *Ambient Intelligence*, Weber, W. et al., Eds., Springer-Verlag, New York, 2004.)

of IDs should be unique, a special unique keyword is used each time a unique identifier is needed.

Now, we can see how a TinyOS application is built by NesC: First, we can use NesC to build a top-level configuration. Then, we define different interfaces in that configuration to wire all needed components together. Figure 6.4 shows an application called `SurgeC`. It consists of the following components: Main, Photo, TimerC, Multihop, LedsC, and SurgeM. Such an application periodically (`TimerC`) acquires light sensor readings (`Photo`) and sends them back to a base station using multi-hop routing (`Multihop`).

NesC is based on the C syntax and executions. However, it is different from C in the following two aspects: (1) NesC does not use function pointers. Its compiler knows the precise call graph of a program. Such a call graph enables cross-component optimizations, which can remove the overhead of cross-module calls. (2) Dynamic memory allocation is not supported in NesC. It *statically* declares all of a program's state. This scheme can prevent memory fragmentation as well as runtime allocation failures.

6.1.3 Execution Model and Concurrency

A WSN can generate many events, such as abnormal sensor data detection, low-battery alert, and sensor sleep/wake-up. The event-centric domain of WSNs requires fine-grained concurrency. And these events can arrive at any time. How do we handle these events?

There could be two approaches: using traditional OS approaches, such as Windows, which atomically enqueues incoming events and runs them at an appropriate time; and executing an event handler immediately in the style of active messages (AMs).

Because many of these events are important to WSN applications (e.g., a detected event needs immediate attention), the second approach is more suitable to WSNs. While the core of the TinyOS execution model consists of run-to-completion tasks that represent the ongoing computation, the event handlers are signaled asynchronously by the hardware.

Tasks are defined as explicit entities. When the program is executed, *tasks* are sent to the task scheduler for execution in any time order (such as FIFO). The execution must obey the *run-to-completion* rule, that is, once it picks up a task to execute, it must complete the execution. We can use "atomicity" to represent the *run-to-completion* nature of a task. However, *tasks* are not atomic if an interrupt handler comes, or if the program needs to respond to *commands* and *events* that an interrupt invokes.

TinyOS defines the synchronous code (SC) and the asynchronous code (AC) as follows: (1) SC—It can only be reached through *tasks*. (2) AC—It is reachable through at least one interrupt handler. *Components* often have a mix of the SC and the AC. TinyOS allows programmers to build responsive, concurrent data structures that can safely share data between the AC and the SC. TinyOS uses non-preemption to eliminate *races* (i.e., competitions of CPU resources) among *tasks*. Unfortunately, there are still potential races between the SC and the AC, as well as between the AC and the SC.

Typically, when there is any update to the shared state that is reachable from the AC, a data race could possibly occur. Then how do we ensure "atomicity" in such cases? We have two options to avoid the races: (1) We could convert all of the conflicting codes to tasks (here, the codes are for the SC only) or (2) we could use *atomic sections* to update the shared state. Here, the *atomic section* is a small code sequence that is guaranteed to run *atomically*. For an atomic section, we cannot use any loops inside it and we cannot turn on any interrupts.

In summary, we need to ensure a race-free program execution using the following approach.

> *Race-free invariant*: Any update to the shared state is either SC-only or occurs in an atomic section. NesC makes sure that the above *race-free invariant* is met during *compile* time, that is, the NesC compiler can prevent nearly all data races.

The data race should be avoided in any program due to the following reasons.

The data race can cause a class of very painful nondeterministic program bugs.
If it is race free, the composition can essentially ignore concurrency, that is, it will not care which components generate concurrency or how those components are wired together, because the compiler will catch any sharing violations during the compile time.

A wide variety of concurrent data structures and synchronization primitives could be enabled by a strong compile-time analysis. NesC has several variations of concurrent queues and state machines to easily handle time-critical actions directly in an event handler, even when they update the shared state. For example, NesC always handles radio events that are in the interrupt handler until a whole packet has arrived, at which point the handler posts a *task*.

6.1.4 Active Messages

An important issue is: How does TinyOS handle wireless communications among sensors? A concept, called *active messages* (*AMs*), becomes the core TinyOS communication abstraction [TVon92]. An *AM* is a small (only 36 bytes long) packet associated with a 1 byte handler ID. When a sensor receives an *AM*, it immediately dispatches the message (using an event) to one or more handlers. And these handlers are registered to receive such *AM*s. Such *handler registration* is accomplished through static wiring and a parameterized interface, as described above.

TinyOS uses *AM* interfaces to achieve an unreliable, single-hop datagram protocol. *AM* interfaces also provide a unified communication interface to both the radio unit and the built-in serial port (for wired nodes, such as base stations).

Multi-hop, reliable communications could be achieved by higher-level protocols above the AM interfaces. The exchange of *AM*s is also event driven. AMs also tightly couple the local CPU computations and radio communications.

6.1.5 Implementation Status

So far, TinyOS has been used in a wide range of hardware platforms. It is suitable to many companies' sensor products. People have extended the TinyOS environment by adding visualization, debugging, and support tools, as well as a fine-grained simulation environment.

By installing TinyOS in both sensors and other machines, such as desktops, we could build proxies between sensor networks and the Internet, allowing WSNs to integrate with server side tools implemented in Java, C, or MATLAB®. TinyOS also allows us to build software interfaces to database engines, such as PostgreSQL.

6.1.6 Main Features

Absolute size: Surprisingly, TinyOS is really a tiny operating system because a base TinyOS environment only needs around 400 bytes. If it includes associated C runtime primitives (such as floating-point libraries), TinyOS can fit in just over 1 kB. If it needs to add some NesC-based applications, in most cases, they fit in less than 16 kB. But, for some extra-large TinyOS applications, such as TinyDB, they still fit in less than 64 kB so far.

Footprint optimization: Besides using standard techniques (such as stripping the symbol table) to reduce the code size, TinyOS also uses a whole-program compilation to prune a dead code. Cross-component optimizations are used to get rid of redundant operations and module-crossing overhead.

As a matter of fact, NesC uses a whole-program analysis to remove many of these boundary crossings and optimize the entire call paths through extensive cross-component optimizations. (Such optimizations include constant propagation and common sub-expression elimination.) Such whole-program optimizations make NesC programs smaller and faster than unoptimized codes and the original handwritten C code that predates the NesC language.

Hardware/software transparency: TinyOS uses flexible component models to easily shift the hardware and software boundary. For instance, components can generate two types of *events*: software upcalls and hardware interrupts.

6.1.7 Low-Power Optimizations

TinyOS possesses a series of features to reduce energy consumption. For example, TinyOS uses split-phase operations and an event-driven execution model to reduce power usage, because these operations avoid spinlocks and heavyweight concurrency (e.g., threads). A TinyOS scheduler can command the microprocessor into a low-power sleep mode whenever the *task* queue is empty. Such a sleep mode further reduces power consumption.

6.2 LA-TinyOS—A Locality-Aware Operating System for WSNs [Huang07]

LA-TinyOS [Huang07] proposes a new WSN OS that uses *locality* to improve event detection performance and, at the same time, to reduce energy consumption. A WSN locality includes two types: *temporal* and *spatial*. These are defined as follows.

Temporal locality: When an event occurs, if it is really a WSN system anomaly, it could be observed again for a limited period of time during its first appearance. This is called *temporal locality*.

Spatial locality: If an anomaly is caused by a mobile object passing through a WSN, such an anomaly is likely to be observed again by neighboring nodes. This phenomenon is known as *spatial locality*.

Good idea

If you could remember the Computer Architecture course, the design of caches also uses the same principle: (1) Based on temporal locality, if an instruction is used in one time, it is likely to be used again in a near future. (2) Based on spatial locality, if an instruction is selected to be executed, its neighboring instructions are likely to be executed too. Therefore, the caches could be used to store such locality-aware instructions to speed up CPU execution.

Sometimes, both temporal and spatial locality could occur in the same anomaly, for instance, in an environmental surveillance application, when a sensor detects an intruder. Such an intrusion event is likely to be continuously raised by the same node for some time, that is, temporal locality occurs. If the intruder moves around, the intruder may be detected by the neighboring nodes shortly, that is, spatial locality occurs.

Typically, a sensor uses a task manager to sense an event. The task manager is activated periodically to sense the event. The detection period could be very long, because the task manager usually senses nothing out of the ordinary.

The longer the detection period, the less the energy consumed by a sensor. However, when an anomalous event occurs, due to temporal and spatial locality, a shorter period is favored, as we need to increase the activation frequency of the task manager to observe the anomaly more closely.

It would be good if a task manager is *locality aware*, that is, if it could *adjust its period automatically*, based on the principle of temporal and spatial locality.

Unfortunately, most of WSN OSs do not provide kernel-level support to facilitate the development of *locality-aware* tasks. Therefore, most WSN applications perform no locality-aware tasks, or construct such tasks in a user mode (i.e., not implemented in the OS). Such a user mode tends to be error prone, less efficient, and redundant.

LA-TinyOS improves TinyOS by considering locality-aware task implementation. It achieves this by adding the `LocalityM` component to TinyOS. `LocalityM` provides an interface called `LocalityControl` for programmers to configure their locality elements. A data structure, as shown in Table 6.1, is maintained by `LocalityM`. Such a data structure can be used to record all *locality* configurations. This table is called the *locality configuration* table.

```
registerEvent(string EventName);
configureLocality (event table entry T e,
uint 8 TimerID,
uint 32 GracefulLength,
uint 8 HopCount,
(void _ ) FuncEnter,
(void _ ) FuncLeave);
triggerEvent (string EventName);
```

Table 6.1 LA-TinyOS Locality Configuration Table

Event	*Timer ID*	*Graceful Length*	*Tasks*	*Hops*	*Adaption Functions*
"A"	1	2000	dataTask	2	enterl()/leavel()
"B"	2	1000	getMax	1	Null/Null

Source: Adapted from Huang, T. et al., LA-TinyOS: A locality-aware operating system for wireless sensor networks, *Proceedings of the 2007 ACM Symposium on Applied Computing (SAC '07)*, Seoul, Korea, March 11–15, 2007, ACM, New York, 1151–1158.

In the above three commands, `registerEvent` registers a new entry in the locality configuration table, `configureLocality` specifies locality configurations, and `triggerEvent` is called when an anomalous event is detected to enter its locality.

An example code that uses a *locality configuration* data structure and commands is shown in Figure 6.5. It is a locality-aware `Oscilloscope` in LA-TinyOS.

In Line 8, we can see that an event named "A" is registered in the locality configuration table. In Line 9, it calls `configureLocality` to specify locality configurations of this event.

In Line 11, the `reg` operator associates `dataTask` with this event. If we look back at Table 6.1, its first row shows the locality configurations of this event.

In Line 14, we can see that when the sensed data is larger than a specific threshold (0x03B0), an anomaly is detected and event "A" is triggered to enter its locality (Line 15).

In the last column of Table 6.1 (called "adaptation functions"), `enterl` and `leavel` are pointers to self-adaptation functions provided by this application. It basically means that when "A" is detected, LA-TinyOS executes `enterl` to enter its locality and it executes `leavel` to leave its locality.

6.2.1 Change Timer to Respond to Temporal and Spatial Locality

Now let us see how LA-TinyOS updates its timer based on locality configurations. A component called `TimerM` maintains a list of software timers, as shown in Table 6.2. The timer ID tells us whether it is a one-shot timer (i.e., it is terminated after it expires) or a periodic timer, its default timer period (a counter's value), and the time left before it expires.

If a timer interrupt is triggered, the interrupt handler of `HWClock` reduces the value in the *Time-to-Expired* field of each software timer. When the *Time-to-Expired* reaches zero, it means that the timer expires; a corresponding handler is then executed.

```
implementation
{
      command result_t StdControl.start(){
      event_table_entry_T*e;
      call SensorControl.start();
      call Timer.start(TIMER_REPEAT. 1500);
      ...
      e = call LocalityControl.registerEvent("A");
      call LocalityControl.configureLocality(e, 1, 2000, 2, enterl,
      leavel);
      reg dataTask() "A";
      }
      async event result_t ADC.dataReady(uint16_t data){
        if (data>0x03B0){ // an anomaly
          call LocalityControl.triggerEvent("A");
      }
      pack->data[packetReadingNumber] = data;
      post dataTask();
      }
}
```

Figure 6.5 Part of locality-aware oscilloscope. (Adapted from Huang, T. et al., LA-TinyOS: A locality-aware operating system for wireless sensor networks, *Proceedings of the 2007 ACM Symposium on Applied Computing (SAC '07)*, Seoul, Korea, March 11–15, 2007, ACM, New York, 1151–1158.)

Table 6.2 List of Software Timers in the TimerM Component

Timer ID	*Type*	*Status*	*Period*	*Time-to-Expired*
0	ONE SHOT	On	300	240
1	REPEAT	On	1500	360
2	REPEAT	Off	500	450

Source: Adapted from Huang, T. et al., LA-TinyOS: A locality-aware operating system for wireless sensor networks, *Proceedings of the 2007 ACM Symposium on Applied Computing (SAC '07)*, Seoul, Korea, March 11–15, 2007, ACM, New York, 1151–1158.

When an event enters its *locality*, LA-TinyOS uses the following data structure to change the period of the software timer:

```
setLocailityTimer (unit8_t TimerID,
unit32_t ReducedPeriod);
```

In the above data structure, LA-TinyOS calculates its *reduced period* to show its adaptation to temporal locality. Its `TimerID` can be easily found by searching the `localityconfiguration` table. The default period will be saved in a kernel data structure. When an event leaves its locality, `setLocailityTimer` is again applied to reset its period.

In Table 6.1, the third column indicates the *graceful length* of each event. Whenever an anomaly is detected, this counter is reset to its full value. When an event enters its locality, LA-TinyOS reduces its *graceful length* counter at each timer interrupt.

Eventually, the *graceful length* counter reaches zero. Then, the "period" of its associated software timer (shown in Table 6.2, column 4) is reset to its default value, to indicate that this event is leaving its locality.

The above description corresponds to the *temporal locality* case. How does LA-TinyOS implement *spatial locality*? It does this by broadcasting alerting messages. The number of broadcasting hops defines the alerting area. The number of hops is also available in the locality configuration table (Table 6.1). When a sensor receives an alerting message, it activates a corresponding anomalous event to enter its locality.

6.2.2 Multiple-Level Scheduler

As we mentioned in Section 6.1, TinyOS uses a non-preemptive FIFO scheduler. Now the issue is as follows: Such a simple scheduler cannot differentiate tasks that are associated with an anomalous event from tasks that are regular and nonurgent.

To solve such an issue, LA-TinyOS proposes a three-level scheduler without changing the TinyOS non-preemptive scheduling, as follows:

Level 1: When a sensor detects an anomalous event, it registers associated tasks in the locality configuration table. These tasks are queued in the first level and are scheduled to be executed with the first priority.

Level 2: For spatial locality, tasks are associated with an event that is triggered to enter its locality by an alerting message. These tasks are queued in the second-level FIFO queue.

Level 3: When level 1 and level 2 tasks do not occur, the nonurgent, normal tasks are served in the third-level FIFO scheduler.

The above three-level FIFO scheduler can make sure that LA-TinyOS performs tasks according to their importance.

Good idea

The multilevel hierarchical tree concept has been used to obtain solutions to many problems. Its basic idea is to avoid flat (i.e., one-level) topology, where all nodes are treated in the same way. By distinguishing among different levels, we have the flexibility to handle different priorities.

6.2.3 LA-TinyOS Code Structure

The code structure of LA-TinyOS is shown in Figure 6.6. We can see that LA-TinyOS enhances TinyOS by adding a *LocalityM* module and a *multilevel scheduler*.

As we discussed before, the *locality configuration* table is used to register and configure the events under either temporal locality or spatial locality. And the LocalityM component in turn uses the original TinyOS kernel components to automatically adjust the detection "period" of a task when it enters and leaves its locality.

Without *LocalityM*, a programmer can still use the original TinyOS components (see Figure 6.7) to program a *locality*-aware application.

Now let us summarize the main advantages of using LA-TinyOS to handle locality-aware applications:

First, the LA-TinyOS kernel component houses all of its locality-aware codes. The kernel execution is much more reliable than the original TinyOS implementation.

Second, LA-TinyOS allows a programmer to easily program locality-aware events, as the programmer only needs to register a locality event during the initialization phase, and then make a method call to enter its locality when an anomaly event is detected.

Finally, when more than one locality event occurs, a programmer can use `LocalityM` to handle the locality-aware code of all events. Therefore LA-TinyOS

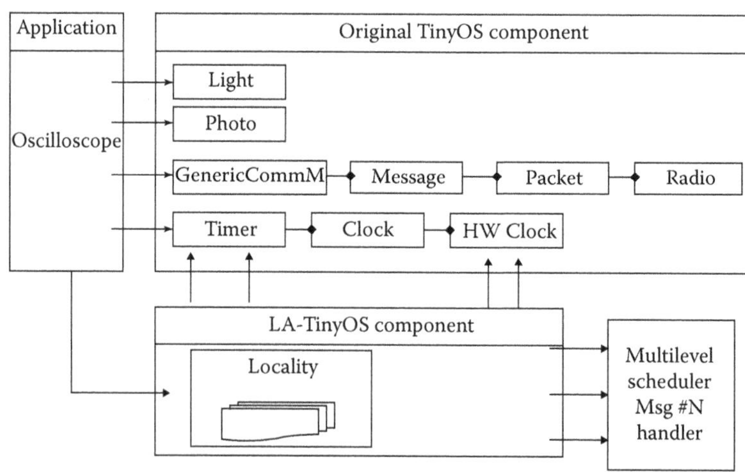

Figure 6.6 The code structure of LA-TinyOS. (Adapted from Huang, T. et al., LA-TinyOS: A locality-aware operating system for wireless sensor networks, *Proceedings of the 2007 ACM Symposium on Applied Computing (SAC '07)*, Seoul, Korea, March 11–15, 2007, ACM, New York, 1151–1158.)

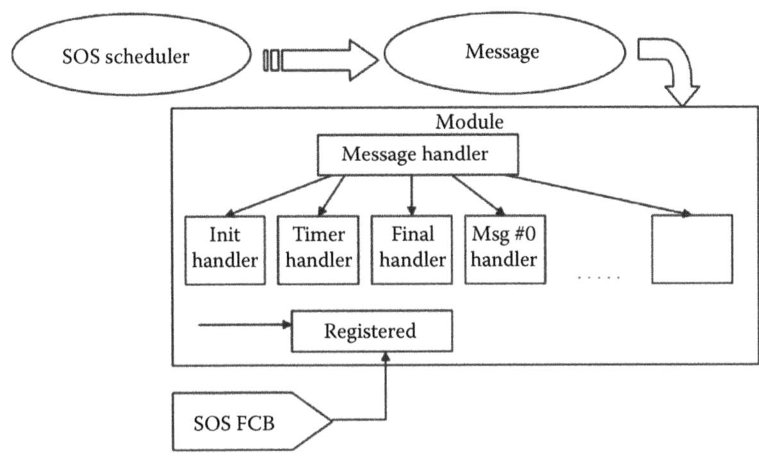

Figure 6.7 Module interactions. (Adapted from Han, C. et al., A dynamic operating system for sensor nodes, *Proceedings of the Third International Conference on Mobile Systems, Applications, and Services (MobiSys '05)*, Seattle, WA, June 6–8, 2005, ACM, New York, 163–176.)

allows a more efficient implementation with multiple locality events. In contrast, a TinyOS implementation needs to provide a redundant locality-aware code for each event.

6.3 SOS [HanC05]

SOS, another improved version of TinyOS, is proposed in [HanC05]. It shows that WSN OSs can achieve dynamic and general-purpose OS semantics without sacrificing significant energy or performance. Its main features include the following:

SOS has a common kernel and dynamic application modules. These modules can be loaded or unloaded at runtime. A system jump table is used by modules to send messages and communicate with the kernel. A module can also register function entry points for other modules to call.

SOS contains no memory protection, which is similar to TinyOS. However, it protects against common bugs. This is an improvement over TinyOS. Dynamic memory is used in SOS for the application modules and the kernel. This makes programming easier, as it decreases complexity and increases temporal memory reuse.

Just as LA-TinyOS uses a three-level task scheduler, SOS also proposes to use priority scheduling to move processing out of the interrupt context and provide improved performance for time-critical tasks.

The SOS kernel has dynamically linked modules, flexible priority scheduling, and a simple dynamic memory subsystem. SOS kernel services provide a higher-level API (applied program interface) to free a programmer from managing underlying services or re-implementing popular abstractions.

6.3.1 Modules

An SOS program uses *modules* (which are position-independent binaries) to implement a specific task or function. SOS consists of multiple interacting modules. From the functionality viewpoint, the *modules* are similar to the concept of *components* in TinyOS. An SOS programmer implements the primary development, including drivers, protocols, and application components, that occur at *the module layer*.

It is challenging to maintain modularity and safety in SOS without incurring high code overhead due to the loose coupling of modules. All SOS modules are self-contained and position independent. They use clean messaging and function interfaces to maintain modularity. Most applications do not need to modify the SOS kernel unless the low-layer hardware or resource management capabilities need to be changed.

6.3.1.1 Module Structure

Figure 6.7 shows SOS module interactions. As we can see, SOS maintains a modular structure by implementing modules with well-defined and generalized points of entry and exit. The flow of execution enters a module either from (1) messages delivered from the scheduler or (2) registered functions (for external use).

A module-specific handler function handles messages between modules. There are two parameters accepted by a handler function: (1) the message being delivered and (2) the state of the module.

When a module is inserted, SOS kernel produces an *init* message. The *init* message handler sets the module's initial state, which includes initial periodic timers, function registration, and function subscription.

When a module is removed, the SOS kernel produces a *final* message. The *final* message handler releases all sensor resources, including timers, memory, and registered functions.

Besides the above *init* and *module* messages, there are also other module-specific messages, including handling of timer triggers, sensor readings, and incoming data messages from other modules or nodes.

SOS handles messages asynchronously (i.e., using a queue to store these messages). Similar to TinyOS, the main SOS-scheduling loop picks up a message from a priority queue and delivers the message to the message handler of the destination module.

Module-specific operations need to run synchronously. SOS uses direct function calls between these modules. A function registration and subscription scheme implements these direct function calls.

The RAM stores the modules' states. Modules are relocatable in memory. The location of inter-module functions is exposed through a registration process.

6.3.1.2 Module Interaction

Messages are used to implement interactions between modules. Messaging enables asynchronous communication between modules. Messaging can also break up chains of execution into scheduled subparts. These subparts are stored into a queue for scheduled execution.

Although the above messaging is flexible, its execution is slow. Therefore, SOS provides direct calls to functions that are registered by modules. These direct function calls can bypass the scheduler to provide lower latency communication between the modules.

SOS uses function registration and subscription to implement direct inter-module communication and function calls from the kernel to modules. A *function control block* (FCB) is used to store crucial information about the registered function. An FCB is created by the SOS kernel and indexed by the tuple {module ID, function ID}. The FCB includes a valid flag, a subscriber reference count, and prototype information.

The module ID and the function ID are used to locate the FCB of interest, and type information is used to provide an additional level of safety. If the lookup succeeds, the kernel returns a pointer to the function pointer of the subscribed function.

A jump table shown in Figure 6.8 is used by modules that need to access kernel functions. Such a jump table also allows each module to remain loosely coupled to the kernel, rather than be dependent on specific SOS kernel versions. It also allows the kernel to be upgraded without the need of recompiling SOS modules. Thus, the same module can run in a deployment of heterogeneous SOS kernels.

6.3.1.3 Module Insertion and Removal

Module insertion: A distribution protocol keeps listening to the advertisements of new modules in the network. The distribution protocol that advertises and propagates module images through the network is independent of the SOS kernel. SOS currently uses a publish–subscribe protocol, similar to a Mobile-Oriented Applications Platform (MOAP). When an advertisement for a module is captured by the protocol, the protocol needs to check whether or not the module is the updated version of a module already installed on the node. It also needs to check if the node is interested in the module and has free program memory for the module.

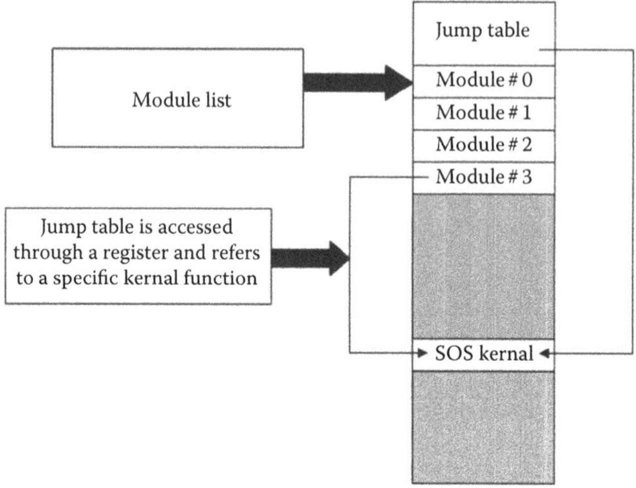

Figure 6.8 Jump table layout and linking in SOS. (Adapted from Han, C. et al., A dynamic operating system for sensor nodes, *Proceedings of the Third International Conference on Mobile Systems, Applications, and Services (MobiSys '05)*, Seattle, WA, June 6–8, 2005, ACM, New York, 163–176.)

To find out whether or not the above two conditions are true, the distribution protocol examines the *metadata* in the header of the packet. Such metadata contains the following information: (1) the unique identity for the module, (2) the size of the memory required to store the local state of the module, and (3) the version information used to differentiate each module version. The protocol will abort the module insertion if the SOS kernel finds that it is unable to allocate memory for the local state of the module.

Module insertion needs to be created as a kernel data structure that is indexed by the unique module ID included in the metadata. Such a kernel stores the absolute address of the handler. It also stores a pointer to the dynamic memory holding the module state and the identity of the module. Finally, the SOS kernel invokes the handler of the module by scheduling an *init* message for the module.

Module removal is initiated by the kernel that dispatches a *final* message. This message commands a module to gracefully release any resources it holds. Such a message also informs other modules that depend on the removed module. After the final message, the kernel performs garbage collection by releasing all resources as follows: dynamically allocated memory, timers, sensor drivers, and other resources owned by the module. FCBs are then used after a module removal to ensure the integrity of the platform.

6.3.2 *Dynamic Memory*

Dynamic memory allocation can use a flexible queue length to adapt to the worst-case scenarios and complex program semantics for common tasks, such as passing a data buffer down a protocol stack. Dynamic memory in SOS is based on a simple, best-fit, fixed-block memory allocation with three base block sizes as follows.

The smallest block sizes are used for most SOS memory allocations, including message headers. Larger block sizes are used for applications that need to move large, continuous blocks of memory, such as module insertion. The largest block sizes are actually a linked list of free blocks and can be used for any complex applications.

All data structures (such as queues and lists) in SOS dynamically grow and shrink at runtime. The dynamic use and release of memory (i.e., dynamic memory) creates a system with effective temporal memory reuse. Dynamic memory can also dynamically tune memory usage to specific environments and conditions.

Modules can transfer memory ownership to reject data movement. SOS annotates dynamic memory blocks by using a small amount of data that is used to detect basic sequential memory overruns. Memory annotations can be used for a post-crash memory analysis to identify suspected memory owners, such as a bad module that owns a great deal of system memory or overflowed memory blocks. SOS memory annotations also enable garbage collection on unload.

6.4 RETOS [Hojung07]

Resilient, expandable, and threaded operating system (RETOS) aims to provide a robust, reconfigurable, resource-efficient multithreaded OS for WSN nodes. Figure 6.9 shows its overall architecture.

Although the *event-driven* approach has been used extensively for sensor OSs due to its efficient implementation in resource-constrained sensors; however, in RETOS, application developers can manage the states of *tasks* and *events* explicitly, via a *program split* process as follows: RETOS explicitly separates applications from the kernel. An application is separately and dynamically loaded into the system (as does the kernel module). RETOS uses a loadable module framework to achieve kernel reconfigurability.

6.4.1 *Application Code Checking*

RETOS uses a software technique, called *application code checking (ACC)*, to perform static and dynamic code checks. The goal of ACC is to prevent user applications from accessing memory outside its legal boundary and direct hardware manipulation. It thus always checks the *destination field* of machine instructions. The *source field* of instructions can also be examined to prevent the application from reading the kernel or another application's data.

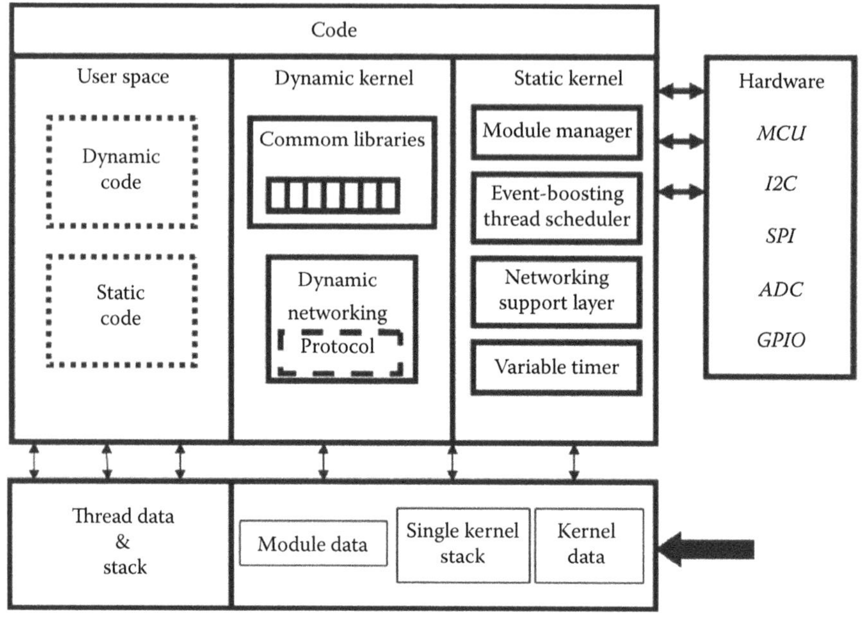

Figure 6.9 RETOS architecture. (Adapted from Cha, H. et al., Resilient, expandable, and threaded operating system for wireless sensor networks, *IPSN '07*, Cambridge, MA, April 25–27, 2007.)

ACC uses *static* code checking to verify direct or immediate addressing instructions and pc-relative jumps during the compilation time. (For details on instructions' addressing modes, please refer to the Assembly Language courses.)

ACC uses *dynamic* code checking to verify the correct usage of indirect addressing instructions during runtime. Dynamic checking is also required for the Return instruction.

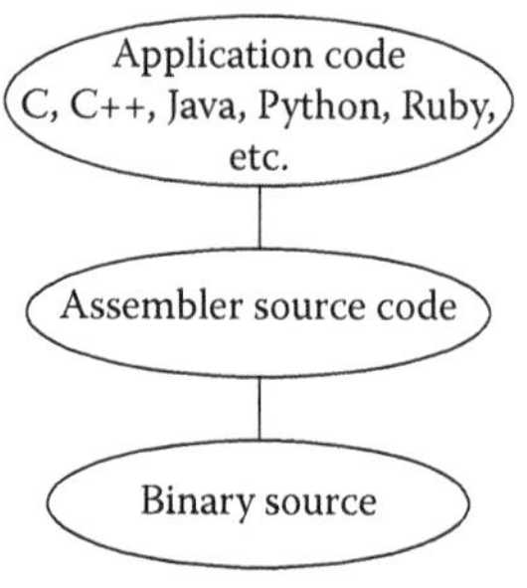

Figure 6.10 Generating trusted code. (Adapted from Cha, H. et al., Resilient, expandable, and threaded operating system for wireless sensor networks, *IPSN '07*, Cambridge, MA, April 25–27, 2007.)

Figure 6.10 shows the procedure of constructing trusted codes. The application source code is compiled to the assembly code. The compiler then inserts checking codes to the place where dynamic code checking is required.

After the dynamic code insertion, *static* code checking is conducted on the binary codes. When the compiler cannot detect some application errors, the missed errors will be reported to the kernel. After receiving these reported errors, the kernel informs users of the illegal instruction address and safely terminates the program.

6.4.2 Multithreading System

We know that TinyOS is an event-driven OS, and TinyOS programmers need to worry about the optimal execution of their programs through explicit concurrency control.

However, RETOS uses a different approach, that is, multithreading, to inherently provide high concurrency with preemption and blocking I/O characteristics of the underlying system. Although the multithreading approach is attractive, it is challenging to implement multithreading in a resource-constrained sensor node environment. In a multithreading environment, each thread needs a stack to maintain state variables. A scheduling scheme is used to perform context switching between the stacks. RETOS has carefully considered memory usage, energy consumption, and scheduling efficiency. RETOS has implemented a single kernel stack and a stack-size analysis, a variable timer, and an event-boosting thread scheduler, respectively.

1. Minimizing memory usage
   RETOS provides two techniques to reduce the memory usage for the kernel.
   a. *A single kernel stack* is used to reduce the size of the thread stack required. The mechanism separates the thread stack into two types: *kernel stacks* and *user stacks*. RETOS has a strict, controlled access to the kernel stack. This is to make sure that the system does not arbitrarily interleave the execution flow during the kernel mode, such as thread preemption. With thread preemption, hardware contexts are saved in each thread's control block based on kernel stack sharing.
   b. A *stack-size analysis* is used to assign an appropriate stack size to each thread autonomously. An accurate thread stack size needs to be estimated to reduce the memory usage. A stack-size analysis has been implemented in RETOS to automatically generate a minimal and system-safe stack for each thread.
2. Variable timer
   Energy is consumed in multithreading computations, which include *timer management*, *context switching*, and a *scheduling operation*.

   Timer management: In multithreading systems, from an energy consumption viewpoint, a variable timer (instead of a fixed periodic timer) could be more energy efficient. Timer requests from threads are processed by the system timer, which then updates the remaining time independent of currently running threads. The timer interrupt interval can be reprogrammed by the variable timer. Such an interval is set to the earliest upcoming timeout among the time quantum of the currently running thread.

The *scheduling operation* does not occur as frequently as passing messages between handlers in an event-driven system. In most WSN applications, the *context-switching* overhead is only a moderate issue.

3. Event-aware thread scheduling
Thread scheduling is based on a priority-aware real-time scheduling interface to enable the kernel's dynamic priority management. Three policies are used to schedule RETOS threads: SCHED_RR, SCHED_FIFO, and SCHED_OTHER.
Event-aware thread scheduling is used to increase the event response time of threads. To handle an important event, the scheduler directly boosts the priority of the thread that handles such a specific event. When an event occurs, the priority-boosted thread will be able to swiftly preempt other threads.

6.4.3 Loadable Kernel Module

Dynamic application loading is supported in RETOS. A memory relocation mechanism is used to support dynamic application loading. Memory relocation cannot be supported by a PIC (position-independent code) approach.

A memory relocation mechanism is shown in Figure 6.11. A RETOS file format consists of a generic portion and a hardware-dependent section. It has compiled codes. If a sensor uses RETOS, its microcontroller needs to support different addressing features, such as relocation type and relative memory-accessing instructions. Therefore, such a file format has hardware-specific information to aid the memory relocation for the corresponding hardware.

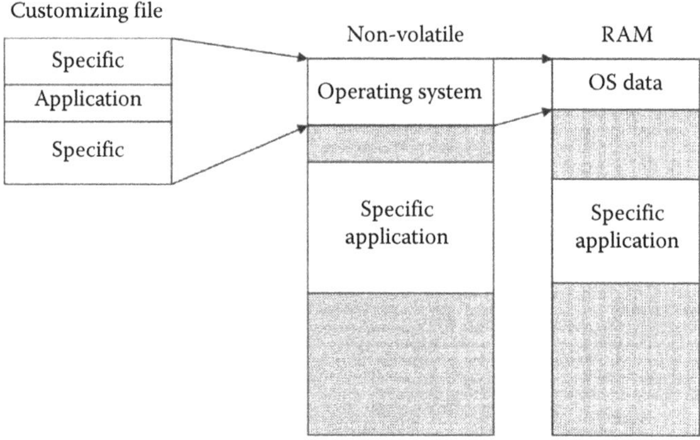

Figure 6.11 Simple RETOS relocation mechanism. Reading from left to right, this mechanism produces the smallest footprint of the OS to free space and resources on the system. (Adapted from Cha, H. et al., Resilient, expandable, and threaded operating system for wireless sensor networks, *IPSN '07*, Cambridge, MA, April 25–27, 2007.)

Problems and Exercises

6.1 Compared to traditional OSs (such as Microsoft Windows), what special characteristics does TinyOS have?
6.2 Explain the TinyOS architecture.
6.3 What enhancements does LA-TinyOS make on the basis of TinyOS?
6.4 Explain the SOS module interaction principle.
6.5 What benefits does RETOS have when using module relocation?

Chapter 7

Middleware Design in Wireless Sensor Networks

This chapter introduces the middleware architecture of a WSN. Our discussions are based on the summarization of [Miaomiao08]. For more details, readers are referred to [Miaomiao08] for a comprehensive survey.

7.1 Introduction

Typically, the network protocol stack can be classified into five layers (from top to bottom): application layer, transport layer, routing layer, MAC layer, and physical layer. As an example, Crossbow Inc. motes (i.e., WSN nodes) allow a user to use NesC (similar to C language) to build sensor network control programs. As shown in Figure 7.1, a user builds these programs in the application layer to control WSN operations, such as performing data aggregation among neighboring sensors. *Note*: The application layer does not deal with WSN routing issues, as they belong to the routing layer. It also does not handle network congestion issues, as they belong to the transport layer.

Although the above direct programming in the application layer can perform many WSN data-processing and other high-level wireless applications, it is still not convenient for a programmer to build these application layer programs due to the following reasons:

1. Most WSN systems do not have convenient programming/compiling tools. For instance, NesC in a TinyOS environment needs a longtime learning curve. A programmer needs to learn dozens of different entity interfaces. TinyOS installation is still a problematic issue today.

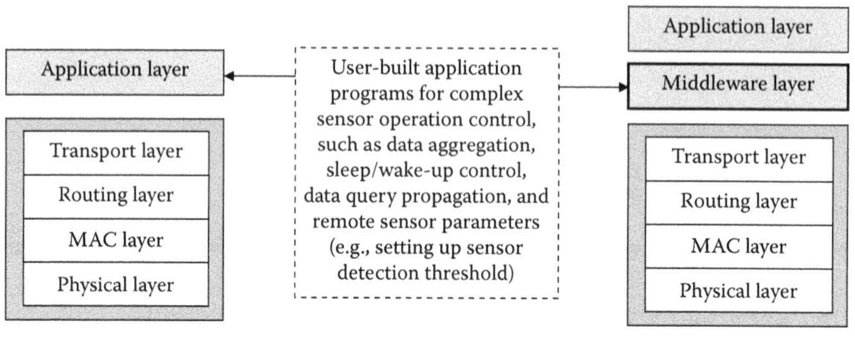

Figure 7.1 Middleware's location.

2. More importantly, a programmer needs to get familiar with many WSN internal operation details to build an efficient, easy-to-use application layer program. For instance, if a programmer wants to build a sensor *data query* software (which is a basic WSN application layer function), she should know the routing layer details, as it is the routing protocol that delivers query commands to each sensor. The programmer should also understand the network topology, as a data query command may need data from certain sensor areas.
3. Although a programmer can build application layer programs to control a sensor's behavior, many WSN operations are built on the collaboration of many nodes instead of just one node. For instance, to save wireless communication energy consumption, a data query command eventually collects data through the data aggregation technology. That is, a program could reside in an individual node, but it needs to control many other nodes. Apparently, this is a challenging task for any programmer.

Therefore, to unburden a WSN programmer from these heavy responsibilities, we need to add a new layer to the traditional network stack. We call this new layer as the *middleware layer*. As shown on the right-hand side of Figure 7.1, the middleware layer should be located between the application layer and the transport layer. Through *the WSN middleware, we could hide the complexity of low-level operations*. A programmer can avoid the troublesome considerations of a WSN's dynamic network topology and low-level embedded OS (operating system) APIs (application program interfaces).

A good WSN middleware provides a programmer with some *reusable code services*, which allow the programmer to access the functionality of network resources, while minimizing the effort of dealing with code dissemination, data aggregation, and power management.

Although traditional middleware schemes (used in distributed computing systems) can also provide transparency abstractions by hiding the context information, they primarily aim to satisfy the interests of *individual* nodes. However, WSN applications are *data centric*, and, therefore, the middleware must be able to operate in all available nodes rather than in individual nodes. Moreover, the WSN middleware should support data aggregation in intermediate nodes along the forwarding path. But the traditional distributed system middleware does not need to support data aggregation, as it uses an end-to-end paradigm [Miaomiao08].

In a middleware, data management is an important task. The middleware needs to provide appropriate abstractions of data structures and operations. Without such abstractions, the application programmer must manage the heterogeneous data and low-level operations [Miaomiao08].

When we design a WSN middleware, we need to make sure that it is lightweight enough for implementation in sensor nodes having limited processing and energy resources.

Good idea

How to minimize a programmer's working load is the goal of many platforms. It is very time consuming if a WSN programmer needs to understand all network operation details before she writes the application layer program. Many WSN companies try to encapsulate the complex sensor/network control into a set of APIs, which is part of middleware tasks. A programmer can then quickly come up with a useful application based on these friendly APIs.

7.2 Reference Model of WSN Middleware [Miaomiao08]

7.2.1 Model Overview

As shown in Figure 7.2, a WSN middleware includes four major components:

1. *Programming abstractions*: A middleware design should first define a set of friendly APIs that hide all complex WSN operations.
2. *System services*: After defining *program abstractions*, a middleware should internally provide concrete implementations of these abstractions. These

WSN middleware components

Programming abstractions	System services	Runtime support	QoS mechanisms

Figure 7.2 WSN middleware components. (Adapted from Wang, M. et al., *J. Comput. Sci. Technol.*, 23(3), 305, 2008.)

implementations are called *system services*, as they belong to part of system codes instead of user codes.

3. *Runtime support*: After we have the above *system service* codes, the sensor OS should be able to run these codes in an optimized way, that is, we need to have *runtime support*.
4. *Quality of service* (*QoS*): In the application layer, people typically use QoS to define some visible application performance metrics, such as data resolution, processing speed, and network delay performance. The middleware should be able to adapt to different QoS requirements.

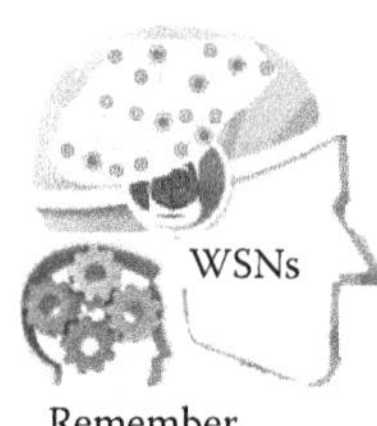

The first three WSN middleware components have a very close relationship among them. The purpose of defining *programming abstractions* is to hide complex WSN operations. As per user's viewpoint, they only require that the middleware provides a set of *system services*. As per middleware designer's viewpoint, they need to write codes to provide *runtime support* for these system services.

Figure 7.3 shows the details of each of the above components. Note that this is just a typical middleware reference model. It does not mean that all WSN middleware implementations should include all of these components.

It is a misunderstanding that the middleware is only implemented in sensors. As a matter of fact, because a user can program the system in different places, the middleware can be located in a sensor node, a sink (i.e., a base station), and a user terminal that communicates with a sink. The distributed middleware components in different places are able to communicate with each other to achieve common goals, such as a data query execution. Figure 7.4 illustrates this point.

7.3 Middleware Example: Agilla [CFok05, Miaomiao08]

A type of middleware implementation is based on the concept of *mobile agent*, which is an execution thread that can migrate from one node to another. Such an agent encapsulates the running codes, the system state, and the application data.

Agilla [CFok05] is an example of the implementation of the agent-based middleware. We can inject a new agent in a WSN to reprogram the network.

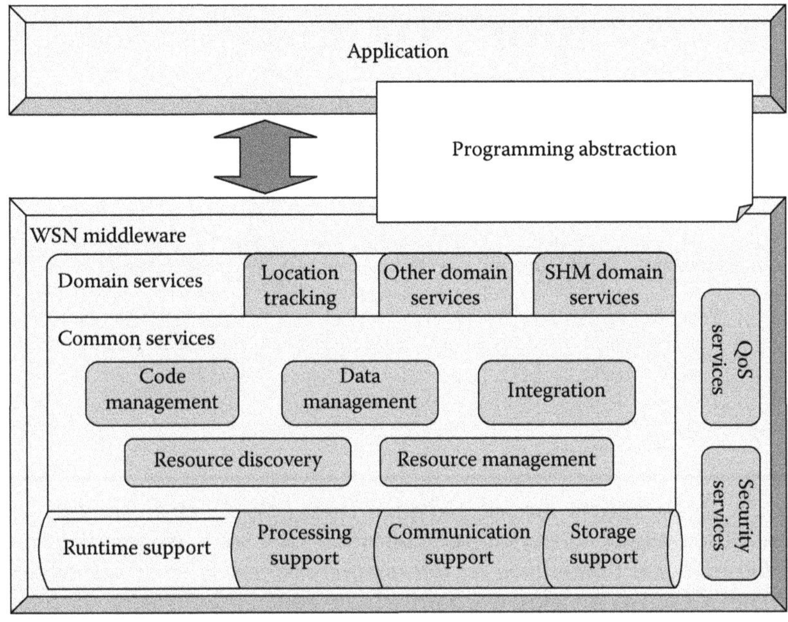

Figure 7.3 WSN middleware reference models. (Adapted from Wang, M. et al., *J. Comput. Sci. Technol.*, 23(3), 305, 2008.)

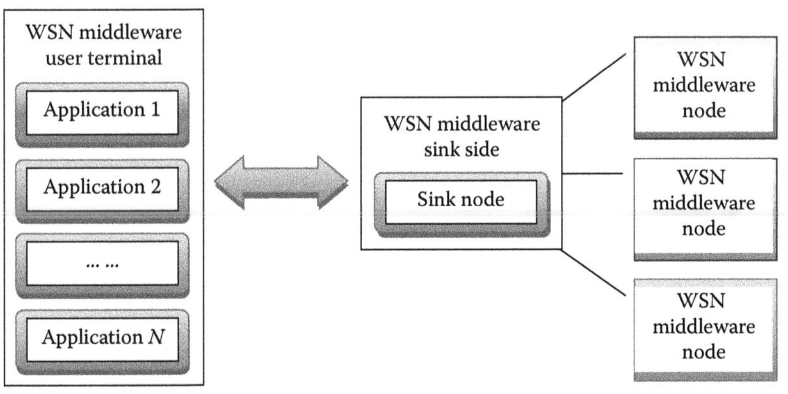

Figure 7.4 System architecture of WSN middleware. (Adapted from Wang, M. et al., *J. Comput. Sci. Technol.*, 23(3), 305, 2008.)

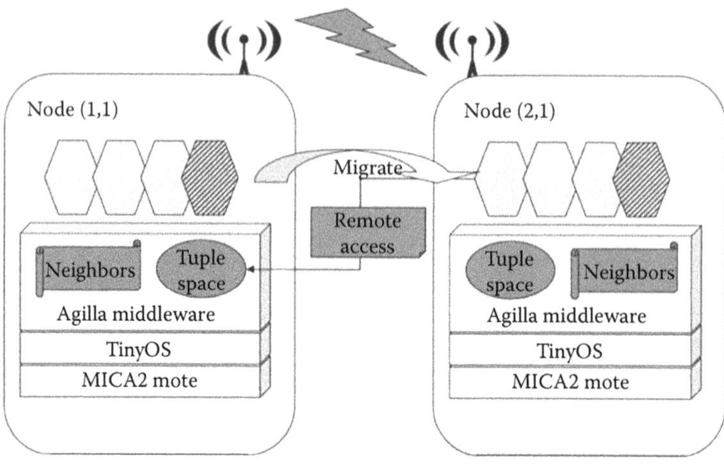

Figure 7.5 Agilla system model. (Adapted from Fok, C. et al., Mobile agent middleware for sensor networks: An application case study, *Proceedings of the Fourth International Conference on Information Processing in Sensor Networks (IPSN '05)*, UCLA, Los Angeles, CA, April 25–27, 2005, 382–387.)

Figure 7.5 shows the *Agilla* system model. Note that each sensor node can support multiple agents. A node also maintains a *tuple space* and a *neighbor list*:

1. The *tuple space* can be shared by all agents residing on the same node. *Agilla* provides special instructions to remotely access the agents in another node's tuple space.
2. The *neighbor list* contains the addresses of all directly adjacent nodes in the WSN. This is for the convenience of agent migration.

Good idea

The mobile agent concept has attracted many researchers' interests. Its basic feature is to allow a physical entity to transfer its unfinished task to another entity. Such a "chain" effect eventually achieves a systematic task. Please note that the mobile agent concept is different from the general multi-agent concept. The latter typically assumes that agents do not migrate/transfer between entities.

As shown in Figure 7.6, a mobile agent in *Agilla* consists of a stack, a heap, and some registers. The heap is actually a memory space to store system variables. Similar to a common CPU architecture, a register consists of the agent ID, the program counter (PC), and

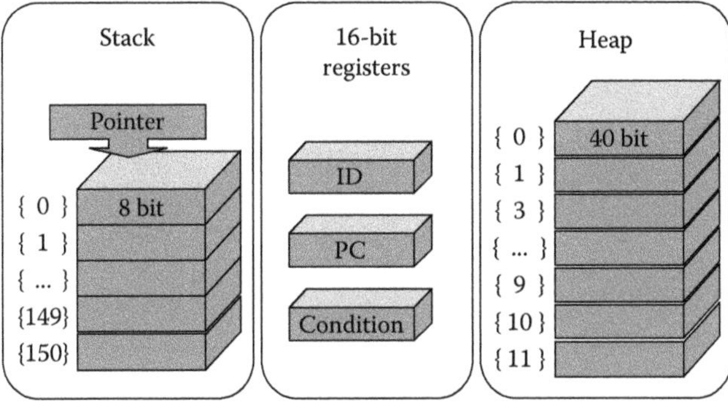

Figure 7.6 Agilla agent architecture. (Adapted from Fok, C. et al., Mobile agent middleware for sensor networks: An application case study, *Proceedings of the Fourth International Conference on Information Processing in Sensor Networks (IPSN '05)*, UCLA, Los Angeles, CA, April 25–27, 2005, 382–387.)

the condition code. The agent ID is unique to each agent. The PC contains the address of the next instruction.

Code migration can be achieved by moving or cloning an agent from one node to another. A *tuple space* can package up all register variables during code migration. When an agent moves, it carries its state variable and runtime code. After an agent reaches a new node, it resumes the code executing. Multi-hop migration is handled by the middleware OS.

7.4 Middleware for Data Acquisition: Mires [ESouto04, Miaomiao08]

A typical task of a WSN middleware is data management, which provides services to applications for *data acquisition*, *data processing*, and *data storage*.

This section uses Mires [ESouto04] as an example of *data acquisition*, which includes a series of functions, such as event definition, event registration/cancellation, event detection, and event delivery. Figure 7.7 shows Mires' middleware architecture.

Mires uses a publish/subscribe paradigm (see Figure 7.8) to implement event-based data acquisition. Such a paradigm supports asynchronous communication and facilitates message exchange between the sensor nodes and the sink node. A publish/subscribe system has two basic components: the event subscriber (in the sink) and the event publisher (i.e., the event broker) (in the sensor nodes).

In Mires, the application layer in the sink *subscribes* the event data of interest. Its subscribe messages are broadcasted down to the network nodes, which *publish* their collected data to the network.

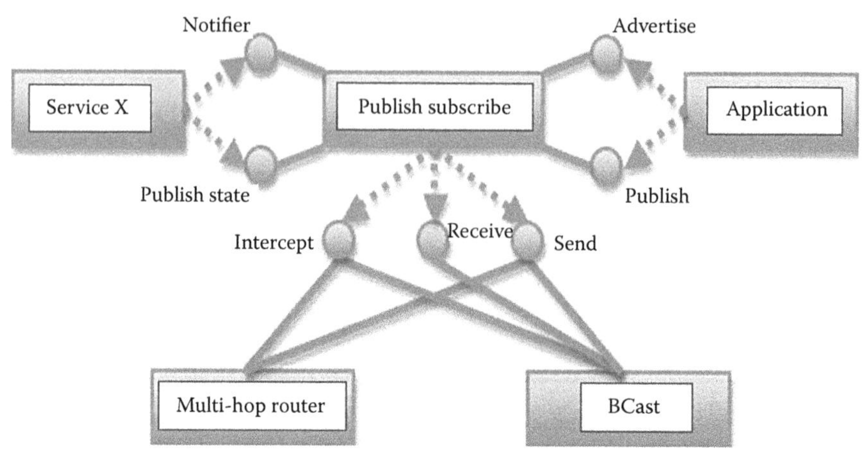

Figure 7.7 Mire's architecture. (Adapted from Souto, E. et al., A message-oriented middleware for sensor networks, *Proceedings of the Second International Workshop on Middleware for Pervasive and Ad-Hoc Computing (MPAC '04)*, Toronto, Ontario, Canada, October 2004, ACM Press, New York, 127–134.)

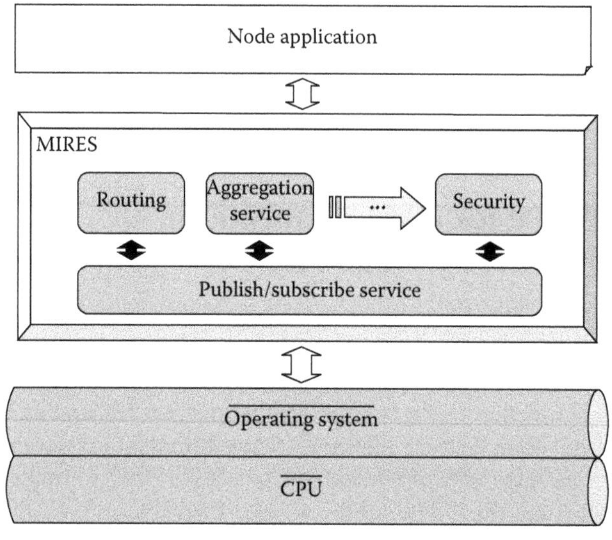

Figure 7.8 Mire's Pub/Sub component. (Adapted from Souto, E. et al., A message-oriented middleware for sensor networks, *Proceedings of the Second International Workshop on Middleware for Pervasive and Ad-Hoc Computing (MPAC '04)*, Toronto, Ontario, Canada, October 2004, ACM Press, New York, 127–134.)

A middleware for *query*-based data models can use *TinyDB*'s [SRM05] flooding approach to disseminate the queries throughout the network.

7.5 Data Storage: DSWare [SLi03, Miaomiao08]

A WSN middleware needs to support one of the most important tasks, that is, data-centric storage. Data service middleware (*DSWare*) [SLi03] is such a middleware. As illustrated in Figure 7.9, *DSWare* implements a database-like abstraction composed of various data services:

1. The *event detection* component actually corresponds to the above-discussed data acquisition service.
2. The *group management* component can implement an important WSN feature, that is, data aggregation.
3. The *scheduling* component can schedule all middleware services based on any one of the two priorities: energy efficiency or delay performance.
4. The data storage component stores data according to the semantics associated with the data. It stores correlated data in geographically adjacent regions to achieve in-network processing.
5. The caching component provides multiple copies of the data that is requested most often. *DSWare* spreads the cached data over the network to achieve high availability and faster query execution.

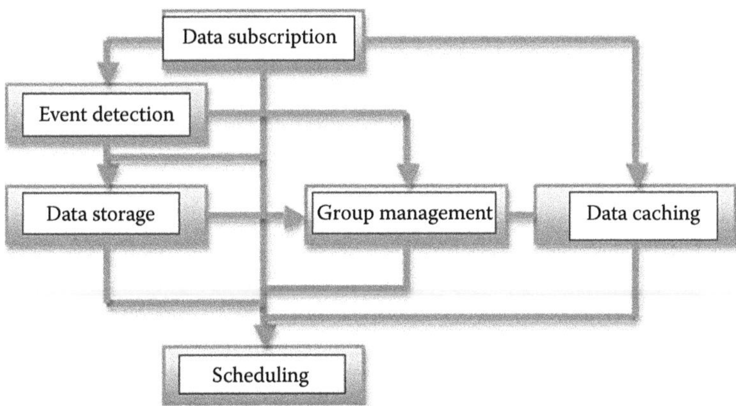

Figure 7.9 DSWare components. (Adapted from Li, S. et al., Event detection services using data service middleware in distributed sensor networks, *Proceedings of the Second International Workshop on Information Processing in Sensor Networks (IPSN '03)*, Palo Alto, CA, April 22–23, 2003, 502–517.)

7.6 WSN Runtime Support Example: Mate [PLevis02, Miaomiao08]

As we mentioned before, all defined middleware services should have some form of runtime support to ensure a well-defined execution environment.

The runtime support has the following basic functions: inter-process communication (IPC), memory control, and power management (in both voltage scaling and component deactivation). These functions are important because they can be used to implement higher-level middleware services, such as multi-thread processing, task scheduling, memory access synchronization, and spread signal spectrum management.

Typically, some kind of *virtual machine* is used to implement runtime support. We can implement a *virtual machine* as a *platform-specific* kernel on top of the embedded OS. *Mate* [PLevis02] is such an example. It is built on top of *TinyOS*. Figure 7.10 illustrates *Mate*'s architecture.

The core of the *Mate* architecture is a scheduler, which maintains a buffer with contexts and interleaves their execution. The *Mate* concurrency model is based on statically named resources, such as shared variables, which should be explicitly specified by any operation.

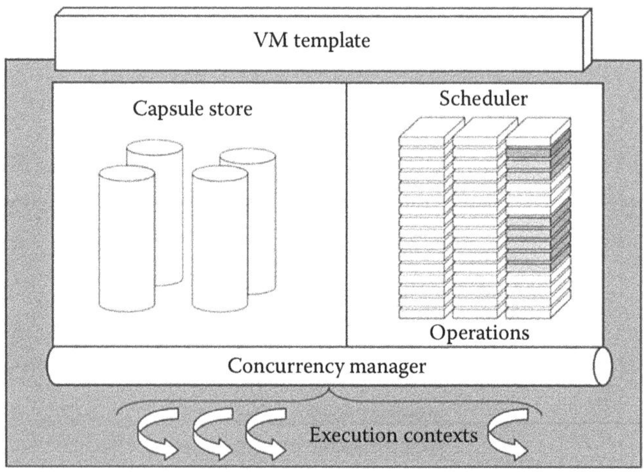

Figure 7.10 Architecture of Mate. (Adapted from Levis, P. and Culler, D., Mate: A tiny virtual machine for sensor networks, *Proceedings of the 10th International Conference on Architectural Support for Programming Languages and Operating Systems (ASPLOS-X)*, San Jose, CA, 2002, ACM Press, New York, 85–95; Wang, M. et al., *J. Comput. Sci. Technol.*, 23(3), 305, 2008.)

7.7 QoS Support Example: MiLAN [WBHeinzelman04, Miaomiao08]

QoS support is important for applications having the following requirements: fault tolerance, reliability, security, and real-time data processing. The following parameters can be used to express QoS in a WSN: packet delay, jitter and loss, throughput, and latency. However, we need more QoS metrics to make a quantitative performance measurement. For instance, we may define some new QoS parameters, including data accuracy, aggregation delay, aggregation degree, coverage, and precision. A WSN middleware that provides QoS support can efficiently support data acquisition.

MiLAN [WBHeinzelman04] has defined a set of QoS support, as shown in Figure 7.11. A WSN application program starts with the conveyance of a set of QoS parameters from the application layer to *MiLAN* (i.e., the middleware layer). Such a QoS conveyance is achieved through a *state-based variable equirements graph* and a *sensor QoS graph*.

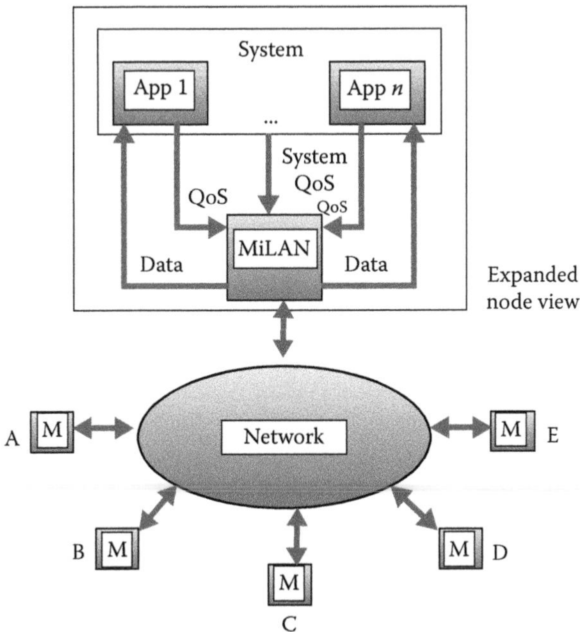

Figure 7.11 QoS support in MiLAN. (Adapted from Heinzelman, W.B. et al., *IEEE Netw.*, 18(1), 6, 2004.)

1. *State-based variable requirements graph*: It specifies the application's minimum acceptable QoS for each performance parameter based on the current running state of a system.
2. *Sensor QoS graph*: It is used to determine which sets of nodes a WSN may support from the QoS requirements viewpoint.

Problems and Exercises

7.1 Explain the roles of a WSN middleware in WSN system design. Point out its components and explain each of them briefly.

7.2 What challenges do we encounter when designing a WSN middleware?

7.3 Why should we use programming abstractions?

7.4 How does *Agilla* implement code management?

7.5 Why does *Mires* use a *publish/subscribe* architecture?

7.6 How does *DSWare* handle data storage?

7.7 Explain in detail the QoS support in MiLAN.

Chapter 8

Sensor Data Management

After sensor data is collected, it needs further processing (such as noise removal) to detect some events. In this chapter, we cover some typical sensor data management issues including sensor data processing (such as data cleaning), sensor database structure and data query strategies, data aggregation, and other issues.

8.1 Sensor Data Cleaning [Elnahrawy 2003]

8.1.1 Background

Sensors differ significantly in precision, accuracy, and their tolerance to hardware and external noises. For example, photovoltaic sensors have large noise distribution [BYCHKOVSKIY03]. The operating environment of sensors also affects the performance of data acquisition. Sensor reading can also be affected by other external and uncontrollable factors, which, in many cases, cause inaccurate measurements. For example, the weight of trucks can be measured by the strain gauge sensors attached to bridges, which can be affected by other objects' vibrations.

Many industries are developing inexpensive sensors that can be distributed everywhere and can be disposed when they run out of batteries. The high sensitivity to internal and external noises, imprecision, and inaccuracies is expected in those inexpensive sensors.

Sensor data has several sources of errors when it comes from the measurements of physical or modeled phenomena. Errors can be classified into two main types: *systematic* (*bias*) or *random* (*noise*). *Systematic errors* come from change of operating conditions such as temperature, humidity, or aging of the sensors, which can be corrected by calibration [Bychkovskiy03]. *Random noise* may come from the

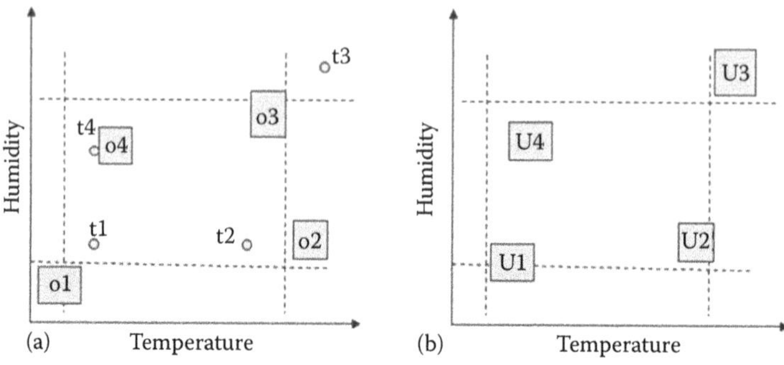

Figure 8.1 (a) Based on the observed readings items 1 and 4 will be thrown away and (b) based on the uncertainty regions; only item 3 will be thrown away. (Adapted from Elnahrawy, E. and Nath, B., Cleaning and querying noisy sensors, *Proceedings of the Second ACM International Conference on Wireless Sensor Networks and Applications*, San Diego, CA, September 19, 2003.)

following sources: (1) random hardware noise, (2) measurement inaccuracy, (3) environmental effect and noise, and (4) imprecision in computing a derived value from measurements (i.e., inconsistency in measuring the same phenomenon under the same conditions). Elnahrawy and Nath (2003) have discussed the reduction of random noise part, as well as their serious effect on sensor data [Elnahrawy2003].

The random errors given by cheap sensors strongly affect the sensor data accuracy. They may give imprecise or even misleading answers, which may yield high loss in immediate critical decisions or the activation of actuators. Therefore, the sensor data errors cannot be neglected.

To support this argument, we use an example from [Elnahrawy 2003]. Bacteria growth is monitored over the time by inexpensive wireless temperature and humidity sensors. If the temperature and humidity go beyond some given thresholds, the item should be discarded. As shown in Figure 8.1, based on the sensor data acquired, items 1 and 2 should be discarded and items 3 and 4 should be kept. However, based on the true conditions, items 1 and 2 should be kept and item 3 should be discarded.

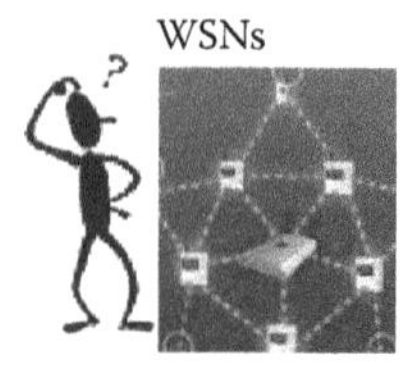

One may wonder why traditional databases do not need data cleaning. Data cleaning is not necessary in traditional data sources that are from either an explicit data-entry operation or a transaction activity with trustworthy steps. As a matter of fact, such data is typically used in banks, companies, or by personnel. Clean data models are assumed. Any noisy data is assumed to be cleaned *off-line* by separate database functionality.

However, the situation is different for WSN data. This data is typically continuously generated and thus forms a data stream. Moreover, we cannot use an off-line approach to handle this data as we typically need such data in a *real-time* way.

Although WSNs have some important performance metrics to consider, such as network bandwidth and energy consumption, data errors could have the same importance as those metrics because they cause uncertainty in determining the true reading (measurement) of the sensor.

Elnahrawy and Nath (2003) have introduced a good approach to cleaning and querying of noisy sensors [Elnahrawy2003]. Its scheme can reduce the uncertainty in sensor readings that arises due to random noise. Specifically, its scheme is based on a Bayesian approach for reducing the uncertainty in an online fashion. Such a scheme is called *Bayesian-based cleaning* (BayC). BayC can be operated in either individual sensors or in the base station. BayC cleans data in each sensor because it assumes that the reading of each individual sensor is important.

BayC makes some assumptions as follows: A set of n sensors, $S = \{s_i\}$, $i = 1,\ldots n$, are deployed in the space and form a WSN. Some networking techniques such as routing, topology maintenance, and communication are already implemented. We also think of each sensor s_i at a specific time instance t as a tuple in the sensor database. The database has some attributes that correspond to the sensor readings. Each sensor has one or more readings corresponding to each measurement. The same sensor can sense different phenomena. It is interesting to see that in the same location we may have some specialized sensors. We can combine their values, which is just like the data from one "virtual" multi-attribute sensor.

All the attributes are assumed to be real valued. We can extend BayC framework to the case of discrete-valued attributes. Note that we always specify the time stamp when each piece of sensed data is collected. Later on, we just drop the time index t when we talk about sensor readings.

8.1.2 General Model

As shown in Figure 8.2, the overall BayC framework is composed of two major modules:

1. The first one is called *cleaning module*, which is used to clean the noisy sensor data online, through computing uncertainty models of data that are unable to predict. This cleaning module has three inputs: (a) noisy sensor data reported from sensors, (b) noise character metadata of every sensor (error model), and (c) true reading distribution at each sensor (prior knowledge). The following is a brief discussion of the latter two inputs. Cleaning module's output is probability models of the reading of "unknown" sensors, that is, a probability density function. We will provide the details of the computation of this model in the next section.

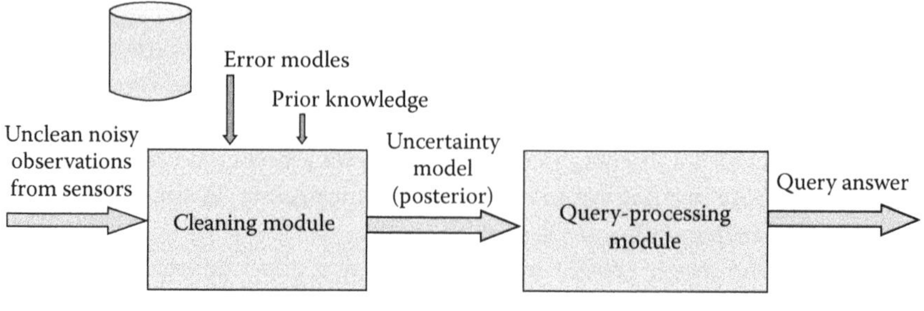

Figure 8.2 The overall framework of sensor data cleaning. (Adapted from Elnahrawy, E. and Nath, B., Cleaning and querying noisy sensors, *Proceedings of the Second ACM International Conference on Wireless Sensor Networks and Applications*, San Diego, CA, September 19, 2003.)

2. The *query-processing* module can produce answers to any posted query to the system based on the current readings' uncertainty model. A traditional query, which assumes a single value for every reading, cannot be used here because the uncertainty models are probabilistic (i.e., statistical distribution). The new query-processing module is based on the algorithm that uses statistical approaches to compute a function over random variables.

Sensors' error model should reflect the noise distribution. We assume that it follows a Gaussian distribution with mean value zero. To determine the error model, we need to calculate the variance, which is based on the specification of each sensor (including accuracy, precision, etc.) and calibration test under normal deployment conditions. The calibration test can be done either by manufacturers or by users after installation and before usage. We also need to consider the environmental factors or characteristics of the field. Error models may change with time and be replaced by new models. They should be stored as metadata in the cleaning module. As sensors' noise characteristics are not homogeneous, each sensor type or even individual sensor should be calibrated to get its own error model.

Prior knowledge is the distribution of the true sensors' readings. Several methods are available to obtain prior knowledge. We can compute it based on the facts about the sensed phenomenon, which is learned over history, or from less-noisy readings, or even from expert knowledge and subjective conjectures. It can be computed dynamically at the same time if the sensed phenomenon is known to follow a specific parametric model. For instance, if the temperature of perishable items is known to drop by a factor of x percent from time $t - 1$ to time t, then the (cleaned) reading at time $t - 1$ is used to obtain the prior distribution at time t. The resultant along with the error model and the observed noisy readings at time t are the input to the cleaning module to obtain the uncertainty model of the sensor at time t. The approach used in this case is similar to Kalman filters [LEWIS86].

The straightforward approach to model uncertainty in noisy sensor readings is to assume that the reading of each sensor's distribution is the Gaussian model that centers around the observed readings and has a variance equal to the noise variance. It is common knowledge that the use of prior knowledge gives more accurate estimations [KAY93]. Thus, our motivation here is to use the prior knowledge to reduce the noisy sensors' uncertainty. Prior knowledge with less variance is more useful than others, as it reduces the uncertainty and enhances the overall accuracy.

If the prior knowledge is not strong enough to give a narrow distribution (compared with the noise distribution), BayC scheme will still be superior. Fortunately, in most cases, a strong prior knowledge can be easily computed even though the noise may have a wide distribution (i.e., in situations where we have noisy sensors scattered in a large area to collect data from a well-modeled phenomenon such as temperature).

Data cleaning and query processing can occur at the sensor level or the database level (or base station). Each option has certain communication and processing costs (i.e., energy consumption, memory storage cost). The decision of which approach to use is made on the sensor capabilities and applications.

Sensor level: When the data cleaning is performed at the sensor level, we need a certain memory space to store the prior knowledge and the error models at the resource-constrained sensors. Furthermore, there could be a significant communication cost to send the prior knowledge from the base station to the sensors.

Database level: We could assume that any processing or storage at the database level has no cost because the base station has enough computing capability to handle database operations. This is also the major advantage of performing the cleaning and the query processing in the base station. Furthermore, we could save communication cost because it is not necessary to send dynamic priors to the sensors in this case.

8.1.3 Reducing the Uncertainty

This section illustrates the method of reducing uncertainty associated with noisy sensor reading, that is, computing more accurate uncertainty models of each sensor. The proposed approach in [Elnahrawy2003] is to use online data cleaning. It combines the prior knowledge of the true reading, the error model of the sensor, and its observed noisy reading together in one step. This step is performed using Bayes' theorem. The likelihood is the probability that data x would fall into a given value of the parameter (θ), and is denoted by $p(x|q)$, which gives the posterior probability density function of θ, $p(\theta|x)$.

$$p(x \mid \theta) = \frac{\text{likelihood} \times \text{prior}}{\text{evidence}} = \frac{p(x \mid \theta)\, p(\theta)}{\int_y p(x \mid y)\, p(y)\, dy} \tag{8.1}$$

Assume that a sensor has only one attribute. The attribute o is noisy, that is, it will be higher or lower than the true value, t. As discussed, the true value t follows a Gaussian distribution with mean value $\mu = t$ and with variance δ^2, that is, $p(o|t) \sim N(t, \delta^2)$. Then the Bayes' theorem is applied to obtain more accurate uncertainty model (posterior probability density function) for t, $p(t|o)$. Finally, the observed value o is combined with error model $\sim N(0, \delta^2)$, and the prior knowledge of the true reading distribution $p(t)$ is as follows:

$$p(t \mid o) = \frac{p(o \mid t)\, p(t)}{p(o)} \tag{8.2}$$

The following two equations are valid when the reading of some specific sensor "s" is known to follow a Gaussian distribution with mean μ_s and standard deviation σ_s, that is, $t \sim N(\mu_s, \sigma_s^2)$ (prior). Generally, there is no restriction of the prior distribution of true reading t, to a specified distribution. However, it is tempting to use Gaussian distribution as it has attractive properties that are convenient for modeling priors.

$$\mu_t = \frac{\delta^2}{\sigma_s^2 + \delta^2} \mu_s + \frac{\sigma_s^2}{\sigma_s^2 + \delta^2} O \tag{8.3}$$

$$\sigma_t^2 = \frac{\sigma_s^2 \times \delta^2}{\sigma_s^2 + \delta^2} \tag{8.4}$$

Case study

This example [Elnahrawy2003] shows how to obtain the uncertainty model of a temperature sensor at a specific time instance. Assume that our prior knowledge is that the temperature r follows a Gaussian distribution, and the temperature is 9 centigrade degrees (in average) with standard deviation of 4, that is, $r \sim N(\mu_s = 9, \sigma_s^2 = 4^2)$. We further assume that the noise at this sensor is known to have a standard deviation of 10, that is, noise $\sim N(0, \delta^2 = 10^2)$. If the reported noisy temperature is 15, using Equations 8.3 and 8.4, we obtain a mean ≈ 9.8 and a standard deviation ≈ 3.7 of the posterior distribution for the true unknown temperature, $p(t|o) \sim N(9.8, 3.7^2)$. This is shown in Figure 8.3.

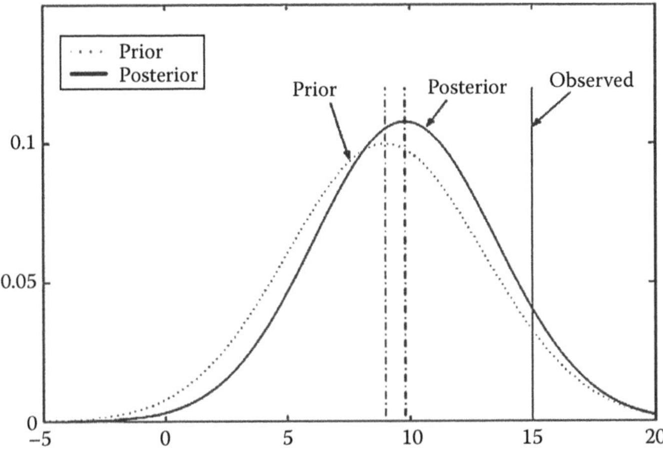

Figure 8.3 The resultant uncertainty model of the true temperature (posterior) and the observed erroneous reading. (Adapted from Elnahrawy, E. and Nath, B., Cleaning and querying noisy sensors, *Proceedings of the Second ACM International Conference on Wireless Sensor Networks and Applications,* San Diego, CA, September 19, 2003.)

Obviously, due to the use of prior knowledge, BayC can greatly reduce the uncertainty. Moreover, when the variance of the prior knowledge becomes very small as compared to the variance of the noise, that is, when the prior data becomes very strong, the error of the posterior becomes smaller and the uncertainty is further reduced. Consequently, BayC's uncertainty model becomes far more accurate than the no-prior case. Equation 8.3 also illustrates an interesting fact: BayC approach makes a good balance between the prior knowledge and the observed noisy data. When the sensor becomes less noisy, its observed reading becomes more important and the model depends more on it. At very high noise levels, the observed reading could be totally ignored.

8.2 TinyDB: An Acquisitional Query-Processing System for Sensor Networks [SRM05]

TinyDB aims to address a few questions on query processing of sensor networks:

1. What is the proper time of sending out a particular sensor data query?
2. Which sensor nodes have data related to the issued query?

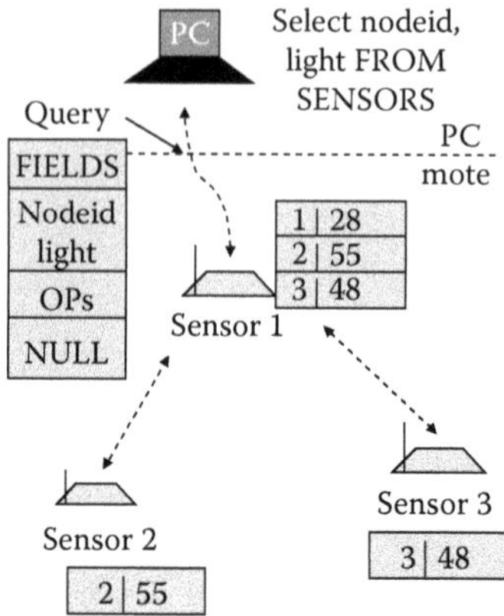

Figure 8.4 A query and results propagating through the network. (Adapted from Madden, S.R. et al., *ACM Trans. Database Syst.*, 30(1), 122, March 2005.)

3. What is the right order of taking sensor data samples, and how to interleave sampling with other WSN operations?
4. Is it worth wasting computational power or bandwidth to process and relay a particular data sample?

Among these issues, (1) is the only one that is acquisitional. The other three questions can be answered by modifying methods used to solve traditional query processing. Questions (2) and (3) can be solved with notions of indexing and optimization, and question (4) bears some similarity to issues that arise in stream processing and approximate query answering.

Figure 8.4 shows the basic principle of data query: After the query is parsed and optimized, it is sent into the sensor network, and then is disseminated and processed. The results are sent back to the base station in the routing tree that is initialized at the time when queries start their propagation.

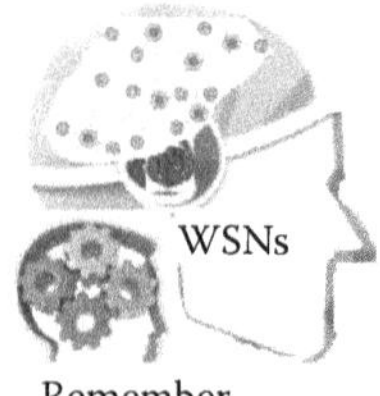

Remember a few important facts on WSN data query. (1) It is built on a certain routing topology, for instance, a hierarchical tree structure with the root as the base station. Such a tree structure allows the efficient propagation of query commands and quick data finding. (2) Although TinyDB data query commands have similar syntax to Microsoft SQL, its internal implementation (i.e., how we send out query commands to sensors and how sensors feedback the required query results) is very challenging. (3) Data query belongs to application layer issues. But it needs help from routing layer.

8.2.1 Data Model

In TinyDB, the sensor data is stored in a table with the following structure: every row is a sensor's data in each time instant, and each column is one attribute (e.g., light and temperature) produced by the sensor.

8.2.2 Basic Language Features

Data queries in TinyDB should be quick to satisfy real-time response requirements. A query command consists of a SELECT-FROM-WHERE-GROUPBY clause to support selection, joins, projection, and aggregation. Its syntax is similar to Microsoft SQL. The FROM clause points out the sensor table or stored tables, which are called *materialization points.* Materialization points are created through special logging queries described below, which provide basic support for subqueries and windowed stream operations.

To define a query, we must first define the sample intervals, which are an important parameter of the query. The time period between the start of each sample period is known as an epoch, which provides the method to structure computation and minimize power consumption. Consider the following query:

```
SELECT nodeID, light, temp
FROM sensors
SAMPLE PERIOD 5 s FOR 15 s
```

This query can be used to report each sensor's identifier (ID), light, and temperature readings once per second for a total duration of 10 s. The query's result is sent via multi-hop topology to the root of the network tree (i.e., the base station), where it can be logged or output to the user. The query result consists of a stream of tuples, clustered into 1 s time intervals. Each tuple has a time stamp to indicate the time it was produced.

At the beginning of each epoch, the data collection is initiated by sensors, as specified in the SAMPLE PERIOD clause. There is a simple time synchronization protocol in TinyDB to ensure that all sensors have correct time stamps.

Note that each query has an ID. This ID can be used to explicitly stop a query via a "STOP QUERY ID" command. Other ways to stop a query include setting a specific time period via a FOR clause, which is shown above (when the time period ends, the query ends), or using a stop condition.

The concept of *windows* is defined in TinyDB based on the materialization points over the sensor streams. A small buffer of data accumulated in these materialization points can be used in other queries. Consider the following example:

```
CREATE
STORAGE POINT recentlight SIZE 8
AS (SELECT nodeid, light FROM sensors
SAMPLE PERIOD 20 s)
```

This statement shows the use of a local (i.e., single-node) location to store a streaming view of recent data.

8.2.3 Event-Based Queries

As an alternative to the continuous- and polling-based mechanism for data acquisition, TinyDB provides *event-based* data collection. TinyDB events are generated explicitly either by another query or by the operating system (in which case the code that generates the event must have been compiled into the sensor node).

Consider the following query:

```
ON EVENT animal-detect(loc):
SELECT AVG(light), AVG(temp), event.loc
FROM sensors AS s
WHERE dist(s.loc, event.loc) < 10m
SAMPLE PERIOD 2 s FOR 30 s
```

This query could be used to extract the average light and temperature data at sensors near a bird's nest where a bird has just been detected. Each time a bird-detect event occurs, the detecting sensor issues the query to collect the average light and temperature from nearby sensors every 2 s for a total duration of 30 s.

Such events allow the system to be dormant until some external conditions occur, instead of continually polling or waiting for some data to arrive. As many CPUs (i.e., microprocessors) include hardware interrupt lines than can wake up a sleeping sensor to begin data processing, events-triggered query can provide significant reductions in power consumption.

8.2.4 Other Queries Defined in TinyDB

We will discuss more queries defined in TinyDB. Those queries are suitable for different special conditions. For example

Network health queries: They are meta-queries over the network itself. For example, we may query parents and neighbors in the network tree, or find nodes with battery life less than some specified threshold. The following *network health query* reports all sensors whose current battery voltage is less than k:

```
SELECT nodeid,voltage
WHERE voltage < k
FROM sensors
SAMPLE PERIOD 10 minutes
```

Actuation queries: In this type of query, users will take actions in response to a data query. `OUTPUT ACTION` clause could be used for this purpose. For example, in a building, a fan could be turned on when temperature is higher than a threshold:

```
SELECT nodeid,temp
FROM sensors
WHERE temp > threshold
OUTPUT ACTION power-on(nodeid)
SAMPLE PERIOD 10 s
```

Note that this OUTPUT ACTION clause tells that a sensor control command should be invoked in response to results satisfying the query. In this case, the power-on command is triggered to pull an output pin on the microprocessor "high," which closes a relay circuit and gives power to sensor. We could also use OUTPUT ACTION to power-off the fan when the temperature falls below a threshold.

Off-line delivery: Sometimes users want to log some phenomenon that happens so fast that we cannot obtain the data in a real-time way. In this case, TinyDB supports the logging of results to EEPROM for off-line, non-real-time delivery.

8.2.5 Power-Based Query Optimization

The above sections introduced the data query syntax. This section discusses some internal implementation (in a sensor) of those queries. This section is primarily focused on optimization for data acquisition, selection, and aggregation.

TinyDB parses queries at the base station in a simple binary format, and then disseminates into the sensor network that will instantiate and execute them. A simple query optimization is performed before the queries are disseminated. The optimization procedure chooses the order of sampling, selections, and joins.

A cost-based optimizer is adapted to choose the query plan with the lowest overall power consumption. This optimization allows us to consider some issues including CPU-processing cost and radio communication, both of which contribute to power consumption.

To start, we look at the metadata types stored in the optimizer. Each sensor in TinyDB maintains a catalog of metadata that describes its local attributes, events, and user-defined functions. Through routing protocols, such metadata is periodically sent to the base station for use by the query optimizer.

Good idea

What is metadata? Let us take a database (say, Microsoft Access) as an example. MS Access uses a table to store raw data. However, the database operations (such as indexing and sorting) should be controlled by a series of commands. Where do we store such commands? Using metadata. Therefore, metadata is "data of data," that is, extracting/building a small amount of control information from large amounts of raw data.

Such metadata can be maintained in the nodes via static linking done at compile time using the TinyOS C-like programming language. If any *event* or *attribute* is needed in query metadata, we can declare it in an interface file and then use a handler function to refer to such an interface file. For example, to make sensor network topology available to the query processor, the TinyOS program components adopt an attribute called "parent node." Such an attribute can be accessed by using a handler that returns the ID of the node's parent in the query "tree."

Event-based metadata has the following structure: a name, a signature, and a frequency estimate that is used in query optimization. For any user-defined predicates, we also use a name and a signature, along with a selectivity estimate that is provided by the author of the function.

Attribute-based metadata in TinyDB is shown in Table 8.1. Attribute-based metadata typically includes the power cost, time to fetch data, and the range of an attribute.

Some examples of power and sample time values are shown in Table 8.2. We can see that the power consumption and sample time can differ across sensors by several orders of magnitude.

Here we use a simple case to show the use of metadata. Suppose we monitor the microclimates created by plants and their biological processes [DELIN00]. Table 8.2 shows the big difference between the order of magnitude in per-sample costs for the accelerometer and the magnetometer. This also means that the power costs of data query plans with different sampling and selection orders could vary substantially. For instance, the following three query strategies could lead to very different power costs: (1) the magnetometer and accelerometer are sampled before either selection is applied, (2) the magnetometer is sampled and the selection over its readings is applied before the accelerometer is sampled, (3) the accelerometer is sampled first and its selection (S_{accel}) is applied before the magnetometer is sampled.

Table 8.1 Metadata Fields Kept with Each Attribute

Metadata	*Description*
Power	Cost to sample this attribute (in J)
Sample time	Time to sample this attribute (in s)
Constant	Is this attribute constant-valued (e.g., id)?
Rate of change	How fast the attribute changes (units/s)
Range	Dynamic range of attribute values (pair of units)

Source: Adapted from Madden, S.R. et al., *ACM Trans. Database Syst.*, 30(1), 122, March 2005.

Table 8.2 Summary of Power Requirements of Various Sensors

Sensor Examples	*Time Per Sample (ms)*	*Start-Up Time (ms)*	*Current (mA)*	*Energy Per Sample (mJ)*
Weather board sensors				
Solar radiation [TAOS, Inc. 2002]	500	800	0.350	0.525
Barometric pressure [INTERSEMA2002]	35	35	0.025	0.003
Humidity [SENSIRION02]	333	11	0.500	0.500
Surface temp. [MELEXIS02]	0.333	2	5.6	0.0056
Ambient temp. [MELEXIS02]	0.333	2	5.6	0.0056
Standard mica mote sensors				
Accelerometer	0.9	17	0.6	0.0048
(Passive) Thermistor	0.9	0	0.033	0.00009
Magnetometer [Honeywell08]	0.9	17	5	0.2595
Other sensors				
Organic by-products	0.9	>1000	5	>5

Source: Adapted from Madden, S.R. et al., *ACM Trans. Database Syst.*, 30(1), 122, March 2005.

8.2.6 Summary of TinyDB Strategies

To conclude, Table 8.3 lists the important techniques used in TinyDB.

8.3 Data Aggregation: AIDA [Tian04]

Without using data aggregation scheme, sensors are mutually independent, and each of them sends its collected data to the end nodes (sink). Such a strategy does not utilize an important feature in WSNs, that is, sensors in a neighborhood typically have redundant data because they can detect the same event in

Table 8.3 Summary of Acquisitional Query-Processing Techniques in TinyDB

Technique	*Summary*
Event-based queries	Avoid polling overhead
Lifetime queries	Satisfy user-specified longevity constraints
Interleaving acquisition/predicates	Avoid unnecessary sampling costs in selection queries
Exemplary aggregate pushdown	Avoid unnecessary sampling costs in aggregate queries
Event batching	Avoid execution costs when a number of event queries fire
Semantic Routing Trees (SRT)	Avoid query dissemination costs or the inclusion of unneeded nodes in queries with predicates over constant attributes
Communication scheduling	Disable node's processors and radios during times of inactivity
Data prioritization	Choose most important samples to deliver according to a user-specified prioritization function
Snooping	Avoid unnecessary transmissions during aggregate queries
Rate adaptation	Intentionally drop tuples to avoid saturating the radio channel, allowing most important tuples to be delivered

Source: Adapted from Madden, S.R. et al., *ACM Trans. Database Syst.*, 30(1), 122, March 2005.

the same area. It is not necessary to send out data by each sensor individually. Such a scheme can also waste much energy due to redundant, long-distance data transmissions.

In-network data aggregation can be used to overcome these shortcomings. As shown in Figure 8.5, each sensor transmits data only to its neighbors. An algorithm is used to select proper sensor to be the data aggregator. The aggregator removes all redundant data. Or the aggregator could generate a new value (such as *average* value) based on the received inputs from nearby sensors. Thus eventually the network traffic is greatly reduced compared to no data aggregation case.

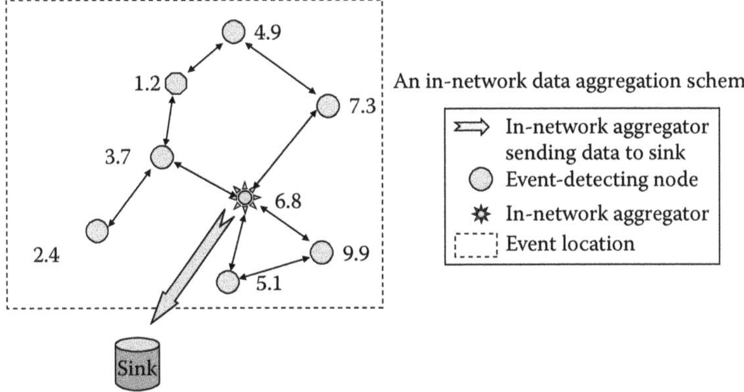

Figure 8.5 In-network data aggregation scheme. (Adapted from Vaidyanathan, K. et al., Data aggregation techniques in sensor networks, Technical Report OSU-CISRC-11/04-TR60, Department of Computer Science and Engineering, The Ohio State University, Columbus, OH.)

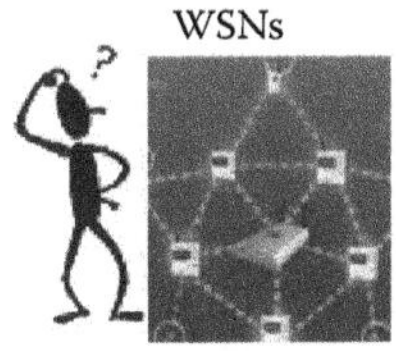

Data aggregation is an important concept in WSNs due to the strong redundancy and correlation among sensors' data. However, data aggregation concept is different from "data fusion." Data aggregation typically uses routing layer protocols (such as network topology discovery) to perform a high-level data analysis. Data fusion typically operates in physical layer, that is, from the signal-processing viewpoint. For instance, how do we deduce a new signal from two spatially correlated time series? This book will not cover data fusion because it belongs to the signal-processing issue.

Grid-based data aggregation [Karthikeyan] is suitable to many sensor network applications such as military surveillance and weather forecasting. In such a scheme the sensor network environment is divided into pre-determined grids, each of which has a grid center (i.e., the data aggregator) that observes and reports data to the sink nodes.

AIDA [Tian04] is an adaptive application-independent data aggregation in WSNs. AIDA's functionality includes two components. One is the *functional unit* that aggregates and deaggregates network packets (units). The other is the AIDA *aggregation control unit*, which is used to adaptively control timer settings and fine tune the desired degree of aggregation.

AIDA protocol [Tian04] works as follows:

For outgoing traffic (i.e., data to be sent out): It puts packets from the network layer into an aggregation pool. Based on the number of packets to be concatenated (in one aggregate session) and also based on the next-hop destinations of those packets, AIDA's aggregation function unit finishes an aggregate and passes it down to the MAC layer for next-hop transmission.

AIDA aggregation control unit determines the number of packets to aggregate and the time to invoke such aggregation. The control unit is a feedback-based, adaptive component. It can make online decisions based on local current network conditions.

For incoming traffic (i.e., data to be received): When data is received at the MAC layer, it is passed up to the AIDA function unit. Within the AIDA function unit, the received aggregates can be refragmented into their original network units. Then each piece of the fragments is passed up to the network layer for next-step routing.

One may argue that it may not be a good idea to perform aggregation if many aggregates are to arrive at the same ultimate destination as we could waste network resources if we deaggregate and reaggregate at every intermediate node. However, to ensure the modularity of layers, and also to allow the networking component to determinate routes independently, AIDA still uses data aggregation. This is because that the aggregation of multiple network data units into a single AIDA aggregate could reduce the MAC layer overhead of channel contention (by using wait/backoff operations in MAC layer) and the communication overhead of control packets (such as RTS/CTS/ACK in 802.11, ACK in regular reliable MAC). By using data aggregation, these costs are incurred only once per aggregate.

Although data aggregation has significant benefits, it is a challenging issue to design an adaptive AIDA control unit that can set up appropriate aggregation timing and parameters online. For instance, what is the good time to perform data aggregation—periodically or wait for enough data arrivals? What data compression scheme should be used? How do we kick out "strange data" (deviating from other data) before we perform aggregation?

An interesting issue is as follows: where (in network layers) should we implement aggregation? To avoid the change of traditional network layers, AIDA uses a *delegation* approach to intercept all function calls and to establish direct communications between the MAC and routing layer (also called network layer). Through *delegation* approach, AIDA data aggregation layer becomes one of the interfaces between the MAC and routing layer.

He et al. [Tian04] have designed different styles of AIDA. Those styles include the fixed, on-demand, and dynamic feedback schemes. The aggregation decisions in those schemes could be based on static thresholds or achieve an ultimate solution that incorporates a dynamic, online feedback control mechanism. In the following discussion, a baseline without aggregation is also mentioned for comparison purpose.

1. *No aggregation*: When no aggregation is used (i.e., the baseline scheme), we simply use the traditional network stack that does not use direct communications between routing layer and the MAC protocol.
2. *Fixed scheme*: Here "fixed" means that AIDA aggregates a fixed number of network data units into one AIDA payload. After a routing layer finishes the data aggregation for those fixed number of data units, it passes the AIDA payload down to the MAC layer for transmission. To ensure that AIDA does not wait for an indefinite amount of time before finishing the collection of fixed number of data units, a sender sets up a timeout value (say, T_{fixed}). If timer expires, it will not wait for more data units and just begin to aggregate what it collects.
3. *On-demand scheme*: In this scheme, data aggregation is just an optional operation. This is because that a sensor will try to be "always busy." That is, a sensor will not waste a long time to collect enough data units (for aggregation). Instead, when the MAC layer is available for transmission, no matter how many data units it has collected so far, it just finishes aggregation and immediately sends out the data. AIDA-layer data aggregation only takes place when a sensor really has nothing to send and gets "stuck" there. (For instance, the outbound message queue has built up or the radio medium is busy preventing the MAC layer from accessing the channel). This on-demand scheme avoids *message delay*.
4. *Dynamic feedback scheme*: This is the ultimate solution based on a combination of on-demand and fixed aggregation. The scheme works by monitoring two parameters: (a) the AIDA output queue size: if the queue has space, it will aggregate more data units and (b) current queuing delay: if the delay is large, it will reduce the aggregation size (i.e., aggregating less data units). AIDA uses control theory to dynamically adjust the degree of aggregation to converge the MAC delay to a certain set point.

8.4 Sensor Data Storage: Tiered Storage ARchitecture (TSAR) [Peter05a]

The data generated by sensor networks must be processed and stored, as typical sensor applications require access to both live and past sensor data. Access to past data is required for applications such as sensor data mining to detect unusual patterns, analysis of historical trends, and off-line analysis of particular events. The essential design considerations of the storage system that stores past sensor data are the location of data storage, indexing or not, and the method for application to access data with energy-efficient manner and low latency.

People have proposed variable approaches on sensor data storage. The simplest scheme allows sensors to stream data or events to the base station for long-term archival storage [PBonnet01]. The data is indexed to ensure efficient access at a later time. The advantage of this approach is that the storage is centralized, and the access to the storage is efficient and inexpensive. The disadvantage is that writes to the storage are expensive and inefficient.

An alternative scheme is to allow each sensor to store its own data locally (e.g., using its flash memory), so all writes are local and efficient. Then a simple read request is handled by a particular sensor, and more complex read requests are handled by flooding to the network. This approach has distributed storage and inexpensive write, but reads are inefficient and expensive.

Some other sensor storage solutions exist between the two extremes discussed above. One of these solution is *geographic hash table* (*GHT*) *approach* [RATNASAMY01, RATNASAMY02], in which every data item has a key associated with it, and a distributed GHT maps keys to different sensors. To read the storage, in-network hash table is looked up and the nodes that store the data item are located. Thus flooding is not needed in this approach.

Compared with the flat and homogeneous architecture where every sensor node is energy-constrained, a new storage architecture named TSAR is proposed in [Peter05a]. TSAR organizes a WSN into a multi-tier architecture. It is a predictive storage architecture, and combines archival storage with caching and prediction. TSAR exploits the resource-rich sensor tiers for caching and prediction.

TSAR stores data on flash storage in each sensor. Sensors send metadata (concise identifying information) to a nearby proxy (a special sensor with network control capabilities). The metadata may be an order of magnitude smaller than the data itself, which helps to reduce the communication cost. Then the resource-rich proxies mutually interact to create a distributed index of the data stored in sensors. This index can be used by applications to query and read past data efficiently, for instance, data that matches a read request can be pinpointed by the index and then retrieved from the corresponding networks. The separation of data stored at the sensors and metadata stored at the proxies gives TSAR the ability to leverage the tethered proxy resources and reduce sensor energy consumption.

TSAR organizes a WSN into three tiers. The bottom one is untethered remote sensor nodes, these nodes are low-power sensors; the middle one is tethered, power-rich sensor proxies; and the upper tier has applications and user terminals.

The middle tier plays a crucial role. Its sensor proxies have significant computation, memory, and storage resources. In a typical WSN application, the proxy tier may comprise tethered base-station class nodes (e.g., crossbow stargate). Each of those nodes is equipped with multiple radios. For instance, it may have an 802.11 radio to communicate with a wireless mesh network, and another radio (e.g., 802.15.4) to connect to the low-tier sensor nodes.

The middle-tier proxies could use solar power cell to have longer lifetime. Each proxy can manage hundreds of lower-tier sensors in its vicinity. A typical WSN deployment will contain multiple geographically distributed proxies.

The upper tier has WSN applications that query the network through a query interface [MADDEN02a]. TSAR aims to design a storage system that exploits the relative abundance of resources at middle-tier proxies to mask the scarcity of resources in the bottom-tier sensors.

TSAR uses the following principles to design the sensor storage system for multi-tier networks:

Principle 1: Store locally, access globally: The cost is lower and the efficiency is better in the case of local storage (compared with networked storage), and this trend will continue for the next few years. To maximize the network lifetime, we could store data locally in a sensor's flash memory, which can save much more energy than exchanging storage message through expensive radio transmissions. TSAR uses an efficient information retrieval mechanism based on local storage.

Principle 2: Distinguish data from metadata: Metadata uses special data fields with pre-determined syntax. It uses identifiers such as location, time, or summarized data values. Metadata is accompanied with each data record to reduce the searching and retrieving time. Then it is indexed by proxies to provide efficient database lookups. TSAR system has a unified logical view for all data, and can exploit the idiosyncrasies of multi-tier networks to improve its performance and functionality.

Principle 3: Support data-centric query: In sensor applications, it is important to build interfaces that allow TSAR to locate data by value or attribute (i.e., location or time). Thus indexing metadata can lower lookup costs.

The key feature of TSAR system design is based on the principles listed above. It uses a distributed index at the proxies. In this system, sensor nodes write data that is composed of opaque data and application-specific metadata. Metadata can be searched on and compared by TSAR. One example is a camera-based sensing application. In this case, the metadata can include coordinates describing the field of view, average luminance, motion values, and other basic information such as time and sensor location. The size of metadata varies depending on the application, as it can be much smaller than the raw data extracted from image or acoustic data.

Not only do sensors store data locally, but sensors also report metadata summary to nearby proxies periodically. This summary carries the information such as the sensor ID, the time interval (t_1, t_2) over which the summary was generated, a handle identifying the corresponding data record (e.g., its location in flash memory), and a coarse-grain representation of the metadata associated with the record. The data representation in the summary depends on the application. One example is the temperature sensor. In this case, the summary consists of maximum and minimum temperature values observed in an interval.

The summary reported is used by the proxy to construct an index, which is global because it collects information from the entire system. The index provides a unified view of distributed data such that it can be queried by the application to access data stored at any sensor. In each query, the lookup is triggered in the distributed index, and the matching results are used to retrieve data from sensors.

The TSAR summarization makes sure no missing summaries (which include the value being searched for) would occur, or no *false negatives*. However, false positive is possible whereas a matching summary does not yield matching value in a remote sensor, which is a waste of network resources.

TSAR uses a novel index structure called the *interval skip graph* that combines interval trees [CORMEN01], an interval-based binary search tree, with skip graphs [ASPNES03], an ordered, distributed data structure for peer-to-peer systems [HARVEY03] to find all intervals containing a particular point for a range of values. It has two advantages, which are ideal for sensor networks. The first advantage is that it has only *O(logn)* search complexity to access the first matching interval, and has a constant complexity to access successive interval. The second advantage is that it uses interval indexing instead of value indexing, which makes it more convenient to index summaries over time. Interval indexing is also suitable for energy constrained nodes, as it saves the energy in transmitting summaries rather than transmitting all sensor data.

Interval skip graphs efficiently look up sensor nodes that contain data relevant to a query. When these queries are sent to the network, the sensors quickly locate the relevant data records in their local archive and respond back to a middle-tier proxy. To enable such lookups, each sensor maintains an archival store of sensor data.

Although it is straightforward to implement such an archival store in resource-rich devices (such as a laptop), sensors have serious resource constraint. Consequently, TSAR archiving subsystem has fully exploited sensor data characteristics. For instance, a distinct characteristic of sensor data is that sensors generate time-series data streams, which tells us that we could achieve data in temporal order.

As a matter of fact, many signal-processing schemes can utilize such a temporally ordered store to perform operations. An example is digital signal processing that has many time-series operations such as fast Fourier transform (FFT), wavelet transforms, clustering, similarity matching, and target detection.

As mentioned before, each raw data record has an associated *metadata field* that includes a time stamp, sensor settings, calibration parameters, etc. Raw sensor data is stored in the *data field* of the record. Note that such a data field is *opaque* and application-specific as the storage system is not aware of this field or does not care about such a field. For example, a video sensor may store binary images in this data field.

8.5 Multi-Resolution Data Processing [GANESAN03a]

In [GANESAN03a], an interesting concept is proposed, that is, extracting sensor data in a *multi-resolution* manner from a sensor network. Multi-resolution means that we could observe the data from different levels, for instance, a coarse level or a fine level. It allows users to look at *low-resolution* data from a larger region quickly and cheaply, before deciding to obtain high-resolution data, which are more detailed and potentially more expensive datasets. In some cases, it is sufficient to use compressed low-resolution sensor data for spatio-temporal querying to obtain statistical estimates of a large body of data [DAI04].

To reflect different resolution levels, a concept called data "dimensions" is proposed in [GANESAN03a]. We have noticed that there are correlations in sensor data along multiple axes such as temporal, spatial, and between multiple sensor modalities. These correlations can be exploited to reduce data dimensionality. Although we can exploit *temporal* correlation locally, the routing protocol needs to be tailored to *spatial* correlation between sensor nodes for maximum data reduction.

The temporal correlation exists in many applications. For example, in a video sensor network, a new captured image has strong correlations with previous image as most background pixels in those two images do not change much.

To obtain different resolutions (i.e., data dimensions), Ganesan et al. (2003) [GANESAN03a] use wavelet subband coding, a popular signal-processing technique for multi-resolution analysis and compression [CORMEN01, Shanmugasundaram04]. Wavelets have many advantages over other signal-processing techniques when viewing a *spatiotemporal* dataset. For instance, the data can be decomposed at multiple spatial and temporal scales. Important features in the data, such as abrupt changes at various scales, can be extracted to obtain good compression. When wavelet thresholding is applied to compression in typical time-series signals, we only need a few coefficients for reasonably accurate signal reconstruction.

Problems and Exercises

8.1 Multi-choice questions:

1. Due to which of the following reasons could the sensor data have noise?
   a. Hardware/circuit noise
   b. Operation environments
   c. Measurement inaccuracy
   d. All of the above
2. Data cleaning generally is not necessary in traditional databases because
   a. Noisy data, if any, is assumed to be cleaned *off-line* by separate database functionality.
   b. The source of data is either an explicit data-entry operation or a transaction activity.
   c. They occupy too much space and data cleaning is hard.
   d. Both a and b.
3. Which of the following is not true on *Bayesian-based cleaning* (BayC)?
   a. It consists of cleaning module and query-processing module.
   b. Traditional query, which assumes a single value for every reading, can be used in its query-processing module.
   c. Prior knowledge is the distribution of the true sensors' readings.
   d. Online cleaning combines the prior knowledge of the true reading, the error model of the sensor, and its observed noisy reading together in one step.

4. Which of following feature(s) does TinyDB have?
   a. Its data query command format is similar to SQL.
   b. Its event-based queries can avoid polling overhead.
   c. It cannot perform data aggregation query.
   d. Both a and b
5. Which of the following features do data aggregation schemes not have?
   a. It is typically used for in-network data reduction.
   b. AIDA is to separate AIDA functionality into two components. One is the functional unit that aggregates and deaggregates network packets (units). The other is the AIDA aggregation control unit, employed to adaptively control timer settings and fine tune the desired degree of aggregation.
   c. AIDA can dynamically adjust the degree of aggregation.
   d. AIDA is not transparent from other network protocol layers.

8.2 Explain the BayC general model and point out how it can clean data noise online.

8.3 How does BayC use prior knowledge to remove uncertainty?

8.4 Explain major features of TinyDB.

8.5 What is the relationship between data aggregation and in-network routing protocol?

ADVANCED TOPICS

Chapter 9

Sensor Localization

In this chapter, we discuss about wireless sensor network (WSN) localization schemes, that is, how the sensors can use message exchanges to find approximate locations of object sensors. First, we discuss about the basic knowledge of sensor localization based on the reference [Xiang04]. Then, we exemplify some typical localization algorithms.

Node localization is an important and interesting topic in many wireless networks. This chapter has introduced seven good sensor localization schemes that have elegant math models. To keep the original meanings of those algorithms, we have kept the original math notations and algorithm procedures in our citations.

9.1 Introduction [Xiang04]

We have addressed localization problems in many fields such as the autonomous robot and vehicle navigation for mobile robotics, virtual reality systems, and user location and tracking in cellular networks. However, the crucial issue of determining the locations of sensors for wireless sensor network operations is far-reaching.

Sensor networks typically form a layered network protocol stack. In the application layer, location-aware applications necessitate sensor localization. Sensor position information is often integral to the use of data collected by sensors. For example, to detect and track objects with sensor networks, the physical position of each sensor is needed for identifying the positions of detected objects. In the network layer, many communication protocols of sensor networks are built upon the knowledge of the geographic positions of sensors. For example, the knowledge of location information and transmission range enables geographic routing algorithms that propagate information through multi-hop sensor networks to operate.

In most cases, location information is unknown upon deployment, and there is no infrastructure available to locate them. It is, therefore, necessary to find some approaches to identify the location of each sensor in WSNs after their deployment.

One of the most popular technologies for localization is the global positioning system (GPS). Many applications have been developed based on GPS. It is quite possible to locate sensors if each is equipped with a GPS; however, this method is impractical for three main reasons. First, GPS is not always available due to the line-of-sight (LOS) conditions. For instance, it does not work indoors, under water, or in a subway. Second, a typical GPS receiver costs approximately $100 so far, so it is often too expensive to equip each sensor with a GPS receiver, considering that these sensors are usually designed to be low cost and disposable. Finally, GPS receivers consume large amounts of power (from a tiny sensor viewpoint).

Based on the previous discussion, alternative sensor localization systems are required. Considering the application scenarios of sensor networks, designing localization systems for sensor networks is more challenging than designing localization systems for applications in many other domains. Sensors are designed to be small and have low power. They are usually randomly and densely deployed within a large region. After being deployed, these sensors self-organize into a distributed ad hoc sensor network. The ideal sensor localization system is also required to have a low computation and a low power cost. The localization system should be able to tolerate ad hoc deployment without infrastructure support for localization, and should be able to perform self-localization. The localization system is expected to scale to include a large number of sensor nodes, and must accommodate a dynamic environment.

9.2 Elements of Localization [Xiang04]

Most localization methods first approximate distances or angles between unknown sensors and anchor sensors (those with known locations), then the location of unknown sensors are calculated with geometry algorithms. *Thus, the most important elements for sensor localization are distance measurement, angle measurement, and geometric constraints.* In the following section, we discuss available techniques for ascertaining each of the prerequisites.

9.2.1 Received Signal Strength Indication

An important characteristic of radio propagation is the attenuation of the radio signal as the distance between the transmitter and the receiver increases. The strength of the received radio signal decreases exponentially relative to the increase in distance. The receiver can measure this attenuation based on *received signal strength indication*

(*RSSI*). RSSI estimates the distance to the sender by measuring the power of the received signal. Based on the transmitted power, the propagation loss is calculated, and the loss can be translated into an estimated distance. This method has been used mainly for radio frequency (RF) signals. In [Rappaport96], radio propagation models are well researched, and they are used to predict the average RSSI at a given distance away from the transmitter. The ideal radio propagation model is

$$P_r(d) = \frac{P_\lambda G_t G_r \lambda^2}{4\pi^2 d^n L} \tag{9.1}$$

Equation 9.1 predicts the received signal power as a function of the distance between the transmitter and the receiver. In the ideal model, P_λ is the transmitted power, G_t is the antenna gains of the transmitter, G_r is the receiver, L is the system loss, and λ is the system wavelength. Usually G_t, G_r, and L can be taken out of the equation, as they are set to one. In [ASavvides01], they have shown the distance estimation with received RF signal strength using the wireless integrated network sensors (WINS) sensor nodes [WINS]. In the experiments, different configuration strategies, including different power levels in transmitters and the deployment strategies of sensors, are used to estimate the relation between the received signal strength and the distance between the transmitter and the receiver. The power of the received radio signal strength attenuates exponentially with the increase in distance as shown in Figure 9.1.

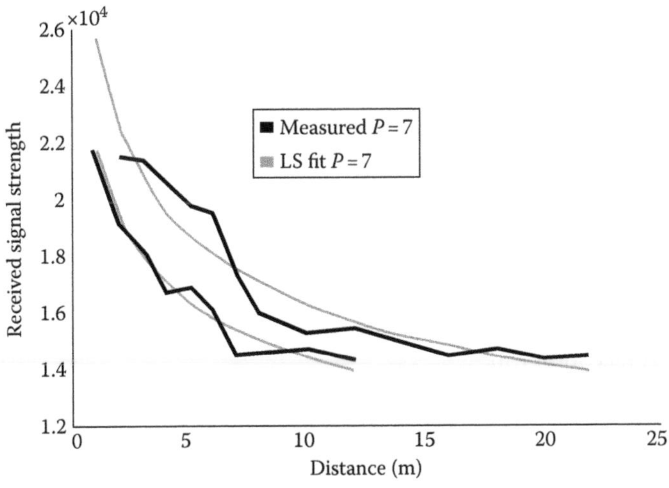

Figure 9.1 The power of the received radio signal strength attenuates exponentially with the increase of distance between the transmitter and the receiver. (From Ji, X., Localization algorithms for wireless sensor network systems, PhD thesis, Department of Computer Science and Engineering, The Pennsylvania State University, Philadelphia, PA, 2004.)

In theory, the power of a radio signal diminishes in relation to the square of the distance from the source of the signal. As a result, a node listening to a radio transmission should be able to use the strength of the received signal to calculate its distance from the transmitter. RSSI suggests a practical solution to the hardware-ranging problem: use the radios present in most sensor nodes to calculate ranges for localization [Jonathan08]. *In practice,* however, RSSI ranging measurements contain noise to the order of several meters. This noise occurs because radio propagation tends to be nonuniform in real environments. For instance, radio signals propagate differently over asphalt than they do over grass. Physical obstructions, such as walls or furniture, reflect and absorb radio waves. As a result, distance predictions using signal strength have been unable to demonstrate the precision obtained by other ranging methods such as time difference of arrival (TDoA).

A more cautious physical analysis of radio propagation and increased precision sensor radio calibration may allow for better use of RSSI data. Thus, it is likely that a more sophisticated use of RSSI could prove to be a superior ranging technology from a price/performance standpoint. Regardless, the necessary technology does not presently exist.

9.2.2 Time of Arrival

The distance between the transmitter and the receiver can be estimated based on the speed of the wave propagation and the measured time for a radio signal to travel between two sensor nodes. This method can be applied to several types of signals, such as RF, acoustic, infrared, and ultrasound. The implementation of this technique depends on the measurement of *time of arrival* (*ToA*). The ToA may be measured with advanced timing techniques. The GPS utilizes one such advanced timing technique for distance estimation [BHW97]. In GPS, each satellite (transmitter) transmits a unique code. The receiver replicates the code. The receiver gradually shifts its internal clock to correspond to the received code; this process is called lock on. Once a receiver has locked on to a satellite, the receiver determines the exact time of the reception of the radio signal from the satellite. Based on that time, the ToA can be determined by subtracting the known transmission time from the calculated reception time.

Although ToA offers a high level of accuracy, such precise measurements require sensor nodes to have relatively fast processing capabilities to resolve small timing discrepancies.

9.2.3 Time Difference of Arrival

The distance from the transmitter to the receiver can be measured by the *TDoA* of a variety of communication media at diverse speeds. For example, the measurement for *ToA* is based on two different modalities of communication, ultrasound and radio, in sensor nodes. The propagation speeds for ultrasound and radio are notably different. Due to this discrepancy, the radio signal is used for synchronization between the transmitter and the receiver, and the ultrasound signal is used to estimate the distance between them. The TDoA technique is used in projects of Active Bat [BWarneke01] and AHLoS [ASavvides01].

In TDoA schemes, each node is equipped with a speaker and a microphone. Some systems use ultrasound while others use audible frequencies. However, the general mathematical technique is independent of particular hardwares. In TDoA, the transmitter first sends a radio message (see Figure 9.2). It waits for a fixed interval of time, t_{delay} (which might be zero), and then generates a fixed pattern of "chirps" with its speaker. When listening nodes captured the radio signal, they record the current time, t_{radio}, and turn on their microphones. When their microphones detect the chirp pattern, they again record the current time, t_{sound}. Once they have t_{radio}, t_{sound}, and t_{delay}, the listeners can compute the distance *d* between themselves and the transmitter, given *the fact that radio waves travel substantially faster than sound waves in air.*

$$d = (s_{radio} - s_{sound}) * (t_{sound} - t_{radio} - t_{delay}) \tag{9.2}$$

TDoA methods are impressively accurate under LOS conditions; however, they perform best in areas that are free of echoes and in situations where the speakers and microphones are calibrated to each other. Several groups are working to compensate for these issues, which will likely lead to improved field accuracy.

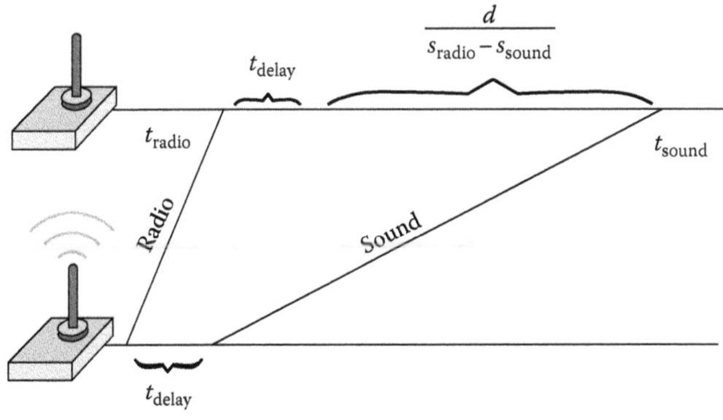

Figure 9.2 TDoA illustrated. (Adapted from Bachrach, J. and Taylor, C., Localization in sensor networks, Computer Science and Artificial Intelligence Laboratory, Massachusetts Institute of Technology, Cambridge, MA.)

9.2.4 Angle of Arrival

Angle of arrival (*AoA*) refers to the angle at which signals are received by the receiver from the transmitter. An AoA system is able to estimate the angle at which signals are received and to use simple geometric relationships to estimate the relative locations of the transmitter and the receiver. Angles of arrival may also be combined with distance estimates to derive relative locations.

The implementation of the AoA system relies on *smart antenna* with antenna arrays to measure the angle at which the signal arrives. A *smart antenna* is an array of antenna elements connected to a digital signal processor. Such a configuration will not only enable AoA estimation but also dramatically enhance the capacity of wireless links through the combination of diversity gain, array gain, and interference suppression. There are two major disadvantages of the AoA techniques, which make it inapplicable to sensor networks. First, the cost of the complex antenna array is high. Second, the AoA techniques will not scale well for systems with a large number of such nodes.

9.2.5 Triangulation

Triangulation is a geometric technique that uses the angles of arrival to determine the location of sensors. With the angle of each anchor sensor, with respect to the unknown sensor node in some reference frame, the locations of the unknown sensor node are calculated with the trigonometric laws of sine and cosine. The computation of triangulation is illustrated by Figure 9.3 [CSavarese02].

9.2.6 Trilateration

Trilateration is a geometric technique that uses distances between three anchor sensors and one unknown sensor to determine the location of the unknown sensor.

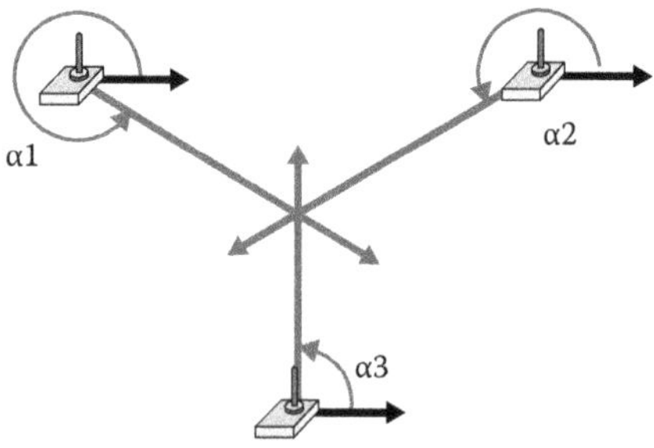

Figure 9.3 Triangulation. (From Savarese, C., Robust positioning algorithms for distributed ad hoc wireless sensor networks, Master's thesis, University of California at Berkeley, Berkeley, CA, 2002.)

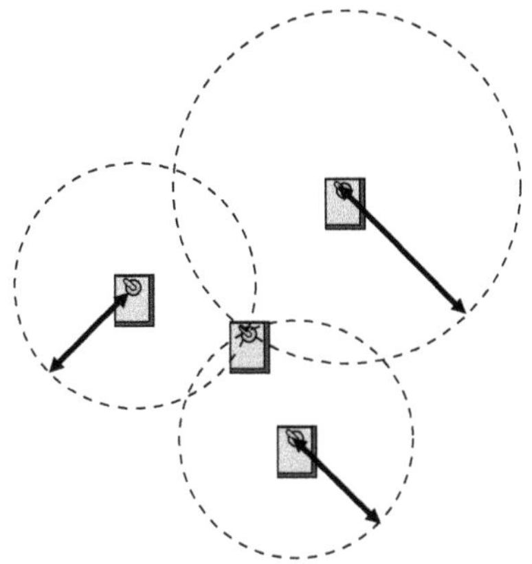

Figure 9.4 Trilateration. (From Ji, X., Localization algorithms for wireless sensor network systems, PhD thesis, Department of Computer Science and Engineering, The Pennsylvania State University, Philadelphia, PA, 2004.)

An unknown sensor is uniquely located when at least three reference points are associated with it in a two-dimensional (2D) space. The location of the unknown sensor is estimated by calculating the intersection of three circles. Figure 9.4 illustrates the computational geometric constraint [ASavvides01].

9.2.7 Multilateration

The location of an unknown sensor may also be estimated with multilateration utilizing its distances to *more than three anchor sensors*. In [JBeutel99], Beutel studied the multilateration with the least square algorithm.

Given n anchor sensors, in a three-dimensional (3D) space, and their distances to the unknown sensor, we have

$$\begin{bmatrix} d_1^2 \\ d_2^2 \\ \vdots \\ d_n^2 \end{bmatrix} = \begin{bmatrix} (x_1 - u_x)^2 + (y_1 - u_y)^2 + (z_1 - u_z)^2 \\ (x_2 - u_x)^2 + (y_2 - u_y)^2 + (z_2 - u_z)^2 \\ \vdots \\ (x_n - u_x)^2 + (y_n - u_y)^2 + (z_n - u_z)^2 \end{bmatrix} \tag{9.3}$$

where

d_i is the distance between the ith anchor sensor and the unknown sensor

(x_i, y_i, z_i) is the location of the ith anchor sensor in 3D space

(u_x, u_y, u_z) is the location of unknown sensor in 3D space

The equation can be converted into the following relations through linear operations:

$$Au = b \tag{9.4}$$

$$\begin{bmatrix} d_1^2 \\ d_2^2 \\ \vdots \\ d_n^2 \end{bmatrix} = \begin{bmatrix} (x_1 - u_x)^2 + (y_1 - u_y)^2 \\ (x_2 - u_x)^2 + (y_2 - u_y)^2 \\ \vdots \\ (x_n - u_x)^2 + (y_n - u_y)^2 \end{bmatrix} \tag{9.5}$$

$$u = \begin{bmatrix} u_x \\ u_y \\ u_z \end{bmatrix} \tag{9.6}$$

$$b = \begin{bmatrix} d_1^2 - d_n^2 - x_1^2 + x_n^2 - y_1^2 + y_n^2 - z_1^2 + z_n^2 \\ d_2^2 - d_n^2 - x_2^2 + x_n^2 - y_2^2 + y_n^2 - z_2^2 + z_n^2 \\ \vdots \\ d_{n-1}^2 - d_n^2 - x_{n-1}^2 + x_n^2 - y_{n-1}^2 + y_n^2 - z_{n-1}^2 + z_n^2 \end{bmatrix} \tag{9.7}$$

The u can be derived with [GGolub96]

$$u = (A'A)^{-1} * A'b \tag{9.8}$$

Figure 9.5 illustrates the computational geometric scenario [ASavvides01].

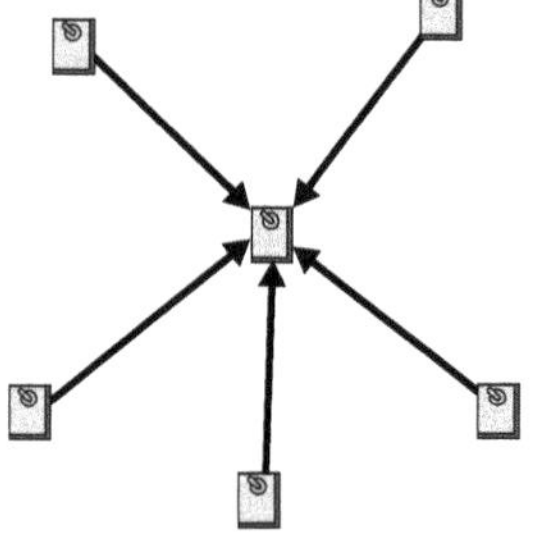

Figure 9.5 Multilateration. (Adapted from Savvides, A. et al., Dynamic fine-grained localization in ad-hoc networks of sensors, *Proceedings of the Seventh ACM International Conference on Mobile Computing and Networking (Mobicom)*, Rome, Italy, July 2001, ACM, New York, 166–179.)

9.3 Using Mobile Robots for Sensor Localization [Pubudu05]

9.3.1 Delay-Tolerant Sensor Networks

A delay-tolerant network (DTN) architecture was proposed in [Kavek04] for sensors deployed in mobile environments lacking an *always-on* infrastructure. These sensors are envisioned to monitor the environment over an extended time period. Rather than using packet switching, communication is based on an abstraction of *message switching*.

A concept called "bundles" could be defined to describe the moderate-length messages for noninteractive traffic. Such a concept is good for network management because it allows the network path selection and scheduling functions to know beforehand about the size and performance of requested data transfers.

Pathirana et al. [Pubudu05] propose a novel localization scheme for DTNs using RSSI measurements from each sensor device at a data gathering mobile robot. It uses one or more mobile robots to perform node localization in a DTN, eliminating the processing constraints of small devices. The mobility of the robots is exploited to reduce localization errors and the number of *static* reference location beacons.

An extended Kalman filter (Robust Extended Kalman Filter [REKF])–based [IPetersen99] state estimation algorithm is proposed for node localization in DTNs. Localization is defined as an online estimation in a nonlinear dynamic system. Its model incorporates significant uncertainty and measurement errors.

Let us take a look at its system dynamic model and the nonlinear measurement model.

9.3.1.1 System Dynamic Model

Assume that sensors are randomly distributed in an environment. The dynamic model for n sensors and the mobile robot can be given in 2D cartesian coordinates as [ASavkin03]

$$\dot{x}(t) = Ax(t) + B_1 u(t) + B_2 \omega(t) \tag{9.9}$$

where

$$A = \begin{bmatrix} \Theta & & 0 \\ & \ddots & \\ 0 & & \Theta \end{bmatrix}, \quad -B_1 = \begin{bmatrix} \Phi \\ \vdots \\ \Phi \end{bmatrix}, \quad B_2 = \begin{bmatrix} \Phi & & 0 \\ & \ddots & \\ 0 & & \Phi \end{bmatrix},$$

$$\Theta = \begin{bmatrix} 0 & 0 & 1 & 0 \\ 0 & 0 & 0 & 1 \\ 0 & 0 & 0 & 0 \\ 0 & 0 & 0 & 0 \end{bmatrix}, \quad \Phi = \begin{bmatrix} 0 & 0 \\ 0 & 0 \\ -1 & 0 \\ 0 & -1 \end{bmatrix} \tag{9.10}$$

The dynamic state vector with $x(t) = [x_1(t) \ldots x_i(t) \ldots x_n(t)]'$ and $x_i(t) = [X_i(t) Y_i(t) \dot{X}_i(t) \dot{Y}_i(t)]'$, where $i \in [1 \ldots n]$, $X_i(t)$ and $Y_i(t)$ represent the position of the ith sensor (Sensor$_i$) with respect to the mobile robot at time t, and their first-order derivatives, $\dot{X}_i(t)$ and $\dot{Y}_i(t)$, represent the relative speed along the X and Y directions.

If $x_c(t) = [x_c(t)\ y_c(t)\ \dot{x}_c(t)\ \dot{y}_x(t)]'$ represents the absolute state (position and velocity in the X and Y directions, respectively) of the mobile robot, and $x_s^i(t) = \left[x_s^i(t)\ y_s^i(t)\ \dot{x}_s^i(t)\ \dot{y}_s^i(t)\right]'$ denotes the absolute state of the Sensor$_i$ in the same order, then $x_i(t) \triangleq x_c(t) - x_s^i(t)$.

Assume that $u(t)$ is the 2D driving/acceleration command of the mobile robot from the respective accelerometer readings, and $\omega(t)$ denotes the unknown 2D driving/acceleration command of the sensor if it is moving. Assume that the sensors are stationary and set $\omega(t) = 0$.

Then we can represent the system in the form of an input ($u(t)$) and measurement (y) system, as shown in Figure 9.6. We omitted B_2 as we only consider the case of stationary sensors. Now the issue is to estimate state x from measurement y.

As the sensor locations are unknown, in the beginning of the algorithm we simply assume an arbitrary location (0, 0) for the sensor to be located. The algorithm can ensure that this assumed state converges to the actual state and, hence, the unknown sensor location can be estimated (as the position/state of the mobile robot is known) within the prescribed time frame.

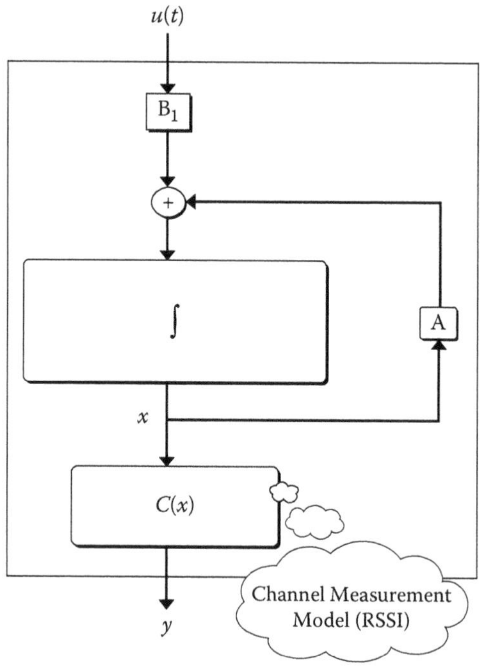

Figure 9.6 Location estimation system. (Adapted from Pathirana, P.N. et al., *IEEE Trans. Mobile Comput.*, 4(3), 285, May/June 2005.)

9.3.1.2 RSSI Measurement Model

As discussed before, we can observe the distance between two communicating entries by using the forward link RSSI of the receiver. The data association is unambiguous when multiple transmitters are present. We can precisely determine which measurement comes from which transmitter by examining the source (transmitter) identifier in the data packet.

For our case, RSSI is measured in decibels at the mobile robot. If we denote the *i*th sensor as Sensor_i (Figure 9.7), we could determine the RSSI from the $Sensor_i$ $p_i(t)$ as [HXia96]

$$p_i(t) = p_{oi} - 10\varepsilon \log d_i(t) + v_i(t) \tag{9.11}$$

where

- p_{oi} is a constant determined by the transmitted power, wavelength, and antenna gain of the mobile robot
- ε is called path loss ratio (typically 2–4)
- $v_i(t)$ is the logarithm of the shadowing component, which is considered as an uncertainty in the measurement
- $d_i(t)$ is the distance between the mobile robot and the Sensor_i, which can be further expressed in terms of the position of the *i*th sensor with respect to the location of the mobile robot, that is, $(X_i(t), Y_i(t))$

$$d_i(t) = (X_i(t)^2 + Y_i(t)^2)^{1/2} \tag{9.12}$$

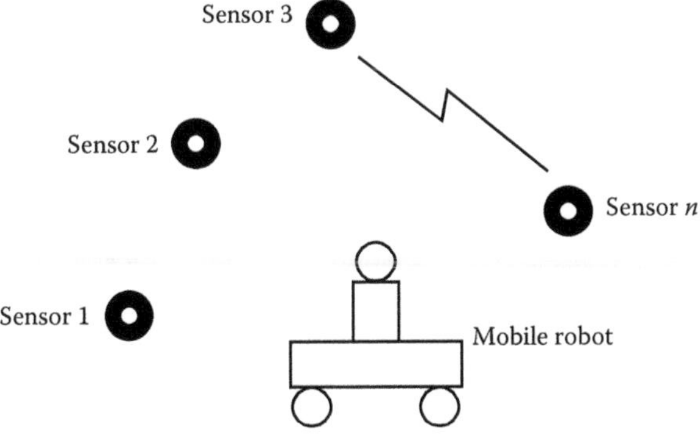

Figure 9.7 Network geometry. (Adapted from Pathirana, P.N. et al., *IEEE Trans. Mobile Comput.*, 4(3), 285, May/June 2005.)

The observation vector

$$y(t) = \begin{bmatrix} p_1(t) \\ \vdots \\ p_n(t) \end{bmatrix} \tag{9.13}$$

is sampled progressively as the mobile robot moves in the coverage area. The measurement equation for the measurements made by the mobile robot for the n number of sensors is in the form of

$$y(t) = C(x(t)) + v(t) \tag{9.14}$$

where $v(t) = [v_1(t) \cdots v_n(t)]'$ with

$$C(x(t)) = \begin{bmatrix} p_{oi} - 10\varepsilon \log(X_1(t)^2 + Y_1(t)^2) \\ \vdots \\ p_{oi} - 10\varepsilon \log(X_n(t)^2 + Y_n(t)^2) \end{bmatrix}$$

REKF can be simply explained as follows: first, we use the state space model with a set of differential equations derived from simple kinematic equations. Such a model has two noise inputs: (1) measurement noise (this is standard with any measurement), v in $y = C(x) + v$ and (2) w-acceleration, which is also considered noise as it is unknown. In this application, the initial condition errors are quite significant as no knowledge is available regarding the sensor locations. If we use REKF in a DTN, the ith system (the mobile robot and the Sensor$_i$), during a corresponding time interval, is represented by the nonlinear, uncertain system, together with the following *integral quadratic constraint* (*IQC*):

$$(x(0) - x_0)' N_i (x(0) - x_0) + \frac{1}{2}\int_0^s (\omega(t)' Q_i(t) w(t)) + v(t)' R_i(t) v(t) dt \le d + \frac{1}{2}\int_0^s z(t)' z(t) dt \tag{9.15}$$

Here, $Q_i > 0$, $R_i > 0$ and $N_i > 0$ ($i \in \{1,2,3\}$) are the weighting matrices for each system i. The initial state (x_0) is the estimated state of respective systems at start-up. Note that we could derive the initial state from the terminal state of the previous system, together with other data available in the network (i.e., robot position and speed). With an uncertainty relationship of the form of this equation , the inherent measurement noise, the unknown mobile robot acceleration, and the uncertainty in the initial condition, are considered as bounded deterministic uncertain inputs. In particular, the measurement equation with the standard norm-bounded uncertainty can be written as

$$y = C(x) + \delta C(x) + v_0 \tag{9.16}$$

where $|\delta| \leq \xi$, with ξ being a constant indicating the upper bound of the norm-bounded portion of the noise. By choosing $z = \xi C(x)$ and $v = \delta C(x)$,

$$\int_0^T |v| \, dt < \int_0^T z'z \, dt \tag{9.17}$$

Considering v_0 and the corresponding uncertainty in w as w_0 satisfying the bound

$$\Phi(x(0)) + \int_0^T [\omega_0(t)'Q\omega_0(t) + v_0(t)'Rv_0(t)]dt \leq d \tag{9.18}$$

Any noise model assumptions in algorithm development are removed by this more realistic approach. This approach also guarantees the robustness. Pathirana et al. [Pubudu05] conducted experiments based on these algorithms and showed that it could converge to the actual sensor locations (see Figure 9.8).

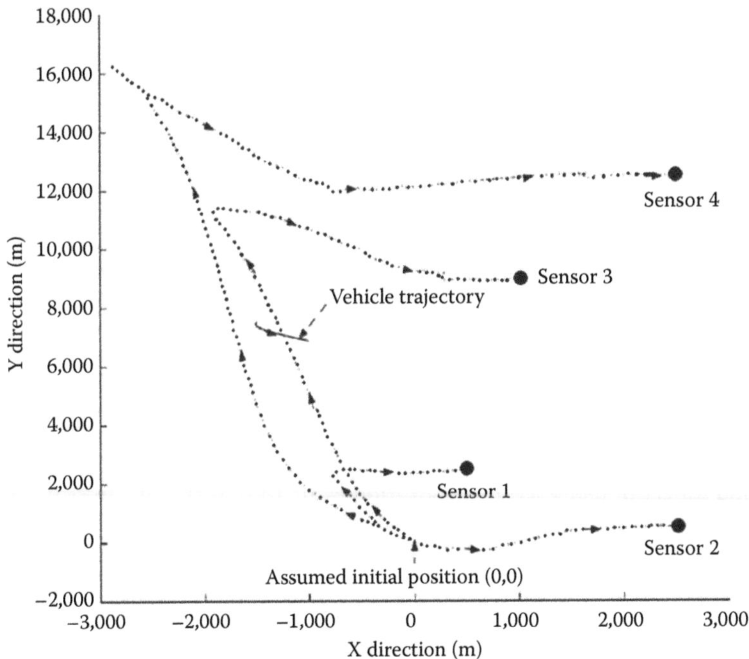

Figure 9.8 Location estimation trajectories converging to the actual sensor locations. (Adapted from Pathirana, P.N. et al., *IEEE Trans. Mobile Comput.*, 4(3), 285, May/June 2005.)

9.4 Sensor Localization with Multidimensional Scaling [Xiang04]

Most existing localization algorithms make use of trilateration or multilateration based on range measurements obtained from ToA, TDoA, and RSSI. In [Xiang04], they explore the idea of using dimensionality reduction techniques to estimate the coordinates of sensors in two- (or three-) dimensional space. They present a centralized sensor localization algorithm based on a dimensionality reduction technique—multidimensional scaling (MDS). It utilizes pair-wise sensor distances to recover locations of sensors in two (or three) dimensions. If pair-wise distances between all sensors are known, a simple *eigen-decomposition* will generate the locations of the sensors.

To estimate the locations of all sensors in a distributed wireless ad hoc sensor network, a small percentage of sensors have their location information known either through manual configuration or equipped with GPS. These sensors with known location information are referred to as anchor sensors, and other sensors without location information are defined as unknown sensors. We hope to estimate locations of all sensors with the assistance of anchor sensors. In general, the anchor sensors broadcast their locations to their neighbors. Neighboring unknown sensors measure their spatial relation from their neighbors, and use the broadcasted anchor sensor locations to estimate their own positions. For an unknown sensor, once an unknown node estimates its position, it becomes an anchor sensor and is able to assist other unknown sensors to estimate their locations.

MDS has been popularly used in the analysis of dissimilarity of data on a set of objects. It can disclose the structure in the data [IBorg97]. We can use MDS as a data-analytic approach to discover the dimensions that underlie the judgments of distance and model data in geometric space.

MDS usually begins by assigning objects to arbitrary coordinates in a 2D space. Next, it computes Euclidean distances among all of the coordinate pairs of points to form a distance matrix. Then, MDS compares the matrix with the measured distances. Finally, the coordinates of each object are adjusted in the direction that best minimizes stress.

A good aspect of using MDS for location estimation is that it can still generate accurate location estimation even when we have erroneous distance information. Although there are numerous varieties of MDS, here we use classical MDS and its iterative optimization.

9.4.1 Classical Multidimensional Scaling

We use $T = [t_{ij}]_{2\times n}$ to denote the *true* locations of n sensor nodes in 2D space. And we use $d_{ij}(T)$ to represent the distance between sensors i and j, based on their location in T, and we have

$$d_{ij}(T)=\left(\sum_{\alpha=1}^{2}(t_{\alpha i}-t_{\alpha j})^2\right)^{1/2} \quad (9.19)$$

The collected distance between nodes i and j is denoted as δ_{ij}. We ignore the errors in distance measurement, δ_{ij} is equal to $d_{ij}(T)$. Let $X = [x_{ij}]_{2\times n}$ denote the estimated locations of n sensor nodes in a 2D space. If all pair-wise distances of sensors in T are collected, we can use the classical MDS algorithm to estimate the locations of sensors:

1. Compute the matrix of squared distance D^2.
2. Compute the matrix J with $J = I - e * e^T/n$, where $e = (1, 1,\ldots, 1)$.
3. Apply double centering to this matrix with $H = -(1/2)JD^2J$.
4. Compute the eigen-decomposition $H = UVU^T$.
5. If we seek for the i dimensions of the solution ($i = 2$ in 2D case), we denote the matrix of largest i eigen-values by V_i, and U_i is the first i columns of U. The coordinate matrix of classical scaling is $X = U_iV_i^{1/2}$.

9.4.2 Iterative Multidimensional Scaling

If we do not know the distances between certain pairs of sensors, we could use the iterative MDS to compute the relative coordinates of adjacent sensors. The iterative MDS is an iterative algorithm based on multivariate optimization n for sensor-location estimation in a 2D space. As only part of the pair-wise distances are available, δ_{ij} is undefined for some i, j. To assist computation, we define weights w_{ij} with value 1 if δ_{ij} is known and 0 if δ_{ij} is unknown, and also assume

$$\delta_{ij} = d_{ij}(T) \quad (9.20)$$

X is randomly initialized as $X^{[0]}$ and will be updated into $X^{[1]}, X^{[2]}, X^{[3]}\ldots$ to approximate T with the iterative algorithm.

We could make a location matrix X approximate T through the minimizing of the following equation:

$$\sigma(X)=\sum_{i<j} w_{ij}(d_{ij}(X)-\delta_{ij})^2 \quad (9.21)$$

We could reach the minimum value of such function when its gradient is equal to 0. The update formula for the iterative algorithm is thus

$$X = V^{-1}\left(\frac{w_{ij}\delta_{ij}}{d_{ij}(T)}A_{ij}\right)T \quad (9.22)$$

where A_{ij} is a matrix with $a_{ii} = a_{jj} = 1$, $a_{ij} = a_{ji} = -1$, and all other elements zeros, and

$$V = \sum_{i<j} w_{ij} A_{ij} \tag{9.23}$$

If V^{-1} does not exist, we replace it with Moore–Penrose inverse of V given by Equation 9.24 [IBorg97].

$$V^{-1} = (V + 11')^{-1} - n^{-2}11' \tag{9.24}$$

We can summarize the iteration steps as follows:

1. Initialize $X^{[0]}$ as random start configuration, set $T = X^{[0]}$ and $k = 0$, and compute $\sigma(X^{[0]})$.
2. Increase the k by 1.
3. Compute $X^{[k]}$ with this updated formula and $\sigma(X^{[k]})$.
4. If $\sigma(X^{[k-1]}) - \sigma(X^{[k]}) < \varepsilon$, which is a small positive constant, then stop; Otherwise set $T = X^{[k]}$ and go to Step 2. (The ε is an empirical threshold based on our accuracy requirement. We usually set $\varepsilon = 5$ percent of the average hop distance. This algorithm generates the relative locations of sensor nodes in $X^{[k]}$.)

These MDS techniques are used in a distributive manner by estimating a local map for each group of adjacent sensors. These maps are then stitched together. Next we will present the details of the distributed-sensor localization method.

9.4.2.1 Hop Distance and Ranging Estimation

In [Xiang04], distance measurement model is based on RSSI. The hop distance is defined as the RF communication range of a sensor. A receiver can estimate the distance to the sender by measuring the attenuation of RF signal strength from the sender to the receiver. For example, there are four sensor nodes A, B, C, and D in Figure 9.9. Hop distance is r_h. The distance between A and D, r_{ad}, can be inferred from A's signal strength at the location of D and r_h.

Other distance measure approaches, such as ToA, TDoA, AoA, and ultrasound, can also be applied here. Although they may generate more accurate distance measure than RSSI, more complex hardware may be needed in each sensor.

9.4.2.2 Aligning Relative Location to Physical Location

After the pair-wise distances of a group of adjacent sensors are estimated, a local map of their relative locations can be calculated using the MDS techniques. As we hope to utilize our distributed localization method to compute the physical

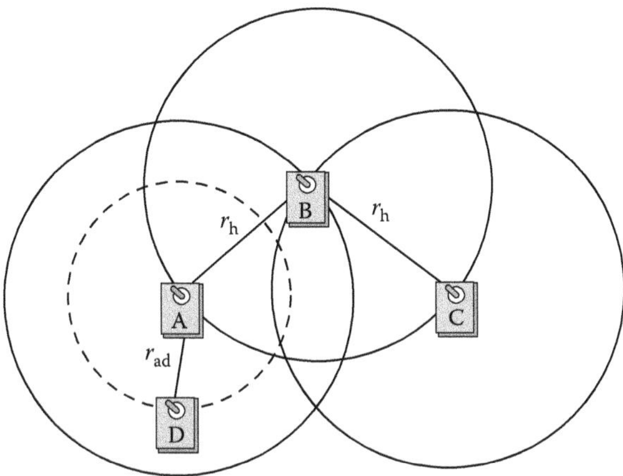

Figure 9.9 Hop distance and signal strength. (Adapted from Ji, X., Localization algorithms for wireless sensor network systems, PhD thesis, Department of Computer Science and Engineering, The Pennsylvania State University, Philadelphia, PA, 2004.)

locations of all sensors, it is necessary to align the relative location to the physical location. This is accomplished via the aid of sensors with known locations. It is known that, in an adjacent group of sensors, the physical locations of at least three sensors are required to locate the remainder of the nodes in the group, in 2D cases. Thus, each group of adjacent sensors must contain at least three nodes with known physical locations; these sensor nodes can be either anchors or nodes with physical locations that were previously calculated.

The alignment procedure includes shift, rotation, and reflection of coordinates. Let us denote $R = [r_{ij}]_{2\times n} = (R_1, R_2, \ldots, R_n)$ as the relative locations of the set of n sensor nodes in 2D space. $T = [t_{ij}]_{2\times n} = (T_1, T_2, \ldots, T_n)$ denotes the *true* locations of the set of n sensor nodes in 2D space. In the following explanation, we assume the nodes 1, 2, and 3 are anchors. A vector R_i may be shifted to $R_i^{(1)}$ by $R_i^{(1)} = R_i + X$. It may be rotated counterclockwise through an angle α to $R_i^{(2)} = Q_1 R_i$, where

$$Q_1 = \begin{bmatrix} \cos(\alpha) & -\sin(\alpha) \\ \sin(\alpha) & \cos(\alpha) \end{bmatrix} \tag{9.25}$$

It may also be reflected across a line

$$S = \begin{bmatrix} \cos(\beta/2) \\ \sin(\beta/2) \end{bmatrix} \tag{9.26}$$

To $R_i^{(3)} = Q_2 R_i$, where

$$Q_2 = \begin{bmatrix} \cos(\beta) & \sin(\beta) \\ \sin(\beta) & -\cos(\beta) \end{bmatrix} \tag{9.27}$$

Before alignment, we only know R and physical locations of three or more other sensors T_1, T_2, and T_3. Given the locations of the other sensor, we can compute T_4, $T_5, \ldots, T_n$. Based on these rules, we have

$$(T_1 - T_1, T_2 - T_1, T_3 - T_1) = Q_1 Q_2 (R_1 - R_1, R_2 - R_1, R_3 - R_1) \tag{9.28}$$

With R_1, R_2, R_3, T_1, T_2, and T_3 known, we can compute

$$Q = Q_1 Q_2 = \left(\frac{R_1 - R_1, R_2 - R_1, R_3 - R_1}{T_1 - T_1, T_2 - T_1, T_3 - T_1} \right) \tag{9.29}$$

Then, $(T_4, T_5, \ldots, T_n)$ can be calculated with

$$\begin{aligned} (T_4 - T_1, T_5 - T_1, \ldots, T_n - T_1) &= Q(R_4 - R_1, R_5 - R_1, \ldots, R_n - R_1) \\ (T_4, T_5, \ldots, T_n) &= Q(R_4 - R_1, R_5 - R_1, \ldots, R_n - R_1) + (T_1, T_1, \ldots, T_1) \end{aligned} \tag{9.30}$$

9.4.2.3 Distributed Physical Location Estimation

An anchor node labeled as the "starting anchor" initializes flooding to the entire network. When other anchor nodes, called "ending anchors," receive the flooding message, they send their location information back to the starting anchor, along with the reverse routes from the starting anchor to themselves. Now, the starting anchor knows the locations of the ending anchors and the corresponding route to each of them. The starting anchor uses the routes to estimate the locations of those sensors that are one hop away. Figure 9.7 illustrates the procedure: A is the starting anchor, and D and G are the ending anchors. A knows the locations of D and H as well as the routes to them, which are (A, B, C, D) and (A,E, F, G,H), respectively. A estimates that the location of B is B′ on the dashed line AD and that the location of E is E′ on the dashed line AH. A also estimates the average hop distances in the direction of AD and AH, respectively.

With the collection of pair-wise distances among neighboring nodes by RSSI sensing, MDS can be performed to calculate the local map, or the relative locations, for neighboring sensor nodes. In Figure 9.10, the relative locations of neighboring nodes A, B, E, J, and K are calculated by A. Through the alignment of the relative locations of A, B, and E with their physical locations, the physical locations of J and K can be calculated as well. In the same manner, localized mapping and alignment are performed for sensor nodes along a route from the starting anchor to an ending

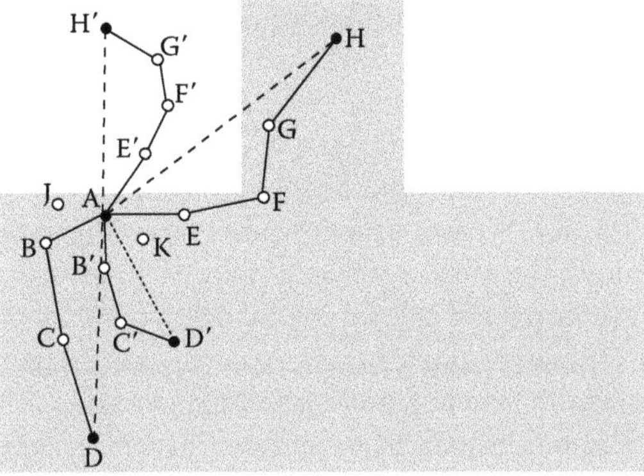

Figure 9.10 Location estimation for a neighborhood. (Adapted from Ji, X., Localization algorithms for wireless sensor network systems, PhD thesis, Department of Computer Science and Engineering, The Pennsylvania State University, Philadelphia, PA, 2004.)

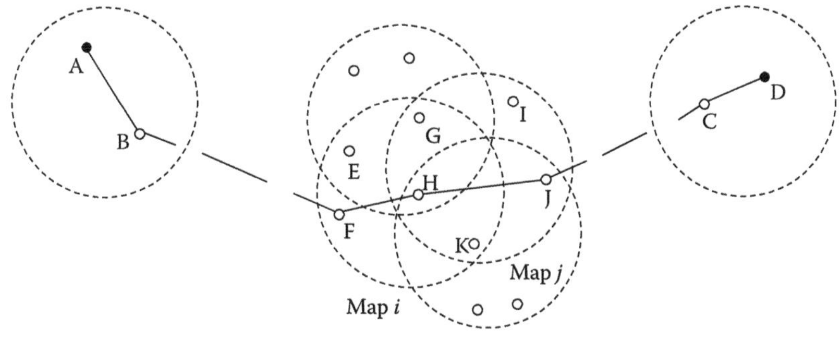

Figure 9.11 The propagation of position estimation. (Adapted from Ji, X., Localization algorithms for wireless sensor network systems, PhD thesis, Department of Computer Science and Engineering, The Pennsylvania State University, Philadelphia, PA, 2004.)

anchor. Figure 9.8 illustrates the procedure of propagated-location estimation from the starting anchor to the ending anchor.

In Figure 9.11, A is the starting anchor and D is the ending anchor. The remainder of the sensor nodes is spread throughout the route of flooding from A to D, and each local map is represented with a dash ellipse. Map *i* contains adjacent sensors E, F, G, H, and K. As the physical positions of E, F, and G have been previously calculated, the physical positions of H and K can be computed with the aforementioned MDS and alignment techniques. Then, H, K, I, J, and G are adjacent sensors that build map *j* to further estimation of I and J's positions.

We can estimate the locations of all nodes surrounding a route from a starting anchor to an ending anchor, as well as the ending anchor itself. For example, in Figure 9.11, the estimated location of nodes E, F, and G are E′, F′, and G′, respectively. Given the physical location of G, we can compare the location of G′ and G. If they are not equivalent, we can align them by rotating ∠G′AG about a center A, and then scale AG″ to AG. We can also apply the same alignment to the coordinates of all sensors along the route, such as E′ and F′. In general, the locations of E′ and F′ are effectively corrected and approximated to their true locations, respectively. This location-estimation procedure is executed iteratively on a route from a starting anchor to an ending anchor, until estimated locations converge.

The experimental results in [Xiang04] indicate that this procedure usually generates highly accurate estimations of location for sensors along a route. Those nodes with accurately estimated location can then be viewed as anchor nodes, and they, in turn, initialize more location estimation for sensors along different routes. This estimation method can be performed on a different portion of sensors in an ad hoc sensor network simultaneously, until all sensors are accurately located.

9.5 Localization in Wireless Sensor Networks [Masoomeh07]

In [Masoomeh07], the authors consider a network with *a small fraction of sensors equipped with hardware such as GPS*, which allows them to be aware of their locations at all times. Aside from this, the sensors are all identical.

We model irregularity in radio range by assuming that the sensor radio range is normally distributed with mean r and standard deviation σ. The simulator uses σ to randomly determine for each packet whether the sender and the receiver are within radio range. Initially, nodes are spread randomly throughout the network area.

The one-hop neighbors of sensor p are those sensors that can communicate with it directly. The algorithms do not require very tightly synchronized clocks. Each node and seed is capable of moving a distance v in a time step in any direction where $0 \le v \le v_{max}$. The nodes know v_{max}, but they do not know the value of v or the direction of movement in any time step.

9.5.1 *The Monte Carlo Method*

In the event that a system state needs to be estimated from some observations, the Bayesian model can formulate the system in which the posterior distribution of the state depends only upon the current observations and state [ADoucet01]. Observations arrive sequentially in dynamic systems, requiring updates for the posterior distribution to take place with the arrival of new observations. The Monte Carlo method estimates the state of the distribution with a set of samples, and updates these samples as new observations arrive.

Although different approaches for the Monte Carlo method have been proposed, we focus on the particle filtering approach [ADoucet01]. This technique is used for estimating robot locations, and is fully distributed and easy to implement [DFox99]. The objective of this method is to represent data distribution of the system as a set of N weighted samples:

$$p(S_t \mid Q_{0\ldots t}) \approx \left\{s_t^{(i)}, w_t^{(i)}\right\}_{i=1,\ldots,N} \tag{9.31}$$

where

$p(S_t|Q_{0\ldots t})$ is the distribution representing the state of the system at time t
$s_t^{(i)}$ is a sample of the state of the system at time t
$w_t^{(i)}$'s are nonnegative numerical weights that sum to one

We need a minimum number of samples so that the set of samples converges to the posterior distribution of the system (see Doucet et al. [ADoucet01] for details).

The steps of the Monte Carlo method are as follows:

Initialization: N samples chose from initial system distribution, $p(S_0)$.
Sampling: N samples, $\tilde{s}_t^i$ for $i = 1,\ldots,N$, are drawn from the distribution $p(S_t|S_{t-1})$ where $p(S_t|S_{t-1})$ is the transition equation or motion model. Compute each sample weight and normalize this value, $w^{(i)} = \eta \tilde{w}_t^i$, where η is the normalizing factor.
Resampling: N samples are chosen (with replacement) from the current sample set according to their weights.

Each node denotes its possible locations with a set of weighted samples. Using the Monte Carlo method, each node updates its samples with each observation. In the upcoming discussions, $d(a, b)$ denotes the distance between locations a and b, and r denotes the ideal radio range. We describe algorithm (1) first.

9.5.2 Algorithm (1)

In algorithm (1), a set of probable locations (samples) is maintained for each node. These samples are given different weights that provide an estimate of their quality. Conceptually, this value represents the likelihood of truly representing the nodes location, given the estimated locations of its neighbors. The steps of the algorithm are

Step 1. Initialization: Nodes have no information about their locations, thus the first set of samples is chosen randomly from the entire field of sensors, using only the seeds within the neighborhood to assign weights to the samples.
Step 2. Sampling: Based on the following transition equation, nodes generate new samples:

$$p(S_t \mid S_{t-1}) = \begin{cases} \dfrac{1}{\pi(v_{\max} + \alpha)^2} & \text{if } d(S_t, S_{t-1}) \le v_{\max} \\ 0 & \text{if } d(S_t, S_{t-1}) > v_{\max} \end{cases} \tag{9.32}$$

where

v_{max} is the maximum speed of a node

$d(S_t, S_{t-1})$ denotes the distance between the locations of a sample at time t and $t-1$

With each time step, new samples are generated from each current sample by randomly selecting a point inside the circle centered at the current location of the sample and with a radius of (v_{max} + α). If α is too small, there will not be enough variability in the new sample selections when the speed of these sensors is low. The value of α, that is, $\alpha = 0.1r$, was determined empirically.

After choosing a sample, neighborhood information can be used to generate its weight as follows. The weight of a sample s chosen for node p, $w_s(p)$, is computed as follows: corresponding to each neighbor q of node p, we find a partial weight for sample s, $w_s'(q)$. The weight of the sample is the product of the partial weights obtained from each neighbor node p. That is,

$$w_s(p) = \prod_{q=1}^{k} w_s'(q) \tag{9.33}$$

where

k is the number of one-hop and two-hop neighbors of node p

q is a neighbor of node p

The partial weight of sample s corresponding to a one-*hop* seed neighbor q is

$$w_s'(q) = \begin{cases} 1 & \text{if } d(s,q) \le r \\ 0 & \text{otherwise} \end{cases} \tag{9.34}$$

The partial weight of sample s corresponding to a two-*hop* seed neighbor q is

$$w_s'(q) = \begin{cases} 1 & \text{if } r \le d(s,q) \le 2r \\ 0 & \text{otherwise} \end{cases} \tag{9.35}$$

The partial weight of sample s corresponding to a one-hop node neighbor q is computed using the weights $w(q_i)$ of samples q_i of node q as follows:

$$w_s'(q) = \sum_{q_i} w(q_i), \quad \text{where} \quad d(s,q_i) \le r + v_{max} \tag{9.36}$$

Similarly, for two-hop neighbors q, $w_s'(q)$ is computed as follows:

$$w_s'(q) = \sum_{q_{ii}} w(q_i), \quad \text{where} \quad r - v_{max} \le d(s,q_i) \le 2r + v_{max} \tag{9.37}$$

Sample s is kept if $w_s(p)$ is greater than a threshold value, β. Parameter β is a real number in the interval [0, 1] and its value depends on the number of neighbors of a node. Therefore, different nodes have different β values. β should be chosen such that its value decreases with an increasing count of neighbor nodes. The reason for this is that $w_s(p)$ is the product of numbers, which is at most 1. Partial weights corresponding to nodes usually retain a value less than 1, and the partial weights corresponding to the seeds are 0 or 1. Here we use $\beta = (0.1)^t$, where t is the number of one-hop and two-hop neighbors of a node.

After computing $w_s(p)$ the weights are normalized to ensure that the sum is equal to one. Therefore, if N samples are chosen for node p, the weight of the ith sample is normalized as

$$\frac{w_i(p)}{\sum_{j=1}^{N} w_j(p)} \tag{9.38}$$

Step 3. Resampling: this step gradually removes samples with lower weights, gradually reducing the set to only those with the highest weights. Each node computes a new sample set from its current set, with the samples of the new set containing all the old samples, but updating them to have a probability proportional to their weights. As the number is fixed, a sample with a small weight has a lower chance of being selected, thus higher weighted samples are likely to have duplicates in the new sample set.

Pseudocode for MSL* is given in Figure 9.12. The algorithm has the same basic structure as *the distance vector* algorithm used for propagating router information.

```
If (node not localized or number of samples are zero)
If (node has first-hop or second-hop neighbors)
find N samples with weights greater than β
Normalize the weights of the samples
Else
closeness = ∞
keep the last set of samples
Else
Sample (α) (Sampling step with parameter α)
If no sample found
closeness = ∞
keep the last set of samples
Normalize weights
Resample the sample set (Re-sampling step)
Send locations and closeness to first- and second-hop neighbors.
```

Figure 9.12 Algorithm (1) in every node. (Adapted from Rudafshani, M. and Datta, S., Localization in wireless sensor networks, *IPSN '07*, Cambridge, MA, April 25–27, 2007.)

Each node weights its own samples using the location estimates of its neighbors. However, by only using neighbors with highly accurate estimates of their locations, we gain an advantage of reducing communication costs. The qualities of these estimates are measured using a parameter called *closeness*. The formula for *closeness* value for a node p with N samples is

$$\text{closeness}_p = \frac{\sum_{i=1}^{N} w_i \sqrt{(x_i - x)^2 + (y_i - y)^2}}{N} \tag{9.39}$$

where
N is the number of samples of node p
(x_i, y_i) denotes the coordinate of the ith sample ($i = 1, \ldots N$)
w_i denotes the weight of the ith sample
(x, y) is the current location estimate of node p

Seeds always have a closeness value of 0, and the closeness of a node is always greater than 0. Lower closeness values indicate more accurate location estimates, making this a good measure of accuracy of the location estimate of a node.

At the start of algorithm (1), closeness values for seeds are 0 and ∞ for nodes. Thus, in the first time step, only the seeds can provide information for the one-hop and two-hop neighbors. As the process proceeds, nodes update their estimates and their closeness, and send this information to their neighbors.

A node will not receive new location information in the event it moves to a new position in which it has no neighbors. If this occurs, the previous sample set is used to estimate the location of the node. The node must then be re-localized and the current sample set must be re-initialized.

This algorithm is performed at a high communication cost due to the fact that each node uses information of all of its first- and second-hop neighbors. We now describe algorithm (2).

9.5.3 Algorithm (2)

Algorithm (2) is communication intensive, due to the transfer of samples between nodes. We assign a weight to each node, which then uses the weights of only its neighbors (as opposed to samples and neighbors) to assign weights to its samples. After computing these weights, MSL computes a single location estimate and a closeness value. Each node broadcasts its estimate and closeness to its neighbors. This way reduces communication costs significantly because it does not transmit the value of its samples.

In this approach, the weight assigned to each node depends on the quality of the location estimate. To achieve this, we define the weight of a node as a function of its closeness value:

$$w_p = b^{-\text{closeness}_q} \tag{9.40}$$

The performance of algorithm (2) was not sensitive to the choice of b; we use $b = 7$ here. Similar to algorithm (1), in algorithm (2) a node uses the locations of only those neighbors with lower closeness values.

In algorithm (2), we compute the weights of seed neighbors just as in algorithm (1), but the weights of first-hop, non-seed neighbors are computed by

$$w_s'(q) = \begin{cases} w_q & \text{if } d(s,q) \le r + v_{\max} + v_{\text{extra}} \\ 0 & \text{otherwise} \end{cases} \tag{9.41}$$

We need to account for extra uncertainty because nodes are using less information than in algorithm (1) (a single location estimate of neighbors rather than a set of weighted samples). Using the parameter v_{extra}, we are able to achieve this affect. *Algorithm (2)* was not sensitive to the choice of $v_{\text{extra}} \in [0.2r, 0.5r]$; we use $v_{\text{extra}} = 0.3r$ here.

The weights of second-hop non-seed neighbors are computed as follows:

$$w_s'(q) = \begin{cases} w_q & \text{if } r - v_{\max} - v_{\text{extra}} \le d(s,q) \\ & \quad \le 2r + v_{\max} + v_{\text{extra}} \\ 0 & \text{otherwise} \end{cases} \tag{9.42}$$

This algorithm uses the Monte Carlo method, like Monte Carlo Localization (MCL) [LHu04a], but we improve on MCL and generalize it in several ways. Modifying the sampling procedure allows our approach to work in static networks and enables it to outperform MCL, even when using only information from seed neighbors.

Second, in both algorithm (1) and algorithm (2), nodes use information from neighbors with more accurate estimates than their own, yielding improved performance in networks with low-speed nodes or low seed densities.

Third, modifying the sampling procedure and permitting samples to have weights greater than a threshold value β, we are able to generate faster convergence of the localization algorithm. This produces faster execution time and better estimation of locations in mobile networks.

For algorithm (2) to function as directed, sensors must move with some predetermined minimal speed, but it cannot work below this speed and does not work in static networks. Algorithm (1) can estimate locations with high accuracy even when sensors are static, and move with low speeds or very high speeds.

9.6 GPS-Free Node Localization in Mobile WSN [Akcan06]

Consider a fire search mission inside a building where a set of mobile nodes explore a floor. The goal is to locate the source of the fire. In a semirigid swarm, the nodes move collaboratively. The swarm follows a path covering the area while taking temperature measurements. A number of issues must be taken into consideration to solve the problem of localization management in GPS-free environments where nodes are mobile. Most importantly, the additive error in the estimated location can amass to very high values caused by mechanical errors in evaluating the direction and distance of movement, which can occur in all measurements. The source for this type of error is due to changes in the environment or manufacturing defects. This means that as the motion changes, the uncertainty of the position and direction of a node decreases.

Akcan et al. [Akcan06] proposed a solution to the problem of *directional localization* in GPS-free sensor networks with mobile nodes. It proposed a motion-based algorithm for node position and direction calculation with respect to the local coordinate system of each individual node. The algorithm is very fast and does not require additional memory. In addition, cumulative position errors do not affect it. More specifically, the algorithm is unaffected by the speed of nodes.

The GPS-free localization algorithm assumes sensors can measure the distance to their neighbors using a well-known range measurement method (e.g., ToA). It also needs motion actuators that allow each node to move a specific distance in a specific direction (with respect to North).

First, the core localization algorithm will be described with two neighbors, n_1 and n_2, that generate two possible relative positions. Later, a verification algorithm that uses a common third neighbor to select the correct solution will be discussed.

Core localization algorithm. The core localization algorithm works on rounds, and each round essentially consists of three steps:

1. Measure distance between neighbors
2. Continues with individual node movement
3. Ends with an exchange (between neighbors) of direction and distance values for that round

Whenever nodes need localization, rounds are initiated. Any other continuity or pattern between rounds are not required. Also, no assumptions are made about the temporal duration of the rounds. However, it is assumed that nodes do not change their directions within a round.

A typical movement of two nodes n_1 and n_2 in a round is shown in Figure 9.13. At time t_1, n_1 is at position (x_0, y_0) and n_2 at (x_2, y_2), and the nodes measure the initial inter-distance d_1. Between times t_1 and t_2, each node $\{n_i \mid i = 1, 2\}$ moves in a direction α_i and covers a distance v_i. At time t_2, the nodes, now at positions (x_1, y_1) and (x_3, y_3), calculate their inter-distance d_2 and exchange v_i and α_i. Each node selects itself as the origin and calculates the position and direction of the other node in its local coordinate system, only after receiving all the information. We choose the position (x_0, y_0) of n_1 as the origin and write the following to solve the equations in the local system of n_1:

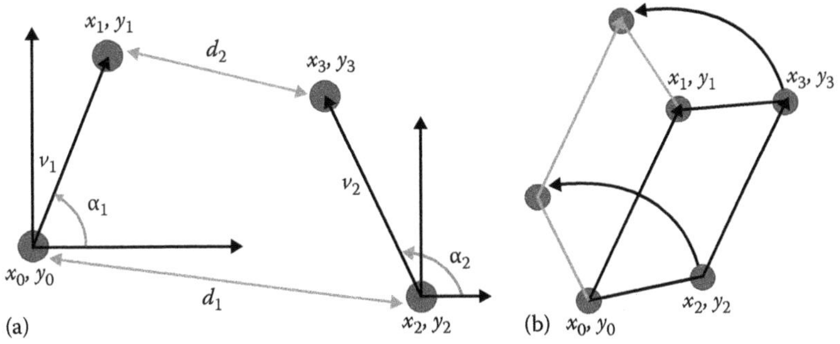

Figure 9.13 Typical movements of two nodes, with angles and distances. (Adapted from Akcan, H. et al., GPS-free node localization in mobile wireless sensor networks, *Proceedings of the Fifth ACM International Workshop on Data Engineering for Wireless and Mobile Access (MobiDE '06)*, Chicago, IL, June 25, 2006, ACM, New York, 35–42.)

$$
\begin{aligned}
&x_1 = v_1 \cos \alpha_1, && y_1 = v_1 \sin \alpha_1 && \text{(i)} \\
&x_3 = x_2 + v_2 \cos \alpha_2, && y_3 = y_2 + v_2 \sin \alpha_2 && \text{(ii)} \\
&(x_3 - x_1)^2 + (y_3 - y_1)^2 = d_2{}^2 && x_2{}^2 + y_2{}^2 = d_1{}^2 && \text{(iii)}
\end{aligned}
\tag{9.43}
$$

Substituting Equations i and ii into Equation iii, we get

$$x_2 A + y_2 B = C \tag{9.44}$$

With the appropriate definitions

$$
\begin{aligned}
&A = v_2 \cos \alpha_2 - v_1 \cos \alpha_1, \quad B = v_2 \sin \alpha_2 - v_1 \sin \alpha_1 \\
&C = \frac{1}{2}(d_2^2 - d_1^2 - v_1^2 - v_2^2 + 2v_1 v_2 \cos(\alpha_1 - \alpha_2))
\end{aligned}
\tag{9.45}
$$

Substituting

$$x_2 = (C - y_2 B)/A \quad \text{and} \quad y_2 = (C - x_2 A)/B \quad \text{into } x_2^2 + y_2^2 = d_1^2 \tag{9.46}$$

We get

$$x_2^2 D - 2x_2 E + F = 0, \quad y_2^2 D - 2y_2 G + H = 0 \tag{9.47}$$

Again with the appropriate definitions

$$D = A^2 + B^2, \quad E = AC, \quad F = C^2 - d_1^2 B^2,$$

$$G = BC, \quad H = C^2 - d_1^2 A^2 \tag{9.48}$$

Note that the coefficient of x_2^2 and y_2^2 is the same in both equations, namely, D.

Using (9.48), each variable solves independently to

$$x_2 = \frac{E \pm \sqrt{E^2 - DF}}{D}, \qquad y_2 = \frac{G \pm \sqrt{G^2 - DH}}{D} \qquad \text{(vi)} \tag{9.49}$$

and solutions can be paired up by using Equation 9.49, as long as $D \neq 0$. In practice, one would compute either x_2 or y_2 using Equation 9.49 and deduce the other variable using (9.48). When $A = 0$ but $B \neq 0$, one would compute x_2 using (9.46), and when $A \neq 0$ but $B = 0$, one would compute y_2 using (9.44) instead.

Figure 9.14 is the core localization algorithm to calculate the position from n_2 to n_1. Each node finds out possible positions for each of its neighbors by solving the equations. Each node has to complete a verification step using an additional common neighbor (n_3) because only one of these solutions is realistic (the other one is due to "symmetry").

If the entire WSN moves in a specific direction to accomplish a goal, the *directional localization* algorithm is very useful. To reflect practical mobility features, the

```
CoreLocalization (n_1, n_2,, n_1, α_1)
d_1 ← inter-distance(n_1, n_2)
Move node n_1 by v_1 and α_1
d_2 ← inter-distance(n_1, n_2)
Retrieve v_2 and α_2 from n_2
Calculate positions of n_2 using equations (4), (5) and (6)
Verification (NeighborList NL)
for each neighbor pair (m, n) in NL do
        if m and n are neighbors then
          d_m,n ← measured inter-distance(m, n)
          for each position pair {m^i, n^j\i, j = 1, 2} do
          Compute Euclidean distance D between m^i and n^j
          if D = d_m,n then
            mark m^i and n^j as exact positions
```

Figure 9.14 Core localization algorithm. (Adapted from Akcan, H. et al., GPS-free node localization in mobile wireless sensor networks, *Proceedings of the Fifth ACM International Workshop on Data Engineering for Wireless and Mobile Access (MobiDE '06)*, Chicago, IL, June 25, 2006, ACM, New York, 35–42.)

reference point group mobility (RPGM) model is used in [XHong99]. Many other mobility models have been surveyed in [TCamp02]. But RPGM is used due to the generality of the model. In RPGM, the movement of an individual sensor is modeled in relation to a randomly chosen directional motion of the entire group. Each individual moves randomly around a fixed reference point, and the entire group move along the logical center of the group. The localization algorithm computes the location/orientation of each node. In RPGM, a sensor does not need to know the center of the group.

A sensor's one-hop neighbors determine the context of random motion of that sensor within the group. Thus we could remove the reference points and use the neighbors to represent the reference points of motion.

The adapted mobility algorithm is presented in Figure 9.15. The network moves with respect to a direction vector. The algorithm imposes a minimum neighbor count k that each node strives to attain to maintain a semirigid formation without disconnecting the network. This is a best-effort k-connected algorithm where a sensor attempts to maintain a neighbor distance that is less than its RF communication range. And it adjusts such a distance dynamically with the number of neighbors so that the neighbors within k hops stay closer while still moving with the network. This idea can avoid network partitioning. The range function returns the wireless range of the given node.

Obviously, as long as we know an initial direction of motion for the entire group, we can keep the entire network cohesive. This idea is similar to "swarming" concept in bio-inspired computing. For example, an oil-search WSN could move in a zigzag pattern, with the goal to discover an oil spill and cover the contaminated area once it is found. In this example, we can just specify a virtual boundary, and the network of sensors will maintain sufficient proximity to communicate, while covering the area.

The mobility algorithm requires only *local* position information. This is good for scaling up to large network (>1000 sensors) because it requires a sensor to communicate with its direct neighbors without message flooding.

```
MoveNode(Node N, NeighborList NL, DirectionVector D⃗, INT k, RangeFactor
 RF)
V⃗ ← 0
count ← 0
for each localized neighbor n in NL do
◁u →N,n n is the vector from N to n
V⃗ ← V⃗ + ūN,n
count ← count + 1
if count < k then
RF ← RF /2
V⃗ ← (RF + range (N) + V⃗ + D⃗)/(count + 1)
Move node N by V⃗
```

Figure 9.15 *k*-Neighborhood mobility algorithm.

9.7 A High-Accuracy, Low-Cost Localization System for WSN [Radu05]

The difficulties of traditional localization approaches are twofold. First, the effective ranges of such devices are very limited, because there are restraints of form factor and power supply. For example, the operation range of the ultrasonic transducers in Cricket [NPriyantha05] is just a few meters. Second, it is expensive to equip these sensors with special circuitry just for a one-time localization because most sensor nodes are stationary.

Many range-free localization schemes have been proposed to overcome these limitations. Most of them estimate the node location by exploiting the radio connectivity information among neighboring nodes. Although they do not need high-cost, specialized hardware, its accuracy may not be satisfactory.

Stoleru et al. [Radu05] achieve a high accuracy in sensor localization while not involving high cost (from communication and calculation complexity viewpoint). It uses a concept called *Spotlight*. Sensor nodes do not need new hardware for localization purpose. All the sophisticated, costly hardware and computation are in a single Spotlight device, which can issue a steerable laser light to illuminate the sensor nodes placed within a known terrain.

Its localization is more accurate (i.e., <1 m) than the range-based localization schemes. It has a much longer effective range (i.e., >1000 m) than the solutions based on ultrasound/acoustic ranging. Because all complicated hardwares/softwares are in a single sophisticated device, the cost is much lesser than the case with additional hardware components in each individual sensor.

Spotlight is a typical *range-free* localization scheme that also works well in an outdoor environment. An LOS is required between a single device and the sensor nodes.

Generating controlled events in the field where the sensor nodes were deployed is the main idea of the Spotlight system. Using the time when an event is perceived by a sensor node, as well as the spatiotemporal properties of the generated events, we can infer spatial information (i.e., location) of a sensor node.

In Figure 9.16, a sensor network deployment and localization scenario is depicted as follows: from an unmanned aerial vehicle wireless sensor nodes are randomly deployed. A time synchronization protocol should be executed among sensors after sensor deployment. An aerial vehicle such as a helicopter, equipped with a device called Spotlight, flies over the network and generates light events. Each sensor detects the events and reports back to the Spotlight device with time stamps when the events were detected. The Spotlight device computes the location of the sensor nodes.

Such a spotlight system assumes the following conditions:

1. The sensor nodes can communicate with the Spotlight device.
2. The aerial vehicle knows about its own position and orientation very well; it also possesses the map of the sensor field.

Figure 9.16 Localization of a sensor network using the Spotlight system. (Adapted from Stoleru, R. et al., A high-accuracy, low-cost localization system for wireless sensor networks, *SenSys '05*, San Diego, CA, November 2–4, 2005.)

3. The Spotlight device can generate spatially large events to be detected by the sensor nodes, even in the presence of background noise (daylight).
4. There exists an LOS between the Spotlight device and sensor nodes.

The Spotlight localization system uses the following definitions:

Let us assume that the space $A \subset R^3$ contains all sensor nodes N, and that each node N_i is positioned at $p_i(x, y, z)$. A Spotlight localization system needs to support three main functions to obtain $p_i(x, y, z)$, namely, an *event distribution function* (EDF) $E(t)$, an *event detection function* $D(e)$, and a *localization function* $L(T_i)$. The following are their formal definitions:

Definition 9.1: An event $e(t, p)$ is a detectable phenomenon that occurs at time t and at point $p \in A$. Examples of events are light, heat, smoke, and sound. Let $T_i = \{t_{i1}, t_{i2}, \ldots, t_{in}\}$ be a set of n time stamps of events detected by a node i. Let $T' = \{t'_1, t'_2, \ldots, t'_m\}$ be the set of m time stamps of events generated in the sensor field.

Definition 9.2: The event detection function $D(e)$ defines a binary detection algorithm. For a given event e

$$D(e) = \begin{cases} \text{true,} & \text{Event is detected} \\ \text{false,} & \text{Event is not detected} \end{cases} \tag{9.50}$$

Definition 9.3: The EDF $E(t)$ defines the point distribution of events within A at time t:

$$E(t) = \{p \mid p \in A \wedge D(e(t,p)) = \text{true}\} \tag{9.51}$$

Definition 9.4: The localization function $L(T_i)$ defines a localization algorithm with input T_i, a sequence of time stamps of events detected by the node i:

$$L(T_i) = \bigcap_{t \in T_i} E(t) \tag{9.52}$$

As shown in Figure 9.17, the sensor nodes support the *event detection function* $D(e)$, which determines whether an external event happens or not. The detection algorithm can be implemented by either a simple threshold-based detection algorithm or other advanced digital-signal-processing (DSP) techniques.

A Spotlight device implements the event distribution $E(t)$ and localization functions $L(T_i)$. The localization function is an aggregation algorithm that calculates the intersection of multiple sets of points.

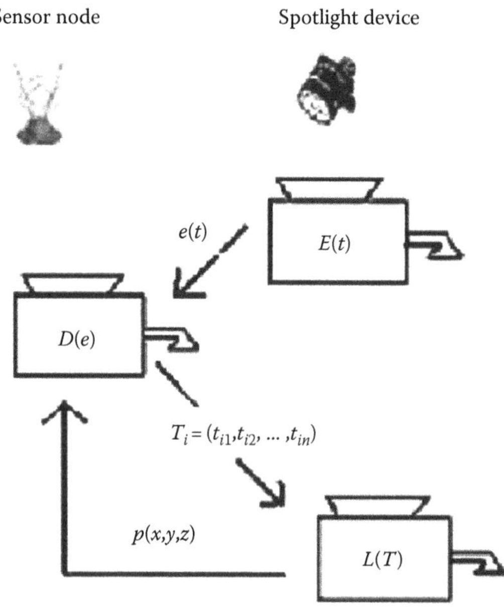

Figure 9.17 Spotlight system architecture. (Adapted from Stoleru, R. et al., A high-accuracy, low-cost localization system for wireless sensor networks, *SenSys '05*, San Diego, CA, November 2–4, 2005.)

The EDF $E(t)$ can also describe the distribution of events over time. It is the core of the Spotlight system, and it is more sophisticated than the other two functions. $E(t)$ is implemented by the Spotlight device (not in sensors).

Based on these three functions, the localization process is as follows:

1. *Events distribution*: A Spotlight device distributes events in the space A over a period of time.
2. *Events detection*: During the event distribution, sensor nodes record the time sequence $T_i = \{t_{i1}, t_{i2}, \ldots, t_{in}\}$ at which they detect the events.
3. *Events report*: After the event distribution, each sensor node sends the detection time sequence back to the Spotlight device.
4. *Location estimate*: The Spotlight device estimates the location of a sensor node i, using the time sequence T_i and the known $E(t)$ function.

In the Spotlight system, the core technique is the EDF $E(t)$. For simplicity, we assume that a set of nodes are placed along a straight line ($A = [0, 1] \subset R$). The Spotlight device generates point events (e.g., light spots) along this line with constant speed s.

The set of time stamps of events detected by a node i is $T_i = \{t_{i1}\}$. The EDF $E(t)$ is

$$E(t) = \{p \mid p \in A \wedge p = t * s\} \tag{9.53}$$

where $t \in [0,1/s]$. The resulting localization function is

$$L(T_i) = E(t_{i1}) = \{t_{i1} * s\} \tag{9.54}$$

where $D(e(t_{i1}, p_i)) = \mathit{true}$ for node i positioned at p_i. The implementation of the *EDF* $E(t)$ is straightforward. As shown in Figure 9.18a, when a light source emits a beam

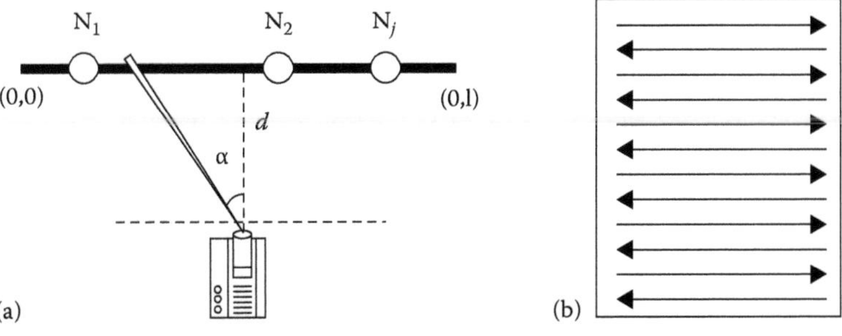

Figure 9.18 The implementation of the point scan EDF. (Adapted from Stoleru, R. et al., A high-accuracy, low-cost localization system for wireless sensor networks, *SenSys '05*, San Diego, CA, November 2–4, 2005.)

of light with the angular speed given by $S_a = d\alpha/dt = (s\cos^2(\alpha))/d$, a light spot event with constant speed s is generated along the line situated at distance d.

Besides this simple one-line case, the point scan EDF can also be extended to the case where nodes are placed in a 2D plane R^2. In this case, the Spotlight system progressively scans the plane to activate the sensor nodes. This scenario is depicted in Figure 9.18b.

- *Line scan EDF*
  Some devices such as diode lasers can generate an entire line of events simultaneously. They can then support the *line scan event distributed function* easily. We assume that the sensor nodes are placed in a 2D plane ($A = [l \times l] \subset R^2$) and that the scanning speed is s. The set of time stamps of events detected by a node i is $T_i = \{t_{i1}, t_{i2}\}$. The line scan EDF is defined as follows:

$$E_x(t) = \{p_k \mid k \in [0,1] \wedge p_k = (t * s, k)\} \tag{9.55}$$

  For $t \in [0, l/s]$, $E_y(t) = \{p_k | k \in [0,1] \wedge p_k = (k, t * s - l)\}$,
  For $t \in [l/s, 2l/s]$, $E(t) = E_x(t) \cup E_y(t)$

  Obviously, we can use the intersection of the two event lines to locate a sensor, as shown in Figure 9.19. More formally

$$L(T_i) = E(t_{i1}) \cap E(t_{i2}) \tag{9.56}$$

  Where $D(e(t_{i1}, p_i))$ = true, $D(e(t_{i2}, p_i))$ = true for node i positioned at p_i.
- *Area cover EDF*
  Besides "line" coverage, we can also perform "area" coverage. Other devices, such as light projectors, can generate events that cover an area. They can implement *area cover EDF*. Area cover EDF partitions the space A into multiple parts and assigns a unique binary identifier, called code, to each section. Let us suppose that the localization is done within a plane ($A \subset R^2$). Each section S_k within A has a unique code k. The area cover EDF is then defined as follows:

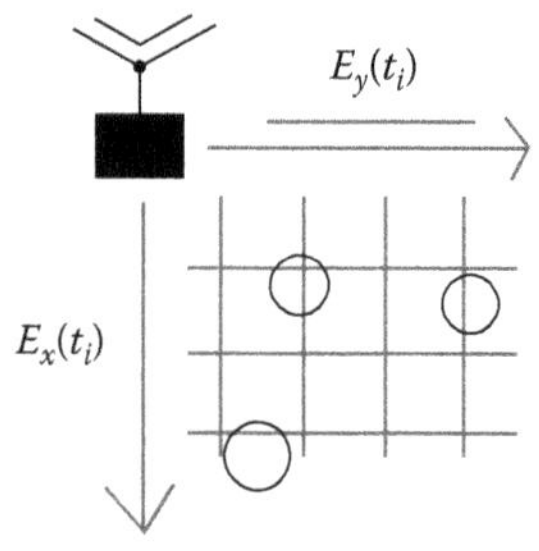

Figure 9.19 The implementation of the line scan EDF. (Adapted from Stoleru, R. et al., A high-accuracy, low-cost localization system for wireless sensor networks, *SenSys '05*, San Diego, CA, November 2–4, 2005.)

$$\text{BIT}(k, j) = \begin{cases} \text{true}, & \text{if } j\text{th bit of } k \text{ is } 1 \\ \text{false}, & \text{if } j\text{th bit of } k \text{ is } 0 \end{cases} \tag{9.57}$$

$$E(t) = \{p \mid p \in S_k \wedge \text{BIT}(k,t) = \text{true}\}$$

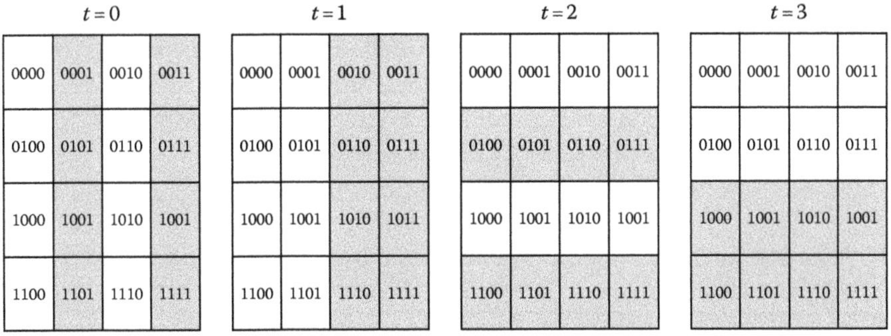

$t = 0$

0000	0001	0010	0011
0100	0101	0110	0111
1000	1001	1010	1001
1100	1101	1110	1111

$t = 1$

0000	0001	0010	0011
0100	0101	0110	0111
1000	1001	1010	1011
1100	1101	1110	1111

$t = 2$

0000	0001	0010	0011
0100	0101	0110	0111
1000	1001	1010	1001
1100	1101	1110	1111

$t = 3$

0000	0001	0010	0011
0100	0101	0110	0111
1000	1001	1010	1001
1100	1101	1110	1111

Figure 9.20 The steps of area cover EDF. The events cover the shaded areas. (Adapted from Stoleru, R. et al., A high-accuracy, low-cost localization system for wireless sensor networks, *SenSys '05*, San Diego, CA, November 2–4, 2005.)

and the corresponding localization algorithm is

$$L(T_i) = \{p \mid p = \text{COG}(S_k) \wedge (\text{BIT}(k,t) = \text{true if } t \in T_i) \wedge (\text{BIT}(k,t) = \text{false if } t \in T - T_i)\} \tag{9.58}$$

where $\text{COG}(S_k)$ denotes the center of gravity of S_k.

As shown in Figure 9.20, the plane A is divided into 16 sections. Each section S_k has a unique code k. The Spotlight device distributes the events according to these codes: at time j a section S_k is covered by an event (lit by light), if jth bit of k is 1. A node residing anywhere in the section S_k is localized at the center of gravity of that section. For example, nodes within section 1010 detect the events at time $T = \{1, 3\}$. At $t = 4$ the section where each node resides can be determined.

9.8 LOCALE: Collaborative Localization Estimation for Sparse Mobile Sensor Networks [Zhang08]

Zhang and Martonosi [Zhang08] present *low-density collaborative ad hoc localization estimation* (LOCALE). It has the following features:

First, it is a distributed localization algorithm. It does not need a center control.

Second, it is built on collaborative localization. That is, a few sensors work together to find a special location.

Third, it works best in sparse, mobile sensor networks. It may not work well in high-density scenarios. But it can be used in mobile cases.

LOCALE can actively predict and maintain the location estimation even during disconnection periods. It uses dead-reckoning (DR) system to achieve such a goal. When a sensor meets a neighbor, they swap position estimates and then refine the location of the nodes by a linear combination of the two estimates, weighted by the variances. The final effect is that we can smoothly average the sensors' movement and give each sensor a distribution that describes its location. Sensors use such distributions to get a good prediction of their actual locations. Moreover, "confidence estimate" of prediction accuracy can be obtained.

LOCALE has the following main features:

- A sensor readjusts its location estimation after swapping information with neighbors it encountered.
- Its localization accuracy is much higher than commonly used beacon-tracking method.
- Fast error correction when sensors do not have accurate estimations.
- Suitable to WSNs with sparse and heterogeneous systems.
- Low-power design.

9.8.1 Collaborative Location Estimation

LOCALE is a delay-tolerant, collaborative localization policy that is effective for sparse mobile sensor networks. LOCALE includes three major phases, shown in Figure 9.21, to maintain and refine location estimations. The first one is called *local phase*, which uses the movement tracking information of the sensor to maintain coarse location estimation. This phase allows the sensor to maintain location information during long periods of disconnection, although it may not be sufficiently accurate. In the *transform phase*, a sensor uses the location estimation of its neighbor to estimate its own location. In the *update phase*, a sensor combines the estimation obtained from the neighbor and the existing estimation. This phase refines location estimation of a sensor.

Next let us see how LOCALE represents the position of a node.

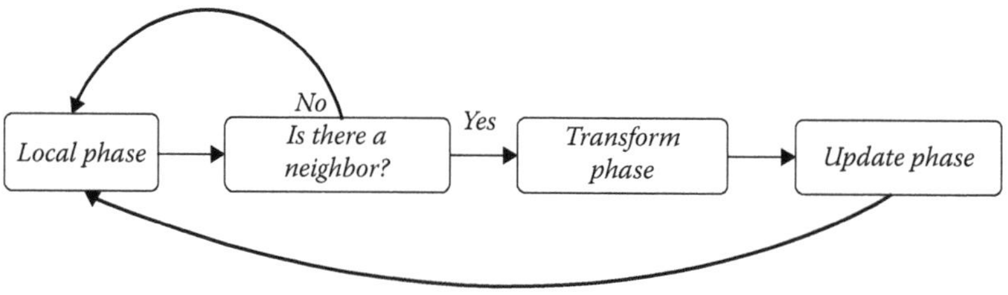

Figure 9.21 LOCALE overview: Location error increases during the local phase and decreases with collaboration in the update phase. (Adapted from Zhang, P. and Martonosi, M., LOCALE: Collaborative localization estimation for sparse mobile sensor networks, *Proceedings of the 2008 International Conference on Information Processing in Sensor Networks (IPSN 2008)*, St. Louis, MO, April 22–24, 2008, Information Processing in Sensor Networks, IEEE Computer Society, Washington, DC, 2008, 195–206.)

9.8.2 Location in LOCALE

If LOCALE needs to predict and merge localization information from multiple estimations, it needs both absolute location estimation and estimation certainty (i.e., confidence), and the location of a node is actually a smooth average of location estimations of neighbors.

A normal distribution can describe the location estimation (mean) and the certainty (variance). Although location estimation of an individual node may not follow a normal distribution, based on the central limit theorem, the averaged estimation should approach a normal distribution. We then use the probability density function as follows:

$$p(X) = \frac{1}{2\pi\sqrt{|C|}} * e^{-\frac{1}{2}(X-\bar{X})^{T}C^{-1}(X-\bar{X})} \tag{9.59}$$

This equation represents the probability of the *true* location for node (X) relative to the *estimated* location ($\bar{x}$). To define the equation we need only the estimated location ($\bar{x}$) and the covariance matrix C. Here we focus on a 2D case; however, it can be easily extended to 3D case with altitude information.

$$C = \begin{pmatrix} \sigma_x^2 & \rho\sigma_x\sigma_y \\ \rho\sigma_x\sigma_y & \sigma_y^2 \end{pmatrix} \quad X = \begin{pmatrix} x \\ y \end{pmatrix} \tag{9.60}$$

In this matrix, the diagonal has the variances along the axes of its coordinate system, and the other values are the covariances between the two axes. When the nodes move and meet neighbors, those values will be updated.

LOCALE keeps three variables: the location estimation $\bar{X}$, the covariance matrix C, and the angle θ between the local coordinates and the global coordinates. Figure 9.22 illustrates the relative angle of the neighboring nodes θ_o, where each node has its own local coordinate (x_h, y_h) (x_n, y_n), and an angle (θ_h, θ_n) relating it to the global coordinate (x, y).

9.8.3 Local Phase

In this phase, each node maintains local position estimation based on any movement tracking methods. LOCALE uses low-cost, low-accuracy dead-reckoning sensors to track their movement relative to their last measured location.

Thus we can obtain a new distribution through the combination of relative measurement distribution and the existing estimation distribution:

$$N = N_{\text{old}}(X_1, C_1) + N_{\text{delta}}(X_2, C_2) \tag{9.61}$$

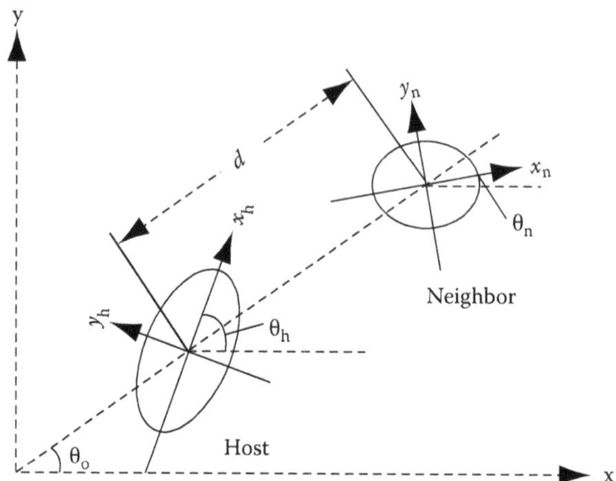

Figure 9.22 Representation of two neighboring nodes with different orientations. (Adapted from Zhang, P. and Martonosi, M., LOCALE: Collaborative localization estimation for sparse mobile sensor networks, *Proceedings of the 2008 International Conference on Information Processing in Sensor Networks* (*IPSN 2008*), St. Louis, MO, April 22–24, 2008, Information Processing in Sensor Networks, IEEE Computer Society, Washington, DC, 2008, 195–206.)

Together this gives the new distribution with mean and variance as

$$N_{\text{combined}} = N(X_1 + X_2, C_1 + C_2) \tag{9.62}$$

The movement covariance matrix is oriented in the movement direction. The covariance matrix in local coordinate C_L is described as

$$C_L = \begin{pmatrix} \sigma_{x'}^2 & 0 \\ 0 & \sigma_{y'}^2 \end{pmatrix} \tag{9.63}$$

The local covariance matrix is rotated to the global coordinate by

$$C = R(-\theta)^T C_L R(-\theta) \tag{9.64}$$

where θ is the direction the node moved, and the rotational matrix is defined as

$$R(\theta) = \begin{pmatrix} \cos(\theta) & -\sin(\theta) \\ \sin(\theta) & \cos(\theta) \end{pmatrix} \tag{9.65}$$

Note that the mean and covariance matrix of the new estimated location distribution is simply the summation. By incorporating the relative movement tracking information, a sensor can merge the movement information along with the covariance information.

9.8.4 Transform Phase

Although we can use the location of a neighbor to refine the location of a node, we cannot simply use the position estimate of a neighbor and merge it with its own. This is because there is distance between any two nodes. Therefore, LOCALE transforms the neighbor's estimate to a format suitable for merging, and that format should be an observation on the location of the host. The transform information needs the information on the "relative" location between the two nodes. Figure 9.23 illustrates the transform principle.

In terms of "relative" location information, LOCALE allows multiple formats, such as the relative measurement of range and direction of the neighbor, or simple information indicating that the neighbor is somewhere within communication range. And it is not difficult to obtain the range and direction information [BKusy07].

In Figure 9.23, we can see the procedure of merging a neighbor's observation into the local frame of the host. It mainly includes the following few steps:

In Step 1, we rotate the observations to comply with the relative coordinate so that the X-axis of the two observations coincide:

$$\begin{aligned} C_h &= R(\theta_o - \theta_h)^T C_{Lh} R(\theta_o - \theta_h) \\ C_n &= R(\theta_o - \theta_n)^T C_{Ln} R(\theta_o - \theta_n) \end{aligned} \tag{9.66}$$

In Step 2, based on the angle uncertainty caused by the location uncertainty of the host, we can calculate the y-component of the transformed covariance matrix.

In Step 3, we can then determine the transformed observation distribution through the x component of the covariance matrix. Note that the matrix should consider the variability of distance, which is the sum of x components of the host variance and $Range^2(1 - 2\sqrt{2}/3)$.

The covariance of the error observations should be 0 because all nodes are oriented in the relative coordinates. The mean value of the observation to be merged is moved by distance d, and the expected distance vector between the two neighboring nodes is $Range/\sqrt{2}$ in the direction $\theta_o - \theta_n$.

$$\begin{aligned} C_{L\,\text{Observed}} &= \begin{pmatrix} \sigma_{\text{radio}} + \sigma_n & 0 \\ 0 & \sigma_h + 2\sigma_n \end{pmatrix} \\ X_{\text{Observed}} &= \begin{pmatrix} x_n + d * \cos(\theta_o - \theta_n) \\ y_n + d * \sin(\theta_o - \theta_n) \end{pmatrix} \end{aligned} \tag{9.67}$$

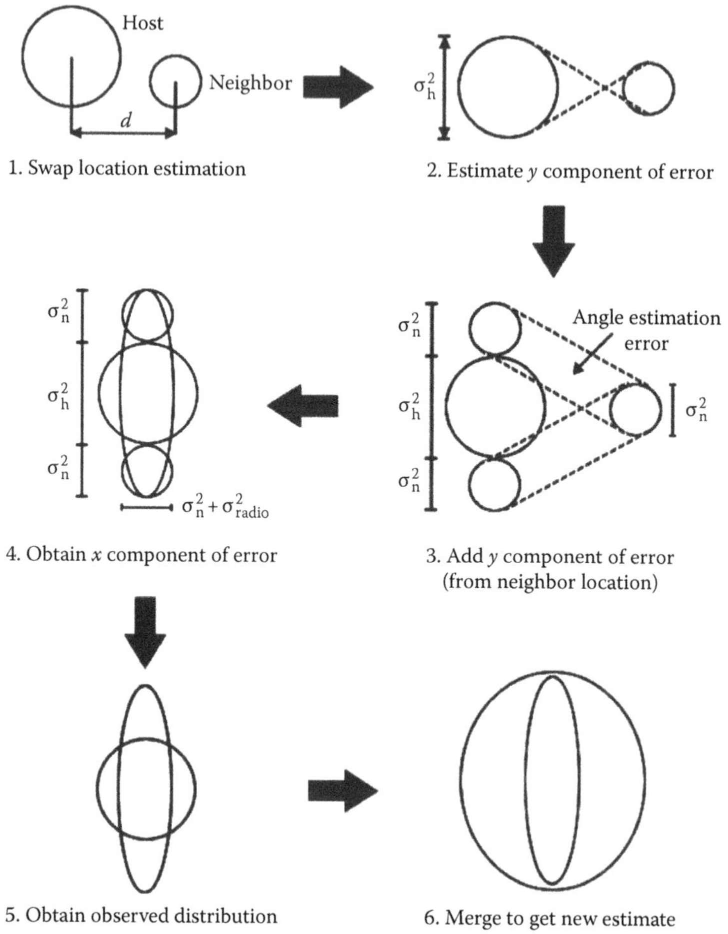

Figure 9.23 Relative location estimation with no direction measurement. (Adapted from Zhang, P. and Martonosi, M., LOCALE: Collaborative localization estimation for sparse mobile sensor networks, *Proceedings of the 2008 International Conference on Information Processing in Sensor Networks (IPSN 2008)*, St. Louis, MO, April 22–24, 2008, Information Processing in Sensor Networks, IEEE Computer Society, Washington, DC, 2008, 195–206.)

Finally, to achieve final merging, the transformed observation and the host distribution need to be rotated to the global coordinate system:

$$C_{\text{Observed}} = R(-\theta_o)^{\mathrm{T}} C_{\text{L Observed}} R(-\theta_o) \tag{9.68}$$

As we can see, a good thing on the transformation phase is that the system can project neighbor observation to a self-observation, allowing for more accurate merging in the next phase. Only probabilistic measurements are needed and no particular radio profile or special hardware are required.

9.8.5 Update Phase

In the two phases given earlier, we can see that LOCALE improves localization accuracy by merging observations from neighbors. This way we can average out measurement errors. On the other hand, if sensors have different movement patterns, the location estimations will have different certainties. We can use variances to represent the certainties. Eventually we merge the estimations weighted by their respective variances. The location distributions are merged as a weighted linear combination.

In the "update phase," the system performs self-estimation preparation and the final merging process, which is shown in Figure 9.24. Because the number of observations can increase, we should calculate the combination of these distributions as the harmonic mean.

The merge factor represents the weight each distribution has on the location estimation result. The merge factor is defined as

$$K = C_{\text{h}} * [C_{\text{h}} + C_{\text{observed}}]^{-1} \tag{9.69}$$

The merge factor is used to calculate both the new covariance matrix and the new location estimation as follows:

$$\begin{aligned} C_{\text{merged}} &= C_{\text{h}} - KC_{\text{h}} \\ \hat{X}_{\text{merged}} &= \hat{X}_{\text{h}} + K(\hat{X}_{\text{observed}} - \hat{X}_{\text{h}}) \end{aligned} \tag{9.70}$$

The new angle of the covariance matrix is

$$\theta = \frac{1}{2}\tan^{-1}\left(\frac{2b}{a-d}\right), \quad C = \begin{pmatrix} a & b \\ b & d \end{pmatrix} \tag{9.71}$$

Finally, we can rotate the merged distribution back to the local coordinate.

$$C_{\text{L new}} = R(-\theta_{\text{merged}})^{\text{T}} C_{\text{merged}} R(-\theta_{\text{h}}) \tag{9.72}$$

This merged location and the covariance matrix will be recorded in the memory in each round as the new self-location estimate. This merging algorithm is a linear combination. Therefore the process can be repeated as long as more neighbors are coming.

The update phase, in conjunction with other parts of LOCALE, achieves the delay-tolerant, collaborative localization in extremely sparse sensor networks.

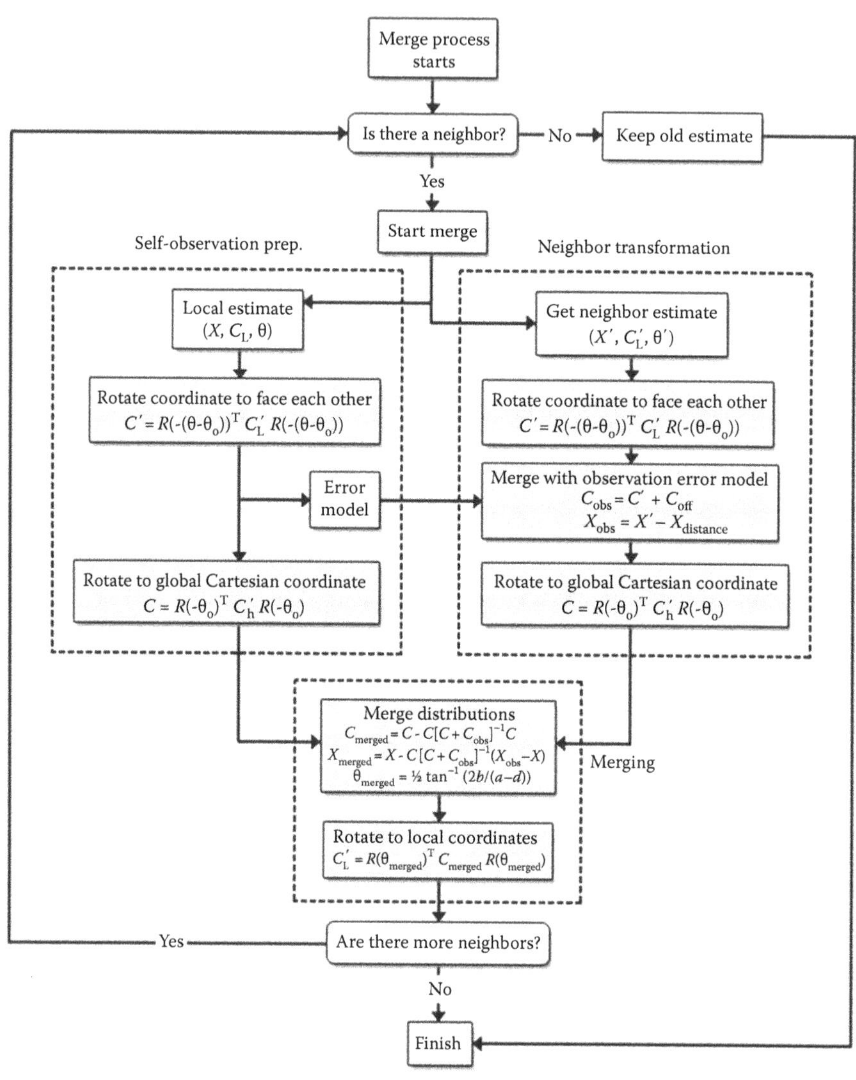

Figure 9.24 Block diagram of the merging process when neighbor is encountered. (Adapted from Zhang, P. and Martonosi, M., LOCALE: Collaborative localization estimation for sparse mobile sensor networks, *Proceedings of the 2008 International Conference on Information Processing in Sensor Networks (IPSN 2008)*, St. Louis, MO, April 22–24, 2008, Information Processing in Sensor Networks, IEEE Computer Society, Washington, DC, 2008, 195–206.)

9.9 On the Security of WSN Localization [ASrinivasan08]

There are four important metrics associated with localization: energy, efficiency, accuracy, and security. WSNs are designed to operate in harsh environments. With hostile environments encountered by military equipment like landmine detection, battlefield surveillance, or target tracking, the conditions can become much more severe. In these unique environments, WSNs have to work autonomously and are thus faced with unique challenges. An adversary can now capture and compromise one or more sensors physically, thus rendering it to the mercy of its assailant. By injection malicious code, the adversary is able to manipulate the workings of the device, extract cryptographic information, or destroy it completely. If sensitive information is extracted from the device, security hurdles such as authentication routines can be bypassed, enabling the assailant to launch an attack from within the system, which would cause most systems to fail.

To understand this, consider a beacon-based localization model, and as sensor nodes are not capable of determining their own location, they are unable to decipher whether or not beacon nodes are transmitting truthful, accurate location information. A malicious node could be present transmitting falsified data, thus causing the receiving node to develop false location information. This is referred to as information asymmetry, where one entity has more information than the other. This model in beacon-based localization has been addressed in [ASrinivasan06], and also presents an effective way of resolving insider attacks. The attacker can also launch sybil, wormhole, or replay attacks to disrupt the localization process.

In this section, we review the existing secure localization techniques, throwing light on their strengths and weaknesses.

9.9.1 SeRLoc

In [LLazos04], Lazos and Poovendran propose a novel scheme for the localization of nodes in WSNs in un-trusted environments called SeRLoc. This is a range-free, distributed, resource-efficient technique in which there is no communication requirement between nodes for location discovery. This method is also robust against wormhole attacks, sybil attacks, and sensor compromise. In this method, we consider two sets of nodes: N, which is the set of sensor nodes equipped with omnidirectional antennas, and L, which is the set of locator nodes possessing directional antennas. The sensors are able to use the location information transmitted by the locators to determine their own location. Each locator transmits different beacons at each antenna sector.

In SeRLoc, an attacker must impersonate a multitude of nodes to compromise the localization process. Also, the adversary has no incentive to impersonate sensor nodes, as each node computes its own location without any assistance from the other sensors. There exist two techniques in this model to prevent wormhole attacks: sector uniqueness property and communication range violation property.

To improve localization accuracy, either more locators must be deployed or more directional antennas must be used. The entire process is made using the assumption that no jamming of the wireless medium is feasible. This is a very strong assumption for a real-world setting.

9.9.2 Beacon Suite

In [DLiu05], Liu, Ning, and Du present a suite of techniques for detecting malicious beacon nodes that provide incorrect information to sensor nodes providing location services in critical applications. This includes the detection of malicious signals, the detection of replayed signals, the identification of malicious nodes, the avoidance of false detection, as well as the revoking of malicious beacon nodes. Beacon nodes serve two purposes: providing location information to sensor nodes and performing detections on the beacon signals they hear from other beacon nodes. It is not necessary for a beacon note to wait idly for beacon signals. It can request location information. The node performing the detection is called the detection node, and the node it is listening from is called the target node. They suggest that the detecting node should use a non-beacon ID when requesting location information from a target node to observe the true behavior of the target node. The revocation scheme works on the basis of two counters being maintained for each beacon node, representing an alert and report counter, respectively. The alert counter is responsible for recording suspiciousness of the corresponding beacon node, and the report counter tallies the number of alerts this node reported.

In the event that a detecting node determines that a target node is misbehaving, this report is sent to the base station. Reports from alert counters are only accepted from detecting nodes whose report counter is below a threshold and against nodes that have not yet been revoked. Meeting these criteria increments the alert and report counter of this respective node. The two counters function on a discrete scale, and the revocation mechanism is centralized. This has been improved to be more robust in [ASrinivasan06] by employing a continuous scale and a reputation and trust-based mechanism.

9.9.3 Attack-Resistant Location Estimation

In [DLiu05a], Liu, Ning, and Du put forward two range-based robust methods to tolerate malicious attacks against beacon-based location discovery in sensor networks. The first method is the attack-resistant minimum mean square estimation, which filters out malicious beacon signals. By examining the inconsistency among location references of different beacon signals, which can be represented by the mean square error of estimation, we can defeat attacks by removing harmful data. The second method, voting-based location estimation quantifies the deployment field into a grid of cells and has each location reference "vote" on the cells in which the node may reside. This method tolerates malicious beacon signals by adopting an iteratively refined voting scheme. Both methods can survive malicious attacks, even if the attacks bypass the authentication process.

9.9.4 Robust Statistical Methods

In [ZLi05], Li, Trappe, Zhang, and Nath introduced the idea of being tolerant to attacks rather than trying to eliminate them by exploiting redundancies at various levels within wireless networks. Two classes of localization are examined: triangulation and RF-based fingerprinting. They propose two statistical methods for securing localization in sensor networks, both based on the simple idea of filtering outliers from the data in the range estimates used for the location estimation.

In the triangulation model, an adaptive least squares and least median squares estimator is used. This adaptive estimator switches to the robust mode with least mean squares estimation when attacked and exhibits the computational advantage of least squares in the absence of attacks. In the fingerprinting model, a Euclidean distance metric is not secure enough, thus they propose a median-based nearest neighbor scheme that defends against location attacks. The authors also discussed attacks unique to localization in sensor networks. The statistical methods proposed in [ZLi05] are based on the assumption that benign observations at a sensor always outnumber malicious observations.

Problems and Exercises

9.1 Use any software (tools) to study and verify the efficiency of one of the discussed WSN localization algorithms.

9.2 Compare different localization schemes from algorithm complexity, accuracy, and practical implementation (in distributed sensors) viewpoints.

9.3 Why should we consider security in WSN localization?

Chapter 10

Time Synchronization in Wireless Sensor Networks

In this chapter, we discuss the basic concepts of WSN clock synchronization schemes. Our discussions are based on the summary of a comprehensive survey paper [Sundararaman05]. Readers may refer to [Sundararaman05] for more details (such as the comparisons between different WSN synchronization schemes).

Remember

Do not think that it is simple to define "time" in wireless sensors. How does a sensor guarantee that its claimed local time is correct (i.e., equal to world standard time)? Maybe one argues, "just asks a server to broadcast a standard time to all sensors." However, wireless transmission delays cannot be ignored. Moreover, when a server gets a message (which has the right time) ready for transmission, it needs to pass through a series of local CPU operations to generate such a message. Such local delay cannot be ignored either. Anyway, eventually a sensor receives a message from a server, and such a message says "9AM now." Can this sensor just set up its local time to "9AM"?

10.1 Introduction

Time synchronization is one of the most important issues in WSNs, because all sensor events need to have accurate time-stamp records. Especially in an object-tracking application, if the timing information is not accurate, we cannot determine the object's trajectory. This is because we form a trajectory by linking all navigated locations at different times.

In *wired* networks, such as the Internet, researchers have created successful clock synchronization protocols, such as NTP (network time protocol). However, these are not appropriate for a WSN environment due to some reasons.

First, these *wired* network synchronization protocols do not work well in a wireless environment that has high error rate due to wireless interference.

Second, a WSN could have thousands of resource-constrained sensors. The synchronization protocol needs to be highly scalable in large WSNs. Moreover, it should be able to achieve a self-organized, robust synchronization without a central control.

Third, these synchronization protocols need to consider energy conservation as a major concern. Power sources cannot be provided to each sensor, and their small sizes limit the amount of energy that can be stored or collected.

Therefore, we need a brand-new clock synchronization protocol for WSNs having unreliable wireless links, large sensor density, and very limited energy and memory.

Before we design a clock synchronization protocol, we need to understand the notion of a computer clock. It has the following basic features:

1. A computer clock could be generated by an electronic device that counts oscillations in an accurately machined quartz crystal, at a particular frequency.
2. Or, a clock could be determined by an ensemble of hardware and software components. The hardware (quartz crystal) and software (timer control program) work together to provide an accurate, stable, and reliable time-of-day function to the operating system and its clients.
3. A computer clock is essentially a timer. The timer counts the oscillations of the crystal. Two registers, that is, a *counter register* and a *holding register*, work together as follows: The counter register decreases by one for each oscillation in the crystal. When the counter reaches zero, the sensor generates a timer interrupt, which can be used to perform a specified timing task, and the counter is reloaded from the holding register (to the original counter value).

How do we provide a time stamp to a sensor event? Such a time stamp is obtained from the system's clock value, which is acquired from the reading of the timer (described above). For instance, each time the counter reaches zero, the timer adds 1 to a system clock.

Although we expect that all sensors have exactly the same pace of *internal timer counters,* unfortunately, in practice, the quartz crystals in each sensor could run at slightly different frequencies, causing the clock values to gradually diverge from each other. We call such a divergence a *clock skew* (defined later), which can lead to an inconsistent notion of time in different sensors.

The clock value generated by an internal timer is also called a *software clock.* It is determined by crystal oscillations. Unfortunately, most crystal oscillations are not very accurate because the frequency that makes time increase is never exactly right. Even a frequency deviation of just 0.001 percent can bring a clock error of about 1 s per day.

Now you see the purpose of clock synchronization—to correct the clock skew in a distributed sensor system. There are two general methods to correct the clock skew:

1. *Absolute synchronization*: Clocks of all sensors should be synchronized to an accurate real-time standard like *UTC* (*universal coordinated time*). In other words, all local clocks must not only be synchronized with each other but also have to adhere to a *physical time.*
2. *Relative synchronization*: In some applications, we do not require that all clocks synchronize with a global time. Instead, *clocks are relatively synchronized to each other* because the requirement is only to provide an ordering of events, and not the exact real-world time at which each event occurred.

Let us further define some important concepts to be used in clock synchronization:

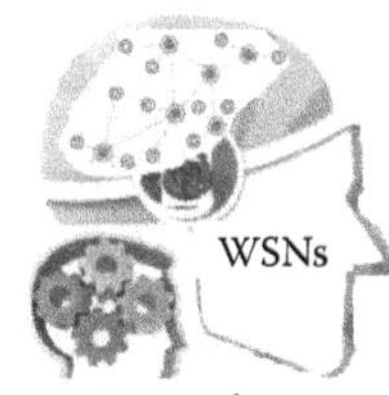

Time: In a sensor p, the clock's reading (i.e., its claimed time) is determined by the function $C_p(t)$. If t is the global standard time, $C_p(t) = t$ represents a *perfect clock* (*i.e., no clock skew*).

Clock offset: We define the difference between the time reported by a clock (i.e., $C_p(t)$) and the *real time* (i.e., t) as *offset*. The *offset* of the clock C_a is given by $C_a(t) - t$. The offset of a clock C_a relative to a clock C_b at time t is given by $C_a(t) - C_b(t)$.

Clock frequency: Frequency is the *rate* at which a clock progresses. The frequency at time t of the clock C_a is the *derivative* of its *time function* (see the above definition), that is, $C_a'(t)$. A perfect clock has a frequency of 1.

Clock skew: The *clock skew* is the difference in the *frequencies* of the clock (its *frequency* is $C_a'(t)$) and the perfect clock (its *frequency* is 1), *that is, the clock skew in a sensor is* $C_a'(t) - 1$. The skew of a clock C_a relative to a clock C_b at time t is $(C_a'(t) - C_b'(t))$.

Clock drift: The drift of clock C_a is the *second derivative* of the clock value with respect to time, namely, $C_a''(t)$. The drift of a clock C_a relative to a clock C_b at time t is $(C_a''(t) - C_b''(t))$

Now let us consider the physical clock synchronization in a WSN to a perfect time—UTC. Assume that the time at UTC is t, certainly we wish $C_p(t) = t$ for all p and all t, that is, the clock frequency $dC/dt = 1$.

However, due to clock skew, the time in the clock of sensor p is $C_p(t)$, which is not equal to t. For this case, a timer (clock) is said to be working *within its specification* if its *clock frequency* is within a scope:

$$1-\rho \le \frac{dC}{dt} \le 1+\rho$$

where constant ρ is the maximum skew rate specified by the manufacturer.

Or, we can say its *clock skew* (relative to a perfect clock) is within a scope:

$$-\rho \le \left(\frac{dC}{dt}-1\right) \le \rho$$

Refer to the previous definition, $C(t)$ is the local time. It can be any function, that is, it may not be equal to perfect time (its function is $C(t) = t$). We also know that clock *frequency* (rate) is its derivative, that is, $C'(t) = dC/dt$. For perfect time, $C'(t) = 1$. *Clock skew* is the difference between the frequency of a local time and that of a perfect time, that is, $(dC/dt) - 1$.

Figure 10.1 illustrates the behavior of fast, slow, and perfect clocks with respect to UTC.

There are some basic requirements for *clock synchronization protocols* that use network message exchange between nodes to achieve a synchronized clock:

1. The synchronization protocol should be robust to unbounded message transmission latencies and unreliable wireless communications.
2. If a node wants to synchronize with another one, it must be able to *estimate the local time on the clock of the other node*, which is not a trivial issue due to network latency.

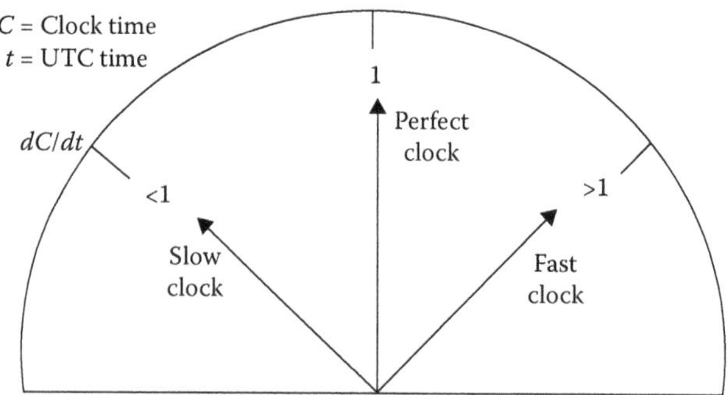

Figure 10.1 Behavior of fast, slow, and perfect clocks with respect to UTC. (Adapted from Sundararaman, B. et al., *Ad Hoc Netw.*, 3, 281, May 2005.)

3. We cannot run time backward, that is, we cannot set back clocks. All clocks must be gradually and gracefully advanced until the correction is achieved.
4. We should minimize the synchronization overhead from network communication viewpoint. For instance, we cannot use too many message exchanges between nodes.

10.2 Synchronization in General Networks (Non-WSN)

Before we discuss the synchronization issues in WSNs, let us take a look at some issues already solved in general networks (i.e., non-WSNs).

10.2.1 Remote Clock Reading

As mentioned earlier, message exchanges are used to accomplish clock synchronization between any two nodes. Because a node does not know local clock values of the other nodes, it can only *estimate* the time in the clock of the other node. Such an estimate should consider the effect of network delay. After it gets an estimated clock value, it can compute the time difference between the clocks of the nodes, and adjust its local clock.

However, there exist nondeterministic and unbounded message delays that make synchronization very difficult. Therefore, *the effectiveness of a synchronization protocol lies in its ability to prevent nondeterministic message delays from affecting the quality of synchronization.*

The *remote clock reading* method [FCristian89] handles unbounded message delays between processes (a process is the clock estimate program in a node). By using the remote clock reading method, it synchronizes several clients to an accurate time service, UTC.

Figure 10.2 shows its procedure to *remotely* read the time of the other node:

1. At a local time point T_0, a client sends a message to the server requesting a time stamp.
2. The server then returns a message holding the time stamp S_{time}. *Note*: S_{time} is the local time at the server.
3. The client receives this message at its local time T_1.
4. The client then sets its local time to S_{time} (accurate time from the server) + $(T_1 - T_0)/2$ (time required to transmit the message).
5. To enhance accuracy, Steps 1–4 will be repeated and the average will be used.

10.2.2 Offset Delay Estimation Method

The most popular clock synchronization method used in the Internet, called *NTP* [DLM92], adopts the *offset delay estimation* method to estimate clock offset (see definition before).

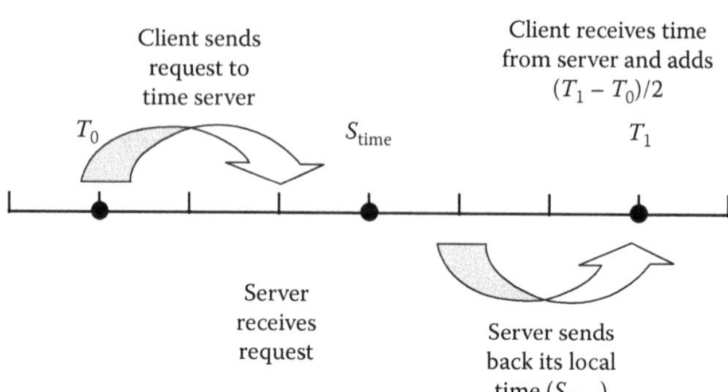

Figure 10.2 Cristian's synchronization protocol. (Adapted from Cristian, F., *Distrib. Comput.*, 3, 146, 1989.)

A hierarchical tree of time servers is implemented in its design. The tree root is the primary server, which synchronizes with the UTC. Secondary servers, which act as a backup to the primary server, are contained in the next tree level,. The clients are located at the lowest tree level. These clients need to synchronize with the tree root that has UTC time.

Because a client node cannot accurately estimate the local time of the target node due to varying network delays during message transmission, NTP performs several round-trip trials and chooses the trial with the minimum delay. This is similar to the above-mentioned Cristian's remote clock reading method [FCristian89], which also relied on the same strategy to estimate message delay.

As shown in Figure 10.3, assume that nodes A and B exchange NTP time stamps. Node A sends a message at T_3; Node B gets it at T_1, and Node B feedbacks a message at T_2, which is received by Node A at T_4. Assume that clocks in A and B are stable and running at the same speed. Then

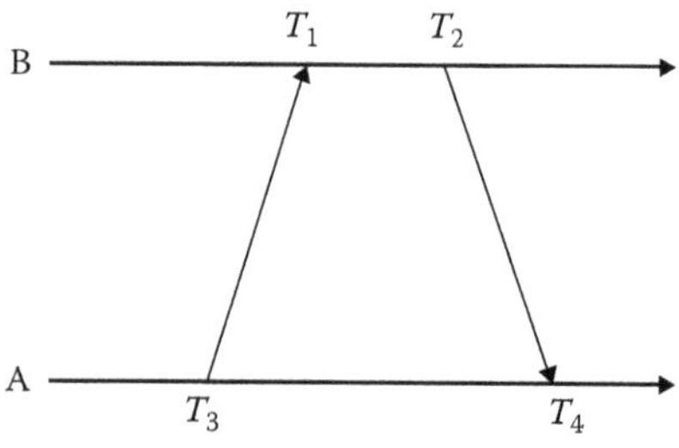

Figure 10.3 Offset and delay estimation. (Adapted from Mills, D.L., *IEEE Trans. Commun.*, 39(10), 1482, October 1991.)

$a = T_1 - T_3$ is the message transmission delay from A → B;
$b = T_2 - T_4$ is the message transmission delay from B → A.

Although a and b could be different due to the asymmetric nature of many communication links, the differential delay (= $a - b$) is small in most cases. We define the *clock offset* θ and *roundtrip delay* δ as

$$\theta = \frac{a+b}{2}, \quad \delta = a - b$$

Note that we could record the values of three time stamps T_1, T_2, and T_3 in the message header when it travels between A and B. However, the value of T_4 cannot be put in the message as it can only be determined upon arrival. Thus, both peers A and B can independently calculate *clock offset* θ and *roundtrip delay* δ using a single bidirectional message stream, as shown in Figure 10.4.

Based on Figure 10.4, we can describe the NTP protocol as follows:

Assume that two servers (A and B) exchange timing messages to achieve time synchronization. A server calculates a pair of parameters (O_i, D_i) during the calculation round i, where O_i is the offset value in that round (i.e., θ) and D_i is the transmission delay (i.e., δ). For all rounds of (O_i, D_i), we select the offset corresponding to the minimum delay.

Specifically, the delay and offset pair (O_i, D_i) are calculated as follows. Assume that message m (from $A \rightarrow B$) takes time t to transfer and m'(from $B \rightarrow A$) takes time t' to transfer. We know that O_i is the offset between A's clock and B's clock. If A's *local* clock time is $A(t)$ and B's local clock time is $B(t)$, we have

$$A(t) = B(t) + O_i$$

Then,

$$T_{i-2} = T_{i-3} + t + O_i$$

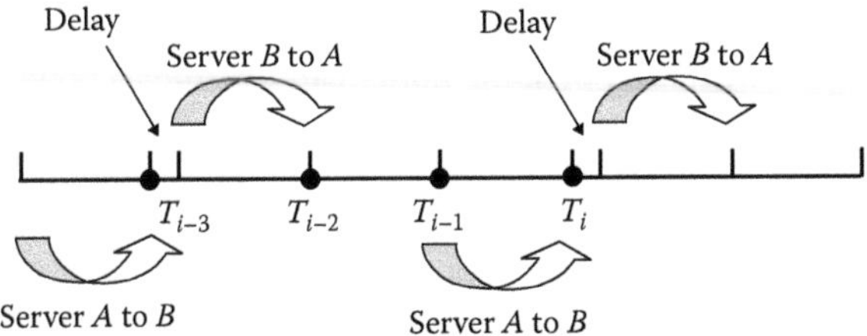

Figure 10.4 Timing diagram for the two servers. (Adapted from Mills, D.L., *IEEE Trans. Commun.*, 39(10), 1482, October 1991.)

$$T_i = T_{i-1} - O_i + t'$$

Typically, we can assume that $t = t'$ in Internet (but not in a WSN). We then subtract these two equations from each other, the offset O_i can be estimated as

$$O_i = \frac{T_{i-2} - T_{i-3} + T_{i-1} - T_i}{2}$$

We can also estimate the *round-trip delay* D_i as

$$D_i = (T_i - T_{i-3}) - (T_{i-1} - T_{i-2})$$

NTP calculates eight most recent pairs of (O_i, D_i). The value of O_i that corresponds to minimum D_i is chosen to estimate the whole network average offset value O.

Due to its averaging approach (it uses eight round trips), the offset/delay estimation protocol is similar to the above-mentioned Cristian's method [FCristian89]. But both methods have a high synchronization overhead in terms of message complexity. However, as delays are partly compensated through forward and backward messages, the accuracy of NTP is better than that of Cristian's protocol.

10.3 Clock Synchronization in WSNs

Traditional clock synchronization protocols work well in wired networks. But they may not be suitable for WSNs due to the following concerns in WSN environments:

1. *Energy efficiency*: Energy conservation is very important in WSNs because all sensors are battery driven. Traditional protocols like NTP [DLM91] use an external standard like GPS (global positioning system) or UTC (universal time) to synchronize the network to an accurate time source. However, the use of GPS poses a high demand for energy that is usually not available in WSNs. This makes it difficult to maintain a common notion of time.
2. *Network dynamics*: An initial set-up is usually required for a stationary sensor network, without any mobility, to begin operation. However, when new sensors keep adding to the WSN, or some nodes can die due to power drainage, we can see the change in the neighborhood of each node and the configuration of the network. A worse case is when the nodes are mobile, the network topology becomes more dynamic. Hence, the WSN synchronization protocols should adapt to both stationary and dynamic network topology, and must ensure *self-configuration* (by the use of suitable neighborhood definition or leader election protocols).
3. *End-to-end latency*: Internet can use NTP to achieve synchronization very well because Internet is based on *wired* networks that are fully connected networks. In such networks, the message transmission delay is relatively

stable (i.e., we can get a constant end-to-end delay throughout the network). However, WSNs use high error rate, wireless transmission over a shared medium. It is not practical to assume a single latency bound between the ends of the network. Therefore, synchronization protocols that assume a fully connected network with constant delay cannot be applied to multi-hop WSNs.

4. *Wireless loss*: Traditional wired networks have very few data loss events. But in a WSN we need to use *multiple rounds* of message exchange to figure out clock parameters (such as offset) due to frequent wireless loss.

- *Classification of WSN synchronization protocols*
  Dozens of different WSN clock synchronization protocols have been proposed. These protocols can be classified as follows:

1. *Sender-to-receiver versus receiver-to-receiver synchronization*
   Sender-to-receiver synchronization. We have already mentioned the NTP protocol that belongs to a "sender-to-receiver" message exchange approach. Such a scheme typically involves three steps. (a) The sender sends a message (with its local time as a time stamp) to the receiver. (b) The receiver sends back a message with its local time stamp. (c) The message delay between the sender and the receiver is calculated by measuring the total round-trip time.

   These steps may need to run multiple rounds to get an average value.
   Shortcomings: Depending on the distance between a sender and a receiver, there could be highly variable latencies in each round of message exchange. When a message travels many hops to reach a receiver, the delay could be huge and very different between rounds. Although we can compute the average message delay after performing many rounds, the time parameters cannot be accurately estimated, and too many rounds of message exchange adds significant network overhead. Also, when we calculate time offset, we must consider the optimization of the time taken by the receiver to process the message, and the time taken by the sender to prepare and transmit the message.

   Receiver-to-receiver synchronization. To overcome the above issues, we may use a *receiver-to-receiver*–based synchronization. This is based on the following principle: if we ask a sender to send out messages to any two receivers that are close to each other, these two receivers will receive the message at approximately the same time. This approach exploits this property of the radio broadcast medium. Instead of sending out messages between the sender and the receiver for multiple rounds, the receivers can exchange the time at which they received the same message and then compute their offset based on the difference in reception times.

   Obviously, this approach can reduce the message-delay variance. We just need to be concerned about the propagation delay to the various receivers and the differences in receive time.

2. *Clock correction versus untethered clocks*

   Clock correction. Today people typically handle synchronization issues by correcting the local clock in each node to run on par with an atomic clock or a global timescale (such as a UTC). Such a *clock correction* scheme provides a convenient reference time. A node corrects its local clock either instantaneously or continually to keep the entire network synchronized.

   Untethered clocks. The above clock correction needs continuous synchronization, which can waste lots of energy. *Untethered* scheme does not correct local clocks based on a global clock. Instead, it just maintains a table with the comparisons of local clock and other nodes' time. Thus, the relative clock difference is monitored. For example, reference broadcast synchronization (RBS) [JElson02] (to be explained later) builds a table of parameters that relate the local clock of each node to the local clock of every other node in the network. Local time stamps are then compared using the table. In this way, a global timescale is maintained while letting the clocks run untethered.
3. *Internal synchronization versus external synchronization*

   Internal synchronization. In this approach, a WSN does not have a global time base. Therefore, our goal is to minimize the local time difference from the readings of sensors.

   External synchronization. In external synchronization, the system may rely on a standard source of time such as UTC. NTP [DLM91] synchronizes Internet nodes in this fashion. But WSNs do not perform external synchronization unless the application demands it, because energy efficiency is a primary concern, and employing an external time source typically induces high energy requirements.

   Internal synchronization can typically give us a more correct operation; external synchronization is primarily used to give the system convenient reference time. While internal synchronization can be performed in a peer-to-peer (i.e., no central server) or a master–slave fashion, external synchronization can only be performed in a master–slave fashion because it requires a master node that communicates with a time service (such GPS) to synchronize the slaves and itself to the reference time.
4. *Probabilistic versus deterministic synchronization*

   Deterministic synchronization. This is a typical way to achieve synchronization. It uses deterministic synchronization algorithms/protocols to guarantee an upper bound on the clock offset with certainty, that is, it can guarantee a certain clock precision for sure.

   Probabilistic synchronization. It cannot provide an absolute clock precision. Instead, it can only use a probabilistic value to show its control of the clock offset, that is, it has a failure probability in terms of absolute probability. Although a probabilistic approach has worse clock precision than a deterministic approach, it does not need to force the synchronization protocol to perform many message transfers. Thus it avoids extra processing. This can help to save energy.

5. *Stationary networks versus mobile networks*

   Mobile networks. The sensors can move. Moreover, a sensor communicates with other sensors only when it enters the geographical scope of those sensors. We need a robust synchronization protocol to handle the frequent network topology changes due to the mobility of the nodes.

   Stationary networks. Most WSNs have fixed network topology, that is, sensors do not move. It is easier to design synchronization protocol for stationary WSNs.

10.4 Evaluation of Synchronization Performance

Time synchronization protocol should be driven by the characteristics and requirements of each application. For instance, most WSN applications can use low-cost, low-precision synchronization protocols. However, some safety-critical applications, such as aircraft navigation or intrusion detection in military systems, will demand high-precision synchronization protocols for nodes to correctly identify events occurring at a certain time.

In [Sundararaman05], some performance metrics are identified to measure the quality of a synchronization protocol.

10.4.1 Precision

Hardware clock: The initial time signals are generated from the internal hardware (oscillator circuits) in a sensor. As mentioned before, the hardware clock can have clock skew. Therefore, we cannot directly use the time generated from such a hardware clock.

Logic clock: As hardware clock is not accurate, sensors generally use a logical notion of clocks and time. We could use software to modify such a logical clock during synchronization protocol. All of our discussions here refer to this logic clock concept.

Based on this logic clock concept, we could define two types of synchronization precision:

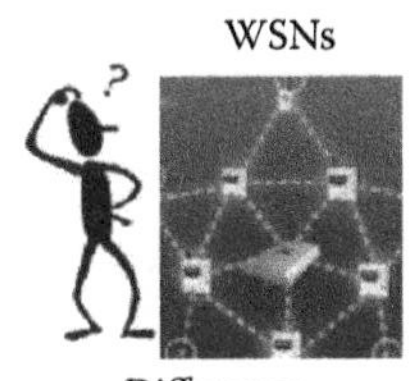

Absolute precision. Using an external standard (such as UTC) to measure the maximum clock skew/offset error of the logical clock of a node.

Relative precision. Without comparison with a standard clock, we only measure the maximum clock skew/offset deviation among logical clock readings of the nodes belonging to the same WSN.

Obviously, the goal of any WSN synchronization protocol is to achieve an absolute or relative precision. However, such a high synchronization precision comes at the expense of increased computational complexity and communication overhead (i.e., the number of messages exchanged among nodes).

A metric similar to the concept of precision is called *accuracy*, which measures how well the time maintained within a WSN is true to the standard time. A synchronization protocol with high accuracy thereby guarantees high precision.

10.4.2 Protocol Overhead

To reduce protocol overhead (i.e., the number of messages exchanged between nodes), we may use piggybacking, which is the process of combining the application data acknowledgment messages with messages that carry time synchronization information, that is, we do not use independent messages to transmit time information. Such control information can be embedded into general raw sensor data packets. This can not only reduce communication overhead but also save memory storage.

10.4.3 Convergence Time

Some synchronization protocols only perform clock control infrequently. And they do not use many message exchanges. Thus those protocols do not have a long *convergence time*, which refers to the total time required to synchronize a network. However, some high clock precision protocols require a large number of message exchanges per synchronization, and thus result in a longer convergence time.

10.4.4 Energy Efficiency

Energy efficiency is always the top concern for any WSN protocol design. A low-complexity protocol may save more energy.

10.4.5 Scalability

A synchronization protocol may need to make a large-scale WSN converge to the same time signal, that is, it should have the capability of synchronizing hundreds of sensors.

10.4.6 Robustness

Because the WSN uses low-bandwidth radio communications under harsh environments, a synchronization protocol should be able to tolerate high packet loss rate and large wireless interference.

10.5 Examples of WSN Synchronization Protocols

10.5.1 Reference Broadcast Synchronization [JElson02]

The RBS protocol [JElson02] is a receiver-to-receiver–based scheme. Two receivers receive the same message (from the same sender) at approximately the same time. RBS is so named because a message (that is broadcast from the same sender) will arrive at a set of receivers with very little variability in its delay.

If each receiver records the *local time* as soon as the message arrives, all receivers can achieve a clock synchronization with a high degree of precision by comparing their local clock values (for the same received message).

RBS broadcasts messages through a time-critical path, which delivers a message that contributes to *nondeterministic* clock errors in a protocol. In Figure 10.5, we can see the difference between sender-to-receiver time-critical path and receiver-to-receiver time-critical path (used in RBS).

It is important to remove/reduce the effects of *nondeterministic* transmission delays as they are detrimental to the accuracy of a synchronization protocol. They also make it difficult for a receiver to estimate the time at which a message was sent and vice versa.

The following four time factors are nondeterministic when a sender sends a message to a receiver in a WSN:

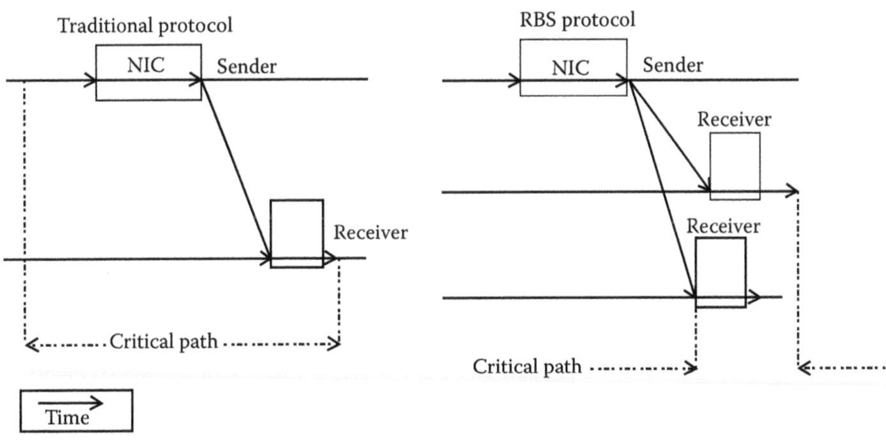

Figure 10.5 Time-critical path for traditional protocols (left) and RBS protocol (right). (Adapted from Elson, J. et al., Fine-grained network time synchronization using reference broadcasts, *Proceedings of the Fifth Symposium on Operating Systems Design and Implementation (OSDI 2002)*, Boston, MA, December 2002, 147–163.)

- *Send time*: This time covers all operations before leaving the sender's side: "send time" includes (1) the time spent by the sender for message construction (in its local machine) and (2) the time spent to transmit the message from the sender's host to the network interface (then ready to leave the sender).
- *Access time*: In a WSN, a sender or a receiver may need to wait for some time until the RF channel is ready for use (i.e., no other nodes use the channel).
- *Propagation time*: This is the real "air fly" time. It is the time for the message to reach the receiver, once it has left the sender.
- *Receive time*: The time spent by the receiver to locally process the message, once it has received the message.

Compared to sender-to-receiver–based approach, RBS, a receiver-to-receiver scheme, considers only the times at which a message reaches different receivers. Therefore, it directly removes two of the largest sources of nondeterminism involved in message transmissions, namely, the *send time* and the *access time*. Thus, this protocol can provide a high degree of synchronization accuracy in sensor networks.

To estimate the phase offset between the clocks of two receivers, the following simple steps are adopted:

1. A sender broadcasts a reference packet (i.e., message) to two receivers.
2. Each receiver records the local time at which the packet was received.
3. (Important step) The two receivers exchange their recorded local times at which they received the same packet.
4. We can then calculate the clock offset between two receivers by computing the difference of the local times at which the receivers received the same message.

To use RBS protocol to generate a high clock precision, it is important for each receiver to record its local clock reading as soon as the message is received. However, a receiving node may not be able to record the time of message arrival promptly, for instance, if the node was busy with other computations when the message arrived.

Obviously, RBS cannot just use a single message transmission to alleviate these nondeterministic time factors. In practice, RBS protocol uses a sequence of reference messages from the same sender. Let parameters i and j denote two receivers. Suppose total m messages are sent. Receiver j will compute its offset relative to any other receiver i as the average of clock differences for each packet received by nodes i and j:

$$\text{Offset}[i,j] = \frac{1}{m}\sum_{k=1}^{m}(T_{i,k} - T_{j,k})$$

where $T_{i,k}$ is node i's clock when it receives broadcast k.

By using receiver-to-receiver time comparisons, RBS removes the largest sources of error (send time and access time) from the critical path by decoupling the sender from the receivers. Clock offset and skew are estimated independent of each other. In addition, clock correction does not interfere with either estimation because local clocks are never modified.

However, it may have high communication overhead, because for a single-hop network of n nodes, this protocol requires $O(n^2)$ message exchanges. Convergence time, which is the time required to synchronize the network, can be high due to the large number of message exchanges.

10.5.2 Time-Diffusion Synchronization Protocol [WSu05]

Time-diffusion synchronization protocol (TDP) is a scalable protocol. It could guarantee all the sensors in a WSN to have a synchronized time within a small bounded time deviation from the networkwide "equilibrium" time. The protocol is comprised of several algorithms.

To guarantee high-precision clock synchronization, TDP runs periodically. It thus has alternating *active* and *inactive* phases. Within each *active* phase, there are multiple *cycles*, each cycle lasting a duration τ. In each cycle, some nodes are elected as the *masters* by an *election/reelection procedure* (*ERP*).

Those *masters* selected by ERP concurrently initiate a diffusion of timing messages. A *tree*-like propagation structure can be dynamically created through those timing diffusion messages.

ERP also selects some *non-leaf* nodes in this tree as "diffused leaders," which also propagate the timing messages. It may happen that a node does not qualify to be a diffused leader node, and hence will not propagate the diffusion.

ERP has two important purposes:

1. ERP can use *Allen variance,* a variance calculation algorithm, to eliminate outlier nodes whose clock variance is above some threshold function. *Allen variance* algorithm determines the variance by exchanging messages and calculating deviations between pairs of adjacent nodes using a *peer evaluation procedure* (*PEP*).
2. ERP can achieve network traffic (also called *load*) distribution among the nodes because the roles of masters and diffused leaders put a greater demand on the energy resource. It achieves load distribution by taking turns at being the master, based on factors such as the available energy level being above a tunable threshold.

TDP is a typical *external* synchronization (see Section 10.3) protocol, that is, it uses the diffusion of timing messages to converge to the local times, and eventually reaches a common notion of the *system-wide time.*

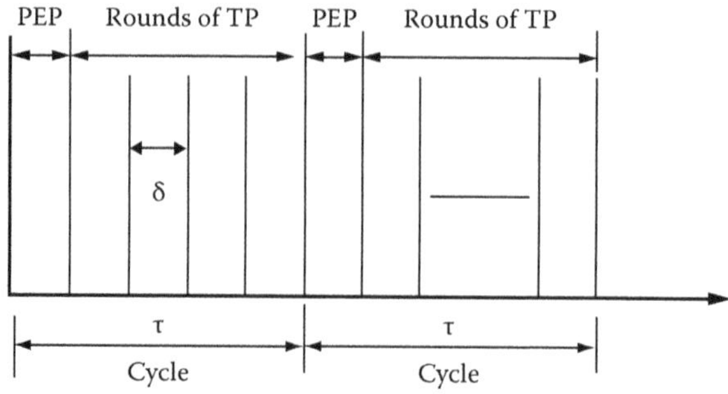

Figure 10.6 Illustration of timing relationships between the TP *rounds* (each with duration δ) and the PEP duration within each *cycle*. Each *cycle* has a duration of τ. (Adapted from Su, W. and Akyildiz, I.F., *IEEE/ACM Trans. Netw.*, 13(2), 384, 2005.)

As mentioned before, TDP has alternating *active* and *inactive* phases. Each *active* phase consists of multiple *cycles*. Each *cycle* has duration τ. *Each cycle* (τ) *consists of two serially executed tasks:* (1) *use PEP to determine master nodes and diffused leader nodes and* (2) *run the main time diffusion procedure* (*TP*), and each TP consists of multiple rounds (each round has a duration of δ). Figure 10.6 shows the relationship between "cycles," PEP, and TP rounds.

After ERP selects masters, each master initiates a concurrent time message broadcast that gets diffused along a tree-like structure, as illustrated in Figure 10.7. A master's time can be coordinated to an external precise time server that does periodic broadcasts of a reference time. If no time servers are available, the protocol may use UTC.

Referring to Figure 10.7, let us see how a master diffuses time message (let us start from tree *Level* 1): First, a master sends many time messages to its neighbors. The neighbors send back acknowledgments containing the two-sample Allen variance of the local clock from the master's clock. Based on the received samples, the master calculates (a) an outlier ratio γ_{yz} for itself and each neighbor, (b) the average of the Allen *variances*, and (c) the average of the *Allen deviations*. Now, values (a), (b), and (c) are sent to each neighbor in a RESULT message.

In each subsequent step $j = 2, 3, \ldots, n$, the *Level* 1 procedure is repeated in each level $j > 1$ between diffused leader node and its neighbors.

During this time message diffusion procedure, all sensors will overhear the outlier ratios and the average *Allen deviation* (with respect to their neighbors). They will use those values to evaluate the quality of their clocks with respect to their neighbors. If a node's average outlier ratio is >1, its local clock deviates

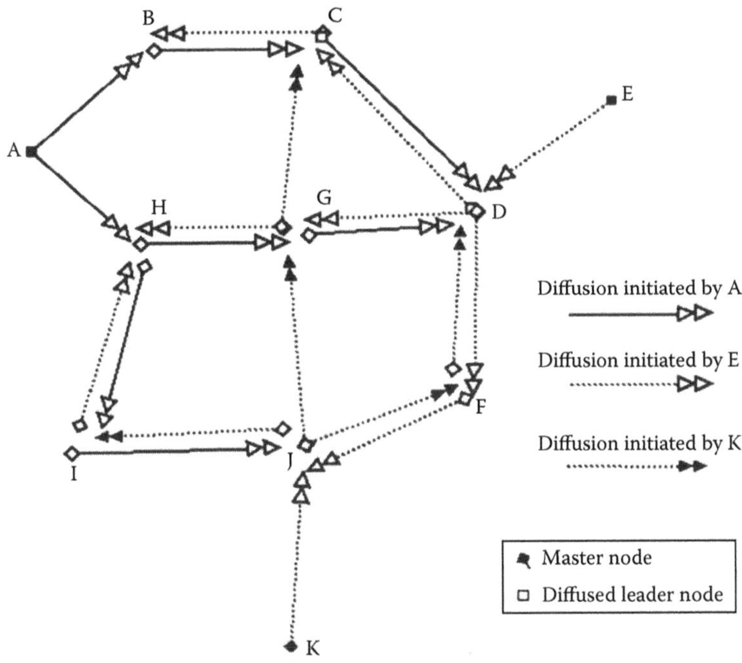

Figure 10.7 Time diffusion with three master nodes and n = 3 hops. In each round, nodes take the hop-weighted (or cumulative deviation weighted) average of the different times received from the masters' diffused broadcasts. (Adapted from Su, W. and Akyildiz, I.F., *IEEE/ACM Trans. Netw.*, 13(2), 384, 2005.)

from the clocks of its neighbors by more than twice the Allen variance. In this case, that node does not become a *diffused leader* during the (TP of the) current cycle, or a master in the next cycle.

Note that a node that is eligible for being a master in the next cycle cannot guarantee to become a real master as TDP considers load distribution (mentioned before). Only when a node has energy availability above a certain (dynamically adjustable) threshold, it will possibly become a master. Load balancing is done by rotating the role of master nodes.

The *TP* procedure performs the main function of diffusing the time from each master in a tree-like manner for n hops, where n is some predetermined parameter smaller than the diameter of the network.

In summary, TDP protocol achieves a system-wide "equilibrium" time across all nodes, computed using an iterative weighted averaging technique, and involves all the nodes in the synchronization process.

10.5.3 Probabilistic Clock Synchronization [SPalChaudhuri03]

Most of the proposed WSN synchronization schemes belong to deterministic algorithms, that is, they can guarantee an upper bound on the error in clock offset estimation. On the other hand, such a deterministic algorithm needs a large number of messages being exchanged during synchronization. This may not be suitable to resource-constrained WSNs.

Therefore, some people propose to use probabilistic algorithms to just provide a reasonable synchronization accuracy at certain probability. Its advantage is low computational and network overhead. PalChaudhuri et al. [SPalChaudhuri03] propose such a probabilistic scheme based on the *extension to RBS*. It provides a probabilistic bound on the accuracy of clock synchronization. It allows to trade off synchronization accuracy for computational and energy resources in low-cost WSN sensors.

As mentioned before, RBS uses multiple messages sent from the sender to a set of receivers. By exchanging messages, all receivers know the differences in actual reception times. Because a sender sends out a set of messages with independent distribution, the difference in reception times can be described by a Gaussian (or normal) distribution with zero mean.

The synchronization error can also be described by the Gaussian probability distribution. We can then easily calculate the relationship between a given maximum error in synchronization and the probability of actually synchronizing *with an error less than the maximum error.*

Let us assume the maximum synchronization error (allowed between two synchronizing nodes) is ε_{max}, then we can use the Gaussian distribution property to derive the probability of synchronizing with an error $\varepsilon < \varepsilon_{max}$:

$$P(|\varepsilon| \le \varepsilon_{max}) = \frac{\int_{-\varepsilon_{max}}^{\varepsilon_{max}} \varepsilon^{(-x^2/2)}\, dx}{\sqrt{2\pi}}$$

From this we can see that when the ε_{max} limit is increased, the probability of failure $(1 - P(|\varepsilon| \le \varepsilon_{max}))$ decreases exponentially.

In [SPalChaudhuri03], the service specification (maximum clock synchronization error) is converted to actual protocol parameters (the number of messages and synchronization overhead). It figures out that the probability of the achieved error (being less than the maximum specified error) is

$$P(|\varepsilon| \le \varepsilon_{max}) = 2erf\,\frac{\sqrt{n}\varepsilon_{max}}{\sigma}$$

where

n is the minimum number of synchronization messages to guarantee the error
σ is the variation of the distribution

In summary, a probabilistic synchronization approach makes a good trade-off between synchronization accuracy and sensor resource cost. In safety-critical applications (e.g., nuclear plant monitoring), a probabilistic scheme may not be enough as it cannot guarantee a certain synchronization accuracy.

Problems and Exercises

10.1 Why do traditional synchronization schemes in wired networks not work well in WSNs? Use NTP as an example to analyze their shortcomings.

10.2 Explain the concepts of *offset*, *skew*, and *drift*.

10.3 Use any software to implement one of the discussed WSN clock synchronization algorithms.

Chapter 11

Security and Privacy in Wireless Sensor Networks

Homeland security is a top issue in any country. The cyber security, especially, plays a very important role in today's society, as many social activities are based on computer communications. WSNs also require security schemes if they are used in critical applications, such as building monitoring.

11.1 Introduction [YangXiao07, Tanya06, Internet07]

11.1.1 General Attack Taxonomy

Tanya et al. [Tanya06] have provided a taxonomy of basic WSN attacks and some corresponding countermeasures. Attacks on sensor networks can be classified into general categories depending on classification standards [CKarlof03] as follows.

1. *Mote-class/laptop-class attackers*: A mote-class attacker typically does not have enough resources to deploy strong attacks. But it can attack low-energy sensors. A laptop-class attacker has access to more powerful devices, such as laptops. Such powerful equipment allows the adversary to launch much more powerful attacks.
2. *Insider/outsider attackers*: An outsider attacker has no special access to the sensor network, because it does not know the WSN keys. But it can use

passive eavesdropping to obtain data. An insider attacker is more difficult to prevent, because it has access to the encryption keys or other codes used by the network. A compromised node, an otherwise legitimate part of the network, can be considered an insider attacker.

3. *Passive/active attackers*: A *passive* attacker compromises the privacy and confidentiality requirement by passively listening to the network data. However, an *active* attacker could damage the function of the network by actively attacking the WSN. For example, the attacker may inject faulty data into the network by pretending to be a legitimate node.

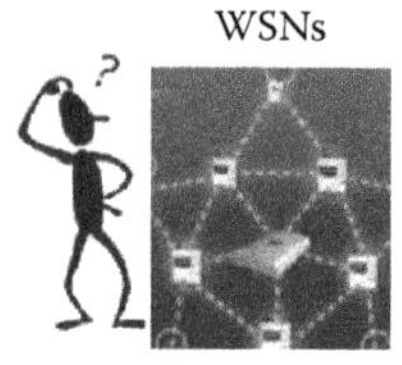

Understanding the difference between different WSN attacks is the prerequisite for designing countermeasure schemes. Please realize that we have many different classification standards of WSN attacks. For instance, if we think of how "strong" an attack could be, an attack may want to save some cost by using low-cost machines to launch attacks. In our society, it may be easy to identify external enemies. But we cannot easily find out spies. Likewise, in WSNs, insider attacks are more threatening than outsider attacks. In the next section, we classify WSN attacks from other perspectives, such as five-layer protocols.

11.1.2 Attacks on Physical Sensor Motes

In WSNs, sensor nodes are vulnerable to intentional physical tampering (e.g., opening a sensor's memory to access the data it contains). Such physical tampering facilitates external attacks on the software running on the sensors. Unfortunately, today's commercial sensor hardware cannot resist physical tampering. If an adversary captures a mote, he or she can easily exploit the shortcomings of the mote's software.

The physical attacks on WSNs include two main types [Tanya06].

1. *Invasive attacks*: An invasive attacker uses reverse engineering followed by probing techniques to study the chip-level components of the device. This attacker has unlimited access to any and all information stored within the components. Such reverse engineering analysis can easily lead to substantial damage to the sensor system.
2. *Noninvasive attacks*: A noninvasive attacker does not open and physically tamper with the embedded device. For instance, a side-channel attack can use the information gathered from the physical implementation of a cryptosystem to get to know some hardware information, such as the power consumption, the timing of the software operation execution, or the frequency of the electromagnetic (EM) waves.

As we can see from the above discussions, to study WSN security, we not only need a good understanding of sensor networking protocols and cryptography knowledge (which is a typical computer science area), but also need to have some electrical engineering knowledge, such as that of EM waves and reverse engineering. Therefore, WSN security is a cross-disciplinary field.

From the above two types of attacks, invasive attacks are more popular. Unfortunately, there is no solution available to make the sensor nodes resistant to physical tampering. The sensor nodes' microcontrollers and memories lack hardware-based memory protection. Although some embedded systems' crypto-processors are physically secure, they do not have a complete set of protection schemes to defend against physical tampering. Therefore, it is important to develop optimized crypto-processors that fit the low-cost, low-energy requirements of sensor networks.

On the other hand, noninvasive attacks, such as side-channel attacks, can also cause serious consequences. For example, a side-channel attack that uses simple power analysis as well as differential power analysis can damage the message authentication codes (MACs) [KOkeya05]. Security key bits can be extracted through the power analysis attack. Power analysis can launch attacks on the block cipher in WSNs. In cryptography, the block cipher uses a symmetric key cipher that operates on fixed-length groups of bits. Linear or differential crypto-analysis is commonly used to launch these attacks. If the block cipher is used as a hash function, an attack can result in the breaking of the hash function.

Another example of side-channel attacks is called timing attacks, which use the nonconstant execution time to leak secret information. Nonconstant execution time can be caused by conditional branching and various optimization techniques. Because the sensor operating system is typically event driven and optimized for low memory consumption, it makes the timing side-channel attack possible. A solution to such an attack could be a constant execution time software. However, it cannot be easily achieved in WSNs. Therefore, the timing attack in sensor hardware is an important subject for future research [Internet07].

A frequency-based attack is also a side-channel attack. It aims to extract secret keys of symmetric cryptographic algorithms.

Some *countermeasures* for side-channel attacks are available, such as power consumption randomization, CPU clock randomization, using fake instructions, and using bit splitting.

11.1.3 Attacks on WSN Communication Stack [Tanya06]

This section classifies WSN attacks from the communication layer viewpoint. We cover attacks in the following categories: physical layer, link layer, network and routing layer, and transport layer.

11.1.3.1 Physical Layer

In the WSN physical layer, jamming, which is one of the most threatening attacks, can launch RF signal interference within the radio channels of sensor communications. The jamming of a few crucial nodes' communication can even disrupt the entire network, because these nodes may be at the intersections of all route paths.

Spread spectrum (SS) communication is a common defense against jamming attacks. SS includes frequency hopping and code spreading. Another solution to the jamming attack has been proposed in [AWood03]. The authors proposed a mechanism to isolate a jammed region through the surrounding nodes. After isolating the jammed region from the network, the rest of the network can be made to function as intended.

In the wireless physical layer, the radio-jamming attack is one of the most challenging issues due to the easy launching of jamming attacks. Just by relying on a radio frequency detector and a strong signal generator, a jammer could make normal data communications in a certain frequency difficult to achieve due to lots of interference signals from the attacker. CDMA (code division multiple access) may be able to achieve certain anti-jamming communications. But CDMA can cause high communication overhead in resource-constrained WSNs.

11.1.3.2 Link Layer

As you may recall, a data link layer protocol defines scheduling schemes for neighboring nodes to access the shared wireless channel. The following are some examples of link layer attacks: an attacker may cause transmission collisions by damaging the scheduling protocols, it could exhaust good nodes' battery by making them have repeated retransmissions, and it could cause unfairness in using the wireless channels among neighboring nodes. Some researchers have proposed a number of solutions for detecting these attacks, such as using collision detection techniques, modifying the MAC (Medium Access Control) protocols so as to limit the rate of requests, and using smaller frames for each packet [AWood02].

Note that MAC could represent different meanings in this chapter. For example, it could mean the *Medium Access Control* protocol, which manages the radio-sharing schedule in wireless access. It could also mean *message-authenticated code*, which is a special binary sequence calculated from the original message data through a function. Such a code is for authentication purposes, that is, find out whether or not a received message is from the right source (not from an enemy's machine).

11.1.3.3 Routing Layer

As we know, routing protocols (also called networking protocols) attempt to find an optimized path from a sender to a destination. Such a path could have higher energy efficiency, or lower latency, or less congestion, or other advantages. The sensors in the path are called relay points, which have a similar function as the routers in the Internet. An attacker could mislead or damage such a path. Here, we mention a few of the attacks on the routing protocols based on the discussion in [CKarlof03, Tanya06].

1. *Spoofed, altered, or replayed routing information*: All data transmissions are controlled by routing protocols. The establishment of a routing path is through the protocol messages among relevant sensors. Therefore, a direct attack against a routing protocol is to target the routing information exchanged between nodes. An attacker could spoof, alter, or replay routing information, and thus create routing loops (i.e., never getting to the destination side), attract or repel network traffic (i.e., misleading routing), extend or shorten source routes, generate false error messages (i.e., reporting wrong error status), partition the network (i.e., making routing difficult in isolated subnetworks), increase end-to-end latency, etc.
2. *Selective forwarding*: WSNs use hop-to-hop routing protocols to relay sensor data. A normal multi-hop routing protocol assumes that all relay nodes will blindly and honestly relay received messages. However, when an attacker uses selective, dishonest forwarding, it can refuse certain messages or drop them altogether. Selective forwarding causes significant data loss and can even disrupt a network. A special form of the selective-forwarding attack is called "black hole." Similar to a universal black hole, an adversary node can refuse to forward every packet it receives. A consequence of this attack is that the neighboring nodes will think that the adversary node has failed and will choose an alternate route.

   In other forms of selective forwarding, an attacker can alter certain nodes' communications and make other nodes transmit as intended. Such an attack could efficiently suppress the data sent from these nodes without suspicion.

   In most cases, selective-forwarding attacks occur when the adversary is within the data path (i.e., becomes one of the relay points). However, an adversary can *overhear* a flow passing through its neighboring nodes, and it then emulates selective forwarding by jamming or causing a collision on each forwarded packet of interest.

Typically, an adversary launching a selective-forwarding attack will select the path with the least resistance as the attack target and then attempt to include herself on the actual path of the data flow.

3. *Sinkhole attacks:* With some similarities to "black hole," a sinkhole attack attracts the nearby traffic through an adversary node that could be an outsider attacker or a native compromised node. The sinkhole attack eventually creates a "hole" around the attacker. By attracting data to its side, it has many opportunities to tamper with application data. In fact, sinkhole attacks can enable many other attacks (such as selective forwarding).

   How does a sinkhole attacker attract traffic to its side? A simple way is to make itself look more attractive than the surrounding nodes, which can be done by spoofing or replaying an advertisement for a much higher quality route to the base station. As we know, WSN routing protocols reply to these advertisements. Once the surrounding nodes see this "attractive" route, they will be much more likely to forward their data to it.

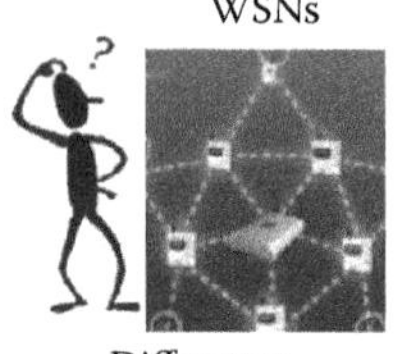

Black hole and sinkhole: Although both of them have some ways to attract data to flow through the enemy's node, black hole makes the incoming data "disappear," while sinkhole does not just simply discard such data. Instead, it may keep the data for further processing, such as content analysis. Thus, a sinkhole may be harder to detect than a black hole.

A good news is that some protocols might actually try to verify the quality of the route with end-to-end acknowledgments (ACKs) that contain reliability or latency information. For instance, a sensor can always ask for a feedback on where its data goes. However, a "strong" sinkhole attacker, such as a laptop-class adversary with a powerful transmitter, can directly (i.e., using one-hop instead of multi-hop) relay the information to the base station or use a wormhole attack (discussed later) to relay data. Due to the "seemingly" high quality route through the compromised node, it is very likely that each neighboring node of the adversary will forward packets (that are supposed to go to a base station) through the adversary. A worse thing is that a good node can advertise a "good" path to its neighbors. Consequently, the sinkhole attacker creates a large "sphere of influence," attracting all traffic from the neighboring nodes.

So why is a sensor network so susceptible to sinkhole attacks? This is due to the WSN routing protocol pattern. In a WSN, typically a base station is the final destination for all sensors' data. If this is the case, why does a compromised node not just simply provide a single high-quality route to the base station? Thus, all nodes will like its route and send data through it.

4. *Sybil attack* [Newsome04]: In a Sybil attack, a single sensor presents multiple instances (i.e., IDs) to other nodes in the network. This is just like a spy who owns multiple countries' passports. Such attacks can significantly reduce the effectiveness of fault-tolerant schemes, such as distributed storage, dispersity, and multipath routing schemes. This is because these schemes rely on some type of redundancy to achieve fault tolerance. However, if different objects (say, routing paths or hard disks) are actually faked by the same node (i.e., a Sybil attack is used), replicas, storage partitions, or routes believed to be using disjoint nodes could be a single compromised node representing several identities.
   In WSN routing schemes, Sybil attacks can seriously damage geographic routing protocols. This is because geographic routing uses a location-aware scheme and requires nodes to exchange coordinate information with their neighbors. By using the Sybil attack, an attacker can make herself appear in multiple places simultaneously. This makes it impossible to efficiently route geographically addressed packets, because we expect that different sensors have really different coordinates.
   If every pair of neighboring nodes uses a *unique* key to initialize frequency hopping or Spread Spectrum (SS) communication, it may be difficult for an adversary to launch such an attack.
5. *Wormhole attack*: The wormhole attack is one of the toughest threats in WSNs. Here, we highlight its main features. In Section 11.2, we will analyze some efficient schemes to overcome wormhole attacks.
   In a wormhole attack, an adversary tunnels the messages received in one part of the network over a low latency link and *replays* them in a different part. Wormhole attacks commonly involve two distant adversary nodes that falsely identify themselves as adjacent nodes.

Good idea

Suppose that a postman has some important mails to be delivered from New York City to San Francisco. He would normally go through many post offices in different states. Although very slow, but such a multi-hop path is safe. However, if somebody says, "Hi, I have built a tunnel for you. This tunnel links a post office (called "A") near New York to another post office (called "B") near San Francisco. From A to B, it takes only 1 hour because I have built a sound-speed train there." Based on the normal mailing service rule, a postman should always find the quickest way to deliver the top-priority mails. Thus, he will take the tunnel to deliver the mails. Haha, this tunnel is totally controlled by the enemy. He can then do anything he wants (such as opening each mail and reading it).

The above analogy can help you understand the wormhole attack.

In a wormhole attack, an attacker should place two machines: one near the source and one *positioned adjacent to a base station* (final destination). There exists a high-quality link (such as a high-speed optical fiber) between these two machines.

By using the wormhole with a high-quality link, an attacker could convince nodes who would normally be multiple hops away from a base station that they are only one hop away.

As we can see, a wormhole can actually create a sinkhole: As the adversary artificially provides a high-quality route to the base station, potentially all traffic in the surrounding area will be drawn through such an "attractive" path.

Of course, if the source is very close to the base station, it is not easy to launch a wormhole attack.

6. *HELLO flood attack*: Many WSN routing protocols require nodes to broadcast HELLO packets to announce themselves to their neighbors. This is called neighbor discovery. Upon receipt of such a packet, a node can assume that the sender is at an appropriate reception distance. However, an adversary with a powerful transmitter can convince all the nodes within a network that it is their neighbor. An adversary who uses the HELLO flood attack could trick every node in the network into believing that the adversary was its neighbor. If the attacker is actually at a long distance, such an attack could effectively cause most of the transmitted data to be lost.

   The WSN could be put into a state of confusion by the HELLO flood attack. Even if one sensor detected a problem with the route, the data still cannot be relayed properly, because all its neighbors are sending data to an attacker.

   Especially, if a WSN routing protocol depends on localized information exchange between neighboring nodes for topology maintenance or flow control, it is subject to this attack.

   An attacker does not need to have the capability of constructing legitimate traffic to use the HELLO flood attack. He can simply use a strong antenna to rebroadcast route-search packets. Such a high-power antenna could make the HELLO packets be received by every node in the network. Therefore, in some sense, HELLO floods can also be thought of as one-way broadcast wormholes.

   Note: When we use the concept of "flooding," we typically mean the epidemic-like propagation of a message to every node in the network over a multi-hop topology. Despite its name, here, we use the HELLO flood attack to mean that an attacker uses a single-hop broadcast to transmit a signal to a large number of nodes.
7. *ACK spoofing*: To achieve route establishment reliability, some sensor network routing algorithms rely on explicit or implicit data link layer ACKs. However, because the wireless links have a broadcast nature, an attacker can spoof data link layer ACKs that are addressed to neighboring nodes.

   The ACK attacker's purpose could be as follows: convincing a neighboring node that a dead node is alive, or claiming a weak signal as a strong one. Such an ACK attack can cause a significant loss of data in networks that determine paths using data link reliability.

An ACK attack reinforces a weak or dead link. This is an effective, but subtle, way of manipulating such a scheme. As packets sent along weak or dead links can be easily lost, an adversary can effectively mount a selective-forwarding attack using ACK spoofing. The consequence is that the target node will transmit packets on those links.

11.1.3.4 Transport Layer

A transport layer, such as the TCP, uses a timer, retransmissions, and end-to-end retransmissions to achieve reliable packet transmissions between a source and a destination. However, transport layer protocols in wired networks cannot be directly used in sensor networks due to resource constraints. Chapter 5 has discussed WSN transport layer protocols.

Some WSN transport layer attack examples are flooding and desynchronization attacks. A flooding attack sends out multiple end-to-end connection establishment requests, effectively exhausting the memory of a node. The desynchronization attack tries to forge packets to one or both ends of a connection using different sequence numbers on the packets. It triggers the end points of the connection to request retransmission of the 'perceived' missed packets.

Source authentication and client puzzles are two possible solutions to guard against these attacks [AWood03]. However, we are still unsure whether or not these solutions can be used in sensor networks and what improvements should be made to facilitate such schemes.

11.1.3.5 Traffic Analysis Attacks

As we know, the main purpose of WSNs is to collect sensor data from many remote sensors to a base station. Therefore, the traffic through the network has a pattern of many to one. This gives the adversary a chance to attack the network. For instance, an attacker could analyze the traffic patterns to gather the topology of the sensor network as well as the location of the base station by observing the traffic volume and pattern.

Another traffic analysis attack is to observe the traffic and deduce the "important" nodes that are at the intersections of many paths. Then the attacker can attack and compromise these nodes, and eventually break the network into multiple disconnected subnetworks. Or an attacker might launch a denial-of-service (DoS) attack against the sensors on the vertex cut-set. These DoS attacks could drain the sensors' energy, and thus reduce the lifetime of the network.

Traffic analysis attacks can be launched in many other forms. For instance, the adversary could observe the packet-sending rate of its neighboring nodes, and then move toward nodes with a higher packet-sending rate. Or it could observe the time gap between packet arrivals (in its buffer), and try to follow the path of the packet that is being forwarded until it reaches the base station.

How do we countermeasure traffic analysis attacks? A possible solution is to try to "confuse" the enemies. For instance, between a source and a destination, we may establish random and multiple paths, or use probabilistic routing, or introduce fake messages in the network.

A *probabilistic geographic routing* (PGR) scheme [SSTanya05] selects the next hop based on the link quality and the residual energy of a subset of the neighbors of a node. Their experiments show that PGR is energy efficient and performs well in terms of high network throughput.

Using "confusing" messages could bring high network overhead in terms of energy consumption and in-network traffic. The confusing messages have to look like real messages. Therefore, the fake messages cannot be optimized.

11.2 Attack and Countermeasure Example: Wormhole Attack

11.2.1 Wormhole Defense Scheme—LITEWORP [Issa06]

We first classify the wormhole attack based on the attack-launching techniques.

1. Wormhole Using Encapsulation
   In [Issa06], Khalil has analyzed a generic wormhole attack. A Dynamic Source Routing (DSR)[DSR] routing protocol is used as an example. Here, we simply review DSR protocol ideas.

   In DSR, if a node S needs to discover a route to a destination D, S floods the network with a *route request* (RREQ) packet. Every node that hears the request processes the packet, adds its identity, and rebroadcasts it. To limit the amount of flooding through the network, each node broadcasts only the *first* RREQ it receives and drops any further copies of the same request. For each RREQ that D receives, it generates a route reply (RREP) and sends it back to S. Based on the RREP messages, the source S then selects the best path that could be either the path with the shortest number of hops or the path associated with the first arrived reply.

   Unfortunately, the DSR protocol can be easily attacked. For instance, an attacker who hears the RREQ packet can tunnel it to a second colluding party at a distant location near the destination. The second party then uses a replay attack, that is, rebroadcasts the RREQ. Based on DSR rules, the neighbors of the second colluding party that receive the RREQ will drop any further legitimate requests that may arrive later on legitimate multi-hop paths. Such an attack is in fact a wormhole attack, which makes the packets (to be passed to the base station) travel between the two adversary nodes. The attacker can do anything to these packets in a "shortcut" path. Such a wormhole attack eliminates the possibility of discovering legitimate paths more than two hops away, as the attacker's shortcut typically uses a one-hop high-quality link.

The other way for two colluding malicious nodes to construct a wormhole route is not building a one-hop shortcut by themselves. Instead, they may just discover the shortest path and use it, even though there may be multiple hops between those two colluding nodes. Consider Figure 11.1, in which nodes A and Z try to discover the shortest path between them, in the presence of the two malicious nodes X and Y. After node A broadcasts a *RREQ*, X gets the *RREQ* and encapsulates it in a packet destined to Y through the path that exists between X and Y (6-7-8-9). Node Y de-marshalls the packet, and rebroadcasts it again, which reaches Z. So X and Y successfully include themselves in the route between A and Z. Any routing protocol that determines "good" paths by the shortest route is vulnerable to such an attack.

In the above example, the two colluding nodes (X and Y) do not need to have any cryptographic scheme, nor do they need any special capabilities, such as a high-speed wire line link or a high power source. Therefore, this mode of the wormhole attack is easy to launch.

2. Wormhole Using Out-of-Band Channel
In this type of wormhole attack, the attackers establish an out-of-band high-bandwidth channel between the malicious nodes. Such a high-bandwidth channel could be a long-range directional wireless link or a direct wired link. Because such an attack needs specific and specialized hardware, it is more difficult to launch than the previous case.

Figure 11.2 shows such a scenario. Node A sends a RREQ to node Z; nodes X and Y are malicious nodes with an out-of-band channel between them. Node X tunnels the RREQ to Y, which is a neighbor of Z. Node Y broadcasts

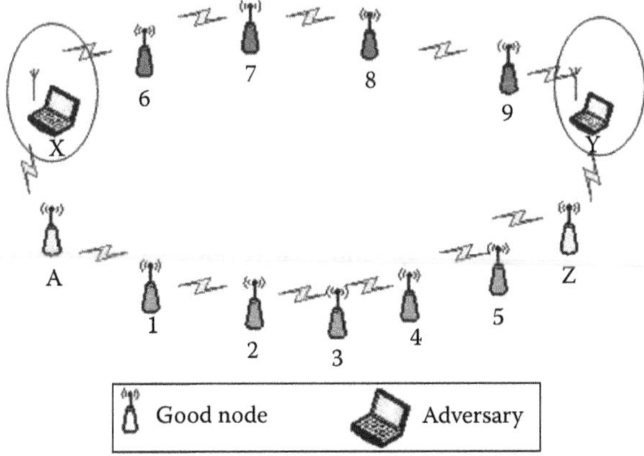

Figure 11.1 Wormhole through packet encapsulation. (Adapted from Khalil, I., Mitigation of control and data traffic attacks in wireless ad-hoc and sensor networks, PhD dissertation, Purdue University, West Lafayette, IN, 2006.)

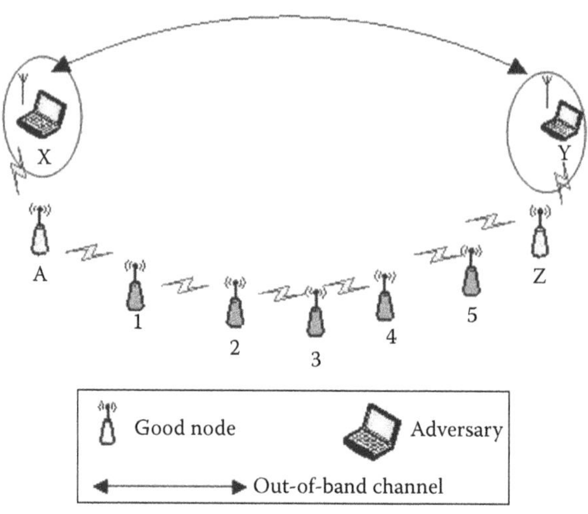

Figure 11.2 Wormhole through out-of-band channel. (Adapted from Khalil, I., Mitigation of control and data traffic attacks in wireless ad-hoc and sensor networks, PhD dissertation, Purdue University, West Lafayette, IN, 2006.)

the packet to its neighbors, including Z. Node Z gets two RREQs—A-X-Y-B and A-1-2-3-4-5-Z-Y. The first route is both shorter and faster than the second one. Z will choose the first one, which results in a wormhole established between X and Y in the route between A and Z.

3. Wormhole Using High Power Transmission

   In this case, an attacker receives a RREQ and rebroadcasts it at a higher power level that is not available in other nodes' antennas. All the nodes that hear the high-power broadcast will rebroadcast it toward the destination. Therefore, the malicious node could easily get involved into the routes established between the source and the destination even without the participation of a second colluding node.

   A way to mitigate this attack is to require each node accurately determine the signal strength of the received signal and use radio propagation models to deduce the distance. As we know, the longer the distance, the weaker the received signal strength (RSS). Thus, each node can determine whether or not the received signal is within the appropriate power threshold. A malicious node that uses high power could be easily detected by such a model, as other nodes will not use such high power.

4. Wormhole Using Packet Relay

   In this case, a malicious node relays packets between two distant nodes (say A and B, assume that there are multiple hops between them) to convince A and B that they are one-hop neighbors. This attack can be launched from a single malicious node, or through the cooperation of a greater number of malicious nodes, which serves to expand the neighbor list of a victim node to several hops.

5. Wormhole Using Protocol Deviations
   This type of wormhole attack tries to disobey the rules in some routing protocols. For instance, Authenticated Routing for Ad Hoc Networks (ARAN) [KSanzgiri02] is a routing protocol that chooses the route with the shortest delay instead of the route with the shortest number of hops. Therefore, an attacker can try to shorten its route-search delay to make its node look more appealing than others.

   How does an attacker shorten its routing delay? In Authenticated Routing for Ad Hoc Networks (ARAN) routing rules, the good nodes will back off for a random period of time before forwarding a RREQ. This is because the MAC layer requires that all nodes carefully access the radio link—try to avoid transmission collisions by waiting for a random time before re-forwarding a RREQ. However, a malicious node will not obey such rules. It can create a wormhole by broadcasting a RREQ without any delay. Through this way, an attacker's RREQ packets can reach the destination first, thus making the entire route delay seem less than the surrounding normal nodes. The malicious node thus has a high probability of being included in the route between the source and the destination.

The above case is in fact a special form of the rushing attack described in [YCHu03].

Table 11.1 summarizes the different modes of the wormhole attack, along with the associated requirements.

Khalil [Issa06] also proposed a wormhole detection and countermeasure scheme called *LITEWORP*. Its basic idea is to isolate the malicious nodes.

Any security scheme has some assumptions. LITEWORP has made some assumptions for it to operate efficiently:

1. The communication links are bidirectional, meaning that if A can send to B, then B can send to A.
2. A finite amount of time is required before a node is compromised. No external or internal malicious nodes exist before the completion of the neighbor discovery.

Table 11.1 Summary of Wormhole Attack Modes

Mode Name	*Min. # of Compromised Nodes*	*Special Requirements*
Packet encapsulation	Two	None
Out-of-band channel	Two	Out-of-band link
High power transmission	One	High-energy source
Packet relay	One	None
Protocol deviations	One	None

Source: Adapted from Khalil, I., Mitigation of control and data traffic attacks in wireless ad-hoc and sensor networks, PhD dissertation, Purdue University, West Lafayette, IN, 2006.

However, this assumption can be removed by using a secure neighbor discovery protocol, such as the one by Hu and Evans using directional antennas [LHu04] or by using trusted and more powerful nodes as in [YTirta06].

3. The WSN nodes are not mobile (this is a reasonable assumption in most WSN applications). However, the network topology can have route changes due to sensor power draining, sensor damage, malicious node isolation, route evictions from the routing cache, or the change in the role of a sensor (e.g., cluster head and data aggregator).
4. Each packet forwarder is required to explicitly announce the immediate source of the packet, that is, the node from which it receives the packet.
5. Finally, LITEWORP assumes a key management protocol, such as Scalable and Energy-Efficient Crypto On Sensors (SECOS) [IKhalil05], used to predistribute pair-wise keys in the network.

11.2.1.1 Building Neighbor Lists

LITEWORP first proposed a neighbor list discovery protocol, which aims to build the data structure of the one-hop neighboring nodes and the nodes surrounding them. A neighbor node is any node that falls within the transmission range. Such a data structure is important to detect malicious nodes and to make a local response to isolate the detected malicious nodes.

A HELLO message is a common way to find neighbors. Immediately after a node (say A) is deployed in the field, it broadcasts a HELLO message for a one-hop distance. Any node, say B, that hears the HELLO message, sends back a reply to A. Node A accepts all the replies that arrive within a predetermined time-out duration.

By collecting these replies, A adds the responder to its neighbor list. Neighbor discovery is not over yet. A will broadcast such a list to all one-hop nodes. When any neighbor (say B) detects the broadcasted list, it stores it.

After we finish the above neighbor discovery process, each node has a list of its direct neighbors and the neighbors of each of its direct neighbors. However, the above process is only performed once per node and is assumed to be secure (which can be achieved through a secure neighbor discovery protocol).

Note: After such a list is built in each node, a node will not forward packets to nodes that are not neighbors. Also, two-hop neighbor information is used to determine if a forwarded packet comes from a neighbor of the forwarder. If a node C receives a packet forwarded by B claiming to come from A in the previous hop, C discards the packet if A is not a two-hop neighbor.

After building its one-hop and two-hop neighbor lists, node A can activate the *local monitoring* procedure to find wormhole attackers.

Here, we show how *local monitoring* can be used to build the detection algorithm individually for each of the first four wormhole attack modes, and also show how existing approaches can be used to detect the fifth mode.

11.2.1.1.1 Detecting Out-of-Band and Packet Encapsulation Wormholes

LITEWORP introduces the concept of sentry (guard) node. Suppose that α is the guard node of another node A. Suppose that α could monitor the wireless link from a node X to A by using the following steps as part of its role in terms of monitoring the sensor network communication:

1. We require that the sentry node, α, stores information from the packet header of each control packet going over the link from X to A and labels it with the deadline τ.
2. Node α overhears every packet going out of A. For all the packets that A claims to come from X, α looks up the corresponding entry in its watch buffer that has the neighbor list.
3. If an entry is found, α discards it, because proper forwarding is assumed to already be accomplished.
4. If an entry is not found, then A is assumed to have fabricated the packet. Therefore, α increments the malicious node's count, *MalC* (α,A), by V_f.
5. If an entry for a packet sent from X to A stays in the watch buffer beyond τ, then A is accused of dropping the corresponding packet. Node α increments *MalC*(α,A) by V_d.
6. If the incoming packet to A is different from the corresponding outgoing packet from A, then A is accused of modifying the packet. Therefore, α increments *MalC*(α,*A*) by V_m.

Let us consider the scenario in Figure 11.3. X and Y are two malicious nodes wishing to establish a wormhole between two nodes (source node, A, and destination node, Z). When X hears the *RREQ* packet from A, it directs the packet to Y. Y then rebroadcasts the *RREQ* packet after appending the identity of the previous hop from which it got the *RREQ*.

Now, Y has two choices for the previous hop—either to append the identity of X or append the identity of one of Y's neighbors, say 9.

In the first choice, all the neighbors of Y will reject the *RREQ* because they all know, from the stored data structure of the two-hop neighbors, that X is not a neighbor to Y.

In the second case, the knowledge of the one-hop and two-hop neighbor lists is not sufficient for all the guards to detect the attack. However, using local monitoring, all the guards of the link from X to Y will detect Y as fabricating the RREQ, because they do not have the information for the corresponding packet from X in their watch buffer.

In both cases, Y is detected, and the guards increment the *MalC* value of Y.

LITEWORP could also use the *RREP* packet to detect the behaviors of X and Y. When the destination node Z gets the *RREQ*, it generates a RREP packet, and sends it back to X. The guard nodes of the link from Z to Y can overhear the *RREP* and save an entry in their watch buffers. Node Y sends the RREP back to X using the out-of-band channel or packet encapsulation.

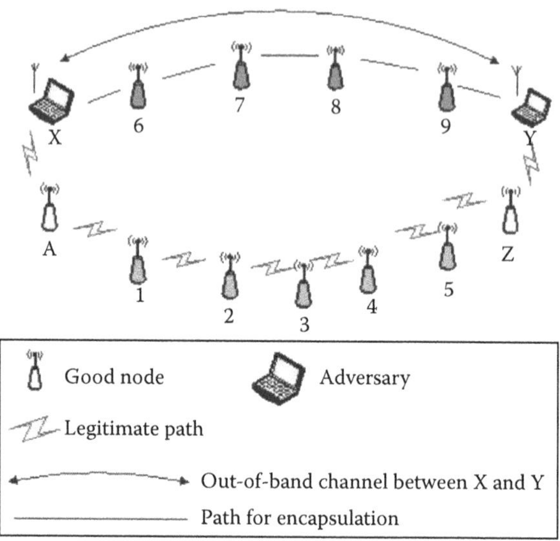

Figure 11.3 Wormhole detection for out-of-band and packet encapsulation modes. (Adapted from Khalil, I., Mitigation of control and data traffic attacks in wireless ad-hoc and sensor networks, PhD dissertation, Purdue University, West Lafayette, IN, 2006.)

After τ time units, the timers in the watch buffers of the guard nodes run out, and thus the guards detect Y as dropping the *RREP* packet and increment the *MalC* of Y. However, if Y is smarter, it can forward another copy of the *RREP* through the regular slower route. In this case, the *MalC* of Y is not incremented. When X gets the *RREP* from Y, X forwards it back to A after appending the identity of the previous hop.

As before, X has two choices—either to append the identity of Y or append the identity of one of X's neighbors, say 6. In the first choice, node A rejects the *RREP* because it knows that Y is not a neighbor to X. Also, all the neighbors of X know that Y is not a neighbor to X. In the second case, all the guards of the link from 6 to X detect X as forging the *RREP*, because they do not have the corresponding entry from 9 in their watch buffers.

11.2.1.1.2 Detecting High-Power-Transmission Wormhole

We could detect this case by using the assumption of *symmetric* bidirectional channels. If a malicious node, X, tries to use high power transmission to forward a packet, P_1, to its final destination or to cross multiple hops to involve itself in the shortest path, all the nodes that do not have X listed as a neighbor realize the fraudulent packet and drop it.

11.2.1.1.3 Detecting Packet Relay Wormhole

We could easily detect this case through the stored neighbor lists at each node. Suppose that a malicious node, X, is a neighbor of two non-neighbor nodes, A and B. If X tries to deceive them by relaying packets between them, both A and B will be able to detect the malicious behavior of X and reject the relayed packet, because A and B know that they are not neighbors to each other.

11.2.1.1.4 Detecting Protocol Deviation Wormhole

LITEWORP cannot detect this case. However, we can use other researchers' ideas on countering selfish behavior in specific protocols. Here, *selfishness* (also called greediness) refers to the property that nodes tend to deny required cooperating services to other nodes to save their own resources, for example, battery power.

The problem of greediness at the MAC layer has been addressed by Kyasanur et al. [Kyasanur03]. Selfishness in routing packet forwarding has been addressed in [SCapkun03]. A solution, called the rushing attack, in which nodes forward information quickly without waiting for the backoff time, is addressed in [YCHu03].

11.2.1.1.5 Response and Isolation Algorithm

The above solution only covers the *wormhole detection* issue. The next step is to use the local response and isolation module to diagnose the attacker and employ the appropriate response to isolate it from the network, thereby dissolving its ability to harm the rest of the network. LITEWORP proposed such an attacker isolation module, which is controlled by the local monitoring module and is only activated upon detection of an adversary node.

LITEWORP uses a local response scheme to propagate the detection knowledge only locally, that is, within two hops from the suspect node. It accomplishes a local response by deleting the suspect node from the first-hop and second-hop lists of all its neighbors.

The following is LITEWORP's local response algorithm. It is activated when a guide node, say *a*, detects a malicious behavior of a node, say A, during the course of local monitoring.

1. When the reputation value, *MalC*(α,A), goes beyond a threshold, *Ct*, the guard α revokes A from its neighbor list and sends to each neighbor of A, say D, an authenticated alert message indicating that A is a suspected malicious node.
   Note: To permanently isolate the bad node, we could use a *shared security key* among nodes to authenticate all nodes for the prevention of false accusations in future. The next section provides more knowledge on key-based WSN security management. Alternately, if the clocks of all the nodes in the network are loosely synchronized, *a* can authenticate local two-hop multicast as in TESLA [APerrig02], or μTESLA [APerrig02], to inform the neighbors of A.

Note that α isolates A without waiting for γ alerts from other nodes, as a node is assumed to trust itself.

2. When D gets the alert, it verifies the authenticity of the alert message and stores the identity of α in an *alert buffer* associated with node A.
3. When D gets enough alert messages about A (we could define a threshold of the alert messages), it isolates the node by deactivating it on all the neighbor lists.
4. After isolation, D does not send any packet to or accept any packet from A.

The above approach can remove the malicious nodes from the network. In addition, it reduces the time between detection and response, as the information can be handled and processed locally. It does not cause much network traffic, as it only sends out messages to each neighbor of A (only in the detection phase). The number of hops each message traverses is at most two hops.

LITEWORP also defined a useful concept, called *detection confidence* (γ), which is useful for reducing the possibility of *framing* with a higher value being favored for this purpose. *Framing* is an attack whereby a malicious node acts as a sentry node and begins sending false accusations about a legitimate node. If γ is set to infinity, then a node only trusts itself and is invulnerable to this attack.

Good idea

From LITEWORP, we can learn many excellent ideas on WSN security. By maintaining an honest neighbor list, we can detect any "bad" nodes that try to get involved into the routing procedure. After we detect these "bad guys," we need to put them into the jail, that is, we must isolate them from good communications.

Fully explore your imagination!

11.3 WSN Security Example: Blom-Based Approach [DuW05]

As we mentioned before, the security keys can be used to achieve authentication (i.e., verify the source) and confidentiality (i.e., encrypt a message). However, managing the keys is a challenge in WSNs, because we need to deal with the key predistribution issue, that is, how do we pre-allocate the keys in different sensors for future security purposes? In [DuW05], Wenliang Du et al. have built a WSN key predistribution scheme based on the enhancement of the Blom scheme [Blom85] (see also [Blundo93]).

Assume that N is the total number of nodes in a WSN. If there exists a secure communication between any two nodes, these two nodes have to share one secret key to encrypt and decrypt the message between them. If we do not use a smart key predistribution scheme to ensure that any two nodes can share at least one key, each node will have to store $(N - 1)$ keys.

In Blom's key predistribution scheme, nodes are required to store only (λ + 1) keys, with $\lambda \ll N$. Obviously, Blom's scheme does not have a perfect flexibility against node capture, because we cannot guarantee that any two nodes share one common key. However, in reality, we do not need to ensure that any two nodes have the same key if those two nodes are not near each other and, thus, never talk with each other.

As a matter of fact, Blom's scheme can ensure a λ-secure property, that is, as long as no more than λ nodes are compromised by an adversary, the communication links between all non-compromised nodes remain secure. Of course, if an adversary compromises more than λ nodes, the entire network of keys becomes compromised.

This threshold λ is a crucial security parameter. By selecting a larger threshold λ, the key-sharing probability is increased, thus allowing better security performance. Therefore, by setting a large λ value, we force an adversary to attack a great portion of the network if it wants to compromise the WSN communications. On the other hand, the increase in λ would require a large memory space to store a lot of key information.

Du's scheme [DuW05] is the enhanced solution to Blom's scheme. It uses a probabilistic approach to increase the pliability of the network against node capture. Unlike Blom's scheme, it does not require too much additional memory.

Blom's scheme uses a *single* key space to ensure that any pair of nodes can compute a shared key, whereas Du proposed a new scheme that uses *multiple* key spaces. It first uses Blom's scheme to construct total ω spaces (where $\omega > 2$), and then it requires that each sensor node carry key information from τ randomly selected key spaces (where $2 \leq \tau < \omega$). Blom's scheme tells us that as long as two nodes carry key information from a common space, these two nodes can compute a shared key.

Although there is only a probabilistic guarantee that two nodes can generate a pair-wise key, Du's analysis shows that when the same amount of memory is used, their new scheme is considerably more resilient than traditional probabilistic key predistribution schemes.

Good idea

Many students/researchers ask a question: How do I propose a good idea to overcome a challenging issue? We can learn the skills from [DuW05]: Blom's algorithm was proposed more than a decade ago. It was "hidden" in millions of papers published by IEEE, ACM, Elsevier, etc. There may be no direct solution to a new issue. However, a new solution could possibly be found by widely reading traditional cryptography papers and by constantly asking ourselves a question, "Even though this paper is not for WSN security, can I borrow some ideas from it and do some extensions/modifications to apply it for the resource-constrained WSNs?" Always ask yourself the above question. Someday you will say "Wow, I could use this idea!"

To understand Du's scheme, let us briefly review Blom's scheme. (Du's scheme has made some slight modifications to Blom's scheme to make it more suitable for WSNs' serious resource constraints, but all Blom's major features remain unchanged.)

Assume that there exists an agreed-upon matrix, G, with a dimension of $(\lambda + 1) \times N$ over a finite field, $GF(q)$ (where $q > N$). Note: The matrix G is not secret. Even adversaries are assumed to know G.

During the *key generation* phase, the WSN base station creates a random $(\lambda + 1) \times (\lambda + 1)$ symmetric matrix, D, over $GF(q)$, and computes an $N \times (\lambda + 1)$ matrix, $A = (D \cdot G)^T$, where $(D \cdot G)^T$ is the transpose of $D \cdot G$.

Note: Matrix D must be kept secret and should not be disclosed to adversaries or to any sensor nodes. On the other hand, as we will discuss next, one row of $(D \cdot G)^T$ should be disclosed to each sensor node. Because D is symmetric, it is easy to see that

$$A \cdot G = (D \cdot G)^T \cdot G = G^T \cdot D^T \cdot G = G^T \cdot D \cdot G = (A \cdot G)^T \tag{11.1}$$

Therefore, $A \cdot G$ is a symmetric matrix. If we let $K = A \cdot G$, we know $K_{ij} = K_{ji}$, where K_{ij} is the element in the ith row and jth column of K. The idea is to use K_{ij} (or K_{ji}) as the pair-wise key between node i and node j. The generation of the pair-wise key, $K_{ij} = K_{ji}$, is shown in Figure 11.4. To carry out the above computation, nodes i and j should be able to compute K_{ij} and K_{ji}, *respectively*. Such a procedure can be achieved from the following key predistribution steps, for $k = 1, \ldots, N$: (1) Node K stores the kth row of matrix A. (2) Node K then stores the kth column of matrix G. We will show later that a sensor does not need to store the whole column, because each column can be generated from a single field element.

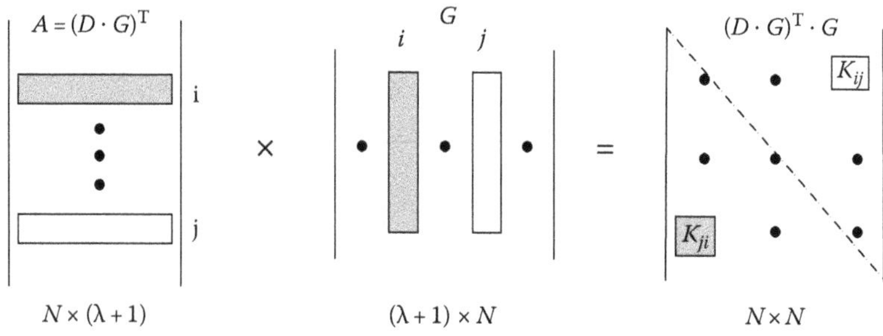

Figure 11.4 Generating keys in Blom's scheme. (Adapted from Du, W. et al., *ACM Trans. Inf. Syst. Secur.*, 8(2), 228, May 2005.)

Then, nodes i and j can generate a shared key (also called a *pair-wise key*) as follows: They first exchange their columns of G, and then use their private rows of A to compute K_{ij} and K_{ji}, respectively. As we mentioned before, as G is not kept private, its columns can be transmitted in plaintext. It has been shown [Blom85] that the above scheme is λ-secure if any $\lambda + 1$ columns of G are linearly independent. This λ-secure property guarantees that no coalition of up to λ nodes (not including i and j) has any information about K_{ij} or K_{ji}.

Du [DuW05] has shown an example of a matrix G. Any $\lambda + 1$ columns of G must be linearly independent. As each pair-wise key is represented by an element in the finite field $GF(q)$, we must set q to be larger than the key size we require. Thus, if we want to generate a 64 bit key, we may choose q as the smallest prime number larger than 2^{64} (or we may just simply choose $q = 2^{64}$).

Assume s is a primitive element of $GF(q)$, that is, each nonzero element in $GF(q)$ can be represented by s^x. We can generate a feasible format of G as follows [Macwilliams77]:

$$G = \begin{bmatrix} 1 & 1 & 1 & \cdots & 1 \\ s & s^2 & s^3 & \cdots & s^N \\ s^2 & (s^2)^2 & (s^3)^2 & \cdots & (s^N)^2 \\ \vdots & \vdots & \vdots & \ddots & \vdots \\ s^\lambda & (s^2)^\lambda & (s^3)^\lambda & \cdots & (s^N)^\lambda \end{bmatrix} \tag{11.2}$$

Because s is primitive, as long as $i = (j \bmod q)$, we have $s^i = s^j$. It can be shown that any $\lambda + 1$ columns of G are linearly independent [Macwilliams77].

Because the matrix G has a nice property, that is, its columns can be generated by an appropriate power of the primitive element s, to store the kth column of G at node k, we only need to store the seed s^k at this node. Matrix G's column can be regenerated when needed.

Du et al. [DuW05] have also provided an interesting theoretical analysis and detailed experimental results. These results have clearly shown the low memory overhead and good security performance of their scheme, which is based on the extension of Blom's scheme.

11.4 Broadcast Authentication: μTESLA [APerrig00, APerrig01]

In this section, we discuss another important security issue: authenticate the source. We are especially interested in broadcast authentication, because, in a

WSN, the base station often broadcasts a command message (such as "tell me the sensor value in area xxx"). Any sensor that receives such a broadcasted command needs to verify the source, as it could be a good base station or an attacker's machine.

Traditional ways for authenticating broadcasts do not work well for sensor networks, because most of them rely on *asymmetric* digital signatures for the authentication. An *asymmetric* digital signature requires a public key and a private key in two nodes, respectively. The source could use its private key to encrypt a message. And any receiver who owns the source's public key could successfully decrypt the message. However, if the source message is sent out from an attacker's machine, the attacker does not have the right public key, and its message cannot be decrypted by any receiver, that is, the digital signature will fail.

Although *asymmetric* digital signatures can authenticate a message, they need public/private keys that require a much larger memory storage overhead than *symmetric* keys (which require only a small key in both nodes). Therefore, *asymmetric* authentication is impractical in sensor networks due to the sensors' small memory space.

The TESLA protocol [APerrig00], an asymmetric mechanism, provides an efficient broadcast authentication method. However, this protocol needs an overhead of around 24 bytes per packet to generate a digital signature key, which exceeds the resources available in common WSNs. As a matter of fact, most WSNs need around 30 bytes for a message. Therefore, disclosing a 64 bit (which is 8 bytes) key and the MAC message-authenticated code with every packet would take up over 50 percent overhead of each packet. Given these facts, it is evident that pure TESLA is not practical for a node to broadcast system.

Therefore, Perrig et al. [APerrig01] proposed a solution, called µTESLA, to overcome the ineptitude of TESLA for sensor networks. The difficult issue is that to achieve strong message authentication, an asymmetric mechanism is preferred over a symmetric one. This is because of the following fact: If we simply use a symmetric scheme (i.e., the same key is used for both the sender and the receiver), a compromised receiver could get such a key and easily forge messages from the sender.

µTESLA solves the problem of the extremely high computation, communication, and storage overhead that occurs in TESLA, by introducing *asymmetry* through a delayed disclosure of *symmetric* keys. Its basic idea is as follows: When a WSN base station sends a packet, it computes a MAC on the packet, but does not yet disclose the MAC key. Packets received by nodes are buffered in nodes' memory until the corresponding MAC key is released by the base station. All sensor nodes know that no adversary could have altered the packet in transit, as the key is only known by the base station. Later on, the node receives the disclosed MAC key and authenticates the packet that was stored in the buffer for some time.

"Use a *symmetric* security scheme to achieve *asymmetric* authentication." This is the main idea of μTESLA. A *symmetric* scheme means only one key is used for each MAC. However, an *asymmetric scheme* requires two keys (public/private keys). μTESLA uses only one MAC key. But the sender (base station) does not give the receiver the MAC key when it sends out the MAC message. Instead, the sender waits for some time (this delay will be larger than maximum round trip delay between the base-station and the sensors) to disclose the previous MAC key. Thus, such a delay achieves the effect of "asymmetry."

A well-known *one-way function* is used to generate each MAC key, and each MAC key is part of a *key chain*. The sender chooses *the last key* (K_n) of the chain randomly and is able to repeatedly apply F to compute all other keys: $K_i = F(K_{i+1})$. Suppose that the last key is K_{100}. Then it would be able to compute all other keys as follows:

$$K_{99} = F(K_{100}), \quad K_{98} = F(K_{99}), \ldots, K_0 = F(K_1)$$

Because $F(.)$ is a *one-way* function, given K_{100}, we can easily figure out K_{99}, K_{98}, …, K_0. However, given K_0, we cannot figure out K_1, K_2, …, K_{100}.

The concept of one-way key chain in μTESLA is shown in Figure 11.5. Its features are as follows.

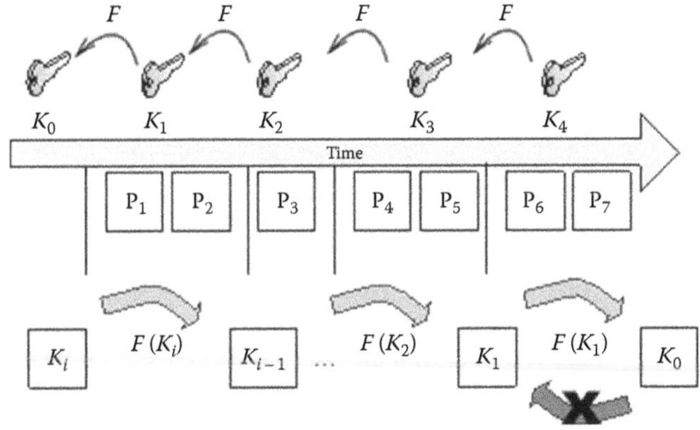

Figure 11.5 The μTESLA one-way key chain. The sender generates the one-way key chain right to left by repeatedly applying the one-way function, *F*. The sender associates each key of the one-way key chain with a time interval. Time runs left to right, so the sender uses the keys of the key chain in the reverse order, and computes the MAC of the packets in a time interval with the key of that time interval. (Adapted from Perrig, A. et al., SPINS: Security protocols for sensor networks, *Proceedings of ACM MOBICOM,* Rome, Italy, 2001.)

μTESLA assumes that the entire WSN has some type of loosely time-synchronized protocol in all nodes. Thus, all nodes will be able to recognize different sending time intervals. When the base station sends out messages (packets), it uses the same key to authenticate all packets sent within one time interval. The receiver knows K_0 (a commitment to the key chain).

In Figure 11.5, packets P_1 and P_2 are sent in interval 1 and contain a MAC with key K_1 (note: without key K_1, a receiver has no way to verify whether or not the MAC is sent from the right base station). Packet P_3 has a MAC using key K_2, and so far the receiver cannot authenticate any packets, because the base station does not disclose the corresponding key of each MAC in that time interval (and it will disclose it until some intervals later).

Note a nice feature of the μTESLA one-way key chain: It can tolerate the loss of previous MAC keys. Suppose that some intervals later, key K_1 (used to verify packets P_1 and P_2) could not be received by a sensor due to wireless loss. However, as long as the sensor can get key K_2 later on, it can verify $K_0 = F(F(K_2))$, and, therefore, know $K_1 = F(K_2)$. So, it can still authenticate all previously received packets.

Case study

Message-authenticated code (MAC): An example of MAC is a keyed hash function. A hash function can map an arbitrary message to a fixed-length message. A hash function is in fact a one-way function, as by giving a hash result you cannot deduce the original message. If we use a key to encrypt the hash result, we will get a MAC. Typically, a sender sends out message M and its MAC to a receiver. The receiver will use the same key to decrypt the MAC and compare the result to M. If they are the same, we say that M is indeed sent from the right source.

11.4.1 μTESLA's Detailed Description

μTESLA consists of a few operation phases, including sender setup, sending authenticated packets, bootstrapping a new receiver, and authenticating packets.

Sender setup: In this phase, the sender (base station) constructs a key chain of secret keys. The key chain is of length n and the sender generates it by choosing the last key, K_n, randomly and using the one-way function, F, to successively generate the remaining values. An example of a one-way function is a cryptographic hash function, such as MD5: $K_i = F(K_{i+1})$. As mentioned before, the one-way nature of function F means that keys can always be computed forward but never backward.

Broadcasting authenticated packets: As shown in Figure 11.5, time is divided into time intervals. And each key of the one-way key chain is associated with one time interval. For each respective interval, the sender uses the key from that interval to compute the MAC message-authentication code of packets in that interval. The sender reveals the key for the respective interval after a preset delay after the respective interval.

Case study

Now the issue is as follows: How much delay should the base station wait for before it discloses the key for that time interval? Suppose that the base station used key K_{76} in interval 76. Definitely, the base station cannot disclose K_{76} in interval 76, because, by doing this, an attacker can immediately get to know K_{76}. As long as the attacker has K_{76}, it can make a MAC using K_{76}. Such a MAC can be used to broadcast a command message. Thus, the attacker can fake any command message to the sensors.

Therefore, the base station should wait for some intervals later. Should it wait for interval 77, 78, or some other interval to disclose K_{76}? Here is µTESLA's solution: The set delay is on the order of a few time intervals and must be greater than any reasonable *round trip time* (RTT) between the sender (i.e., base station) and the receivers (i.e., sensors).

Good idea

Why does the base station wait for at least a RTT to disclose the MAC key? The answer is simple: We do not want to give any attacker the chance to receive the corresponding MAC key and fake a command. If we wait for the maximum RTT value (which can be obtained from empirical data), it will be too late for an attacker to fake a command, because all sensors would have already got the right interval key.

Bootstrapping a new receiver: As mentioned before, each sensor just needs to know K_0, which is the last key of the key chain. We call K_0 the *commitment.* Based on the one-way nature of the key chain, it is obvious that a sensor can verify whether a received MAC key, K_x, is the right one, by constantly applying the one-way hash function as follows:

$$F(\ldots(F(F(K_x)))) = K_0?$$

If it is not equal to K_0, we know such a key does not belong to the right key chain.

The procedure of assigning each sensor the *commitment* K_0 is called *bootstrap.* As we can see, it is easy to bootstrap a new sensor in µTESLA by ensuring that the receiver has one authentic key of the one-way key chain as a commitment to the entire chain.

Loose time synchronization is also important to the correct operation of µTESLA, because the receiver will get to know the beginning of each time interval.

The above-mentioned two requirements, that is, the loose time synchronization and the authenticated key chain *commitment* in each sensor, can be met with a mechanism that ensures *freshness* (i.e., verifying that any message is a new one instead of a replayed one by an attacker) and *point-to-point authentication* (i.e., verifying that the source is a good base station instead of an attacker).

To ensure correct μTESLA operation, the base station needs to securely let the sensors know the following parameters: the current time, T_S (for the time synchronization purpose); a key, K_i, of the one-way key chain used in a past interval, i (disclosed after RTT); the starting time, T_i, of interval i; the duration, T_{int}, of a time interval; and the disclosure delay, δ. We can use the following communications to achieve secure parameter transmissions:

Sensors → *Base Station*: Nonce
Base Station → *Sensors*: $T_S \mid K_i \mid T_i \mid T_{int} \mid \delta$, MAC ($K_{MS}$, $N_M \mid T_S \mid K_i \mid T_i \mid T_{int} \mid \delta$)

Note: We use "nonce" (i.e., a random number that is used only once in the entire session) in the above communications to ensure that each transmitted message is a "fresh" one instead of a replayed one. Also note that the base station does not need to encrypt the data, as the system requires no confidentiality. The MAC uses the secret key shared by the base station and the node to authenticate the data.

Authenticating broadcast packets: If a sensor receives a key, K_j, that was used for the MAC in a previous time interval, it can verify the correctness of the key K_j by checking that it matches with the last authentic key that it knows (say K_i), by applying the *one-way* function F for a few times: $K_i = F_{j-i}(K_j)$. If the verification is successful, the new key, K_j, is authentic, and the sensor can then authenticate all packets that were sent within the time intervals i to j. The receiver also replaces the stored key, K_i, with K_j for the next time check.

11.5 Practical Security Schemes for "Motes"

In this section, we discuss some practical implementations of security schemes in sensor hardware (called "motes"). Especially, we discuss data link layer security, because it is important to achieve security among sensor neighbors.

11.5.1 TinySec [Karlof04]

In conventional networks, such as the Internet, message security (including authenticity, integrity, and confidentiality) is usually achieved by an *end-to-end* security mechanism, such as SSH [TYlonen96], SSL [SSL], or IPSec [IPSec]. This is because the Internet mostly uses end-to-end communication. The routers between the sender and the receiver only need to view message headers. They do not need to access message bodies.

However, WSNs mostly use one-to-many (base station to sensors)/many-to-one (sensors to base station) traffic modes. Moreover, WSNs often have a large number of nodes in an environmental monitoring application. Therefore, neighboring nodes in WSNs often witness the same or correlated environmental events. If each node sends a packet to the base station separately, we will waste a lot of energy and bandwidth. To avoid sending redundant messages, WSNs use *in-network* processing, such as data aggregation techniques, to achieve duplicate data elimination [Samuel02].

Because in-network processing requires intermediate sensors to suppress the contents of messages (or perform other processing), end-to-end security mechanisms may not be as important as hop-to-hop (i.e., data link layer) ones. As a matter of fact, if we just use end-to-end security mechanisms, all message integrity is only checked at the final destination. Then we cannot detect the network attacks in each sensor; for instance, an adversary may inject packets in the middle. Therefore, data link layer security is required to detect unauthorized packets when they are first injected into the network. Some researchers have proposed data link layer security mechanisms for wired networks to resist DoS attacks [Mohamed02].

TinySec [Karlof04] is a WSN data link layer security mechanism that achieves authenticity, integrity, and confidentiality of messages between neighboring nodes, while permitting in-network processing.

TinySec does not need heavy message overhead. It can easily be integrated into other sensor network applications. TinySec can be portably used in a variety of sensor hardware and radio platforms. For more details, please refer to [Karlof04].

11.5.2 MiniSec: A Secure Sensor Network Communication Architecture [Mark07]

MiniSec [Mark07] is also a data link layer security scheme. It consumes lower energy than TinySec; however, it can achieve a higher level of security. It accomplishes this by leveraging three techniques as follows:

1. It uses a block cipher to provide both secrecy and authenticity.
2. It sends only a few bits of the IV (initialization vector); however, it can retain the security of a full-length IV per packet. In contrast, previous approaches (such as TinySec) require two passes over the plaintext (one for encryption and the other for authentication) and transmission of the full-length IV.
3. In the broadcast mode (i.e., from the base station to sensors), MiniSec employs a Bloom-filter-based replay protection mechanism that avoids the per sender state. Such an improvement in energy consumption is achieved at the cost of a modest increase in memory size, which is a desirable trade-off in sensor nodes, as the memory technology is increasing fast.

For TinySec and MiniSec details, please refer to [Karlof04, Mark07].

11.6 Special Case: Secure Time Synchronization in WSNs [Hui07]

WSN security is a wide field with a lot of issues, because security could be implemented in many aspects, such as routing layer, data link layer, and hardware chips. In this section, we introduce the WSN time synchronization security implementations.

Most of the existing time synchronization schemes (in WSNs or other networks) are designed without security in mind and, thus, are vulnerable to malicious attacks. In this section, we focus on a specific type of attack in WSN synchronization schemes, called the *delay attack*, which cannot be addressed by traditional cryptographic techniques. In [Hui07], the author has proposed two approaches to filter the outlier data (caused by the *delay attack*) using the *time transformation* technique and the *statistical* method, respectively.

We have covered WSN synchronization in the previous chapter. We have known that many WSN applications require time to be synchronized among all sensors. Examples of such applications include data link access scheduling, μTESLA, and in-networking aggregation, to name a few. We also know that all WSN time synchronization methods *rely on message exchanges between nodes.*

When a sensor network is deployed in an adversarial environment, such as a battlefield, the time synchronization protocol is an attractive target to the adversaries. For example, time synchronization is the prerequisite of target tracking, because the tracking time needs to be accurately recorded to estimate the object trajectory. Therefore, if an adversary can attack the time synchronization protocol, the estimated direction of a mobile object could be seriously deviant from its actual direction.

Hui [Hui07] defines the delay attack as follows: The attacker deliberately delays some of the time messages, for example, the beacon message in the RBS scheme, so as to fail the time synchronization process. Figure 11.6a shows the normal RBS scheme without the delay attack. Figure11.6b and c shows two ways to launch the delay attack in the RBS scheme. In Figure 11.6b, two colluding nodes act as the reference nodes for nodes A and B, respectively. They send the reference beacon *b* to nodes A and B at different times. As a result, nodes A and B are deceived to believe that they receive the beacon at the same time, although they actually receive it at different times. Figure 11.6c shows that a malicious node can launch the above attacks alone if it has a *directed* antenna (instead of an omnidirectional antenna), so that nodes A and B only hear one beacon message.

Note: If a benign node is synchronizing with a compromised node, the delay attack can also be launched. The compromised node can intentionally add some delay to the beacon-receiving time to mislead the good node to synchronize to a wrong time.

The above example shows the delay attack in a *receiver–receiver*-based synchronization model. A delay attack can also occur in a *sender–receiver*-based model [Ganeriwal03], where the sender and the receiver exchange time synchronization

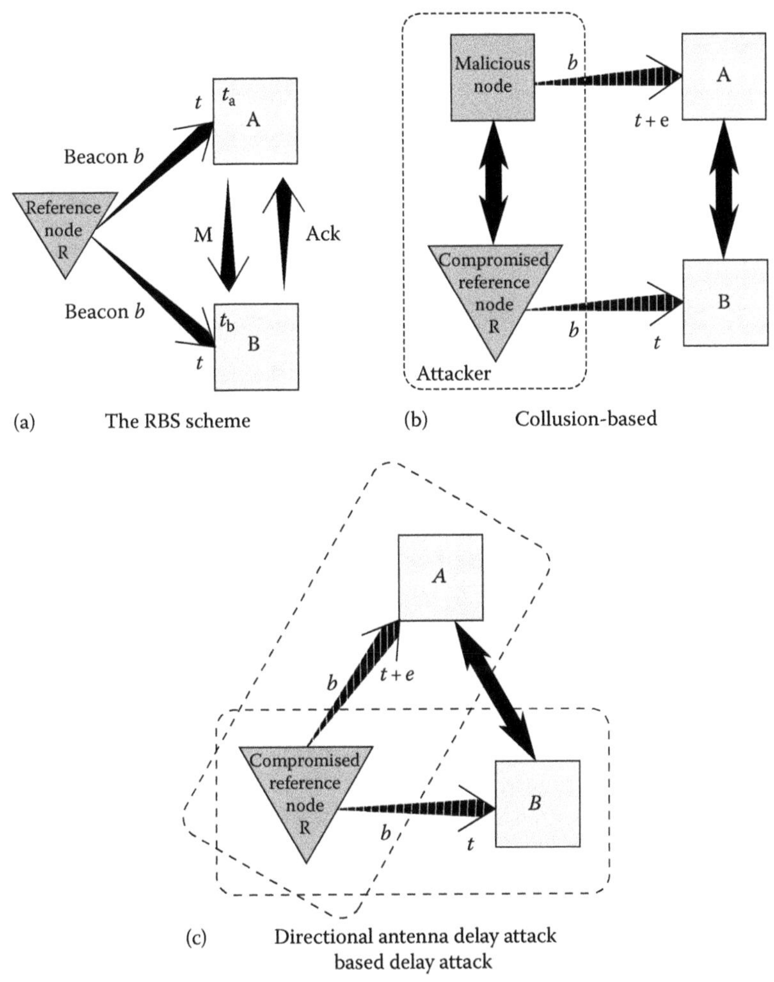

Figure 11.6 The RBS scheme and the delay attacks. (From Hui, S., Secure wireless sensor networks: Building blocks and applications, PhD dissertation, Department of Computer Science and Engineering, The Pennsylvania State University, University Park, PA, 2007.)

messages to estimate the round trip time (RTT) of transmission between them, and synchronize their clocks after finding the clock offset between them. A node can be deceived if it synchronizes with a malicious one. Therefore, these schemes are also subject to the aforementioned delay attacks.

The general idea of defending against *delay attacks* is to find out bad time messages and exclude them. Its basic steps are as follows.

The first step is to collect a set of time offsets from involved nodes.

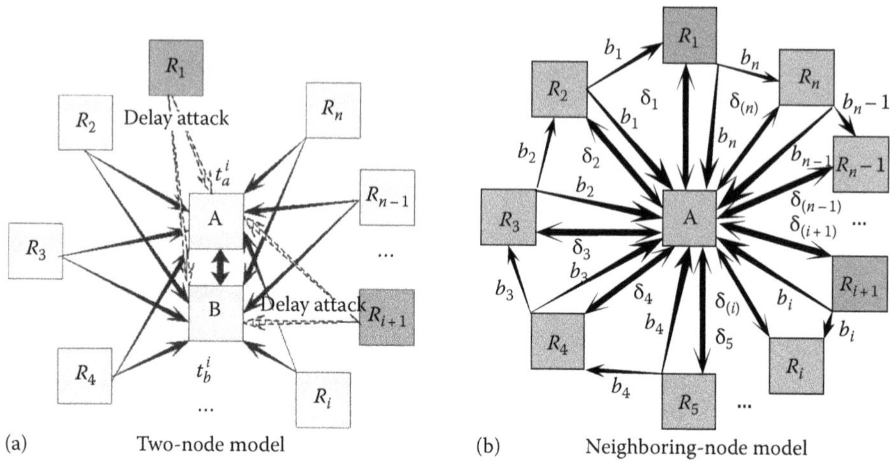

Figure 11.7 Two models for secure time synchronization. (From Hui, S., Secure wireless sensor networks: Building blocks and applications, PhD dissertation, Department of Computer Science and Engineering, The Pennsylvania State University, University Park, PA, 2007.)

Then, some special schemes (such as outlier-detection-based statistical algorithms) are used to identify the malicious time offsets that are under delay attacks.

Finally, the identified malicious time offsets will be excluded and the rest of the time offsets are used to estimate the actual time offset.

Hui [Hui07] has presented two models for *collecting the time offsets*: the two-node model and the neighboring-node model, which are described in the context of the RBS scheme.

Two-node model: In this model, *a node only needs to synchronize with its cluster head*. As shown in Figure 11.7a, suppose that node B is the cluster head, and A is a common node within the cluster. Due to security concerns, node A only trusts the cluster head and not other nodes in the cluster. And node A is required to synchronize only with cluster head B.

To countermeasure delay attacks, node A uses multiple reference nodes (R_1, R_2, …, R_n) to obtain a set of time offsets. If $\langle t_a^i, t_b^i \rangle$ represent the two beacon-receiving times obtained by using a reference node R_i (i.e., the receiving times at node R_i when sending a message from A → i and B → i, respectively), define $\delta_i = (t_a^i - t_b^i)$ as the time offset. Thus, we can obtain a set of n time offsets $\{\delta_1, \delta_2, \ldots, \delta_n\}$. Based on the collected time offsets, we can use some statistical algorithms to detect and exclude the malicious time offsets and obtain a more accurate estimation on the actual time offset between A and B.

Neighboring-node model: In this model, *a node is required to synchronize with its multiple neighbors* (>2) *to detect a delay attack*. The reason of using the neighboring-node model is that the two-node model is not enough when one or multiple neighbors may have been compromised. The good nodes could synchronize with the malicious nodes that generate delay attacks, as illustrated in Figure 11.7b. Suppose that A has n neighbors: $R_1, R_2, \ldots, R_n$. We run the RBS scheme between A and each of its neighbors and each time we use a different node as reference to obtain a time offset. After collecting a set of n time offsets, we can detect the outliers, exclude them, and make a good estimation on the actual time offsets.

Good idea

The delay attack countermeasure is based on outlier detection, which picks up a "strange" value from a lot of data. As you can see, we can achieve security from many different perspectives: Although traditional encryption/decryption could work on most cases, if an internal node is compromised and becomes a "spy," we need other *non-cryptographic ways* to find out the "spy." This section describes the use of math statistics for bad behavior detection. Always remember: All disciplines can be related to each other to generate some "magic" solutions to some challenging issues.

Besides the above two models, other models are also possible to be used in terms of collecting time offsets. However, all of these models have one thing in common: They collect a set of time offsets, which may include the malicious time offsets.

Then the next question is as follows: Given a set of time offsets, how do we identify the outliers and achieve an attack-resilient time estimation?

We could imagine that without delay attacks, the time offsets among nodes follow a similar statistical distribution. The existence of delay attacks makes the malicious time offsets much different from the others. From the statistics viewpoint, these malicious time offsets are referred to as *outliers*, which are defined as "an observation which deviates so much from other observations as to arouse suspicious that it was generated by a different mechanism" [Hawkins80].

Many schemes have been proposed to detect outliers values ([Iglewicz93] has a good survey). Hui [Hui07] introduced generalized extreme studentized deviate (GESD) as an outlier detection algorithm. GESD is built on the extreme studentized deviate (ESD) test (also called Grubb's test). The ESD test can detect one outlier in a random normal sample.

Definition of the ESD test: Given a data set $\Gamma = \{x_1, x_2, \ldots, x_n\}$, the mean of Γ is denoted as $\overline{x}$, and the standard deviation of Γ is denoted as s. Let

$$T_i = |x_i - \overline{x}|/s, \quad \text{where } i = 1, \ldots, n$$

T_i is also called the corresponding T value of x_i. Let x_j be the observation that leads to the largest $|x - \overline{x}|/s$, where $i = 1, \ldots, n$. Then x_j is an outlier when T_j

exceeds a tabled critical value, λ. In principle, if T_j does not exceed the critical value A, we need not single out x_j. Assuming that this test finds an outlier, we then look for further outliers by removing observation x_j and repeating the process on the remaining $n - 1$ observations. *However, the ESD test can only detect one outlier each time.*

The GESD procedure [Hui07] is a modified version of the ESD test, which can *find multiple outliers at a time.* GESD has two critical parameters: (1) r is the estimated number of outliers in the data set and (2) λ_i is the two-sided $100 * a$ percent critical value as follows:

$$\lambda_i = \frac{t_{n-i-1,p}(n-i)}{\sqrt{(n-i-1+t_{n-i-1,p}^2)(n-i+1)}} \tag{11.3}$$

where $i = 1, \ldots, r$, $t_{v,p}$ is the $100 * p$ percentage point from the t distribution with v degrees of freedom, and $p = 1 - [\alpha/2(n - i + 1)]$. Given α, n, and r, the critical values, λ_i, can be calculated beforehand.

Definition (*GESD-based delay attack detection*). Given the time offset set $\Gamma = \{\delta_1, \delta_2, \ldots, \delta_n\}$, all the time offsets δ_i that are identified as outliers by GESD are claimed to be under the delay attack.

In GESD, r is the estimated number of malicious time offsets. Note that it is important to select a proper value of r. If r is set to too small a value and there are more than r malicious time offsets among the m time offsets, some of them cannot be detected using GESD. On the other hand, if r is too large, it wastes time in checking the nodes that are in fact good.

In GESD, because the number of time offsets is small (e.g., 20), r is set to be half of the total number of time offsets. GESD also assumes that the number of malicious time offsets is less than half of the total number of time offsets. Without this assumption, GESD may not work, because it may find the malicious time offsets to be benign and the benign ones to be malicious.

GESD *estimates* r as follows: Let the median of the time offset set Γ be $\hat{x}$ and s be the standard deviation. r is defined as the number of time offsets, x_j, such that

$$\frac{|x_j - \hat{x}|}{s} > 2, \quad \text{where } i = 1,2,3,\ldots,n \tag{11.4}$$

When the number of malicious nodes is small, that is, less than five percent of the total, we can utilize the median of the time offsets to set r. As shown in the above definition, r is the number of time offsets that are two standard deviations away from the median. In most cases, the data and time offsets are normally distributed, which means that 95 percent of the values are at most two standard deviations away from the mean.

Problems and Exercises

11.1 Multi-choice questions

1. With respect to sensor network security, which of the following is not correct?
   a. Key management is a crucial step to achieve sensor network security, because it controls key generation and distribution.
   b. The most important goal of sensor network security is to guarantee the confidentiality of transmitted data. Other security goals are minor.
   c. Traditional Internet security schemes may not be suitable to the sensor network case due to their high calculation overhead.
   d. The security schemes should occupy small memory space (<100 kB) in the sensors.
2. Which of the following statements is not correct on sensor network attacks?
   a. Attackers do not belong to sensor network nodes. They have no access to the encryption keys stored in the sensors' memory.
   b. A side-channel attack refers to any attack that is based on the information gathered from the physical implementation of a cryptosystem. The attacker may analyze the power consumption, the timing of the software operation execution, or the frequency of the EM waves.
   c. A typical network physical layer attack is the jamming attack.
   d. A link layer attack tries to damage the normal medium access control operations.
3. With respect to sensor network routing attacks, we have which of the following fact(s)?
   a. An attacker can mislead a routing control command.
   b. An attacker can attract the network traffic into its machine through the sinkhole attack.
   c. A Sybil node can fabricate a new identity.
   d. All of the above.
4. Wormhole attacks have which of the following features?
   a. An adversary tunnels the messages received in one part of the network over a low latency link and replays them in a different part.
   b. An adversary that had tricked every node in the network into believing that the adversary was its neighbor could effectively cause most of the transmitted data to be lost, provided that the adversary was at a long distance.
   c. Due to the inherent broadcast medium, an attacker can spoof link layer ACKs for "overheard" packets addressed to neighboring nodes.
   d. Wormhole attacks try to hide the enemies' IDs.
5. Which of the following is not correct with respect to wormhole attacks?
   a. Wormholes can be launched by encapsulating the routing packets.
   b. Two attackers can use a worse communication channel (compared to normal sensors' links) to build a wormhole attack.

c. Wormholes can be made by a higher power transmission between enemies' machines.
d. A malicious node can create a wormhole by not following the normal routing protocol and broadcasting without delay. This is an effort to make the packet reach its destination first, thus making the delay seem less than the surrounding nodes.

6. With respect to time synchronization security, we have which of the following fact(s)?
   a. A delay attack can delay some of the time messages to fail the time synchronization process.
   b. Traditional encryption/decryption schemes can solve time synchronization security issues.
   c. Outlier detection aims to use statistical mean values to overcome attacks.
   d. None of the above.
7. MiniSec improves TinySec in which of the following aspect(s)?
   a. MiniSec exploits the fundamental distinctions between unicast and broadcast communication, providing two energy-optimized communication modes.
   b. MiniSec is more energy efficient than TinySec.
   c. MiniSec can run in the routing layer.
   d. Both A and B.
8. μTESLA does not have which of the following features?
   a. It first uses TESLA to filter bad nodes.
   b. The entire time is divided into intervals.
   c. Authentication keys are disclosed after some time.
   d. The keys are generated by a one-way hash function.

11.2 Explain the principle of LITEWORP in terms of overcoming the wormhole attack. Illustrate with the operation diagrams.

11.3 Why does μTESLA wait for some intervals to release an authentication key? How long should the waiting time be?

11.4 Use a broadcast communication scenario to explain the detailed operation principle of μTESLA.

11.5 Explain the delay attack countermeasure in the time synchronization protocol.

11.6 Read the papers on TinySec and MiniSec, and explain the major operation differences between them.

11.7 Explain the Blom-based key generation principle.

VI
SPECIAL WIRELESS SENSOR NETWORKS

Chapter 12

Wireless Sensor and Actor Networks

12.1 Introduction [Melodia07, Akyildiz04]

Wireless sensor and actor networks (WSANs) [Akyildiz04] are distributed wireless communication and control systems consisting of heterogeneous devices referred to as sensors and actors. Sensors have the same characteristics as general WSNs such as low-cost, low-power, multifunctional devices that communicate untethered in short distances [Akyildiz02]. Actors collect and process sensor data and consequently perform actions on the environment. Different from sensors, actors have rich resources, such as high processing capabilities, high transmission power, and long battery life.

The term *actor* is different from the more conventional notion of an actuator. An actuator typically refers to a device that can convert an electrical control signal to a physical action, and may be used for flow-control valves, pumps, motors, etc. An actor has functions of an actuator; more importantly, it is also a single and separated network entity that performs networking-related functionalities, that is, receive, transmit, process, and relay data. For example, a robot may interact with the physical environment by means of several motors (i.e., actuators). However, from a networking perspective, the robot constitutes a single entity, which could be called an "actor" [Melodia07].

As shown in Figure 12.1, sensors and actors are deployed in the wide field. A sink monitors the overall network, and communicates with the task manager node and sensor/actor nodes. Similar to WSNs, a WSAN has a possibly large number of sensor nodes, that is, of the order of hundreds or thousands. Such a dense

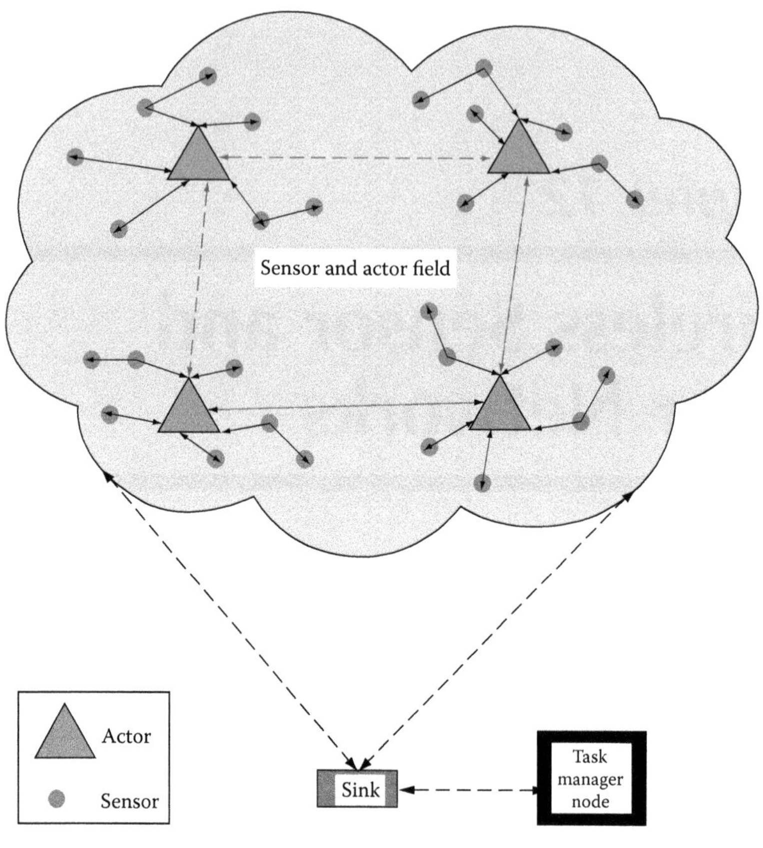

Figure 12.1 The physical architecture of WSANs. (Adapted from Akyildiz, I.F. and Kasimoglu, I.H., *Ad Hoc Netw.*, 2, 351, October 2004.)

deployment is not necessary for actors, because actors are typically more expensive than sensors and have higher capabilities to act on large areas.

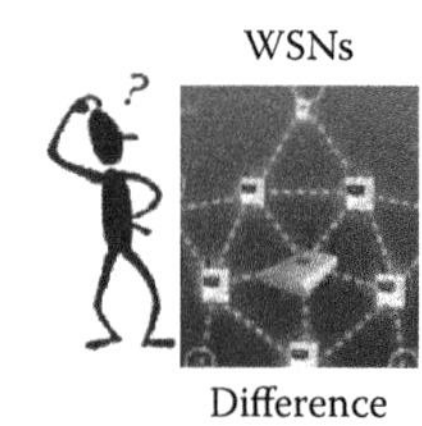

As we can see, WSAN is a special WSN. The biggest difference between them is that WSAN has a small number of actors that are mobile, have better CPU performance, and longer wireless communication range than sensors. Those actors need to coordinate with sensors to determine how to respond to a certain sensed event.

If a sensor detects an event, they could send their readings to the nearby actors that process all incoming data and initiate appropriate actions, or they route data hop-to-hop

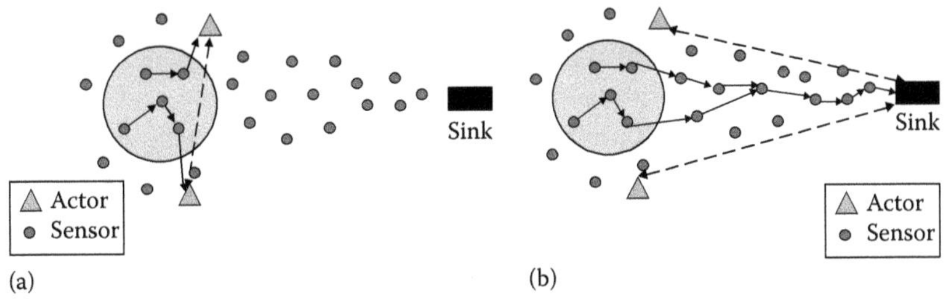

Figure 12.2 (a) Automated versus (b) semiautomated architecture. (Adapted from Akyildiz, I.F. and Kasimoglu, I.H., *Ad Hoc Netw.*, 2, 351, October 2004.)

to the sink, which then finds out an actor to handle the event. We call the former case automated architecture (see Figure 12.2a) due to the nonexistence of central controller, for example, the sink, while the latter case is called semiautomated architecture (see Figure 12.2b) as the sink (central controller) collects data and coordinates the acting process. These two schemes have different pros and cons. The automated mode saves action time; but the semiautomated one has better global management because a sink can check all actors' resource status and make decisions.

After sensors detect a special event, they may directly (without going to a sink first) request actors to respond to the event. These sensors and actors may coordinate with each other to establish efficient routing paths among them. Such a coordination procedure is referred to as sensor–actor coordination [Akyildiz04].

On the other hand, when an actor receives an event-handling request from sensors, it may not be able to handle such an event efficiently due to limited capacity. In this case, it will coordinate with other actors to make a collaborative decision on how to perform the action. This process is referred to as *actor–actor coordination*.

There are some crucial requirements for WSANs:

First, the WSANs need to have real-time coordination and communication among actors (and even between sensors and actors) to guarantee timely execution of the right actions.

Second, energy efficiency is still required, especially for sensors, which are resource-constrained nodes with limited battery lifetime.

Third, as in WSNs, we still have scalability requirements for WSAN protocols and algorithms because the number of sensors can be arbitrarily high.

WSANs have wide applications in battlefield surveillance, nuclear, biological, or chemical attack detection, home automation, and environmental monitoring. Akyildiz and Kasimoglu [Akyildiz04] have a few good examples:

Fire control: When a building has a fire event, the temperature/smoke sensors can detect the exact origin and intensity of the fire, and send those parameters data to water sprinklers (i.e., actors) that will extinguish the fire before it is out of control.

Pollution control: Sensors may detect visible or measurable discharges of contaminants in water or in the air, and actors (such as pollution cleaner) can reactively take countermeasures.

Building surveillance: In a building, motion, acoustic, or light sensors can detect the presence of intruders and request the cameras to track them. Security personnel (also called actors) can move to the area where the intruder has been detected.

12.2 Sensor–Actor Coordination Problem [Melodia07]

As discussed in the previous section, sensor–actor communications have real-time requirements. Melodia et al. [Melodia07] introduce a set of interesting schemes to solve the sensor–actor coordination issues. Besides real-time requirements (i.e., bounded communication delay), it also considers communication reliability issue. Especially, it introduces "delay-bounded reliability," which accounts for the percentage of packets that are generated by the sensors in the event area and are received within a predefined latency bound (which is referred to as reliable packets). Note its reliability concept is related to the real-time delivery of data packets from sources to actors, and is calculated at the network layer.

The concept of *latency bound B* is defined as follows: it is the maximum allowed time between the instant when the sensor samples the physical features of the event and the instant when the actor receives a data packet eventually.

Obviously, if a packet does not meet the *latency bound B* when it is received by an actor, it is useless. Likewise, a data packet received within the *latency bound B* is said to be unexpired and thus, reliable.

We define the event reliability r as the ratio of reliable data packets over all the packets generated in a decision interval. We also define r_{th} as the minimum event reliability required by the application. The lack of reliability is the difference $(r_{\text{th}} - r)$ between the required event reliability threshold r_{th} and the observed event reliability r at a given time.

Now we can formulate the sensor–actor coordination problem as follows:

How do we establish routing paths from each sensor residing in the event area to the actors under the following two conditions?

- The achieved reliability r should be larger than the threshold r_{th} (i.e., $r \geq r_{\text{th}}$).
- The routing paths should have minimum energy consumption.

Based on the above goals, Melodia et al. [Melodia07] solve the sensor–actor coordination problem through event-driven partitioning with multiple actors, and using a math model called *integer linear program* (*ILP*) [Ahuja93]. To describe ILP, we need to first define WSAN network model and energy model.

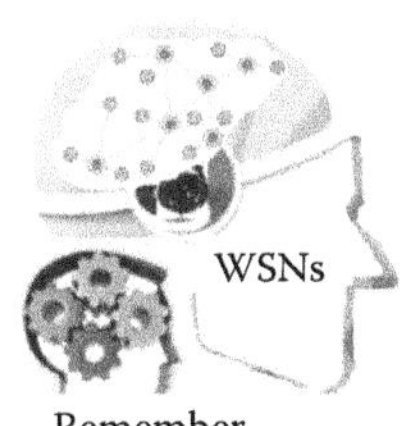

Remember: when you try to minimize or maximize an *object function* under a series of constraints/conditions, and if those conditions can be formulated into math equations (could be the formats of $A > B$, $A < B$, $A = B$, where A and B are functions), then you may consider using *ILP*. ILP is actually a function optimization problem. In MATLAB® (a math tool) you could set up those ILP constraints.

12.2.1 Network and Energy Model

Melodia et al. [Melodia07] use a graph model to describe the network topology. A WSAN is represented as a graph $G(S_V, S_E)$, where $S_V = \{v_1, v_2, \ldots, v_N\}$ is a finite set of *Vertexes* in a finite-dimension terrain, with $N = |S_V|$, and S_E is the set of *Edges* among nodes, that is, $e_{ij} \in S_E$ iff nodes v_i and v_j (also i and j for simplicity in the following) are within each other's transmission range.

Let S_A represent the set of actors, with $N_A = |S_A|$. We refer to an actor that is collecting data from one or more sources as *collector.*

Let S_S be the set of data sources, with $N_S = |S_S|$. This set represents the sensor nodes that detect the event, that is, the sensors that reside in the event area.

We define $P = \{(s, a): s \in S, a \in A\}$ as the set of source–destination connections.

Energy model: Following the model in [Heinzelman02], we can assume that the energy consumption per bit (in physical layer) is $E = 2E_{elec} + \beta d^{\alpha}$, where α is the exponent of the path loss ($2 \le \alpha \le 5$), β is a constant, and E_{elec} is the energy needed by the transceiver circuitry to transmit or receive one bit [J/bits].

12.2.2 ILP Algorithm

The ILP-based sensor–actor routing path search problem is to find *data aggregation trees* (*da-trees*) from all the sensors that reside in the event area (referred to as sources) to the appropriate actors. All tree leaves in a *da-tree* are source sensors (but not all source sensors are necessarily leaves), and each actor is either the root of a *da-tree* or does not participate in the communication.

The ILP-based algorithm aims to construct a set of *da-trees* where each source sensor belongs to one tree only. Each *da-tree* has only one actor as its root. Therefore, each source sensor is associated with an actor to achieve an optimal event-driven partition.

The ILP algorithm aims to build such an event-driven partitioning with the following two major steps: (1) first, we need to select the optimal subset of actors to which sensor readings will be transmitted and (2) second, after we select those actors, we can

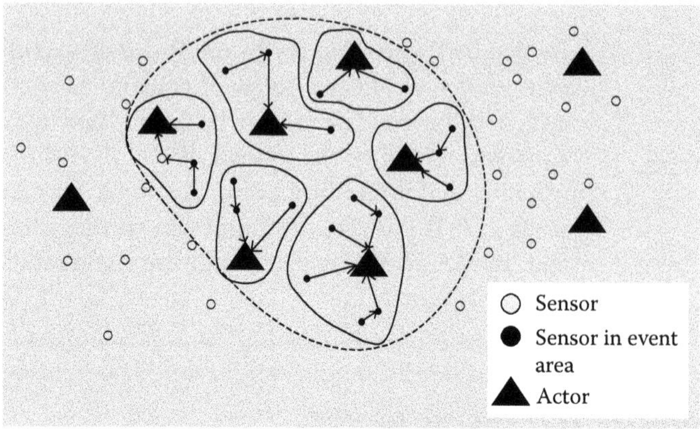

Figure 12.3 Event-driven partitioning with multiple actors. (Adapted from Melodia, T. et al., *IEEE Trans. Mobile Comput.*, 6(10), 1116, October 2007.)

then construct the minimum energy *da-trees* toward those selected actors that meet the required event reliability constraint (i.e., the above-mentioned two conditions).

Therefore, we could partition the set of source nodes in the event area based on the *da-trees* rooted at the actors. Figure 12.3 gives an example of event area partition.

Before we use ILP algorithm to formulate the area partition problem, we introduce the following notations that are used in the ILP model:

e_{ij} is a binary variable (either 0 or 1). It equals to 1 if nodes i and j are within each other's transmission range.

c_{ij} is the energy cost of the link between nodes i and j, that is, $2E_{elec} + \beta d_{ij}$, where d_{ij} is the distance between nodes i and j.

x_{ij}^k is a binary variable that equals 1 if link (i, j) is part of the *da-tree* associated with actor k.

$f_{ij}^{k,s}$ is a binary variable that equals 1 if source sensor s sends data to actor k and link (i, j) is in the path from s to k.

$l_{k,s}$ is a binary variable that equals 1 if sensor s sends data to actor k.

p_{ij} is the propagation delay associated with link (i, j), defined as d_{ij}/v, where v is the signal propagation speed.

$\tilde{d}$ is a parameter that accounts for processing, queuing, and medium access delay at each sensor node.

B is the latency bound on each source–actor flow.

r and r_{th} are the actual event reliability and the required event reliability threshold, respectively.

$b_{k,s}$ is a binary variable that equals 1 if the connection between source s and actor k is not compliant with the latency bound, that is, the end-to-end delay is higher than the *latency bound B*.

Q is the number of noncompliant sources.

The *ILP*-based area partition problem (i.e., *da-trees* construction) can then be formulated as follows:

$P_{\text{Min}}^{\text{Com}}$: event area sensor partitioning with multiple actors

Given: e_{ij}, c_{ij}, p_{ij}, v, $\tilde{d}$, B, r_{th}

Find: x_{ij}^{k}, $f_{ij}^{k,s}$, $l^{k,s}$, $b^{k,s}$,r

Minimize:

$$C^{\text{TOT}} = \sum_{k \in S^A} \sum_{(i,j) \in S^{\varepsilon}} x_{ij}^{k} \cdot c_{ij} + \gamma \cdot Q \tag{12.1}$$

The above equation is called the *object function*: Once all routing paths (from sensors to actors) are established, the entire system should have minimized energy consumption. This constraint imposes a penalty by multiplying the number Q of noncompliant source sensors by a penalty coefficient γ.

Subject to the following constraints:

$$\sum_{j \in S^V} (f_{sj}^{k,s} - f_{js}^{k,s}) = l^{k,s}, \quad \forall s \in S^s, \quad \forall k \in S^A \tag{12.2}$$

(This constraint guarantees that a source sensor generates a data flow on the *da-tree* of the selected actor, and only on that *da-tree*; while non-source nodes do not generate any data flow.)

$$\sum_{j \in S^V} (f_{sj}^{k,s} - f_{js}^{k,s}) = -l^{k,s}, \quad \forall s \in S^s, \quad \forall k \in S^A \tag{12.3}$$

(This constraint requires that data flows generated by each source sensor be collected by one actor only.)

$$\sum_{j \in S^V} (f_{sj}^{k,s} - f_{js}^{k,s}) = 0 \tag{12.4}$$

(This constraint imposes that the balance between incoming and outgoing flows is null for non-source and nonactor nodes.)

$$\forall s \in S^s, \forall k \in S^A, \forall i \in S^V \quad \text{s.t.} \quad i \neq s, \ i \neq k$$

$$f_{ij}^{k,s} \leq e_{ij}, \quad \forall s \in S^s, \quad \forall k \in S^A, \quad \forall i \in S^V, \quad \forall j \in S^V \tag{12.5}$$

(This constraint ensures that data flows are created on links between "adjacent" nodes [i.e., they are within the transmission range of each other]).

$$f_{ij}^{k,s} \le x_{ij}^{k}, \quad \forall s \in S^{s}, \quad \forall k \in S^{A}, \quad \forall i \in S^{V}, \quad \forall j \in S^{V} \tag{12.6}$$

(This constraint forces all data flows from different source sensors but directed toward the same actor to be aggregated in the *da-tree* associated with that actor.)

$$\sum_{k \in S^{A}} l^{k,s} = 1, \quad \forall s \in S^{s} \tag{12.7}$$

(Thus constraint imposes that each source sensor sends data to exactly one actor.)

$$f_{ij}^{k,s} \le l^{k,s}, \quad \forall s \in S^{s}, \quad \forall k \in S^{A}, \quad \forall i \in S^{V}, \quad \forall j \in S^{V} \tag{12.8}$$

(This constraint ensures that all flow variables from a source to a particular actor are zero unless that actor is selected by the source.)

$$\varepsilon \cdot \left[B - \sum_{(i,j) \in S^{\varepsilon}} f_{i,j}^{k,s} (p_{ij} + \tilde{d}) \right] \le b^{k,s}, \quad \forall s \in S^{s}, \quad \forall k \in S^{A} \tag{12.9}$$

[This constraint requires that the binary variable $b_{k,s}$ be equal to 1 if and only if the flow between source sensor s and actor k violates the *latency bound B*. The small negative coefficient ε scales the value in the square parentheses to make it smaller than 1. Hence, when the latency bound is violated, the left side of (12.9) is a small positive value, which forces the binary variable $b_{k,s}$ to be 1. Conversely, when the latency bound is met, the left side of (12.9) is negative and $b_{k,s}$ will assume the 0 value to minimize the objective function in (12.1).]

$$Q = \sum_{k \in S^{A}} \sum_{s \in S^{s}} b^{k,s}; \qquad r = \frac{|S^{S}| - Q}{|S^{S}|} \ge r_{\text{th}} \tag{12.10}$$

(Q is defined as the number of noncompliant source sensors, and the reliability r is calculated as the ratio of compliant source sensors over all source sensors. r should be larger than the required threshold.)

12.2.3 Sensor–Actor Coordination: Distributed Protocol

The next step is to transform the above math models to practical sensor–actor coordination protocol. The objective of the distributed protocol is to build *da-trees* between the source sensors (that reside in the event area) and the actors in such a way as to minimize the objective function in (12.1), that is, to provide the required reliability r_{th} while, in the meantime, minimizing energy expenditure.

As we mentioned before, the result of sensor–actor coordination protocol is a set of *da-trees* that have all the routing paths from source nodes to actors. It is an approximate solution to the event area partitioning with multiple actors problem. Melodia et al. [Melodia07] refer to the protocol as *distributed event-driven partitioning and routing (DEPR) protocol.*

We also know that routing algorithms with localized routing decisions (i.e., based on local topology information) can generate routing paths with energy efficiency close to the global optimum [Melodia05]. Therefore, the objective of the DEPR protocol is to minimize the energy consumption by relying on local information and on greedy routing decisions.

To guarantee predetermined delay bounds in each sensor–actor routing path, we require some form of end-to-end feedback. DEPR relies on collective feedback from the receiving actors. Each actor advertises the observed reliability.

DEPR adopts local behavior control in each individual sensor node to lead to a global network effect with two most important aspects: (1) providing event reliability r above the required threshold r_{th} and (2) minimizing the energy consumption.

However, DEPR controls the reliability through the adjustment of the routing delays, which can be achieved by modifying the average routing path length. In general wireless networks, we could control the energy consumption in each communication link by changing the transmitted power in the sender node. That is, when we decrease the antenna power level of the sender node, we could save more energy. On the other hand, changing the power level of the transmitter can control the signal propagation distance. The larger the transmitter power, the longer distance the signal can propagate, and less delay the routing task takes.

DEPR protocol makes some assumptions in its geographical routing algorithms: Some sensor positioning schemes should be used to ensure that each sensor is aware of its position. Each sensor should also be aware of the position of its neighbors, as every node locally broadcasts its position. It should also be aware of the position of the actors, as each actor periodically beacons its position in the sensor field. The entire network should be synchronized by means of one of the existing time synchronization protocols [Sundararaman05].

12.2.4 Overview of DEPR

Let us repeat the objective of DEPR protocol: it aims to create *da-trees* between the sources and a subset of the actors. Those actors are also called collectors. A *da-tree* is created between each collector and the source sensors that provide sensor data to the collector. The protocol generates a partitioned event area, with each part composed of the source sensors associated with a single collector.

DEPR protocol requires each sensor to operate among four different states, namely, idle, start-up, speed-up, and aggregation states. When the reliability requirement is not met, the main objective of these state transitions is to reduce the number of hops, which results in decreased delay. When the reliability requirement is met, our objective is to save energy.

Besides the state transition, DEPR protocol controls the energy and reliability performance by transmit power control. The transmit power level affects the quality of the received wireless signal and thus impacts packet error rate. Higher transmission power can bring higher reliability.

The transmission power also determines the radio communication range and thus affects the pool of feasible next hops at the routing layer. High transmission power reduces the number of hops needed to reach the intended destination. A high transmission power makes the network more connected by increasing the number of direct links seen by each node.

On the other hand, a lower transmission power reduces the energy consumption and thus increases the network lifetime. However, a low transmission power causes shorter communication range of a sensor, and thus requires more forwarding nodes (i.e., more hops), which results in higher end-to-end latency.

In summary, an efficient power control scheme should make a balance between energy consumption and routing latency. This is also the objective of the DEPR distributed protocol. In the following discussion, we will provide more details on the DEPR protocol.

The latency of each packet is calculated by an actor that checks the time stamp in the packet header. In each decision interval, the actor computes the data arrival reliability r as the ratio of unexpired packets over all generated packets and periodically broadcasts its value to its neighborhood. Sensor nodes (associated with that collector) control their state transitions (among idle, start-up, speed-up, and aggregation states) based on the reliability observed by the collector, which is broadcasted at the end of each decision interval.

A sensor transfers states as follows: it starts in an idle state, where it senses data from the environment and monitors the wireless channel for incoming data packets. A sensor then enters the start-up state when it either senses a special event or receives the first data packet from a neighboring sensor.

Sensor nodes expect feedback messages from the collector (i.e., actor) they are associated with. If the event reliability r is advertised to be below the low event reliability threshold r_{th}^{-}, we need to reduce the sensor–actor routing delay, which can be achieved by reducing the end-to-end path length. Hence, when $r < r_{th}^{-}$, a sensor in the start-up state will enter the speed-up state with probability P_{st-sp}, which can be a monotonically increasing function of the lack of reliability $(r_{th}^{-} - r)$. Note that here we use a probabilistic policy (P_{st-sp}) to prevent state transitions deadlock (i.e., system oscillations), which could occur if all sensors changed state at the same time.

If the event reliability r is above the high event reliability threshold r_{th}^{+} (i.e., $r > r_{th}^{+}$), we may consider to save energy. In this case, a node in the start-up state enters the aggregation state with probability P_{st-ag}, which can be a monotonically increasing function of the excess of reliability $(r - r_{th}^{+})$. In this case, it tries to minimize the energy consumption by relaying data to the closest neighbor that participates in a *da-tree.*

Then, sensors can alternate between the speed-up and the aggregation state to respond to feedback messages from collectors. The objective of the DEPR protocol is to converge to a solution with reliability close to the event reliability threshold with minimal energy consumption. A sensor goes back to the idle state if it does not generate or receive packets for timeout (a parameter) seconds.

12.3 Hierarchical Sensor–Actor Coordination Mechanism [Yuan06]

12.3.1 Hierarchical WSAN Coordination Architecture

A three-level sensor–actor coordination model for WSANs is proposed in [Yuan06], which can be shown in Figure 12.4a through c.

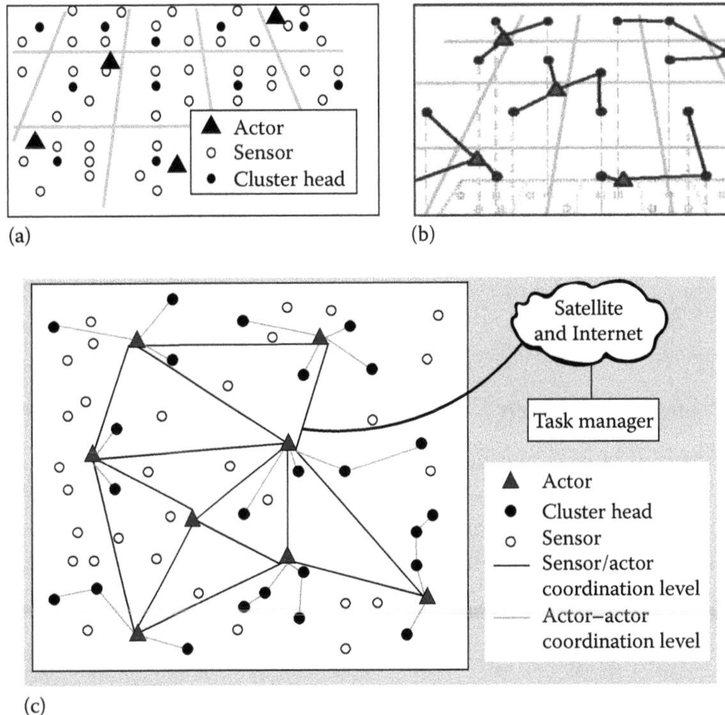

Figure 12.4 Three-level coordination model. (Adapted from Yuan, H. et al., Coordination mechanism in wireless sensor and actor networks, *Proceedings of the First International Multi-Symposiums on Computer and Computational Sciences (IMSCCS '06)*, April 20–24, 2006, IEEE Computer Society, Washington, DC, 2006, Vol. 2, 627–634.)

- *Level 1: Sensor–sensor coordination*: As shown in Figure 12.4a, sensor–sensor coordination is based on the clustering technique with the cluster head collecting data from other neighboring sensors. The sensor–sensor coordination aims to minimize the energy consumption of sensors and maximize the network lifetime.
- *Level 2: Sensor–actor coordination*: As shown in Figure 12.4b, the goal of sensor–actor coordination is to minimize the latency between sensing and acting when the cluster heads transmit data to appropriate actors. Another goal of sensor–actor coordination is to make actors perform most of the energy-consuming tasks such as routing computation and data aggregation.
- *Level 3: Actor–actor coordination*: As shown in Figure 12.4c, the purpose of actor–actor coordination is to control actors to perform effective and reliable actions. The main goal of actor–actor coordination is to maximize their overall task performance by actualizing optimal policy of the task allocation and cooperation.

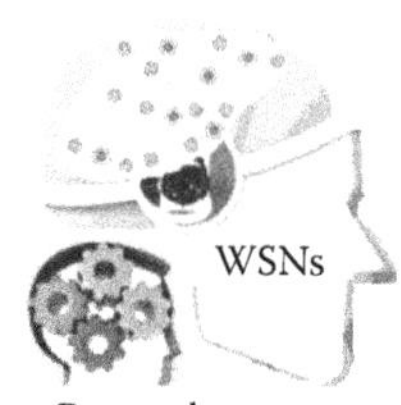

The multilevel coordination scheme has been used in many problem solutions. Its basic feature is to define two or more levels with close mapping relationship between any two neighboring levels. Higher levels typically have less (however, more capable) objects than lower levels. Therefore, there is less communication traffic in higher levels. In lower levels, people concern scalability and energy efficiency as large amount of objects need to communicate with each other.

12.3.2 "Sensor–Sensor" Coordination Level—Use Clusters

Sensor–sensor coordination is based on the cluster-based routing protocol. Both LEACH [WBHeinzelman02] and TEEN [AManjeshwar01] can be used because they all use clusters. However, they did not use geographical location information, which is necessary here because Yuan et al. [Yuan06] assume a grid-based routing architecture. Each sensor/actor needs to know their positions before they know which grid they belong to.

GAF [YXu01] (geographical adaptive fidelity) is a location-based routing algorithm, but is not a cluster-based algorithm. In GAF, the network area is divided into fixed zones and forms a virtual grid. Inside each grid, nodes collaborate with each other to become active nodes or sleeping nodes in turns. GAF conserves energy by turning off unnecessary nodes.

Figure 12.5 shows a clustering algorithm based on GAF. The nominal radio range R of stationary sensors is shown as a dashed line in Figure 12.6. Assume that the virtual

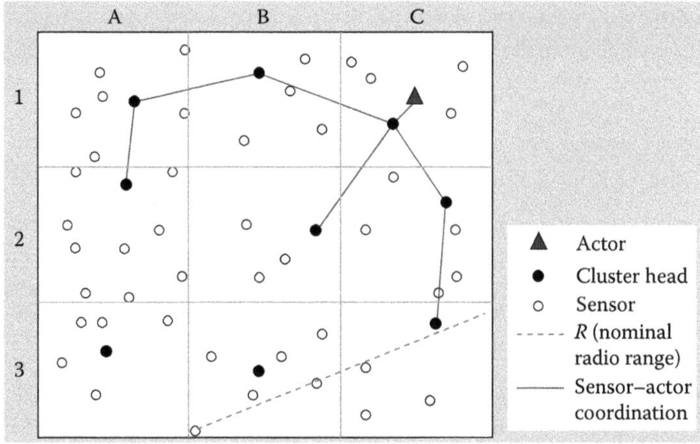

Figure 12.5 R-based clustering and routing. (Adapted from Yuan, H. et al., Coordination mechanism in wireless sensor and actor networks, *Proceedings of the First International Multi-Symposiums on Computer and Computational Sciences* (*IMSCCS '06*), April 20–24, 2006, IEEE Computer Society, Washington, DC, 2006, Vol. 2, 627–634.)

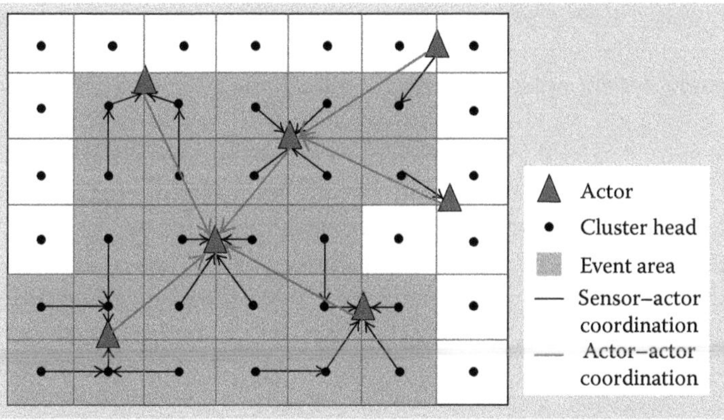

Figure 12.6 Two-level aggregation tree. (Adapted from Yuan, H. et al., Coordination mechanism in wireless sensor and actor networks, *Proceedings of the First International Multi-Symposiums on Computer and Computational Sciences* (*IMSCCS '06*), April 20–24, 2006, IEEE Computer Society, Washington, DC, 2006, Vol. 2, 627–634.)

grid is a square with r units on a side. The distance between two possible farthest sensors in any two adjacent grids must not be larger than R. Yuan et al. [yuan06] require

$$r \le \frac{R}{\sqrt{5}} \tag{12.11}$$

Sensors in adjacent grids can communicate directly with each other. Inside each grid, sensors elect one sensor as cluster head that aggregates the sensing data from each sensor associated with it, and is responsible for monitoring and reporting data to the appropriate actors.

12.3.3 "Sensor–Actor" Coordination Level

To achieve sensor–actor coordination, we require that all actors announce their information (such as their current locations and accurate time) periodically or throughout the duration of actors' movement. After receiving such information, each cluster head (special sensors) achieves clock synchronization, and maintains a routing table that includes nearby actor(s) to deal with the mobility of actors.

As the whole sensor area is divided into location-based grids, the serial number of each grid can be regarded as the ID of each cluster head. Furthermore, every cluster head need not exchange and maintain the ID of other cluster heads, because event information is transmitted through cluster heads to actors only grid by grid; only grids included in the routing path are involved in the communication process.

When an important event is sensed, which actor(s) should respond to it? In Section 12.2, we have used ILP-based event area partition algorithm to make all sensors in the event area coordinate with each other to select the appropriate actors according to different criteria, for example, the distance between the event area and the actor, the energy consumption of sensors, or the acting ranges of the actors. When the actor is closer to the event area, the actor can be informed earlier, thus the actor reacts more quickly. So, the best criterion of selecting the actor is the distance between the cluster head and the actor, that is, the distance between the event area and the actor.

In the sensor–actor coordination scheme proposed by Yuan et al. [Yuan06], the cluster head is responsible for monitoring and transmitting data to the closest actor. In the case of a cluster head far from all actors, the cluster head coordinates with the adjacent cluster head that is closer to the actor (based on geographical location information). Eventually, it can find the closest actor.

After a cluster head finds the closest actor, it maintains some information on the actor such as the current location, accurate time, and maintains a routing table toward it. When an event occurs, each cluster head can relay event information immediately to the closest actor with one hop or multi-hop, without establishing routes that will decrease energy or timing consumption.

If an event area is wide and thus relates to many actors, every cluster head in that event area still communicates with the closest actor. All cluster heads associating with one actor construct a data aggregation tree toward the selected actor, and all actors triggered by the same event construct a second-level aggregation tree in the actor–actor coordination level as shown in Figure 12.6, toward the actor in the center of the event area. Such a strategy can achieve an optimal strategy to provide energy efficiency and meet the required event reliability and timing constraint.

12.3.4 "Actor–Actor" Coordination Level

Depending on the characteristics of the detected event, one or multiple actors may be triggered to perform one or more tasks. To solve the task allocation problems, the actor–actor coordination mechanism can use two approaches: the *action-first* (AF) scheme and the *decision-first* (DF) scheme.

1. *AF scheme*: when actors near an event area receive event information via sensor–actor coordination, these actors perform action immediately without negotiating with farther actors. Each involved actor broadcasts the action information to other actors that are one hop or *m* hop away. Those farther actors can learn the event information and make decision to join or retreat the action independently.

   We do not want too many actors to get involved in the event processing. Therefore, Yuan et al. [yuan06] propose the use of a prespecified action threshold to control the number of actors to join the action. The action expectation can be represented as follows:

$$ex(N,A) = \alpha\, d(N,A) + \beta\, e(N) - \gamma\, n(A) + \delta\, p(A)$$

where

$ex(N, A)$ is the expectation of actor N to join the action A
$d(N, A)$ is the distance between the actor N and the action area A
$e(N)$ is the remaining energy of the actor N
$n(A)$ is the number of actors executing the action A
$p(A)$ is the priority of the action A
α, β, γ, and δ are proportional parameters

Remember

It is a common way to define a series of factors and assign different weights to different factors depending on the importance of each factor. The final equation can be used to represent the combined effect of all factors. How to assign weights to different factors is a difficult issue.

Let us denote the action threshold as TH. Then if $ex(N, A) > TH$, the actor N will join the action A. By this way, latency between sensing and acting can be very low. If a certain actor is not capable of doing the action due to coverage or energy constraints, the actor may deliver the "help" message to other actors instead of forcing itself to perform the action.

2. *DF scheme*: In this scheme, all actors receiving event information coordinate closely to maximize their overall task performance. Section 12.2 discussed the ILP-based scheme that achieves optimal task allocation among actors. It can then be used for this scheme.

   In the ILP-based scheme, an actor–actor coordination model is presented to solve the problem of overlapping area, that is, a certain area where multiple actors can act on. Likewise, in the DF scheme, according to the event features and the location-based grids, the area can be optimally split among different actors. The acting area of each actor can be located efficiently among the nonoverlapping area.

Problems and Exercises

12.1 Explain the differences between general WSNs and WSANs.

12.2 In the algorithm of Melodia et al. [Melodia07], we explained "sensor–actor" coordination algorithm. Read the original paper and explain the "actor–actor" coordination algorithm.

12.3 Why do Yuan et al. [Yuan06] use a three-level hierarchical architecture (instead of a one-level one)?

Chapter 13

Underwater Sensor Networks

13.1 Introduction [Melodia07, Akyildiz04a, Pompili06, Pompili09]

13.1.1 Underwater WSN Applications

In our planet, around 70 percent of the surface is comprised of water. Therefore, it is important to perform underwater communications using underwater devices. Underwater acoustic networks (USNs) use the interconnection of large amounts of underwater sensors and mobile vehicles to perform collaborative monitoring tasks.

USNs can perform adaptive sampling of the three-dimensional (3D) coastal ocean environment. They can carry out important underwater tasks such as pollution monitoring, ocean/wind monitoring, and biological monitoring. Pollution monitoring could help to find the level of metals such as lead in water. Ocean/wind monitoring is important to analyze climate change, weather forecast, or the understanding of the effect of human activities on marine ecosystems. Biological monitoring could be used to keep track of fish or microorganisms.

There could be more useful applications if we integrate underwater sensors with other sensors. For instance, seismic sensor networks can provide tsunami warnings to coastal areas, or study the effects of submarine earthquakes (seaquakes). Seismic monitoring allows reservoir management approaches when dealing with oil extraction. Some underwater navigation sensors can be used to detect seabed hazards, to locate dangerous rocks or shoals in shallow waters, to explore submerged wrecks, or to perform bathymetry profiling. Moreover, various underwater mobile vehicles equipped with sensors could perform mine reconnaissance. Those underwater vehicles have acoustic and optical sensors to perform rapid environmental assessments and detect mine-like objects.

Good idea

Only recently, low-cost, large-scale underwater sensor networks have attracted lots of attention. The U.S. Navy, in particular, has invested lots of effort in practical USN design. Please note that although a USN is a special WSN, it has dramatically different characteristics from terrestrial WSNs. This will be further explained later on. This is also the reason why we use a separate chapter to explain USNs in detail.

Most traditional underwater networks do not use wireless (acoustic) sensors. Instead, they use cables to connect small amounts of high-cost underwater sensors. Such an approach has the following disadvantages:

Because of high wiring fee, it is difficult to perform real-time monitoring. Most times the recorded data cannot be retrieved until the instruments are recovered. It is difficult to achieve online system reconfiguration due to the limited length of the cable and the lack of interaction between onshore control systems and monitoring instruments.

High-cost deployment makes it unsuitable for large water area–monitoring applications.

There are many challenges in the design of USNs. For instance, acoustic links have very limited available bandwidth for radio communications; propagation delay under water is five orders of magnitude higher than that in radio frequency (RF) terrestrial channels; the sensor battery power is limited and usually batteries cannot be recharged because solar energy cannot be exploited; and several other issues.

13.1.2 Differences between USNs and Terrestrial Sensor Networks

In the following, we list a few major differences between terrestrial and underwater sensor networks.

The most important difference is the radio communication frequency: Under water, popular terrestrial radio frequencies (such as 2.4 GHz and 833 MHz) cannot be used due to the special characteristics of water—it can severely weaken RF signals in a short distance. However, acoustic signals (typically <1 MHz) could propagate for a much longer distance than radio signals under the water.

Special acoustic modem and advanced underwater transceivers are needed for USNs, and we need sensor protection in the extreme underwater environment. For instance, water can corrode the sensors. The higher the communication distance is, the more complex signal-processing techniques the receivers need. All the above aspects make the USN design cost higher.

Underwater sensors are typically sparsely deployed compared to the densely deployed terrestrial sensor networks.

The readings from terrestrial sensors are often correlated, and this is more unlikely to happen in underwater networks due to the higher distance among sensors.

Acoustic underwater communications need higher power than that needed by terrestrial radio communications because the acoustic communications in USNs need more energy.

Terrestrial sensor nodes could have very limited storage capacity. However, underwater sensors may need to have the ability to do some data caching as the underwater channel may be intermittent.

13.1.3 Network Topology

Typically a 3D (instead of 2D) network topology is used to detect and observe the objects. Note: a sensor network deployed in ocean bottom cannot perform cooperative sampling of the 3D ocean environment.

As shown in Figure 13.1, in 3D underwater networks, sensor nodes float at different depths to observe underwater parameters. A metal object with a rope can anchor a sensor to the ocean bottom. Also, a sensor can be equipped with a floating buoy that pulls the sensor toward the water surface. By adjusting the length of the wire that connects the sensor to the anchor, we can control the depth level of the sensor.

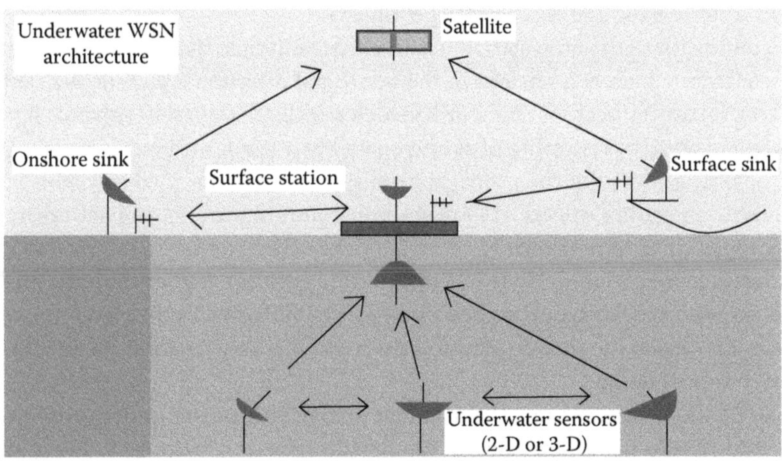

Figure 13.1 Architecture for 3D underwater sensor networks. (Adapted from Melodia, T. et al., *IEEE Trans. Mobile Comput.*, 6(10), 1116, October 2007.)

An underwater WSN protocol design needs to consider the effect of ocean currents on sensor mobility.

Good idea

Please note that most of the WSN routing protocols only assume a 2D, instead of a 3D, structure. In the underwater case, all sensors are located in different depth levels (vertical direction). In each depth level, lots of sensors form a 2D topology. A sensor needs to keep communication connections with both horizontal and vertical sensors.

13.1.4 Acoustic Signals Propagation

As we mentioned before, radio signals cannot propagate well in water. Hence, acoustic signals should be used. However, acoustic communications can be influenced by *path loss, noise, multipath, Doppler spread,* and *high propagation delay.* These factors can tell us the temporal and spatial variability of the acoustic channel.

Path loss, also called signal *attenuation,* is caused by energy absorption during the conversion of acoustic energy into heat. Higher distance or acoustic frequency can bring more signal attenuation. The signal attenuation could be caused by multiple signal propagation phenomena such as wave scattering and reverberation, refraction, and dispersion. Water depth also has an effect on the attenuation.

Communication noise is mostly caused by machinery noise such as pumps and shipping activity (such as hull fouling or cavitations especially) in areas with heavy vessel traffic. Ambient noise comes from tides, current, storms, wind, and rain. It is also related to seismic and biological phenomena.

The multipath signal propagation (i.e., there are multiple transmission paths for a signal sent from a source) is obvious in the horizontal direction (i.e., from sea bottom to surface). Smart transceivers could utilize such multipath signals to enhance signals.

Doppler spread is scattering of sound energy as a result of the expansion of the wave fronts. It grows with the propagation distance. There are two common kinds of geometric spreading: spherical in deepwater communications and cylindrical in shallow water communications.

Acoustic propagation delay is five orders of magnitude higher than that in the radio channel case (the acoustic speed is around 1500 m/s). Such a large propagation delay can lessen the throughput of the system. It is very harmful for an efficient network protocol design.

Most of those factors limit the available bandwidth of the underwater acoustic channel that is mainly dependent on signal range and frequency. Long-range acoustic communications (~tens of kilometers) only have a bandwidth of a few kilohertz. Short-range (~100 m) may have a higher bandwidth (>100 kHz). In both cases, these factors lead to low bit rates.

Table 13.1 Available Bandwidth for Different Ranges in UW-A Channels

	Range (km)	*Bandwidth (kHz)*
Very long	1000	<1
Long	10–100	2–5
Medium	1–10	~10
Short	0.1–1	20–50
Very short	<0.1	>100

Source: Adapted from Melodia, T. et al., *IEEE Trans. Mobile Comput.*, 6(10), 1116, October 2007.

We may classify the underwater acoustic communication links based on their ranges as very long, long, medium, short, and very short links. Table 13.1 [Melodia07] shows typical bandwidths of the underwater channels in different propagation ranges.

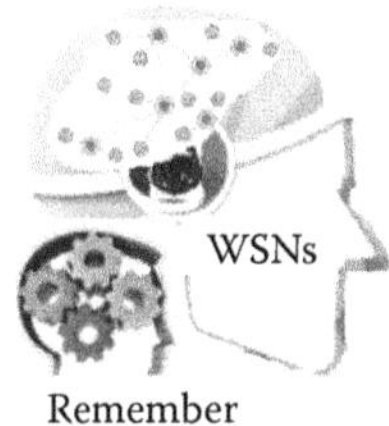

Always remember that the *long, variable* acoustic delay in USNs necessitates a set of new protocols compared to RF-based terrestrial WSNs. An acoustic wave is similar to human sound. It travels very slowly but reaches a long distance in the water. An underwater sensor needs special wireless transceiver, called the acoustic modem, to communicate with other nodes. Commercial acoustic modems are very expensive.

13.1.5 Underwater Sensors

An underwater node has a similar architecture to general WSN nodes. The microcontroller/CPU works along with analog underwater sensors through a sensor interface circuitry or oceanographic instrument (Figure 13.2). The microcontroller accepts the data (such as water pollution level and metal level) from the analog sensor, stores it in the onboard memory, processes it, and delivers it to other network devices by controlling the acoustic modem. The electronics are usually mounted on a frame that is protected by the housing. By housing all components beneath a low-profile pyramidal frame, we could prevent water corrosion.

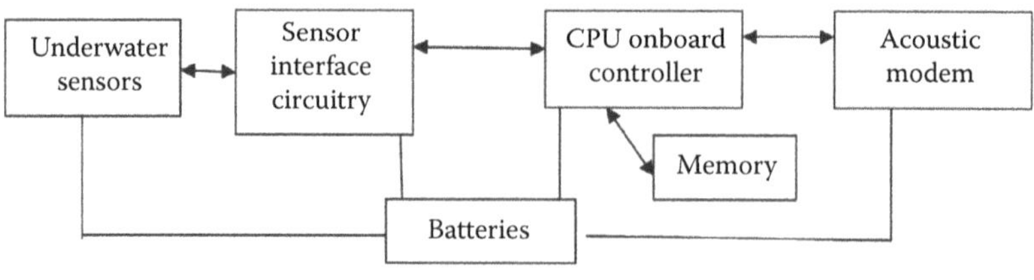

Figure 13.2 Internal architecture of an underwater sensor. (Adapted from Melodia, T. et al., *IEEE Trans. Mobile Comput.*, 6(10), 1116, October 2007.)

The underwater sensors could assess the quality of water by measuring parameters such as temperature, density, salinity, chemicals, conductivity, pH, oxygen hydrogen, dissolved methane gas, and turbidity. We could also use disposable sensors to detect the highly poisonous protein that could be found in castor beans and thought to be a potential terrorism agent. DNA microarray sensors have been designed to monitor abundance and activity-level variations among natural microbial populations. Force/torque sensors can simultaneously measure several forces and moments.

The trend is to develop less-expensive/robust underwater nano-sensors. All underwater sensors need periodical cleaning mechanisms against corrosion and fouling, which may impact the lifetime of underwater devices. Integrated sensors are under research for synoptic sampling of physical, chemical, and biological parameters to better understand the processes in marine systems.

13.2 USN Protocol Stack [Akyildiz04a, Melodia07]

13.2.1 Physical Layer

In terms of modem and modulation schemes, a simple way is to use the frequency shift keying (FSK) modulation scheme. In the FSK scheme, the multipath effects can be tightly bounded by inserting time guards between successive pulses. We could also use dynamic frequency guards between frequency tones to adapt the communication to the Doppler spreading of the acoustic channel.

However, FSK has low bandwidth efficiency and is not stable for high-data-rate communication applications. We could use coherent modulation techniques for long-range, high-throughput applications. For instance, differential phase shift keying (DPSK) encodes information relative to the previous symbol.

Recently, orthogonal frequency division multiplexing (OFDM) spread spectrum technique has become a promising solution for underwater communications. OFDM is also called multi-carrier modulation because it modulates signals across multiple sub-carriers simultaneously. As the symbol duration for each individual carrier is wider than many other modulation schemes, OFDM systems perform robustly in severe multipath environments, and achieve a high spectral efficiency.

Besides modulation design, other physical layer issues need to be addressed to enable underwater acoustic sensor networks. For example, inexpensive transmitter/receiver modems for underwater communications need to be developed.

13.2.2 Data Link Layer

This layer solves multiple neighboring sensors' acoustic channel access issues. People also call this layer as MAC (medium access control) layer because channel access is the main design issue in the data link layer. How do we make sure that no conflict exists when those neighbors try to access the channel at the same time? A good access schedule needs to be developed.

Channel access control in USNs should adapt to the limited bandwidth, and high/variable delay. Frequency division multiple access (FDMA) may not be used due to its narrow bandwidth.

If using TDMA, we need to design a good channel access schedule that can overcome the variable acoustic delay. Such a TDMA-based scheme should be based on a precise clock synchronization because TDMA needs a common timing reference.

Contention-based techniques such as *carrier sense multiple access* (*CSMA*), which uses ready-to-send/clear-to-send (RTS/CTS) to avoid conflicts, may not be practical in underwater due to the large acoustic delays in the propagation of RTS/CTS control packets. The high variability of acoustic delay also makes it very difficult to predict the start and finish time of the transmissions. The result is that the collisions can still occur.

Therefore, we cannot simply use MAC schemes in terrestrial sensor networks for underwater sensor networks because we need to overcome a few challenges in underwater channels: variable and high propagation acoustic delays and very limited underwater communication bandwidth.

Code division multiple access (CDMA) may be a good solution although it needs more complex hardware. It is robust to frequency-selective fading caused by underwater multipaths because it can use orthogonal codes to distinguish among simultaneous signals transmitted by multiple devices. CDMA also reduces the number of packet retransmissions, which results in decreased battery consumption and increased network throughput. If we adopt CDMA, we need to design communication access codes with high autocorrelation and low cross-correlation properties to achieve minimum interference among users. We need to find out the optimal data packet length to maximize the network efficiency.

13.2.3 Network Layer (Routing Layer)

The network layer aims to search a good routing path between a source and a destination node. There are many routing protocols proposed for terrestrial sensor networks. However, we cannot directly use them for USNs due to the high/variable acoustic delay.

The existing WSN routing protocols mainly include three categories, namely, proactive, reactive, and geographical routing protocols.

Proactive protocols always make an up-to-date routing table ready at all times from each node to every other node. It thus avoids the message latency induced by route discovery. However, it has high routing updating overhead when used in underwater networks.

Reactive protocols initiate a route discovery process *only* when a route to a destination is required. It does not maintain an "always correct" routing table.

Geographical routing protocols select each relay node based on the position of sensors. Global Positioning System (GPS) receivers may be used in terrestrial systems to accurately estimate the sensor location. However, they do not work well inside water because USN does not use radio signals (it uses acoustic signals).

It is important to devise routing algorithms that are robust with respect to the intermittent connectivity of acoustic channels.

13.2.4 Transport Layer

We also need a transport layer protocol in USNs to achieve end-to-end reliable transport of event features, and to perform flow control and congestion control. The most popular transport layer protocol, TCP, is not suitable to the underwater environment because it uses a window-based flow control mechanism that relies on an accurate estimate of the *round trip time* (RTT). Unfortunately, the versatility of the underwater RTT would make it difficult to effectively set the time-out of the window-based mechanism.

USNs transport layer protocol should deal with the following issues for reliable data transport: large propagation delays, low bandwidth, energy efficiency, high error probabilities, and highly dynamic network topologies. A good transport layer solution for the underwater environment may use the following design principles such as shadow zones, minimum energy consumption, rate-based transmission of packets, timely reaction to local congestion, and reliability. The following provides more details:

- Proper handling of communication shadow zones (they have very weak signals) requires the parameters from the routing layer such as the delay in each hop.
- Minimum energy consumption should be considered due to the battery-driven nature in underwater sensors.

- Rate-based transmission of packets is better than window-based schemes (such as TCP case) because we could achieve a more accurate traffic control.
- Timely reaction to local congestion should adapt to local conditions immediately, and it should decrease the response time in case of congestion. Thus, rather than relying on the base station, intermediate nodes should be capable of determining and reacting to local congestion. A hop-by-hop reliability control is better than an end-to-end control.

13.3 MAC Design Example [Min07]

As we mentioned before, it is challenging to design USNs MAC protocols due to energy limitations, long propagation delays, low data rates, and the difficulty of synchronization in underwater environments. The MAC protocols used in terrestrial sensor networks are not suitable for the underwater acoustic communication medium that experiences a very large propagation delay of 1 s over 1.5 km. This section will use [Min07] to exemplify the efficient MAC design in underwater environment.

First, let us review some related works in this field. FDMA was used in the underwater network project called SeaWeb [JRice00]. But it was found to be restrictive and inefficient in terms of bandwidth utilization. SeaWeb 2000 [JGProakis01] favored a carrier sense multiple access/collision avoidance (CSMA/CA) solution with RTS/CTS handshake/exchange. However, we will have high energy consumption overhead when using the RTS/CTS packets.

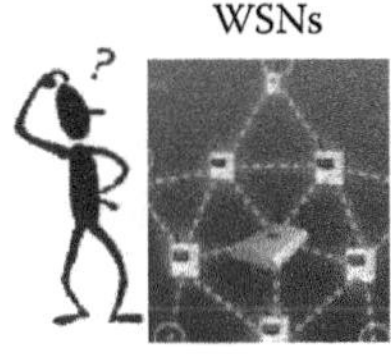

You have learned some MAC schemes for terrestrial WSNs such as S-MAC. Those MAC protocols put "energy efficiency" as a top priority. They try to put sensors into "sleep" status for a long time. They also try to reduce the channel access scheduling complexity to further reduce energy. However, in underwater WSNs, the MAC design should put "adaptation to *long, variable* acoustic delay" as the top priority.

Min and Volkan [Min07] thus designed the UWAN-MAC protocol. Its basic idea is illustrated in Figure 13.3 [Min07], which explains how to achieve a locally synchronized schedule even in the presence of long, unknown propagation delays.

1. *Determination of "Listen" cycles*: As shown in Figure 13.3, assume that sensor A broadcasts an SYNC packet (the shaded rectangle) to its neighborhood in the beginning of its cycle period, and then goes to sleep status (by turning off its transceiver circuits to save energy). This SYNC packet announces A's communication cycle period T_A.

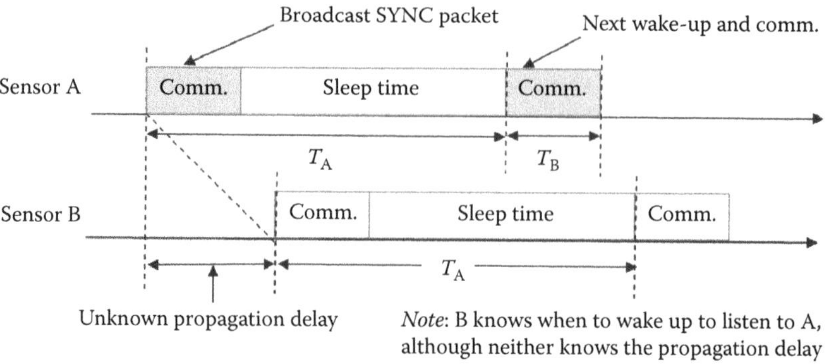

Figure 13.3 Basic idea of the UWAN-MAC protocol. (Adapted from Park, M.K. and Rodoplu, V., *IEEE J. Ocean. Eng.*, 32(3), 710, July 2007.)

Assume that another sensor, B, belongs to neighborhood of A. When node B joins the network, it first tries to capture this SYNC packet to achieve communication synchronization with sensor A. (The white rectangles in the figure indicate node B's receptions of node A's SYNC packets.)

Because B follows the explicit stamping of sensor A's transmission cycle period instead of using absolute wake-up period, B can wake up at exactly the correct time in the next transmission cycle to listen to A without any knowledge of the acoustic propagation delay. (Of course, here we assume that the propagation delay remains almost fixed from one cycle to the next and the clock drift is not significant in one cycle. This is a reasonable assumption if the clock cycle is not so long.)

We should use the above SYNC broadcasting protocol for any neighboring sensors (such as A and B). Further, this scheduling algorithm does not require any adjustments to the sensors' clocks because absolute timing information is not needed.

2. *Determination of transmit start times*: The MAC protocol also needs the help of network topology control protocol that keeps track of the neighbors of a node. Even though the listen times are determined based on the above described scheme, the initial transmission time of a sensor is selected randomly and independently by each sensor. However, once a node chooses a certain transmission start time, it sticks to its schedule by transmitting its data at that time again in the next cycle. As long as the cycle period is much longer than the transmit duration (in Figure 13.3, $T_A \geq T_B$), the probability of channel access collisions will be small.

In UWAN-MAC protocol, each node compares its stored neighbor list (from neighbor discovery protocol) with the list of nodes from which a node has actually received signals. After this comparison, the node generates the "missing node list" and sends the list of "missing" neighbor nodes in the header of the data packet in its next transmission cycle.

During regular operation, every node keeps sending its cycle period (i.e., T_A in Figure 13.3) in its SYNC header. The SYNC message allows a node to change its current cycle period, and its neighbors can decode the modified SYNC message and change their wake-up times. If a node, say B, loses contact with node A during this modification, it will use the missing neighbor list to recover node A as a neighbor.

Note that T_A (cycle period) actually includes three parts: (1) data transmission (including sending and receiving); (2) idle listening; and (3) sleep status. After the *data transmission* phase, a node does not immediately go to *sleep* but rather enters *idle-listening* mode. In the listening mode, the node is still awake but operates at low power. If it hears something, it will go into the receiving mode.

This listen duration can also be used to hear newcomers. The length of the listen duration needs to be chosen carefully: A very long duration decreases the energy efficiency of the protocol due to idle period energy consumption, but a very short duration might not be enough to catch some of the newcomers' messages.

Handle newcomer: When a new node joins the network, as long as it hears from another node, it can send a HELLO packet back to that neighbor to tell about its transmission schedule. Such a HELLO packet should be sent out in "data transmission" phase. Typically, a sensor divides its transmission time slot uniformly among M time slots, and a slot lasts for the duration of a HELLO packet. The sensor randomly selects one out of M slots to transmit a HELLO packet. Such a random HELLO transmission avoids the possible collisions among HELLO packets from different newcomers, in case an existing node has multiple newcomer neighbors who simultaneously enter the network.

Handle node failure: If the channel condition is poor, or if a sender fails, a node may not be able to receive data at a scheduled wake-up time. If the receiver node does not hear from the sender at a scheduled time, the receiver puts the sender in its missing node list, as explained before.

Handle variable acoustic delay: In Figure 13.3, we assume that acoustic propagation delay remains fixed from cycle to cycle. However, in reality it varies due to channel fluctuations caused by sensor motions/water current. We may assume that a node has the knowledge of the *maximum propagation delay*. For example, for densely deployed underwater WSNs, the maximum propagation delay is about 70 ms for distances of up to about 110 m.

Here is a new MAC protocol to handle variable delay: *each node i places a certain "guard time" on both sides of its transmit duration*. To reduce the packet transmission collision rate, we can choose the guard times of each node in a completely localized manner to reduce collision rate.

Figure 13.4a shows an example that does not use guard time. A receive–receive collision occurs in the presence of variable propagation delays. To avoid this

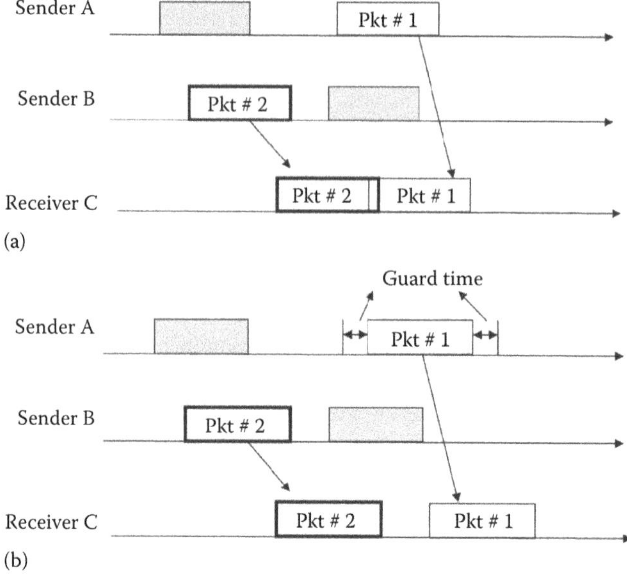

Figure 13.4 Collision with propagation delays and CA using the guard time policy. (a) Receiver–receiver collision due to the variable acoustic propagation delays. (b) Guard time solution. (Adapted from Park, M.K. and Rodoplu, V., *IEEE J. Ocean. Eng.*, 32(3), 710, July 2007.)

collision, node A or B (here we ask A to add guard times) reselects its transmission start time to avoid collisions. Figure 13.4b suggests one possible solution for the case where $\tau_2 < \tau_g < \tau_1$. No collision occurs because node A's newly selected transmission start time has been applied.

"Guard time" is really a good idea! In wireless multichannel communications, people typically add a "guard channel" to a communication band (a narrow bandwidth) to prevent signal interference among neighboring bands. Here we use "guard time" to make the transmission time more "flexible" to avoid receiver–receiver collisions.

13.4 Routing Design Example: Vector-Based Forwarding Protocol [PXie05]

USN routing protocols need to meet two requirements: (1) It needs to be energy efficient because the sensors are battery driven and (2) the underwater sensors can move around due to water current. Thus routing protocol needs to handle

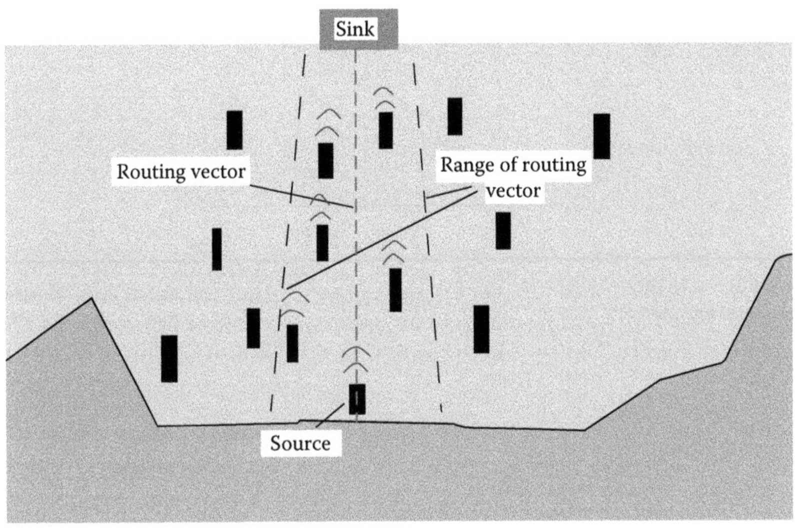

Figure 13.5 VBF protocol scenario. (Adapted from Xie, P. et al., VBF: Vector-based forwarding protocol for underwater sensor networks, UCONN CSE Technical Report, UbiNet-TR05-03 (BECAT/CSETR-05-6), February 2005.)

node mobility. Vector-based forwarding (VBF) protocol [PXie05] meets these requirements.

We have noticed that the USN routing has a direction from the sea bottom to the surface. Therefore, we could use a routing vector to represent such a path. The idea behind VBF is shown in Figure 13.5. Node S_1 is the source and S_0 is the receiver. The routing vector is $\overrightarrow{S_1S_0}$, and the routing pipe is shown with a pre-controlled radius of *W*. VBF does not require state information at each node and is scalable to the size of the network. As only nodes in the forwarding path contribute to the forwarding, it makes the network energy efficient.

In VBF, each data packet needs to have the positions of the sender node, the target node, and the forwarding (relay) node. When a node receives a packet, it computes its position in relation to the forwarder by measuring (1) the distance to the forwarder and (2) the angle of arrival (AOA) of the signal. All nodes that receive the packet compute their positions recursively.

If a node determines that it is close enough to the routing vector (i.e., it could be added to the routing pipe), it adds its position to the packet and forwards it. Otherwise, it discards the packet. All the packet forwarders in this sensor network thus form a "routing pipe" that consists of only the nodes that are eligible for packet forwarding (Figure 13.5).

Routing vector from sender A to receiver B is usually denoted as $\overrightarrow{AB}$. In a 3D space, if A's coordinate is (A_x, A_y, A_z) and B's coordinate is (B_x, B_y, B_z), then the vector $\overrightarrow{AB}$ can be represented by $(B_x - A_x, B_y - A_y, B_z - A_z)$.

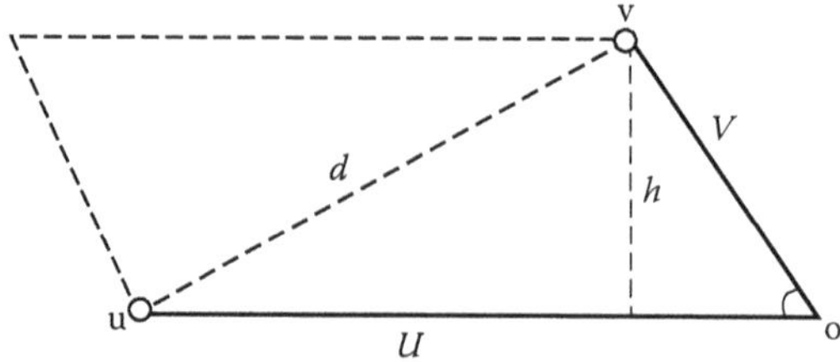

Figure 13.6 An illustration of vector calculation. (Adapted from Xie, P. et al., VBF: Vector-based forwarding protocol for underwater sensor networks, UCONN CSE Technical Report, UbiNet-TR05-03 (BECAT/CSETR-05-6), February 2005.)

Assume that we have a point $V = (v_x, v_y, v_z)$ and another point $U = (u_x, u_y, u_z)$. The distance between points v and u can be calculated by $d = \sqrt{(v_x - u_x)^2 + (v_y - u_y)^2 + (v_z - u_z)^2}$ (Figure 13.6).

VBF routing protocol is built based on *protocol packet* (not *data packet*) exchanges. Each of the protocol packets consists of three position fields, OP, TP, and FP, which are the coordinates of the sender, the target, and the forwarder. When a *data packet* reaches the area specified by its TP, the packet is flooded in an area controlled by the range field.

Each *protocol packet* has a field called *radius*, which is a predefined threshold and can be used by the sensor nodes to determine if they are close enough to the routing vector to be considered in packet forwarding (within the routing pipe).

VBF can execute two types of sensor data queries. The first query is *location dependent*. The sink is interested in a specific area and knows the area's location. The second is *location independent*. The sink wants to know some specific data regardless of positions. An example of this case is if the sink wants to know if any metal pollution exists anywhere in the network.

Good idea

"Routing pipe" is an interesting and useful idea. In some cases, it may not be robust to rely on a line of single nodes to achieve multi-hop wireless communications. Therefore, people suggest to use a "pipe" concept that consists of a thickness of nodes in the path. The thicker the pipe, the more robust the routing scheme is. This is because any of the nodes in the pipe can help to relay data. On the other hand, such routing robustness may bring higher routing complexity due to the maintenance of the routing pipe.

13.5 Hardware Prototype Design [Hu2009e]

In this section, we will discuss a USN prototype designed by us [Hu2009e]. Most commercially available underwater communication systems [DSPComm08] are designed for long-range communications (with link distances of several kilometers). These modems can carry sustained data rates of approximately 1–40 kbps. The implementation of a system is thus cost prohibitive.

In our design, the majority of the sensor node's functionality is software defined, allowing for low-cost reconfiguration of the platform for a variety of tasks. The underwater sensor hardware remains relatively simple and provides only the following functionality: amplify both outgoing and incoming signals, and provide signal conditioning for all environmental sensors and probes. This leaves the remainder of the node's functionality to be defined in software, including both modulation and demodulation.

Although a variety of embedded processors could fulfill the networking and routing needs, a digital signal processor (DSP) is needed for this platform because modulation and demodulation will take place in the software. Figure 13.7 [Hu2009e] shows the connection of DSP chip with other components such as the transmitter (Tx) and the receiver (Rx).

Case study

Using loudspeakers to serve as hydrophones is a simple approach and works for prototype purposes. However, it cannot be used in commercial applications. We still need to design robust acoustic modem to achieve underwater communications.

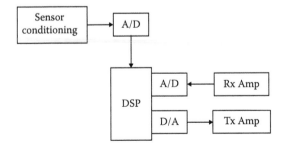

Figure 13.7 Hardware and DSP interaction.

13.5.1 Hardware Design

Hydrophones (used to convert the electrical signals into acoustic waves and vice versa) cost approximately $1000, even for small transducers designed for shorter-range applications [Transducer08]. Therefore, a cheaper alternative than commercially available hydrophones was needed. For prototype purposes, we have used a small loud speaker in which the paper cone was waterproofed. These speakers produced an audibly louder tone at the desired frequency range, and were used as both the transmitting and receiving hydrophones (Figure 13.8).

We have used two types of sensors, pH and temperature. The signal must be conditioned so that it produces a valid signal for the analog-to-digital converter (ADC).

The pH amplifier has a constant gain of 2.4. It consists of a non-inverting amplifier. It should be able to accept high impedance sources because the pH sensor used has an impedance of 50 MΩ, and the underwater node's ADC can only accept at most 5 kΩ sources.

The temperature amplifier has a constant gain of 2, which covers the analog range of the ADC, and also provides greater precision. The temperature amplifier utilizes a Wheatstone bridge and an inverting differential amplifier. The temperature sensor used is a 10 kΩ thermistor.

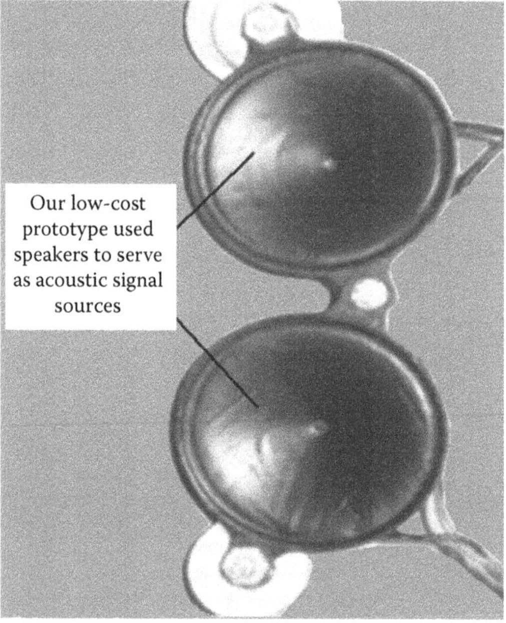
Our low-cost
prototype used
speakers to serve
as acoustic signal
sources

Figure 13.8 Loudspeakers serving as hydrophones.

To execute acoustic communication protocols with a neighboring underwater node, a microcontroller is used to control all sensors. It is a fixed-point DSP with 256 kB of direct memory access (DMA)–capable memory. We use a fixed-point processor because a floating-point processor is more expensive, and requires substantially more power to operate. The board has a RAM of 100 kB, which is good enough for signal sampling and signal processing. The DMA system is used to allow the sampled symbols to be placed into memory directly, thus we do not need to use valuable CPU cycles to move the samples from the ADC to memory. It allows a symbol to be demodulated concurrently with sampling, so that no samples are lost during the long demodulation calculation.

The microcontroller is packaged with an interface board, which contains peripherals that are directly accessible to the processor. These include a variety of digital I/O ports, a digital potentiometer-based DAC, and a 10-bit/12-bit ADC. Having these peripherals integrated directly with the CPU, our system saves substantial development time and cost.

Figure 13.9 shows our fabricated underwater board (it has DSP and acoustic communication modules; analog sensors are not shown).

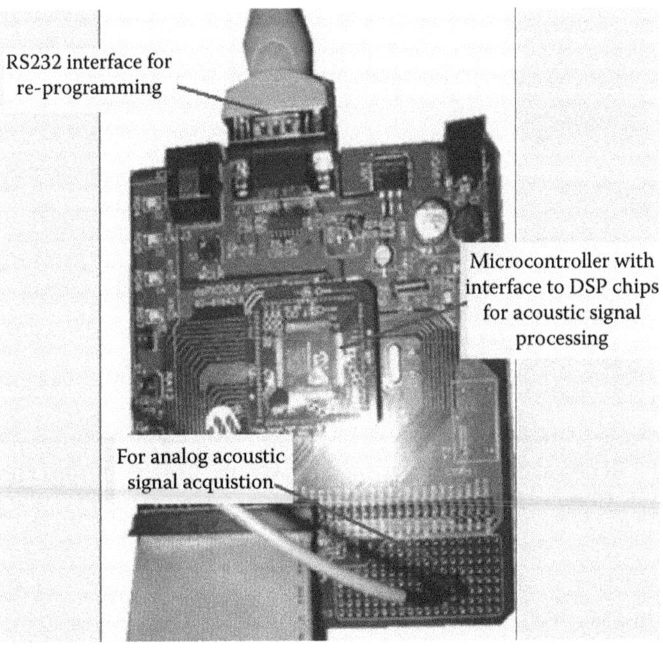

Figure 13.9 Fabricated underwater node with microcontroller and acoustic transceiver. (Adapted from Hu, F. et al., *J. Circuits Syst. Comput.*, 17(6), 1203, 2008.)

13.5.2 Software Design

The software of the system is divided into two sets of components, those associated with transmission and those associated with reception. Figure 13.10 shows our software structure in the receiver side. It includes analog sensor data filtering and acoustic demodulation. Figure 13.11 is the transmitter's software structure that has modulation and CRC (cyclic redundancy check) error control.

13.5.3 System Testing

The water surface sensor (i.e., the sink) is connected to a PC via an RS-232 connection at 2400 baud. The surface node has no sensors of its own in this case. The underwater nodes have the pH and temperature sensors. The underwater nodes are to poll its sensors for information every 30 s. Once the sensor data is retrieved a packet is constructed and the CRC of the packet is computed.

After an underwater sensor sends out a packet to the sink, the sink must send an acknowledgment (ACK) back to the underwater node so it knows the packet was received successfully and will not have to be resent. If the CRC shows that the

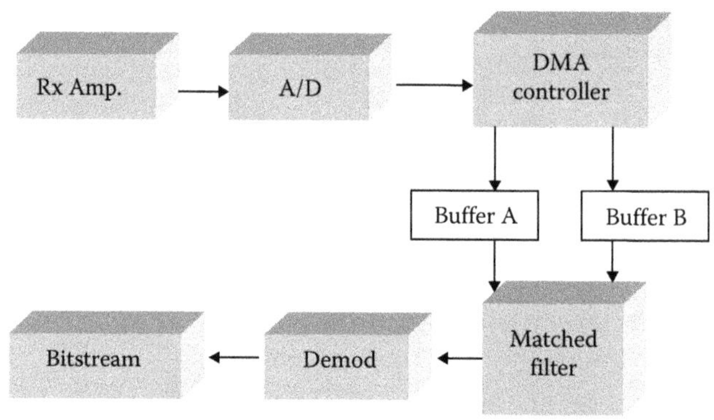

Figure 13.10 Receiver side software block diagram. (Adapted from Hu, F. et al., *J. Circuits Syst. Comput.*, 17(6), 1203, 2008.)

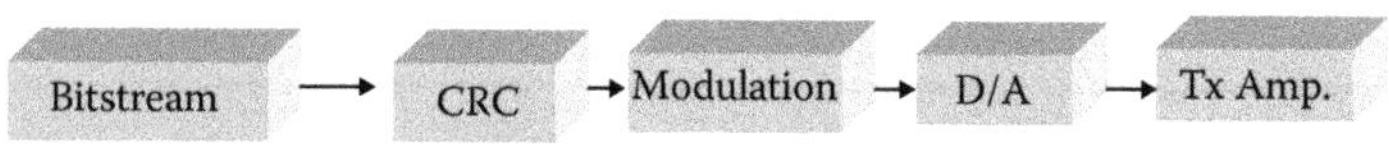

Figure 13.11 Transmitter side software block diagram. (Adapted from Hu, F. et al., *J. Circuits Syst. Comput.*, 17(6), 1203, 2008.)

packet is corrupted, or the packet does not arrive at all, the receiver will send out a negative ACK. Eventually, the sender will time out while waiting for the acknowledgment, and send the packet again. The sequence of error recovery is shown in Figure 13.12. This is known as a stop-and-wait Automatic Repeat reQuest (ARQ) mechanism.

The lab test has a setup as shown in Figure 13.13. The system achieved its theoretical bit rate of 15.625 bits/s. An average bit error rate (BER) of 0.091 was observed. The CRC did not avoid all errors because it is not the strongest error-checking scheme in wireless networks. These can be seen in Figure 13.14 where spikes are present in the pH temperature results. These points are outlier values (when the pH greatly exceeds 14, and it must be a result of bit errors).

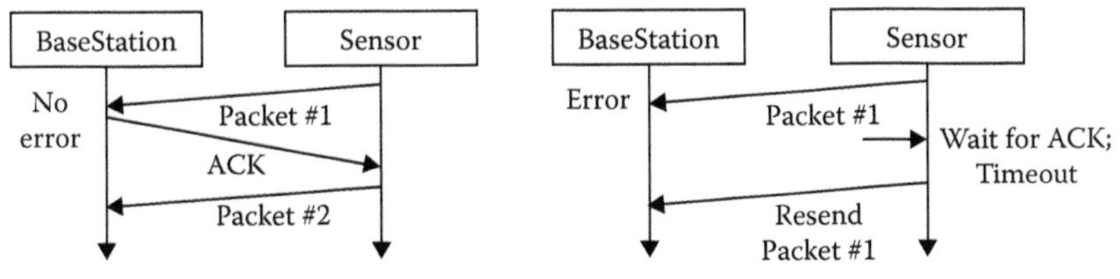

Figure 13.12 ARQ interactions. (Adapted from Hu, F. et al., *J. Circuits Syst. Comput.*, 17(6), 1203, 2008.)

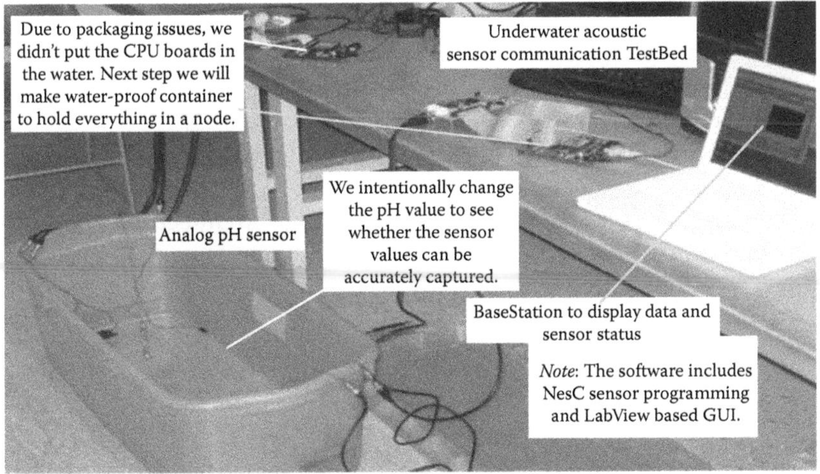

Figure 13.13 Lab test setup. (Adapted from Hu, F. et al., *J. Circuits Syst. Comput.*, 17(6), 1203, 2008.)

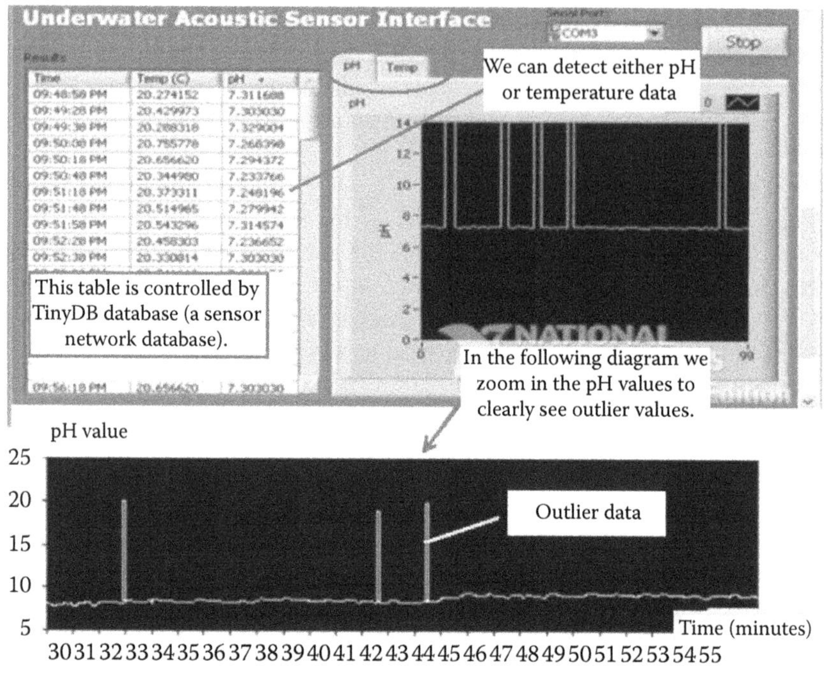

Figure 13.14 Underwater sensor data collection GUI. (Adapted from Hu, F. et al., *J. Circuits Syst. Comput.*, 17(6), 1203, 2008.)

Problems and Exercises

13.1 Explain the differences between underwater and terrestrial sensor networks.

13.2 What specific improvements should we make when changing general WSN MAC layer protocols (see Chapter 3) to underwater WSNs?

13.3 Can we directly use the routing protocols in Chapter 4 to USN cases? Why or why not?

13.4 Draw the underwater sensor network system architecture (hardware/software) based on Section 13.5.

Chapter 14

Video Sensor Networks

14.1 Introduction

Today WSNs can use hundreds of different analog sensors to collect environmental data. Video capture can provide invaluable information that no other sensors are capable of capturing. We can integrate video sensors with other types of sensors to provide multimedia data, or we could use them alone in applications such as video surveillance.

Although it is very exciting to use video sensors, video data requires a tremendous amount of storage space and wireless bandwidth. To overcome such challenges, we can make use of various resource-conserving techniques while maintaining their ability to capture and transmit video data.

In the following, we list some example applications of video sensor networks (VSNs) as well as resource requirements associated with VSNs [Feng05].

Underwater exploration: Oceanographers can use VSNs to observe and study the development of underwater sandbars. They can also use digital image processing to determine the evolution of such sandbars over time. The wireless nature of the application requires self-sufficient video sensors. For instance, we may use solar panels or other dynamic methods (water current) to generate sensor power. VSN protocols should be power saving. To reduce energy consumption, we may establish network connectivity intermittently, that is, only collect data from time to time. Although video sensors operate in a low-power state, the network protocols should transmit the most important video data.

Environment surveillance: For building/outdoor surveillance applications, we may not know beforehand which data would be important. To enhance network scalability, reduce network traffic, and save storage space, the video sensors should try to filter useless data as much as possible. For instance, some image-comparing techniques could be used to find out any newly entered objects in an area. If the video sensors cannot detect image change, they will not send out video data.

Emergency response systems: A video-based sensor network may be deployed for emergency response applications. Video sensors can be used to capture and transmit high-quality video for a specified period of time (i.e., the duration of the emergency). The VSN operation should have quick response and low power adaptation attributes, to provide the emergency response personnel with the video data throughout the incident.

For the above three applications, some common tasks are required, as follows. [Purushottam07]

Object detection: One of the VSN goals is to detect familiar or new objects or scene changes in the observation area. For example, an animal habitat–monitoring application should detect when an animal enters/leaves an area. A building security system should be able to detect intrusion events, such as a person entering the area. To perform object detection, the video sensors can use lots of proposed object detection algorithms. These detection algorithms should be able to spend minimum time to detect each new object that enters the security area.

Object recognition: If *object detection* just finds out whether or not a new object has entered the area, we also require *object recognition* schemes to recognize what exactly that object is. For instance, after a new object is detected, we need to determine its type (e.g., normal personnel or enemies, zebra or deer). Such a recognition procedure can find out whether or not the object is of interest. Recognition is usually achieved by comparing the captured video/images to a database of object images. Good matching algorithms can quickly find out the targeted object.

Object tracking: After we recognize an object of interest, we may want to keep track of such a target as it moves through the environment. In object tracking, we typically first determine the current location and the trajectory of the object; then, we hand off the tracking task when an object moves out of the visual range of one camera sensor and enters into the range of another sensor.

For each of the above three tasks, we need to design both hardware and software to capture the desired information.

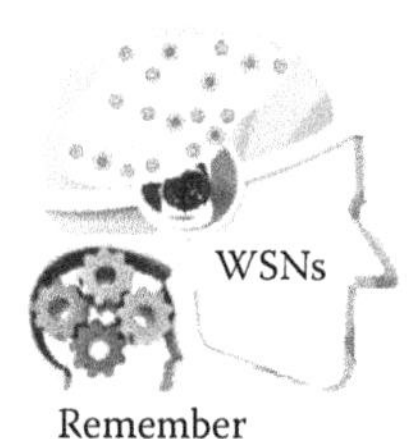

Today, many object detection/recognition/tracking schemes have been proposed. These typically need the knowledge of machine learning that can classify objects efficiently. The neural network is a traditional scheme that performs object recognition. Object tracking typically requires accurate object-positioning techniques.

14.2 Panoptes [Feng05]

Feng et al. [Feng05] have illustrated the hardware design of video sensors. In designing and choosing a video sensor, we must consider power source, memory space, and CPU speed requirements. Although Intel's StrongARM-based PDA (personal digital assistant) was very popular in a number of high-profile research projects at MIT and ISI, it still cannot meet the low power requirements of video sensor design. In addition, traditional video sensor designs have more shortcomings, as follows.

Limited I/O bandwidth: Many of today's embedded sensors use PCMCIA-based devices. They typically require significant power. Some devices use USB interfaces. Unfortunately, low-power, tiny video sensors cannot support USB 2.0 (455 Mb/s) well, as a fairly large processor would be required to store the incoming data.

Lack of floating-point (FP) processing: Today, many embedded devices use StrongARM processors and Xscale processors. However, both of them do not support FP operations. As we know, video compression algorithms are based on FP operations.

Memory bandwidth: Traditional devices are not optimized with respect to memory bandwidth, which is very important for video sensor operations due to a lot of image/video data.

The Panoptes video sensor [Feng05] has been developed to overcome the above shortcomings. It uses the Linux operating system due to Linux's simplicity in controlling the device and its flexibility to modify parts of the system. In Panoptes, the video-sensing task is achieved through several components, including capture, compression, filtering, buffering, adaptation, and streaming. Some of the important components of the Panoptes system are shown in Figure 14.1. Next, we briefly describe these components.

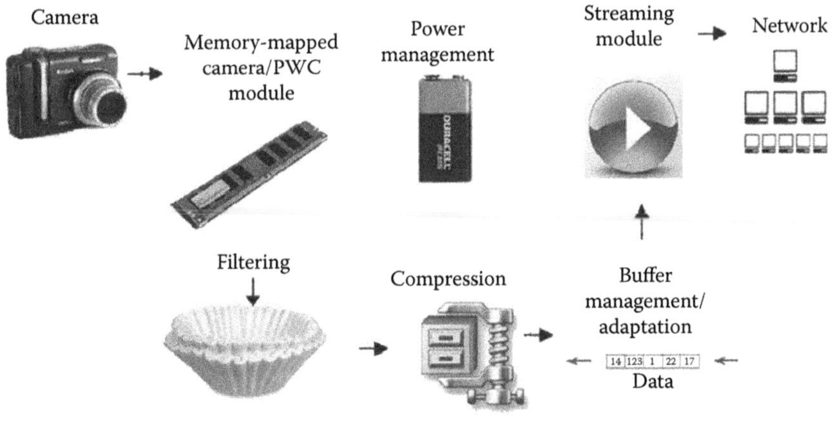

Figure 14.1 Panoptes' sensor software components. (Adapted from Feng, W. et al., *ACM Trans. Multimedia Comput. Commun. Appl.*, 1(2), 151, May 2005.)

14.2.1 Video Capture

Feng et al. [Feng05] use the Philips Web camera interface with video for Linux. A Linux kernel decompresses the video data before passing it to the user space, where the decompressed video data (>10 frames per second) is available via memory mapped access. When a video frame is ready to be read, it is further processed through a filtering algorithm, a compressor, or both.

14.2.2 Video Compression

To reduce memory storage and network traffic, we need to compress the video frames both spatially and temporally. Panoptes can use multiple compression formats, such as JPEG, differential JPEG, and conditional replenishment compression formats. Although JPEG itself does not achieve *temporal* compression of data, it can reduce math computational cost (compared to formats such as MPEG), and thus save sensor power. As compression is performed by the CPU, the quality of video and the level of compression depend on the CPU's processing capability.

14.2.3 Data Filtering

The video sensors must have the ability to filter data at the sensor level to reduce network traffic. Note that filtering data at the sensor level instead of at the network level allows us to reduce the overall network design cost. In many applications, such as video security surveillance, the sensors should be able to filter "uninteresting" data while compressing and transmitting only the desired information (such as new faces). For environmental observation, the filter could create a time-elapsed image to compress image data only, as it is required to reduce the amount of data that needs to be transmitted [Stockdon00].

Panoptes has a filter for the user to specify which video data should be removed. Because of the relatively high cost of Discrete Cosine Transformation (DCT)-based video compression, low-complexity filtering algorithms should be run if they can reduce the number of frames to be compressed.

14.2.4 Data Buffering

Data buffering is a key component in developing a successful video sensor. We can buffer the video data based on some type of priority-control scheme to ensure that all important data is transmitted in the event of network congestion or outages. In a buffering scheme, if the buffer within the video sensor is full, efficient priority-control mechanisms should be used to determine which data should be discarded first. Panoptes uses a priority-based streaming mechanism to support the video sensor.

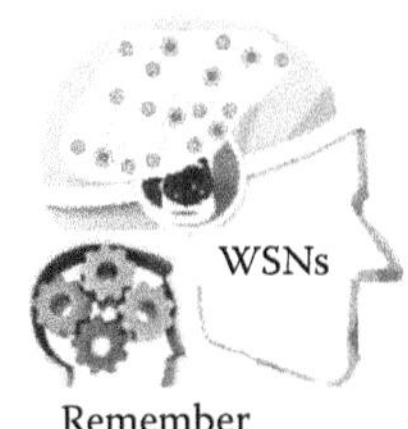

Remember

As we can see, a VSN mainly needs two aspects of knowledge: (1) digital image/video processing (such as data compression, object detection, and tracking) and (2) network multimedia processing (such as buffering and QoS schemes). Distributed data compression is a good example that needs the integration of networking protocols and image processing.

14.3 Cyclops [Rahimi05]

Cyclops [Rahimi05] separates the above-mentioned two aspects: (1) local image capture (in sensors) and (2) wireless network communications. It has programmable logic and memory circuits for high-speed data transfer. It also has a dedicated microcontroller (MCU) to serve as a sensor-to-network interface. By using these hardware components, it could separate high-speed data transfer from the low-speed capability of the embedded MCU. Another benefit of such separation is to make the time-consuming image capturing and interpretation localized and not affected by sensor communications. This is particularly useful for network-enabled sensors that experience asynchronous events (e.g., MAC layer radio access) and, in the meantime, need stringent delay constraints.

But we should note that one of the Cyclops design goals is to make power consumption minimal for large-scale deployment and extended lifetime. Such a goal also makes the platform face serious constraints with respect to computational power and imaging size. Therefore, Cyclops is appealing only for certain classes of applications. It may not be so efficient when an application requires high-speed processing or high-resolution images.

Good idea

In summary, Cyclops hardware has two appealing features: (1) very efficient energy-saving architecture and (2) separate localized video processing from network sensor communications. To achieve these two advantages, Cyclops exploits computational parallelism to isolate prolonged sensing computations, uses on-demand control of clocking resources to decrease power consumption, and uses automatic relaxation of each subsystem to make each of them reach its lower power state.

Cyclops node hardware consists of an imager, an MCU, a complex programmable logic device (CPLD), an external SRAM, and an external flash (Figure 14.2). The MCU can control the video sensor by setting its parameters, instructing it to

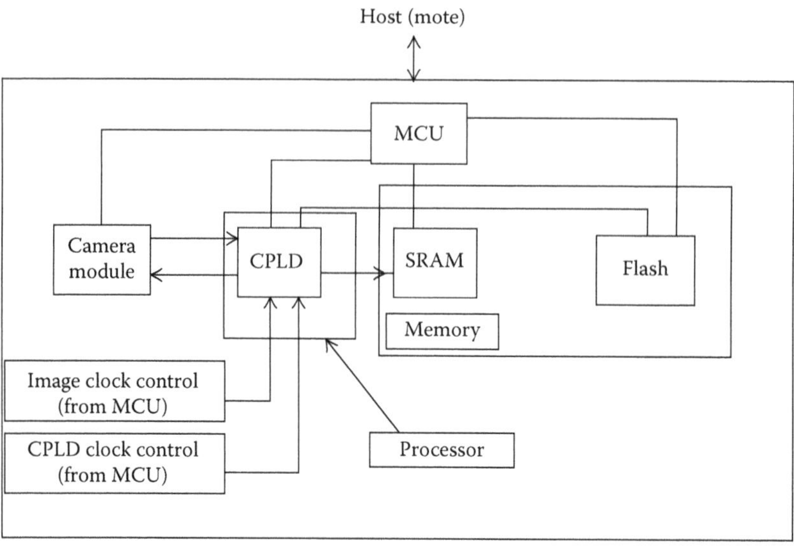

Figure 14.2 Hardware architecture of first generation of Cyclops. (Adapted from Rahimi, M. et al., Cyclops: In situ image sensing and interpretation in wireless sensor networks, *Proceedings of the Third International Conference on Embedded Networked Sensor Systems* (*SenSys '05*), San Diego, CA, November 2–4, 2005, ACM, New York, 192–204.)

capture video frames, and telling it when to run image computation. The complex programmable logic device (CPLD) produces high-speed clock synchronization signals and memory control commands that are required for image capture. The MCU can also work with the CPLD to provide low-power processing as well as access to high-speed clocking that are important in performing image capture. It is interesting to know that the CPLD can also perform some image-processing tasks, such as background subtraction and frame differentiation during video capture. Such a design makes an extremely economical use of hardware resources, as the CPLD is already clocking during the video capture. If the MCU does not require the clocking or processing features of the CPLD at any point of time, it can send out a halt command to stop the CPLD (this conserves power).

Another important feature of Cyclops nodes is their use of an external SRAM, which is useful when the internal memory of the MCU is insufficient for many applications. The external memory has enough space for image storage, computation, and manipulation purposes. Moreover, at the time of capture and computation, the external memory allows us to access memory resources. The SRAM is kept in the *sleep* state when we do not need extra memory resources (i.e., the internal memory is enough). Cyclops also contains external flash memory for permanent data storage that is used in template matching.

Bus architecture: The MCU, CPLD, and memory modules all share the same address bus and data bus. This feature helps fast and easy data transfer between hardware components. Such a common bus structure requires a special mechanism for synchronized data access among components.

Each module in Cyclops works in a few power "states." The lower the power a state uses, the higher the wake-up cost it requires to get back to the "active" state. Therefore, the application cannot just simply use a power state without considering the balance between the power saved (by entering the low power state) and the power used (to bring the module back to full power).

Cyclops also has an asynchronous trigger command wire. It can be used as a paging channel to perform event triggering in applications that require a quick "wake-up" from the sleep state. For example, the trigger wire could be connected to an IR (InfraRed) sensor, a microphone, or a magnet sensor, to trigger image capture when a motion or sound is detected.

The firmware that controls the Cyclops platform should support automatic relaxation to the lowest-possible power state, allow longtime image computations, and support synchronized access by both the MCU and the CPLD to shared resources such as SRAM. These requirements indicate that a network-centric approach is not suitable to primarily asynchronous events. Instead, Cyclops requires a "sequential" approach, which performs sequential image capture and processing. In this type of approach, a frame capture is followed by a series of long synchronous operations with little concurrency.

Cyclops firmware is written in the NesC [Gay03] language and runs in the TinyOS operating system environment. The use of TinyOS allows abstract functionalities to be used in the form of "modules" that can be easily interfaced. In addition, the operating system provides a scheduler and services that can be used for event-timing control.

14.4 VSN Calibration [Purushottam07]

As we mentioned in Section 14.1, VSNs perform several common tasks, such as object detection, object recognition, and object tracking. While *object detection* can detect a new object that appears in the range of the video sensors, *object recognition* determines what exactly the object is, and *object tracking* uses multiple video sensors to continuously track the object.

The VSN needs to be *calibrated* at the initial setup time for all of the above three tasks to be performed. For VSN calibration, we need to determine where each camera is placed (i.e., its location) and how the camera is oriented (i.e., its angle). The *location* of a camera is its position (3D coordinates) in a reference coordinate system, while *orientation* is the direction to which the camera points. Only when these two calibration statistics are obtained will we know the scope each video sensor can view.

By using this calibration information (location and orientation), the entire observation area can be broken down to figure out which parts of that area are covered by more than one sensor. We could also determine the relationship with other nearby sensors, such as the overlap area among neighboring cameras.

After we know these overlapped view areas, we may determine which sensor should sense which areas. We can also triangulate the location of the position through the use of overlapping sensors. We can allocate tracking responsibilities to sensors when the object moves out of a sensor's view.

Calibration of cameras is a well-studied area in the computer vision field. Many techniques can accurately estimate the location and orientation of cameras (such as in [Horn86] and [Tsai87]). Typically, they assume that the coordinates of a few landmarks are known beforehand. By using the projections of these landmarks as well as the principles of optics, we could determine a camera's coordinates and orientation.

Unfortunately, we cannot simply use these camera-based calibration schemes, because *video sensors have serious computational limits and power constraints*. Video sensors' limited calibration capability can lead to imprecise positioning of objects.

Even though we may borrow the concept of landmarks for the calibration of VSNs, we may not have landmarks at all in a VSN due to their high cost (compared to sensors). If we do not use landmarks, we may equip each video sensor with a positioning device, such as a GPS, and a directional digital compass [Sparton08]. These two things enable direct determination of the node location and orientation. Although this idea is extremely useful, today's GPS systems are expensive (compared to microsensors) and are not so accurate (errors up to 5–15 m). Another alternative is to use ultrasound-based positioning and ranging technology [Priyantha00] that provides greater accuracy. But the use of additional hardware in low-power video sensors could be cost prohibitive and power inefficient.

Therefore, accurate calibration is challenging for the initialization of resource-constrained WSNs without infrastructure support. Is it possible to implement calibration in video sensors without the use of known landmarks or without using any positioning technology?

If it is cost prohibitive to achieve highly accurate calibration through an *absolute* position knowledge, determining *relative* relationships between neighborhood nodes may be the only available option. This raises the following questions: (1) How to determine *relative* locations and orientations of video sensors without the use of known landmarks or positioning infrastructure? (2) How accurate are these techniques? and (3) What is the performance of applications based on approximate initialization?

Let us consider a wireless network with randomly deployed video sensors. Each sensor node consists of a low-power imaging sensor (such as Cyclops) and an RF mote (such as the Crossbow mote or the TelosB node). No GPS hardware is used. Our goal is to determine a parameter called *k-overlap*, which is a fraction of a viewable area that overlaps with *k* video sensors. Assume that there is a *reference* object present in the environment at any place. We assume that we know beforehand the reference object's dimensions as well as the focal length of each video sensor. Kulkarni [Purushottam07] describes approximate techniques to determine the degree of overlap and the region of overlap for camera sensors.

14.4.1 Determining the Degree of Overlap

We need to determine the value of *k-overlap*. Let us analyze a general case (i.e., k is an arbitrary value). Thus, 1-*overlap* is the fraction of a sensor's viewable region without overlap with any other sensors, 2-*overlap* is the fraction of region viewable to a sensor itself and one of the other sensors, and so on.

As shown in Figure 14.3, $k1$ is an area that is viewable by a single sensor, $k2$ is an area viewed by two cameras, and $k3$ is an area that all three cameras in the figure can view. Obviously, the union of the *k-overlap* regions of a sensor is exactly the total viewable range of that sensor (i.e., the sum of the *k-overlap* fractions of a sensor is 1).

The next step is to determine the *k-overlap* for each sensor, $k = 1\ldots n$, where n is the total number of sensors in the system.

14.4.2 Estimating k-overlap

Assume that some reference objects are deployed randomly. We call the location of each reference object as a reference point. Let us assume a uniform distribution of reference points in the environment. The video sensors will then take pictures of the environment. Once the pictures are processed, we can then see which reference objects each camera can view.

Suppose that a camera i can see r_i reference points in the total set. Among these r_i reference points, let r_i^k denote the reference points that are simultaneously visible to exactly k cameras. The *k-overlap* for camera i is given by

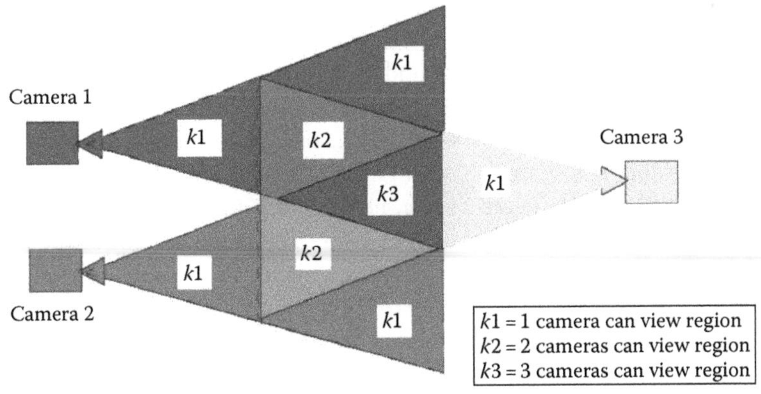

Figure 14.3 Different degrees of overlap (*k-overlap*) for a camera. (From Kulkarni, P., SensEye: A multi-tier heterogeneous camera sensor network, PhD thesis, Department of Computer Science, University of Massachusetts, Amherst, MA, February 2007.)

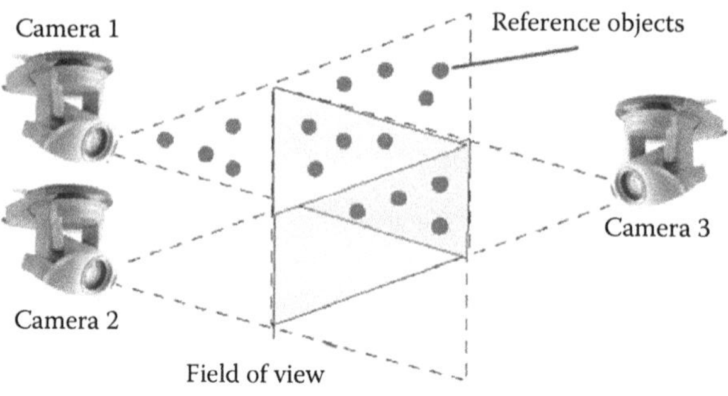

Figure 14.4 Estimation of *k-overlap* with the distribution of reference points. (From Kulkarni, P., SensEye: A multi-tier heterogeneous camera sensor network, PhD thesis, Department of Computer Science, University of Massachusetts, Amherst, MA, February 2007.)

$$O_i^k = \frac{r_i^k}{r_i} \tag{14.1}$$

As shown in Figure 14.4, camera 1 can see sixteen reference points, of which eight are visible only to itself; four are visible to cameras 1 and 3; and another four to cameras 1, 2, and 3. This yields a 1-*overlap* of 0.5, and 2-*overlap* and 3-*overlap* of 0.25 for camera 1. Likewise, we could find the *k-overlaps* for other cameras.

14.5 SensEye [Purushottam07]

SensEye [Purushottam07] is a VSN comprising multiple tiers (see Figure 14.5). As we have seen before, a sensor node consists of an analog video sensor, an MCU, a radio transceiver, and on-board RAM and flash memory.

In each tier, all sensors are homogeneous (i.e., of the same type). However, different tiers have heterogeneous sensors (i.e., these video sensors have different capabilities). Higher-tier sensors have higher capabilities than lower-tier sensors. Here, the capabilities include processing, networking, and imaging. On the other hand, higher-tier sensors consume more power. Therefore, to reduce the system power consumption, we should activate or use higher-tier sensors only when the lower-tier sensors are incapable of effectively capturing the image. Because different tasks will execute on multiple tiers, we need energy-efficient protocols to coordinate among various tiers of video sensors.

SensEye has made a good trade-off when allocating tasks to different tiers. SensEye uses a three-tier architecture (see Figure 14.5).

1. The lowest tier of SensEye consists of Crossbow motes [Crossbow08] (RF = 900 MHz), and low-fidelity Cyclops or CMUcam video sensors.

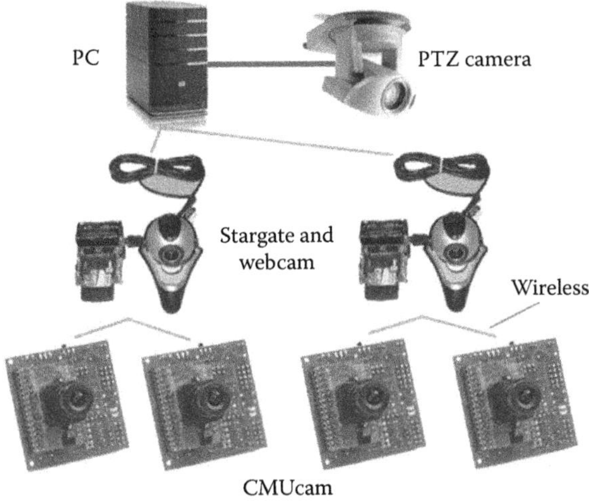

Figure 14.5 Multi-tier SensEye's hardware architecture. (From Kulkarni, P., SensEye: A multi-tier heterogeneous camera sensor network, PhD thesis, Department of Computer Science, University of Massachusetts, Amherst, MA, February 2007.)

2. The second tier consists of Stargate [Stargate08] nodes equipped with webcams. Each Stargate has an embedded 400 MHz XScale processor that runs Linux. Obviously, this tier's webcam can capture higher-resolution images than Tier 1 video sensors. To maintain upstream and downstream communications, each Tier 2 node has two radios—an 802.11 radio for peer-to-peer communications among Stargate nodes and a 900 MHz radio for communications with Tier 1 motes.
3. The third tier has a sparse deployment of high-resolution pan-tilt-zoom cameras that are connected to embedded systems (such as portable PCs). These cameras are programmable. They can be utilized to fill the coverage gaps provided by Tier 2. They can also perform calibration.

The deployment of multi-tier video sensors is an interesting idea. It is not energy efficient to implement all object detection/recognition/tracking algorithms in one tier. The higher tiers have a higher CPU capacity. But they consume much energy. This hierarchical structure has been used in many problems. For instance, the Internet backbone is such a multilevel architecture. Backbone routers are superfast (>40 Gbs). But these are very expensive. Campus LANs use inexpensive routers for edge traffic processing. Such a tree-based multilevel architecture is very similar to the architecture of human society.

The design of the SensEye multi-tier camera sensor network is based on the following three principles.

Principle 1: Assign tasks to the lower tier: Try to assign tasks to the lower tier to reduce power consumption. However, if the lower-tier sensors cannot meet certain requirements for some tasks (for instance, we need to execute some tasks correctly, reliably, and quickly), we need to seek the help of higher-tier sensors.

Principle 2: Wake up nodes only when necessary: To conserve energy, the processor, the radio, and the sensor on each node are duty cycled. SensEye will wake up nodes from sleep states through the use of triggers only when it is necessary. For example, we only wake up a higher-fidelity camera when we need to acquire a high-resolution image after a new object is detected by a lower-tier sensor. Putting these devices into the sleep state as much as possible will drastically improve the network's lifetime.

Principle 3: Exploit redundancy in coverage: Try to exploit overlaps in the coverage of cameras for calibration. For example, we could use two cameras with an overlapping coverage to localize an object and compute its (x, y, z) coordinates. Such data can then be used to intelligently wake up other sensors or to determine the trajectory of the object. Moreover, by exploiting redundancy in coverage, we improve the power consumption performance and maximize system lifetime.

SensEye detects objects with low latency and high reliability, as well as with energy efficiency. These conflicting goals cannot be achieved in a homogeneous, single-tier network. It allocates tasks through the seeking of a point solution in the space of all possible allocation permutations across tiers.

SensEye has used four types of cameras: (1) the Agilent Cyclops (discussed in Section 14.3), (2) the CMUcam Vision sensor [CMUcam08], (3) a Logitech Quickcam Pro Webcam, and (4) a Sony PTZ camera.

The RF communication is achieved through three different platforms—Crossbow motes [Crossbow08], Intel Stargates [Stargate08], and a mini-ITX embedded PC. These motes interface with different tiers of cameras.

Tier 1: It consists of a low-power camera sensor, such as Cyclops, connected to a low-power mote sensor platform. Unfortunately, SensEye can only use a prototype of the Cyclops camera, because mature products are not available. The software in Cyclops provides support for frame capture, frame differencing, and object detection.

Tier 2: It has a more capable platform and camera. Each Tier 2 node has a wake-up circuit to wake the node from a sleep or suspended state once it receives a trigger from a Tier 1 node. In SensEye implementation, an Intel Stargate sensor platform is used along with an attached mote that acts as the wake-up trigger. As the Stargate does not have hardware support for automatic wake-up, a relay circuit described in Turducken [Sorber05] is used for this purpose. The Logitech Webcam connects to the Stargate through the USB port.

Tier 3: A Tier 3 node comprises a Sony SNC-RZ30N PTZ camera connected to an embedded PC running Linux.

Figure 14.6 shows the software framework of SensEye. In this figure, it is assumed that Tier 1 is comprised of motes connected to CMUcam cameras. We could replace a CMUcam with Cyclops. The first two tiers of SensEye have four software components: (1) CMUcam frame differentiator, (2) mote-level detector, (3) wake-up mote, and (4) object recognition at the Stargate.

Tier 1 frame differentiator: Tier 1 nodes can capture an image for differencing. The CMUcam can capture an image and quantize it into a lower-resolution frame. It then performs frame differencing by using the reference background frame. Such a frame-differencing procedure can highlight objects by using nonzero difference values. The CMUcam has two working modes during frame differencing: (1) a low-resolution mode, where it converts the current image (of 88 × 143 or 176 × 255) to an 8 × 8 grid for differencing; and (2) a high-resolution mode, where a 16 × 16 grid is used for differencing.

Mote-level detector: Tier 1 motes report the results of object detection to the higher-level nodes. On start-up, the Tier 1 mote sends initialization commands to its analog video sensor to set up its background and frame-differencing parameters. The video sensor periodically captures an image and performs frame differencing. The mote uses a user-specified threshold and the returned frame difference result to decide whether an object appears or moves. If an event is detected, the mote broadcasts a trigger to the higher-tier motes.

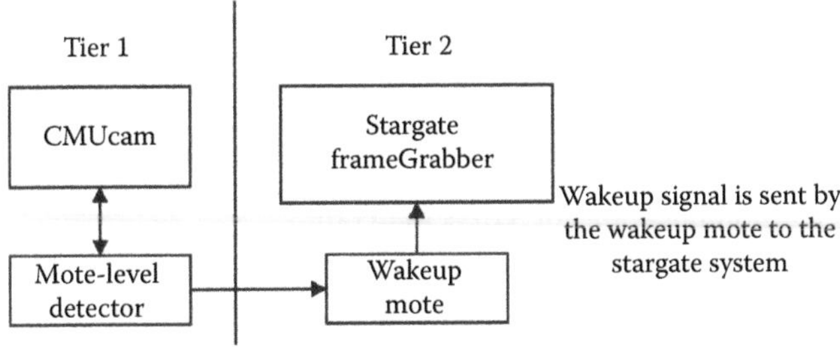

Figure 14.6 ***SensEye's*** **software architecture. (From Kulkarni, P., SensEye: A multi-tier heterogeneous camera sensor network, PhD thesis, Department of Computer Science, University of Massachusetts, Amherst, MA, February 2007.)**

Wake-up mote: The Tier 2 motes (connected to the Stargate) receive triggers from the lower-tier motes and decide whether to wake up the Stargate for further video processing. This procedure needs the localized coordinates. Note that we typically do not compute the object coordinates at a Tier 1 mote, as this would cause significant coordination operations between the Tier 1 nodes. SensEye thus uses a Tier 2 mote to compute these coordinates, and the Tier 1 nodes just simply piggyback parameters such as θ and ϕ, as well as the centroid of the image of the object. The Tier 2 mote then uses calibration algorithms to derive the coordinates. The Stargate is then woken up if the object location is within its field of view; otherwise, the trigger is ignored.

High-resolution object detection and recognition: The Stargate mote can immediately capture an image of the webcam's current view upon being awoken. Frame differencing between the captured image and the reference background image is performed. Frame differencing finds out the pixels and boundaries where the potential objects appear in the image. SensEye removes noise pixels by using smoothing techniques based on color threshold filtering and an averaging of the neighboring region. Next, object recognition algorithms are used to find each possible object. SensEye uses an averaging scheme based on the pixel colors of the object. It produces an average value of the red, green, and blue components of the object. The values of red, green, and blue can then be matched against a library containing many defined objects, to classify the object. SensEye can be extended by adding sophisticated classification techniques, face recognition, and other vision algorithms.

PTZ controller: Tier 3 has some retargetable cameras that can fill the coverage gaps and enhance coverage redundancy. The pan and tilt values for the PTZ cameras use localization techniques to implement calibration. The cameras export an HTTP API for a program-controlled camera movement. SensEye uses one such HTTP-based camera driver [Sony08] to retarget the Tier 3 PTZ cameras.

Problems and Exercises

14.1 Explain a few applications of VSNs in detail.

14.2 What special requirements does a VSN have compared to general sensor networks?

14.3 Illustrate the importance of video sensor calibration.

14.4 Why does SensEye use a three-tier architecture?

14.5 Explain the basic principle of *k-overlap* determination.

MISCELLANEOUS TOPICS

Chapter 15

WSN Energy Model

Because the sensor motes are battery driven, the energy consumption models are needed to quantify the sensor's lifetime. The design of all WSN protocols should keep energy efficiency in mind. In this chapter, we introduce some popularly used WSN energy models.

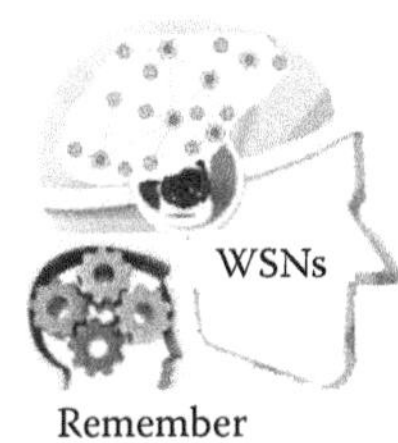

In many cases, we need to use math or simulation models to describe a performance metric. The purpose is to obtain an estimated performance value for any proposed WSN protocol before we test it in the realistic hardware testbed. In practice, we could not easily measure the energy consumption of a sensor except approximate system lifetime measure (e.g., how long a sensor's battery can last). Therefore, model-based energy calculation could give us a good metric on WSN performance.

15.1 Basic WSN Energy Model [Carlos04]

As discussed earlier, the sensor communications are the ones that consume most of the energy on a typical wireless sensor node. Other parts (such as CPU calculations) could consume certain energy, but typically less than communication energy. Let us take a look at a simple wireless link (Figure 15.1) and its energy models. This model has been used in many WSN energy calculations.

The energy consumed when sending a packet of m bits over one-hop wireless link can be expressed as [Carlos04]

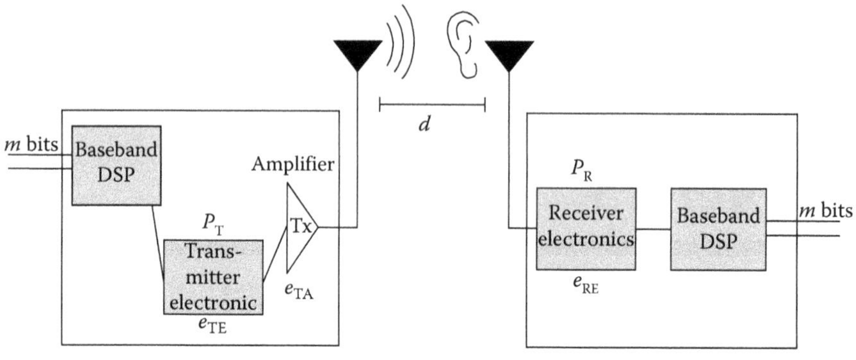

Figure 15.1 WSN energy model. (Adapted from Pomalaza-Ráez, C., Wireless sensor networks energy efficiency issues, Lecture notes, Fall 2004, University of Oulu, Oulu, Finland.)

$$E_L(m,d) = \{E_T(m,d) + P_T T_{st} + E_{encode}\} + \{E_R(m) + P_R T_{st} + E_{decode}\} \quad (15.1)$$

where

- E_T is the energy used by the transmitter circuitry and power amplifier
- E_R is the energy used by the receiver circuitry
- P_T is the power consumption of the transmitter circuitry
- P_R is the power consumption of the receiver circuitry
- T_{st} is the start-up time of the transceiver
- E_{encode} is the energy used to encode
- E_{decode} is the energy used to decode

If we assume that there exists a linear relationship for the energy spent per bit at the transmitter and receiver circuitry, E_T and E_R can be written as

$$E_T(m,d) = m\left(e_{TC} + e_{TA}d^{\alpha}\right) \; E_R(m) = me_{RC} \quad (15.2)$$

where

- e_{TC}, e_{TA}, and e_{RC} are hardware-dependent parameters
- α is the path loss exponent whose value varies from 2 (for free space) to 4 (for multipath channel models)

The effect of the transceiver start-up time, T_{st}, will greatly depend on the type of MAC protocol used. To reduce power consumption as much as possible, the transceiver should be in a sleep mode as much as possible. This can save a lot of power, but there is a caveat in that: constantly turning the transceiver on and off can consume energy.

An explicit expression for e_{TA} can be derived as [Carlos04]

$$e_{TA} = \frac{\left(\frac{S}{N}\right)_r (NF_{Rx})(N_0)(BW)\left(\frac{4\pi}{\lambda}\right)^{\alpha}}{(G_{ant})(\eta_{amp})(R_{bit})} \tag{15.3}$$

where

- $(S/N)_r$ is the minimum required signal-to-noise ratio at the receiver's demodulator for an acceptable E_b/N_0
- NF_{Rx} is the receiver noise
- N_0 is the thermal noise floor in a 1 Hz bandwidth (Watts/Hertz)
- BW is the channel noise bandwidth
- λ is the wavelength in meters
- α is the path loss exponent
- G_{ant} is the antenna gain
- η_{amp} is the transmitter power efficiency
- R_{bit} is the raw bit rate in bits per second

The expression for e_{TA} can be used for those cases where a particular hardware configuration is being considered. The dependence of e_{TA} on $(S/N)_r$ can be made more explicit if we rewrite the previous equation as

$$e_{TA} = \xi * (S/N)_r \quad \text{where } \varsigma = \frac{(NF_{Rx})(N_0)(BW)\left(\frac{4\pi}{\lambda}\right)^{\alpha}}{(G_{ant})(\eta_{amp})(R_{bit})} \tag{15.4}$$

It is important to bring this dependence explicitly as it highlights how e_{TA} and bit error rate p are related to each other. p depends on E_b/N_0, which in turn depends on $(S/N)_r$. Note that E_b/N_0 is independent of the data rate. To relate E_b/N_0 to $(S/N)_r$, the data rate and the system bandwidth must be taken into account, that is,

$$(S/N)_r = (E_b/N_0)(R/B_T) = \gamma_b(R/B_T) \tag{15.5}$$

where

- E_b is the energy required per bit of information
- R is the system data rate
- B_T is the system bandwidth
- γ_b is the signal-to-noise ratio per bit, that is, (E_b/N_0)

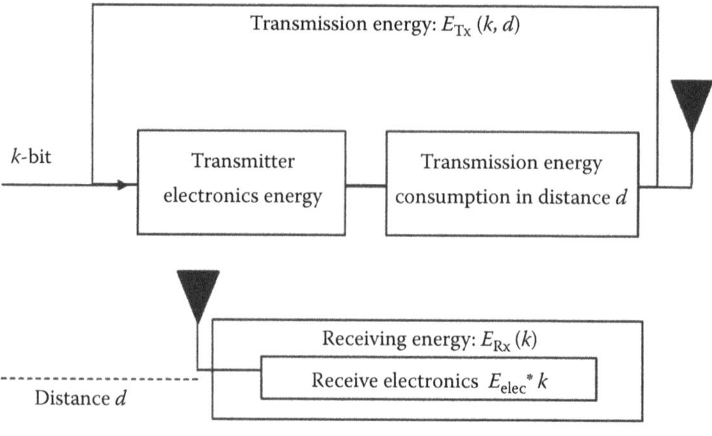

Figure 15.2 Simplified energy model.

Although the above model could accurately calculate the sending/receiving energy, many WSN developers prefer a simplified model (Figure 15.2) as follows [Akyildiz02]:

In the sender side, the energy consumption includes two sources: (1) local electronics and (2) sending out *k* bits for a distance of *d*. In the receiver side, the local electronics consume energy when getting *k* bits.

Example: Suppose each hop has a distance of 5 m. The sender–receiver distance is 100 m. The data amount is 1 M bits. The coefficient E_{elec} = 50 nJ/bit, E_{amp} = 100 pJ/bit. How much does the WSN consume?

Solution: There will be 100 m/5 m = 20 pairs of transmitting and receiving sensors (each sensor will be a receiver in last hop and a sender in next hop).

For each pair of sensors, the energy consumed is as follows:
Sender side energy consumption: $E_{Tx}(k,d) = E_{elec} * k + E_{amp} * k * d^2$
That is: $E_{Tx}(k,d) = (50 * 10^{-9}) * 10^6 + (100 * 10^{-12}) * 10^6 * 5^2 = 0.0525\,J$
Receiver side energy consumption: $E_{Rx}(k) = E_{elec} * k = 0.05\,J$
Each pair thus consumes energy = $E_{Tx}(k,d) + E_{Rx}(k) = 0.1025\,J$
The entire WSN consumes energy = 0.1025 * 20 = 2.05 J

15.2 Simulation-Based Energy Modeling [DSchmidt07]

The above math models could provide quantitative results on WSN energy consumption. Another approach to measuring energy is through simulation models. In [DSchmidt07], a finite state machine (FSM)–based simulation model could accurately measure Crossbow sensor energy.

Good idea

No matter whether you use math models or simulation models, both approaches do not use realistic hardware platforms to measure energy. In reality, it is not easy to use instruments to measure CPU, radio chip, or other circuit energy consumption. In your WSN research, you may create some energy models based on empirical data (such as experimental measurement) or system status analysis (such as FSM models).

The above model describes energy from the system level. Some software tools (such as SPICE) could model the energy consumption of sensor chips in the transistor level or in the register level. Although these simulation tools cover all effects including leakage and switching energy, it is difficult to create the energy models because those tools require a detailed in-depth knowledge of the hardware architecture such as register interfaces. Also, simulation on circuit level is very time consuming. It takes a long time to simulate whole networks with a large number of individual nodes.

As the circuit-level simulations are complex, energy models that simulate the hardware on the *instruction* level (i.e., running codes) have been created for some CPUs. Such an approach is usually done by measurement of synthetic software benchmarks. These benchmarks use a series of program loops that execute only one kind of instruction, so that the energy consumption of every single CPU instruction can be calculated from the measurements. Some special energy measure should be used to model inter-instruction dependencies and the impact of algorithm operands on CPU energy consumption.

The above instruction-based energy models could achieve relatively accurate energy simulations without the need of much chip knowledge as in the hardware circuit simulations. Another benefit of instruction-level simulation is the improved simulation runtime compared to circuit simulation. However, the cost of CPU model creation is relatively high, and the resulting model has too much overhead when used at runtime, because sensor nodes typically have very limited computational power.

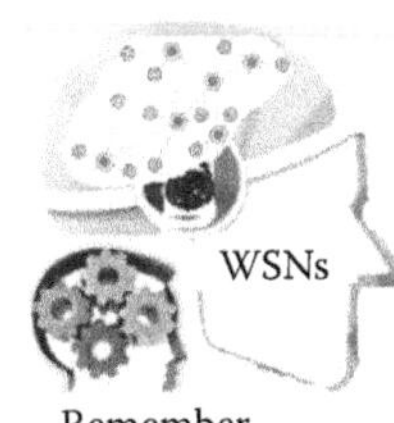

Remember

If you have taken the computer architecture course, you should be able to understand the importance of benchmark programs in CPU performance measurement. Those benchmarks are well recognized by CPU designers in terms of comparing different CPUs' speed and energy consumption. However, here we measure not only CPU energy but also some other parts such as wireless communication energy. Those benchmarks would not be so useful.

To overcome the mentioned problems, Schmidt et al. [DSchmidt07] proposed a *component*-based high-level modeling approach. We all know that a WSN sensor is made up of several hardware components such as a microcontroller (i.e., CPU), a radio transceiver chip, sensor electronics, and various other devices like LEDs, flash RAMs, etc. Each of these components can be operated in different *states*. For example, a transceiver chip would be able to operate in the following four states: power down, idle, transmitting, and receiving. The CPU could also work in idle, interrupt, and calculation states. A natural approach would be the use of an FSM to model the component operation procedure and attach an energy model to each state as follows:

- We call every operational status of a sensor component as one *state* in the FSM.
- We model every possible *change* from one operational state to another as a state transition in the FSM.
- Every state in the FSM is associated with an energy consumed per unit time.
- Every transition in the FSM is associated with a time period needed to switch between two operational states.
- The FSM has a well-defined initial state that corresponds to the stable state a component reaches after power up.

How do we obtain the time and energy consumptions associated with states and state transitions? Typically, we could get those data from experiments or simple measurements. We could start with the FSM model of every component of a sensor node. Then the FSM for the whole sensor node can be constructed.

In most cases, only the microcontroller (CPU) can trigger the state changes. An FSM typically represents the possible states the system could be in. However, it does not reflect constraints on state changes that appear in normal operation. Wireless communication is a good example. After the transceiver has begun frame transmission, it cannot be interrupted. Such a case cannot be represented in the FSM model. Hence, in a second-step, the dynamic behavior of the sensor node in typical scenarios is analyzed to model these constraints.

The SDL (*specification and description language*) model could define a node's dynamic behavior. SDL includes the runtime environment implementing the SDL semantics, and the code transformation patterns. We could represent power saving strategies in the SDL model. We could use either explicit power-saving strategies that are part of the application model, or implicit strategies that are part of the runtime environment. The WSN dynamic behavior can be formalized as a set of communicating state machines. The SDL model is based on state machines that have common actions with the state machine describing the sensor hardware.

Figure 15.3 shows an example for a sensor's dynamic behavior. It represents the task of sending one frame of data via the wireless interface to a distant node. It starts in the state where the microcontroller is the only active component of the sensor

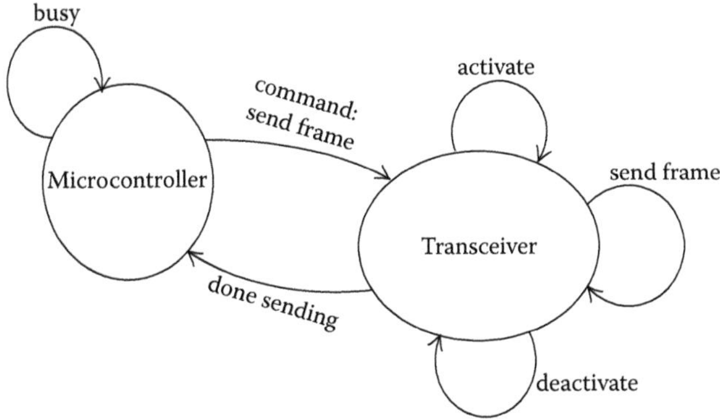

Figure 15.3 Parametric flowchart for frame transmission. (Adapted from Schmidt, D. et al., *Adv. Radio Sci.*, 5, 347, 2007.)

node. Later on, the transceiver is triggered to transmit one frame. The number of data bytes to be transmitted determines the amount of time spent in transmit mode. After transmission, the transceiver is deactivated again to save energy, and the initial state is triggered again.

From the above flowchart (Figure 15.3), it is very easy to calculate the energy consumed by the specified task, conforming to Equation 15.6 below. Here P_{state} is the power consumed in the state and t_{state} is the time spent in this state. P_{trans} and t_{trans} are the power and time for the transitions between two states.

$$E = \sum_{state} P_{state} \cdot t_{state} + \sum_{trans} P_{trans} \cdot t_{trans} \tag{15.6}$$

Such an FSM-based analytic model can be integrated in a software simulator, which makes it possible to perform simulations that predict the energy consumption of a sensor system very accurately.

On the other hand, some environment factors, such as platform resources, network resources, and energy resources, can affect the behavior of the sensor nodes. As one factor's status can significantly affect all of the other simulated factors, and can eventually change the outcome of the simulation, we need to model all of these factors to simulate the energy consumption of an array of sensor nodes accurately.

Let us see an example. The radio transceiver chip is used for communicating with other network sensors. It significantly affects the energy consumption of a sensor node participating in a large sensor network. Some network-level factors, such as network congestion and wireless bandwidth limitations, could cause a large number of transmission errors. This could change the communication pattern of every sensor in the network, and thus affecting the system energy consumption. Furthermore, platform limitations such as inaccurate clocks/timers can make this situation worse by introducing clock jitter into time-synchronized networks. We need to design a number of highly specialized simulators to capture all of these effects.

Some simulators that can simulate network behavior are already available [SAM06]. These specialized simulators can be programmed into different simulation components. To form a system-level simulator, we could use a message-based interface to interconnect those components.

In the above component-system simulators, how do we model energy consumption? A good idea is to use two steps:

Step 1: Use an FSM to describe an individual sensor's energy consumption behavior.
Step 2: Use the message-based interfaces to couple the simulation of energy consumption with other network-level operations (such as congestion control).

We could have two methodologies to support the above idea: (1) If the simulation of the energy consumption in a hardware component (such as CPU) is already implemented in a software simulation module, it would be easy to integrate energy consumption into the existing simulators. In this case, we only need to implement an interface between the energy simulation component and system simulation. (2) If it is difficult to add another interface to an existing simulator, we could implement energy consumption as a new simulation component. In the second approach, we have separate software simulators: one simulation component simulates the behavior of a sensor hardware component and another one simulates its energy consumption by implementing the energy model.

We could easily integrate the above two methodologies into the energy simulator framework. Figure 15.4 shows that the central component, "network node," which is

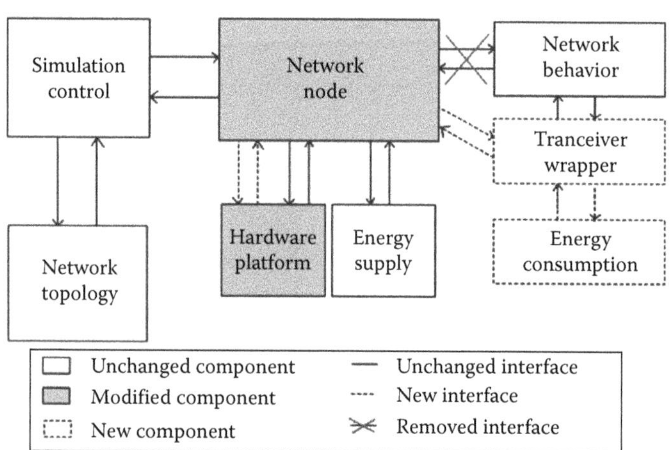

Figure 15.4 Structure of the simulator integration framework. (Adapted from Schmidt, D. et al., *Adv. Radio Sci.*, 5, 347, 2007.)

the core part of the simulator framework, needs to integrate the energy consumption into the simulation software. In fact, such a core component controls all other simulators that simulate one sensor's different operation states.

For the simulated hardware, we could integrate its energy consumption into the already available simulator. This is actually the first methodology mentioned before.

For the simulation of the network-level system behavior, a new component that implements the energy model could be created, and the simulation component for the network behavior was replaced with a wrapper (see Figure 15.4). Such wrapper distributes simulator messages among the original simulation component (for network behavior simulation) and the added component that tracks the energy consumption of each sensor.

Schmidt et al. [DSchmidt07] provide an excellent example on Crossbow MicaZ energy simulation. The MICAz node has an 8-bit Atmel microcontroller with RISC architecture, clocked at 7.3728 MHz, with 4 kB of internal SRAM and 4 kB data EEPROM as well as 128 kB of internal flash memory. Its transceiver chip operates at data rates up to 250 kbit/s. A 512 kB flash memory can be accessed via two SRAM page buffers of 264 bytes each. Three LEDs can be used to show the operational status of the device and each node is equipped with a serial-number chip that gives a node its unique ID. The MICAz has a 51-pin expansion connector as an interface to arbitrary sensors. Figure 15.5 shows the overall architecture.

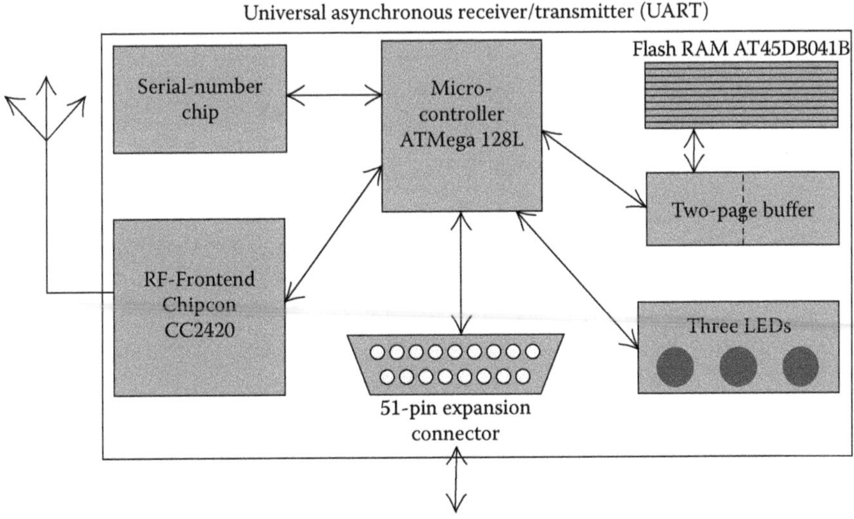

Figure 15.5 Architecture of MICAz. (Adapted from Schmidt, D. et al., *Adv. Radio Sci.*, 5, 347, 2007.)

To simulate MicaZ energy consumption, Schmidt et al. [DSchmidt07] considered the microcontroller, the transceiver chip, and the flash memory. The LEDs can be turned off to reduce energy consumption, and the energy consumption of the serial-number chip is negligible.

Figure 15.6 shows the component models of the transceiver and the microcontroller of the MICAz nodes. Such an FSM model also shows the state transition

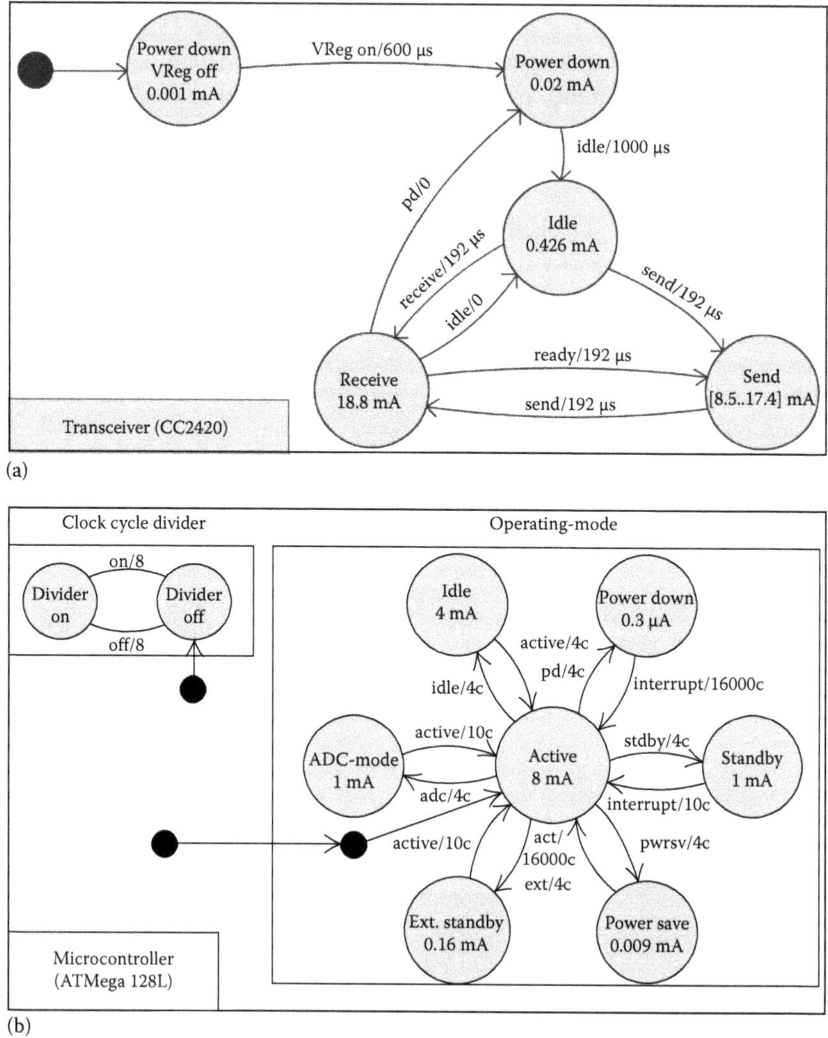

Figure 15.6 Energy model of MICAz (only microcontroller and transceiver). (Adapted from Schmidt, D. et al., *Adv. Radio Sci.*, 5, 347, 2007.)

times. For instance, it uses seconds for the transceiver, and uses clock cycles for the microcontroller.

The microcontroller can be in different power-reduced states. The MICAz operates at a constant input voltage of 3 V (2 AA batteries). Its energy consumption for every state is given in milliamperes. Please note that the transceiver sending state is not a single value, instead, it is a range of energy consumption. The actual consumption in this state depends on the chosen output power of the transceiver chip.

15.3 Battery-Aware Routing [Chi06]

Nickel–cadmium and lithium-ion batteries have been widely used in wireless devices and sensors. In a battery, lots of cells are arranged in series, in parallel, or a combination of both. The active materials of each cell consist of two electrodes (an anode and a cathode) separated by an electrolyte. A continuous reduction–oxidation reaction can transfer electrons from the anode to the cathode after the cell is connected to a load.

Figure 15.7 illustrates this phenomenon through the demonstration of a simplified symmetric electrochemical cell. Figure 15.7a is a fully charged cell, where the electrode surface contains the maximum concentration of active species. When the cell is linked to an external load, an electrical current flows through the external circuit.

Figure 15.7b shows the discharge process. In this case, active species are consumed at the electrode surface and replenished by diffusion from the bulk of the electrolyte. But the diffusion process cannot compensate for the consumption. That is why a concentration gradient can build up across the electrolyte in Figure 15.7b.

A higher current load causes a higher gradient concentration, that is, a lower concentration of active species at the electrode surface [Doyle93]. This low concentration decreases the battery voltage. Eventually the voltage can drop below a predefined cutoff threshold, which means that the electrochemical reaction can no longer be sustained at the electrode surface. In this case, the battery dies (see Figure 15.7e).

Note that we cannot use the electroactive species that have not yet reached the electrode. Such unused charge is called discharging loss. It is not physical "loss." It just means that the species are not available because of the difference between the reaction and the diffusion rates.

Before the battery stops working, if the battery current is very weak or is zero, that is, in the battery recovery status (see Figure 15.7c), we can see that the concentration gradient flattens out after a sufficiently long time, and reaches equilibrium again.

Following the above recovery process, the concentration of active species near the electrode surface makes unused charge available again for extraction (Figure 15.7d).

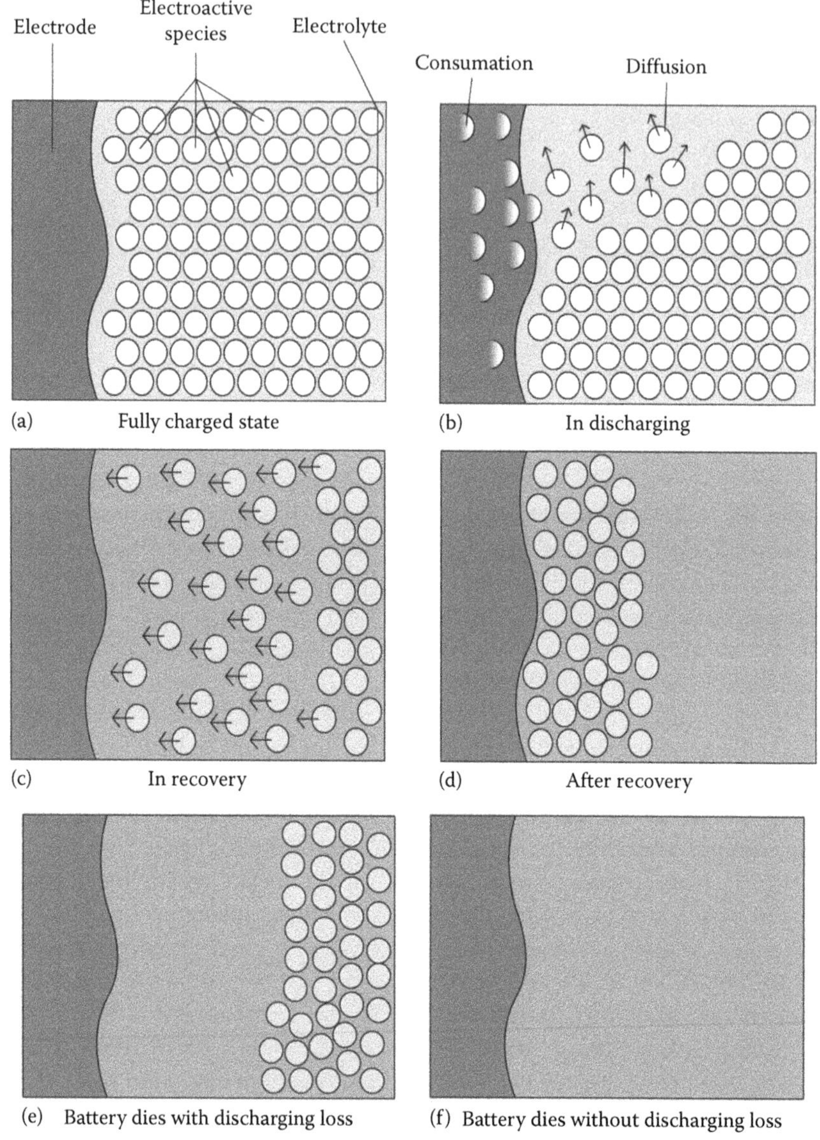

Figure 15.7 Battery operations at different states. (Adapted from Ma, C. and Yang, Y., *Mobile Netw. Appl.*, 11, 757, 2006.)

Therefore, battery recovery can reduce the concentration gradient, recover discharging loss, and hence prolong the battery lifetime (Figure 15.7f).

Some experiments on nickel–cadmium battery and lithium–ion battery have demonstrated that the discharging loss might take up to 30 percent of the total battery capacity [Rakhmatov03]. Hence, it is important to precisely model battery behavior to optimize system performance in sensor networks.

Battery-aware routing (BAR) in streaming data (such as video/audio) transmissions can be simply modeled as streaming packets from a source to its corresponding destination. But how do we maximize the communication lifetime between a source–destination pair? In [Chi06], they propose the concept of BAR. Its basic idea is to choose the "well-recovered" sensors as relay points, and use "fatigue" sensors for recovery. If we could dynamically schedule routing paths to efficiently recover the sensor battery capacity, we can minimize the discharging loss on sensors, and thus maximize the system lifetime and data throughput between a source–destination pair.

Especially in [Chi06] a BAR protocol is proposed. They use BAR algorithm to setup a routing path in a source–destination pair. Before we describe BAR protocol, let us make some assumptions.

Suppose sensors are randomly deployed in a WSN. And each node knows its geographic position (this could be achieved by some accurate sensor localization algorithms, see Chapter 9). The node is powered by AA batteries. Let us target the steaming applications such as video monitoring where transmission is viewed as a stream. If a sensor is on the routing path from the source to the destination, we call it routing node. In each time slot, a routing node can be assigned for a task ("active" status) or in "idle" status. A task may be a routing activity, video displaying, software execution, or any other power-consuming function at this node. Multiple tasks may be assigned in the same time slot.

Let us define some parameters. Assume that C is the battery residual capacity. Also assume that β (a constant) is an experimental chemical parameter. It varies from battery to battery. The larger the β, the less the discharging loss.

Let us observe an example shown in Figure 15.8 [Chi06]. In this sensor network, the source node S transmits packets to the destination node D. The battery residual capacity C and parameter β are indicated in the figure. We compare the following two approaches.

In the first approach, S sends packets to D through multi-hop path S → A → C → F → E → D. Some time (say 45 minutes) later, node A uses up its energy. After that the routing path changes to S → B → C → F → E → D. The total connection lasts around 90 minutes [Chi06].

However, the lifetime can be extended in a simple way by alternating between the above two paths. In the second approach, nodes A and B alternate each other as the router. A recovers its battery while B is routing, and so on. In this way, the total lifetime is around 113 minutes [Chi06]. It is increased by 24.8 percent.

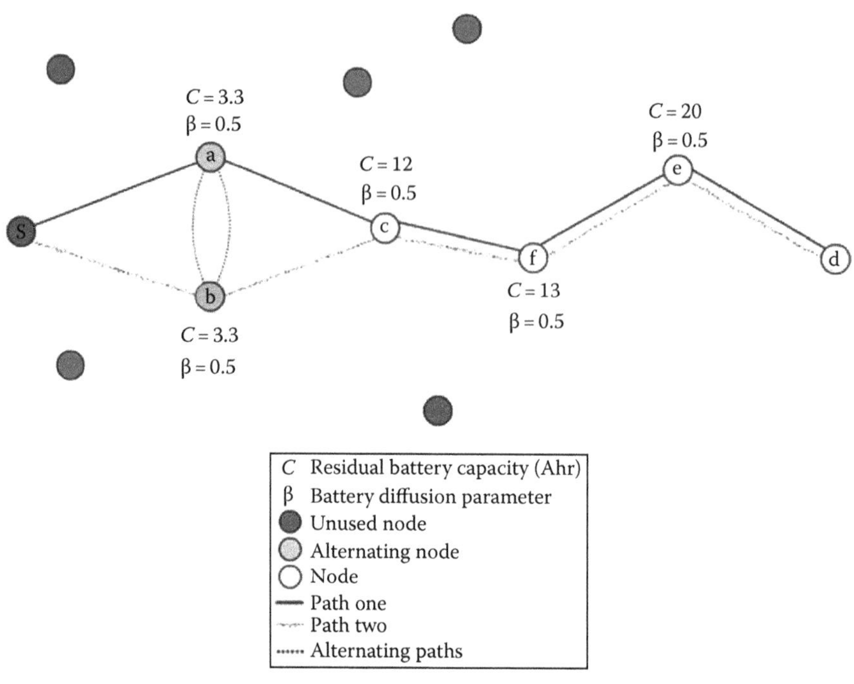

Figure 15.8 BAR in a sensor network. The current at each node is I = 3.5 A. By alternating between nodes A and B, the network achieves longer lifetime. (Adapted from Ma, C. and Yang, Y., *Mobile Netw. Appl.,* 11, 757, 2006.)

In summary, in the battery-aware energy-efficient routing protocol, we could alternatively recover batteries to extend node lifetime. The important idea is that we always choose the most fully recovered nodes as routing nodes.

BAR is an excellent idea. Although there are many energy-efficient WSN routing protocols proposed to extend the system lifetime by choosing the energy-saving path, very few ideas could go deep into the battery itself and explore the cell charging/discharging concept. As you can see, we may solve the same problem from different hardware levels (system level, board level, component level, chip level, transistor level, etc.). The lower the level is, the more accurate model we may define.

Problems and Exercises

15.1 Figure 15.1 is the most typical WSN energy model. Please use such a model to explain that a 100-m, 1-hop communication energy consumption is higher than 10-hop (each hop 10 m) relay-based communication.

15.2 Go to http://www.xbow.com to read the data sheet on MicaZ motes. Explain its power consumption features and search for some good energy models that can simulate MicaZ energy consumption behaviors.

15.3 Besides the BAR example, can you search other energy-aware WSN protocol design examples?

Chapter 16

Sensor Network Simulators

Developing the right simulation tools is especially important for WSN studies. In many cases, we may not have large-scale (>1000 sensors) WSN hardware testbeds. Software-based simulations provide inexpensive design performance tests. Today, many simulation tools can simulate noise, interference, and other uncertain factors in WSNs. These can even analyze energy consumptions in different hardware parts. This chapter introduces some typical WSN simulators.

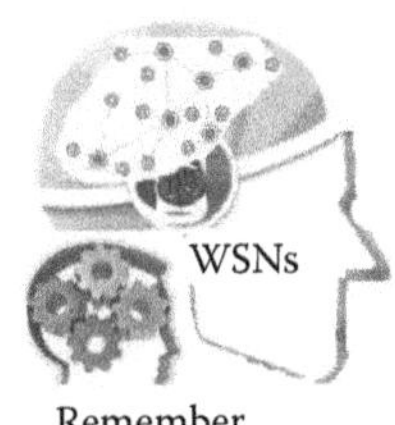

Some engineers may underestimate the role of software simulations in the design of WSNs. As a matter of fact, to save performance test time and cost, we typically use discrete-event-based simulation tools first, to verify the network protocol efficiency in a large-scale WSN. These tools have accurate radio propagation model and energy analysis tools. After we obtain the simulation results, we may be able to avoid some potential engineering design errors.

16.1 GloMoSim [GloMoSim]

GloMoSim can be used to build a scalable simulation environment for wireless and wired network systems, including WSNs. It is designed based on the parallel, discrete-event simulation capability provided by Parsec. Most network systems are built using a layered approach that is similar to the OSI seven-layered network architecture. GloMoSim uses a similar layered approach. Standard APIs (application program interfaces) are used between the different simulation layers. This allows the rapid integration of models developed at different layers by different people. The protocols that are shipped with the current GloMoSim library include the following aspects:

Layers	*Protocols*
Mobility	Random waypoint, random drunken, trace based
Radio propagation	Two ray and free space
Radio model	Noise accumulating
Packet reception Models	SNR bounded, BER based with BPSK/QPSK modulation
Data link (MAC)	CSMA, IEEE 802.11, and MACA
Network (routing)	IP with AODV, Bellman–Ford, DSR, Fisheye, LAR scheme 1, ODMRP, and WRP
Transport	TCP and UDP
Application	CBR, FTP, HTTP, and Telnet

Source: Adapted from Zeng, X. et al., GloMoSim: A library for parallel simulation of large-scale wireless networks, *Proceedings of the 12th Workshop on Parallel and Distributed Simulations,* May 26–29, 1998, Banff, Alberta, Canada.

To run GloMoSim, one will need the latest Parsec compiler (now included with the GloMoSim distribution). If protocol developers write pure C codes, they need to use the Parsec compiler. The Parsec code is used extensively in the GloMoSim kernel. Most users do not need to know how the kernel works.

16.2 SensorSim [SensorSim]

SensorSim is built on the ns-2 simulator and provides additional features for modeling sensor networks. The main features of this platform include (1) sensing channel and sensor models, (2) battery models, (3) lightweight protocol stacks for WSNs, (4) scenario generation, and (5) hybrid simulation.

Figure 16.1 shows its internal simulation modules. It provides a graphical user interface (GUI) for sensor data generation and visualization. Figure 16.2 shows the simulation architecture of an individual sensor node. It includes accurate WSN sending/receiving energy consumption models.

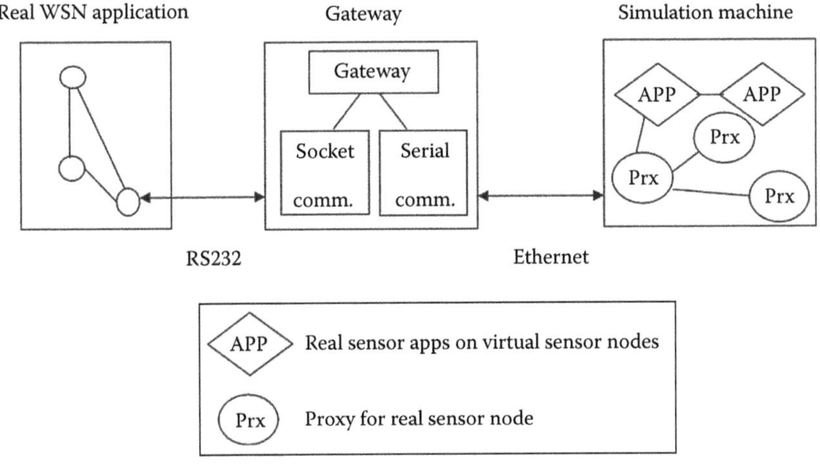

Figure 16.1 System model in SensorSim. (From Park, S. et al., SensorSim: A simulation framework for sensor networks, *Proceedings of 3rd ACM International Workshop on Modeling, Analysis and Simulation of Wireless and Mobile Systems,* August 20, 2000, Boston, MA, 2000.

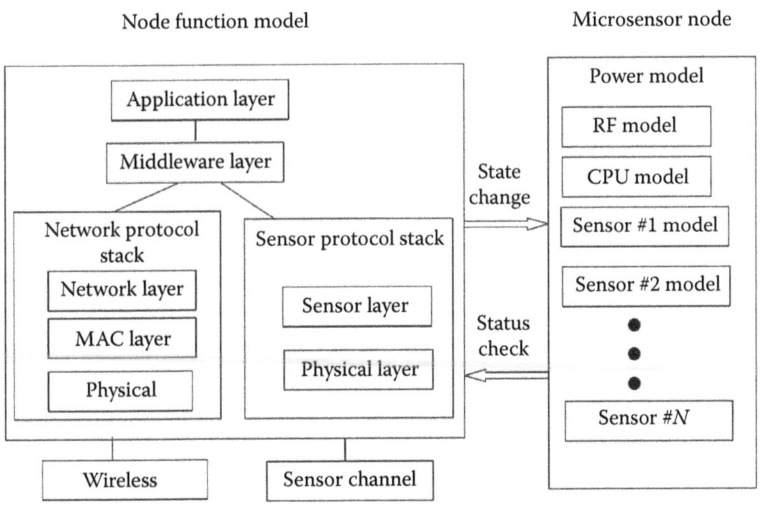

Figure 16.2 Sensor node model in SensorSim. (From Park, S. et al., SensorSim: A simulation framework for sensor networks, *Proceedings of 3rd ACM International Workshop on Modeling, Analysis and Simulation of Wireless and Mobile Systems,* August 20, 2000, Boston, MA, 2000.

16.3 TOSSIM [Philip03]

TOSSIM [Philip03] can be used to describe the behavior and interactions of networks of thousands of TinyOS motes at network bit granularity. Figure 16.3 shows a graphical overview of TOSSIM. It consists of five parts: (1) an interface to TinyOS component graphs, (2) a discrete event queue, (3) some re-implemented TinyOS hardware abstraction components, (4) extensible radio and analog-to-digital converter (ADC) models, and (5) communication services for external programs to interact with a simulation.

In TOSSIM, discrete-event simulations are directly generated from TinyOS component graphs. It runs the same codes that run on realistic WSN hardware. By replacing some low-level components (see the shaded parts in Figure 16.3), TOSSIM can translate hardware interrupts into discrete simulator events.

TOSSIM uses a very simple but powerful abstraction for a sensor network scenario. The sensor state has what it hears on the radio channel. This abstraction allows testing perfect wireless links (i.e., the bit error rate is zero). It also easily captures the hidden terminal problem, and can capture many of the different problems that can occur during packet transmission (such as symbol detection failure and data corruption).

As shown in Figure 16.3, the TOSSIM engine provides a set of communication services to interact with external applications. These services allow users' programs

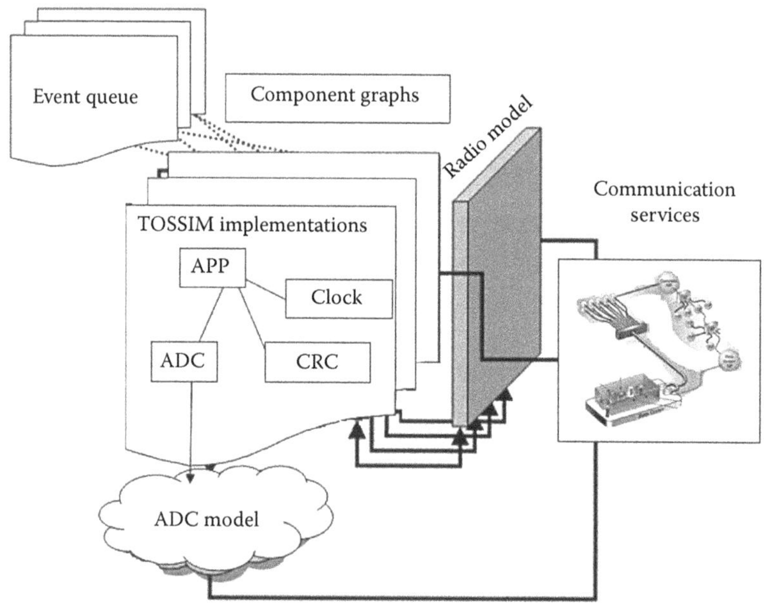

Figure 16.3 TOSSIM architecture: frames, events, models, components, etc. (Adapted from Levis, P. et al., TOSSIM: Accurate and scalable simulation of entire TinyOS applications, *SenSys '03*, Los Angeles, CA, November 5–7, 2003.)

to interface with TOSSIM through a TCP socket to help programmers monitor or actuate a running simulation. A user can also use these services to get to know the specifications of the ADC and radio models, such as sensor readings and packet loss rates.

TOSSIM has the support of the TinyOS tool-chain. This feature simplifies the transitions between simulated and real networks. Compiling to the native code allows developers to use traditional debugger tools in TOSSIM. Users can set debugger breakpoints and step through what is normally a real-time code (such as packet reception) without disrupting the operation.

TOSSIM describes each hardware resource as a component. It can emulate the behavior of the underlying raw hardware, which includes an ADC, a clock, a transmit strength variable potentiometer, an EEPROM, a boot sequence component, and several of the components in the radio stack.

The TOSSIM network model can easily capture the inter-sensor interactions. Each bit transmission activates the model, which changes the state of the channel observed by receiving events on other nodes.

Note that TOSSIM allows users to develop, test, and evaluate physical layer/MAC layer network protocols, in addition to application layer characteristics.

Figure 16.4 is a sample TOSSIM execution procedure.

Time (4 MHz ticks) Action

---

100 Dequeue simulator event at time 100.

The clock interrupt handler is called, signaling the application Timer event.

The application's Timer handler requests a readingfrom the ADC.

The ADC component puts a simulator ADC event on the queue with time stamp.

The interrupt handler completes; the clock event re-enqueues itself for the next tick.

400 Dequeue and handle simulator ADC event at time 400.

The ADC interrupt handler is called, signaling an ADC ready event with a sensor value.

The application event handler takes the top three bits and calls LEDs commands.

The ADC interrupt handler completes.

1000 Simulator event is dequeued and handled at time 1000.

The clock interrupt handler is called, signaling the application Timer event.

... execution continues as above.

Figure 16.4 Sample execution. (Adapted from Levis, P. et al., TOSSIM: Accurate and scalable simulation of entire TinyOS applications, *SenSys '03*, Los Angeles, CA, November 5–7, 2003.)

TOSSIM allows TinyOS developers to choose the accuracy and complexity of the radio model as necessary for their simulations. The radio models are independent of the simulator. Thus, it would be easy for a user to change these models.

In TOSSIM, the network is modeled as a directed graph. As we know, in a graph model, we use a vertex to represent a node, and each edge (u, v) in the graph represents the error rate when mote u sends to v, and is distinct from the edge (v, u). Such a model allows accurate simulation of asymmetric links, as we may have different error rates in back and forth directions. Bit errors are independent. Link probabilities can be specified by the user and changed at runtime. Transmission events propagate to the simulated input channel of each connected node. Each mote has its own local view of the network channel.

For example, assume that sensor T transmits data to mote R on an error-free channel. On each of its bit events, T transmits a 0 or a 1. This transmission modifies the internal state of R, representing what it hears over the air. On each of its bit events, R reads this state, and passes the bit up to a TinyOS component.

A user can use a TCP/IP to allow PC applications to communicate with TOSSIM to drive, monitor, and actuate simulation. Such a simulation-application protocol is a command/event interface based on TinyOS abstractions.

A user can send commands to TOSSIM to actuate a simulation and modify its internal state. These commands could be operations to change radio link error rates/sensor readings, to turn sensors on and off, and to inject network packets.

Users can also write their own systems that interface with TOSSIM in new ways. The monitoring/actuation codes and statements are removed when compiling for a mote.

TOSSIM has a visualization tool, called TinyViz, which is a Java-based GUI for TOSSIM. It allows simulations to be visualized, controlled, and analyzed. It has visual feedback to show the simulation state. It also has mechanisms to control the simulation procedure, such as modifying sensor readings and radio link error rates.

TinyViz also has a plugin interface that allows developers to implement their own application-specific visualization and control codes based on the TinyViz engine. The TinyViz engine manages the event/command interface with TOSSIM. The TinyViz engine sends out TOSSIM events to loaded plugins. This is very useful in some cases, For example, a plugin may be used to visualize network traffic as sensors receive data. TinyViz plugins can also send commands to TOSSIM to invoke a simulation. For example, when a user turns off a sensor in the visualization window, the control plugin sends the corresponding power-off command to TOSSIM.

Besides the network and control plugins described above, TinyViz also has a set of default plugins for basic debugging and analysis purposes. A plugin displays (in a list format) all debug messages. Another one can graphically display the data in radio and UART packets. A sensor plugin displays sensor values in the GUI and allows the user to set individual sensor values during simulation. A radio model plugin can update wireless connectivity based on the distances between motes in the GUI. It can graphically display link probabilities, providing basic mechanisms for experimenting with how networks behave under change.

Based on the built-in models in TOSSIM, a user can write her own powerful TinyViz plugins. For example, the user can model wireless radio obstacles (such as metal barriers) by changing bit error rates. She can also use a plugin to model failures by turning motes off at scripted times. One can also use the plugins and simulation data to examine and analyze the application behaviors. TinyViz can use the communication services of TOSSIM to allow a user to take an omniscient view of a large network, examining the internal mote in a running simulation.

16.4 PowerTOSSIM [Victor04]

In realistic WSN experiments, it is very difficult to accurately measure how much power each chip component (such as the CPU and memory) consumes. A good news is that we can use PowerTOSSIM [Victor04], which consists of the instrumentation of the TinyOS codebase, to track hardware power state transitions. PowerTOSSIM also has an accurate CPU-cycle-counting mechanism based on basic-block-level profiling, and analysis tools to visualize and analyze power consumption results on a per-mote basis.

16.4.1 PowerTOSSIM Architecture

Figure 16.5 illustrates the architecture of PowerTOSSIM. The PowerState module accepts request-energy calls from simulated hardware components (radio, sensors, LEDs, etc.) and emits power state transition messages for each component. With the calculation of a power model, these messages finally produce detailed power consumption data or energy visualizations.

PowerTOSSIM is able to keep track of the power state of each hardware component for any simulated sensors. Its tracking procedure is achieved through specific power state transition messages that are recorded during the simulation run. TOSSIM can issue calls to a component, PowerState, to track hardware power states for each mote, and logs them to a file during the run.

A challenging issue is to estimate the CPU usage if we want to know the CPU energy consumption. Because PowerTOSSIM runs the software as a native binary program on the host machine, it does not know the length of time that a sensor spends using the CPU. But it can accomplish CPU profiling by mapping the basic blocks executed by the simulation code to cycle counts in the corresponding mote binary. PowerTOSSIM combines its generated power state data with a power model to determine per-mote and per-component energy usage. The above tracing procedure can be done off-line to obtain the detailed power consumption for each hardware component of each sensor, or can be output into the TinyViz visualization tool to display the power consumption data in real time. The reason of decoupling the generation and processing of the power state transition data is to achieve efficiency and flexibility.

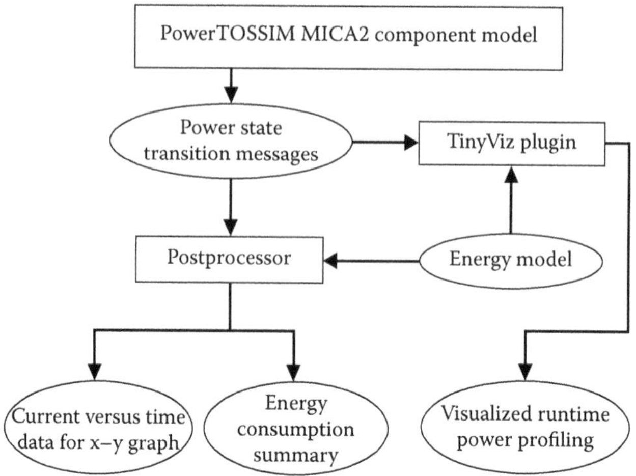

Figure 16.5 PowerTOSSIM architecture. (Adapted from Shnayder, V. et al., Simulating the power consumption of large scale sensor network applications, *SenSys '04*, Baltimore, MD, November 3–5, 2004.)

Efficiency: Just like TOSSIM, PowerTOSSIM can also simulate large networks that scale to thousands of nodes. To preserve this scalability, we should avoid high overheads in the simulations. If we record hardware state transition messages only at runtime, we could achieve very low overhead. Likewise, if we allow the simulation to run as a native binary program, we could avoid the overhead of instruction-level simulation.

Flexibility: PowerTOSSIM provides a high degree of flexibility for capturing and modeling the power state of the mote. But it does not assume a particular hardware platform, because new designs are constantly being developed. Through the decoupled design, we could evaluate the power efficiency of potential hardware designs only by plugging a new power model into the PowerTOSSIM analysis tools. And the simulation software itself need not be re-executed.

Good idea

Remember this good idea: modularity in complex system design. Remember Internet network layers, such as the application layer and the transport layer? The Internet does not use one layer, as it will be much easier to revise each submodule without touching the entire system. As long as the interfaces among modules remain the same, we can easily update each module based on new design requirements.

16.4.2 Component Instrumentation

As we know, for each hardware component of a sensor, TinyOS has a specific software module that is responsible for controlling the hardware component's operations. For example, most aspects of wireless communication from the ChipCon CC1000 radio can be achieved by the CC1000RadioIntM module. TOSSIM (see last section) simulates these TinyOS hardware drivers through its own software modules, which makes it possible to link a TinyOS application to the simulated hardware with very few code changes.

Like TOSSIM, PowerTOSSIM can instrument each of the simulated hardware drivers with power state transition messages that are logged during the simulation. PowerTOSSIM issues call (from each hardware driver) to a new module, called PowerState, which can generate log messages when the power state of each hardware component changes as time goes on. By implementing power state transitions in a separate module, we could allow the interface to be readily extended to support new hardware components, such as new sensor platforms (non-Crossbow products).

16.4.3 CPU Profiling

PowerTOSSIM can compile the TinyOS application codes into a binary file that runs directly on the simulation machine. This design is efficient. However, such a design cannot easily determine how much time a CPU spends in the "active" state (when actively executing instructions) compared with in the "idle" state, or any of the other low-power states. In many cases, we need to track the amount of time that the CPU spends in the active state to get to know the accurate power consumption, especially for CPU-intensive operations (such as security algorithms) or for some special occasions (for instance, a sensor may spend much time in low-power sleep modes, and only wake up and perform computation infrequently).

Today, most sensors' microcontrollers consume approximately constant power while executing instructions. This is because they do not use the sophisticated chip-level power management strategies, as in more advanced processors. In a sensor, most components (such as instruction core, SRAM, ADC, oscillator, timer, and other peripherals) are always on when the controller is in the "active" mode. In Crossbow MicaX sensors, the ATMEL Atmega128L CPU consumes about 8 mA while executing instructions and 3.2 mA while idle. Likewise, the cycle time for each instruction is well documented and usually deterministic, or at least predictable. Therefore, PowerTOSSIM can compute the CPU energy usage easily by tracking the amount of time the CPU spends in each power state. The amount of time that a node spends in the "idle" mode depends on external factors, such as the timing of clock interrupts, which are already modeled by TOSSIM.

Although TOSSIM cannot capture the time spent in executing CPU instructions, PowerTOSSIM can determine the CPU execution time by simulating the execution of each instruction. The detailed strategy includes the following four steps:

1. PowerTOSSIM checks the binary programs to obtain an execution count for each basic program block with no-branch instructions.
2. It maps each program block to the corresponding assembly instructions.
3. It determines the number of CPU cycles for each program block using simple instruction analysis.
4. It combines the simulation basic block execution counts with their corresponding cycle counts to obtain the total CPU cycle count.

When the simulation is finished, PowerTOSSIM can write down basic program block execution counters that are processed *off-line* to obtain CPU cycle count totals. Such a process is fairly accurate and incurs very little overhead during the simulation time.

16.4.4 PowerState Module

If we scatter the *power state* tracking code throughout the simulator, it could incur high overhead. PowerTOSSIM thus uses a single TinyOS module, called PowerState. Other TinyOS components make calls to it to register hardware power state transitions. PowerState has a single interface with one command for each possible state transition. Each function tests if power profiling is enabled, and if so, emits a log message detailing the sensor ID, the specific power state transition, and the current simulation time.

An excerpt from this log is shown below [Victor04]:

0: POWER: Mote 0 LED_STATE RED_OFF at 18677335
0: POWER: Mote 0 LED_STATE YELLOW_OFF at 18677335
0: POWER: Mote 0 ADC SAMPLE RSSI_PORT at 18990479
0: POWER: Mote 0 ADC DATA_READY at 18990679
0: POWER: Mote 0 RADIO_STATE TX at 18993551
0: POWER: Mote 0 RADIO_STATE RX at 19199375

16.4.5 Analysis Tools

PowerTOSSIM also includes several tools to analyze and visualize the power consumption data. These tools accept the input from the log files generated by PowerState, the CPU profiling information, and a hardware power model.

One of the tools is called a postprocessor, which can compute the total energy for the various hardware components for each sensor and output a time-series trace of power consumed by each sensor.

PowerTOSSIM also has a plugin for TinyViz (part of TOSSIM software). Such a plugin can report per-mote power consumption as the simulation runs. For visualization convenience, the plugin assigns different colors to sensors based on how much power it has consumed during the simulation, making it possible to visualize power hot spots in the network.

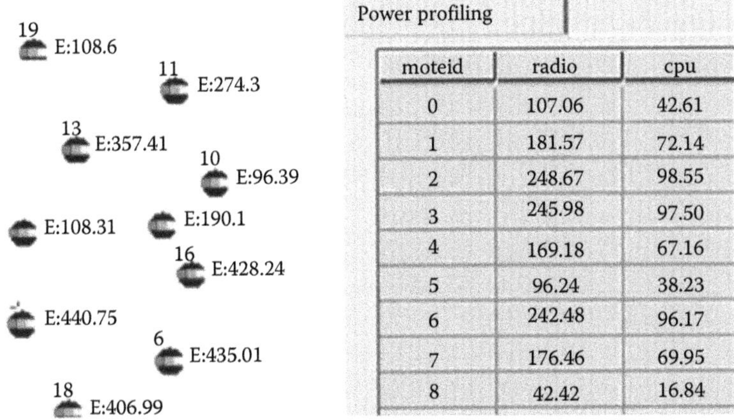

moteid	radio	cpu
0	107.06	42.61
1	181.57	72.14
2	248.67	98.55
3	245.98	97.50
4	169.18	67.16
5	96.24	38.23
6	242.48	96.17
7	176.46	69.95
8	42.42	16.84

Figure 16.6 Screenshot of the PowerProfiling plugin for TinyViz. (Adapted from Shnayder, V. et al., Simulating the power consumption of large scale sensor network applications, *SenSys '04*, Baltimore, MD, November 3–5, 2004.)

Figure 16.6 shows a typical screenshot of the visualization. The table on the right reports a runtime summary of the energy consumed by each component of the simulated network. Each sensor is also assigned a color based on the total amount of energy that it has consumed since the start of the simulation run.

Problems and Exercises

16.1 Explain the pros and cons of simulations compared to a real WSN testbed.

16.2 Refer to [GloMoSim] for software download. Play with a few simple wireless network demos.

16.3 Use PowerTOSSIM to observe a sensor node's power consumption in a CPU and an RF transceiver.

CASE STUDIES

Chapter 17

Case Study 1: Tele-Healthcare

In this chapter, we provide a case study on an important application—tele-healthcare based on wireless sensor networks (WSNs). The results presented here are based on the author's (Hu) research. This chapter mainly refers to the author (Hu) and his colleagues' previous publications, including [Hu08, Hu2009a, Hu2009b, Hu2009c, Hu2009d, Hu2009f, Sunil08a], and others.

17.1 Introduction

Today, in the world, especially in developed countries, cardiovascular diseases are the largest cause of morbidity and mortality [MGHunink97]. Based on the World Health Report 2000, each year the coronary artery disease (CAD) kills an estimated 7 million people representing 13 percent of all male deaths and 12 percent of all female deaths. Thus, low-cost, high-quality cardiac healthcare delivery is a critical challenge.

Many new cardiovascular disease healthcare systems come up, such as primary/secondary prevention and patient empowerment. These promote the development of novel care approaches [LAShort98], in which out-of-hospital monitoring and follow-up are basic aspects [CardioNet08]. Therefore, the development and utilization of tele-cardiology systems that provide new modes of cardiac patient contact is of increasing interest [Istepanian04]. Most tele-cardiology systems use wearable devices (such as portable ECG recorder, sphygmomanometer, and pulse oximetry) to collect remote cardiac patients' physiological data (including ECG, blood pressure, pulse rate, etc.).

In nursing homes or hospitals, the ad hoc interconnection of ECG sensors is a promising approach to perform automatic heartbeat anomaly detection [Martin00]. Today, many ECG machines are claimed as "portable"—but this does not always indicate that they are tiny. As a matter of fact, many of these appliances receive power from an electrical outlet and are so heavy that they have to be mounted on a cart and wheeled from one location to another.

As mentioned in Chapter 1 (Introduction), the biggest advantages of WSNs are the sensors' tiny size, low cost, and low power. If any of these features is lost, we may classify such a network as a common wireless network or an ad hoc network, which has a simpler design than WSNs.

We could interconnect ECG sensors to form a low-power *medical ad hoc sensor network* (*MASN*), which can significantly improve the ECG portability and timeliness. An MASN can also be regarded as a special type of WSN. A simple MASN scenario is shown in Figure 17.1. Each patient's ECG signal can be automatically collected and processed (such as analog-to-digital conversion), and then be wirelessly sent to a remote ECG server for data analysis (such as using data classification to find out arrhythmia). If an ECG sensor reports any abnormal heartbeat signals, an emergency communication channel established between the physician's office and the patient's wireless device (such as a beeper or a cellular phone) will be used to send out alerts to provide the patient some medical suggestions (such as taking drugs or performing other further processing). In a typical MASN, a patient's ECG sensor can use a neighbor sensor to relay its data for multi-hop communications.

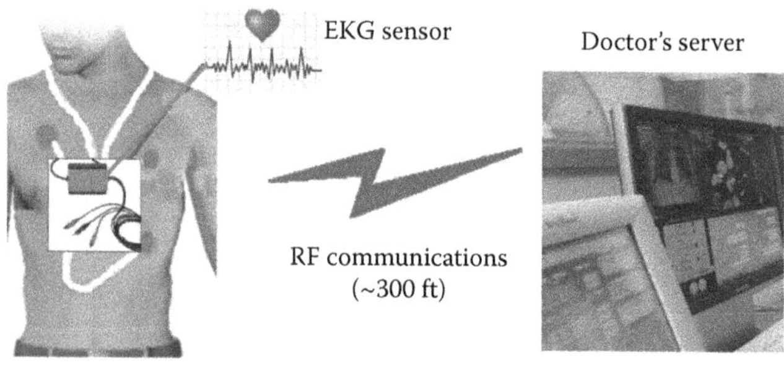

Figure 17.1 Tele-cardiology sensor networks (MASNs).

17.2 MASN Hardware Design

17.2.1 ECG Sensors and RF Communication Hardware

An MASN consists of multiple wireless ECG communication units. Each unit is called a *mobile platform*. As shown in Figure 17.2, each platform is composed of a customized ECG sensor board providing connections to a 3-lead ECG monitoring system, which is housed on a wireless communication board (also called RF motes). While the ECG sensor board gathers useful patient ECG data, the RF mote provides limited local-signal-processing capabilities (such as ECG noise filtering), and also wireless communication to transmit the ECG signals back to the server for feature pattern extraction.

Figure 17.3 shows the *logic* architecture components of the MASN mobile platform. Please note that the sensing chip detects analog inputs from patients' bodies, such as ECG (heartbeat signals), SpO_2 (oxygen level), and temperature.

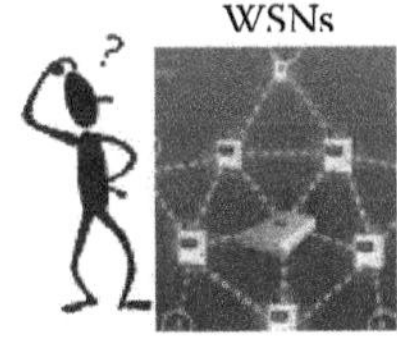

There are many commercial products on ECG sensors. However, if these sensors can only generate analog signals, we cannot interconnect all ECG sensors into a network, as these sensors need to have CPU (central processing unit) and RF chips. Also remember: Even if an ECG sensor can interface with a network, WSNs' definition points out that each sensor has serious resource constraints. And the WSN protocols should be able to adapt to large-scale (>1000 sensors) sensor interconnections. All these WSN features make their design very challenging. In this chapter, medial sensor networks are used to monitor thousands of patients.

Figure 17.2 Mobile platform appearance (includes ECG sensor + RF mote).

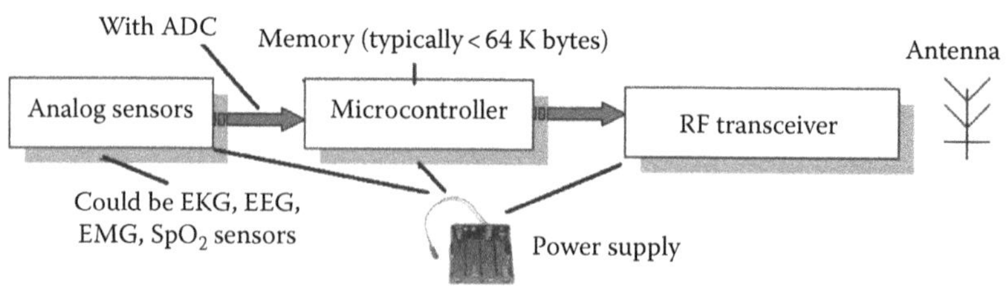

Figure 17.3 MASN mobile platform: logic architecture.

Our original RF mote (see Figure 17.2) was based on TelosB motes from Crossbow Inc. [Crossbow08]. It offers an on-chip RAM of 10 kB and also provides an IEEE 802.15.4 Chipcon radio [Chipcon08] with an integrated on-board antenna providing up to 125 m of range. By using a TI MSP430 microcontroller [Ti08], the TelosB worked fine in this application for its on-board ADC peripherals with expansion bays, to which the customized sensor board is connected. But the TelosB also has a few shortcomings. First, it is expensive when deployed in large-scale networks (its cost was around $200 each in 2009). Second, its power lifetime is around three to six months depending on how often the ECG signal is transmitted back to the server, which is not long enough for most medical applications—one year of lifetime is desirable. Third, its radio components cannot be enhanced or replaced (for instance, we cannot use a better radio transceiver/antenna to reach a longer distance).

Due to the above reasons, we have used Ember CPU-RF chips [Ember08] to build our own RF motes. As shown in Figure 17.4, the core of the RF board is the micro central unit (MCU)/ZigBee [Zigbee08] transceiver unit.

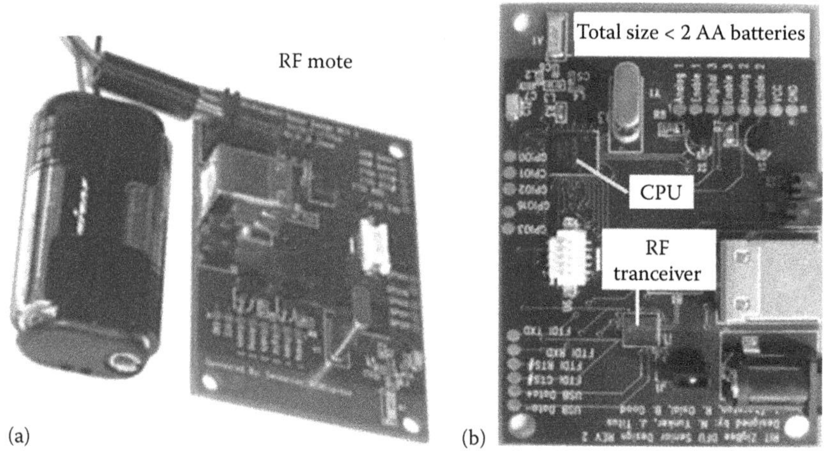

Figure 17.4 Customized RF board with ECG RF communication capabilities.

Multiple options were considered before selecting the final option. For instance, we could (1) use a separate MCU and transceiver or (2) use an SoC (system-on-chip) that incorporates the two devices together. The SoC option was chosen as it would be cheaper to implement, decrease programming complexity, and create an easier printed circuit board (PCB) layout (as there will be fewer parts in circuit layout).

Good idea

An SoC can typically save the manufacturing cost compared to separate chips. The internal chip-to-chip interfaces have been optimized in SoC chips. When using separate chips, an engineer could easily make mistakes in the chip-to-chip pin connections. On the other hand, for separate chips, you can optimize the system performance by easily replacing any of the components.

Figure 17.5 shows the connection between an ECG sensor board and our built RF board. The RF board takes the analog ECG data (sensed from a patient's body), converts it to a digital format, then uses network protocols to form packets, and finally sends the packets out through the RF antenna. Its RF transceiver can also receive ECG data from a neighbor RF board (to achieve patient-to-patient multi-hop communications).

The ECG analog sensor board design is based on the results from the Harvard University CodeBlue team [CodeBlue06]. The ECG lead extensions from the sensor board are pin compatible and color coded to standard 3-lead ECG monitoring systems. Although there are different flavors of physiological chest leads, this system was designed to match any 3-lead ECG snap set leadwires.

If real patients are not available, for the convenience of the test, we could use ECG generator hardware to emulate different heartbeat signals. The generator used in this prototype is the Model 430B, 12-lead ECG generator, as shown

Figure 17.5 Connection between ECG sensor and our built RF mote.

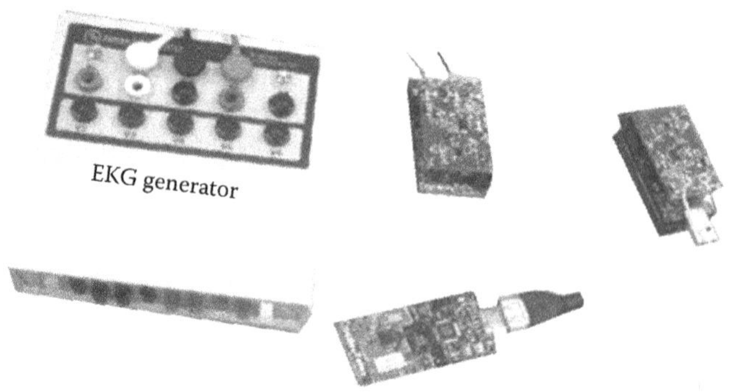

Figure 17.6 Model 430B patient simulator.

in Figure 17.6. This generator provides a complete PQRST waveform at six preset rates (60, 75, 100, 120, 150, and 200 BPM) as well as six preset amplitudes (0.1, 0.2, 0.5, 1.0, 2.0, and 5.0 mV). It is also capable of generating square waves using its five ECG snaps plus ten banana jacks. This provides a good testing interface if this system is adapted into a 12-lead monitoring system in the future. Figure 17.6 also shows the connection between the 430B ECG simulator and the RF communication boards.

Good idea

In many countries, there are strict government policies on the use of real patients or animals for medical tests. It could require a long procedure to get the approval for such real tests. Fortunately, there are many accurate commercial signal generators that can simulate different types of body parameters. For instance, the above-mentioned Model 430B, 12-lead ECG generator could simulate dozens of heart disease signals.

17.3 Reliable MASN Communication Protocols

17.3.1 Enhanced Cluster-Based MASN Data Transmission [Sunil08, Sunil08a]

It is important to achieve a fast and reliable detection of the ECG signals from the patients. We have used a cluster-based, energy-aware ECG collection scheme where the ECG data is reliably relayed to the sink (i.e., a server) in the form of aggregated data packets [Sunil08, Sunil08a].

Good idea

Clustering is a good idea in distributed computing. Its basic idea is to group nodes into different "clusters" based on some common attributes (such as physical location proximity and similar CPU capacity). Typically, each cluster has a member elected as cluster head (CH). Cluster-to-cluster communications are achieved through head-to-head connections. There are some research issues in clustering schemes, such as the forming of clusters under node mobility, CH selection rules, intra-cluster and intercluster routing schemes, cluster size, and reliable clustering.

Our proposed MASN routing scheme is different from LEACH [WBHeinzelman02] and other clustering schemes *due to our consideration of energy-level determination of sensor nodes, event-triggered cluster formation, and dynamic adaptation of reliability based on the cluster member density and event proximity*. The details are as follows.

We assume that the sensor nodes know their maximum energy (E_{max}), residual energy (E_R), and threshold energy (E_{th}). Here, E_{th} is the minimum energy required by the sensor nodes to identify themselves in one of the n energy levels. A sensor node having $E_R \le E_{th}$ belongs to the energy level 0. Initially, the energy of a sensor node is divided into n levels as follows:

$$n = \left\lceil \log_x \frac{E_{max}}{E_{th}} \right\rceil \tag{17.1}$$

where the energy range of level L is defined as the difference between the upper and lower energy values and x is the ratio between the maximum and minimum values of a level. The value of x depends on the requirement of the application. The energy level (L) of a sensor node is determined as follows:

$$\text{If } E_R < E_{th}, \quad L = 0; \quad \text{else } L = n - \left\lfloor \frac{E_{max}}{E_R} \right\rfloor \tag{17.2}$$

A sensor node decides to participate in the cluster formation process if the amplitude of the event parameter crosses a predetermined threshold, Δ. Here, the value of Δ depends on the measured event parameter.

While forming clusters, the sensors with the highest energy level are given an opportunity to become the CHs, to ensure longer cluster lifetime. In areas lacking high-energy sensor nodes, the lower-energy sensor nodes take the initiative to form CHs. This is mainly to ensure that the primary purpose of reliable event detection at the sink is achieved. The sensor nodes then elect their CHs based on the energy level and the AMRP value. Here, the AMRP is defined as the average minimum power level required by the r neighboring nodes to reach the sensor node claiming to become the CH, as follows [OYounis04]:

$$\mathrm{AMRP} = \frac{\sum_{i=1}^{r} \mathrm{Min\,PWR}_i}{r} \tag{17.3}$$

where

MinPWR_i denotes the minimum power level required by a node v_i, $1 \leq i \leq r$, to communicate with the CH
r is the number of neighbor nodes

The sensor nodes advertise themselves as CHs based on their energy level. The sensor node claiming to be a CH broadcasts the advertisement message to its neighbors using maximum power (MaxPWR). The normalized AMRP is defined as the ratio of AMRP to that of the MaxPWR.

Other sensor nodes receiving the advertisements decide to join a CH based on a function of CH energy level and communication power. Every sensor node waits for a random time before advertising itself to other sensor nodes to become a CH. This delay time for sending the advertisement message is based on a function of the energy level (L) of the sensor node and the normalized average minimum reachability power (nAMRP).

The sink assigns a reliability value, REL, for an event in terms of the total number of packets of the event required to be reported in a time T at the sink. This reliability factor is distributed among the clusters formed in the event area based on (1) the number of sensor nodes in the cluster and (2) the cluster-event proximity. Every CH of the event area transmits the number of its cluster members in the aggregated data packet header to the sink through multi-hop communications. By analyzing the values of the measured event parameters in the aggregated data packets, the sink knows which of the CHs are closest to the event. The sink assigns a reliability value to each cluster as follows:

$$\mathrm{CR}_i = \frac{\mathrm{REL} * (J_i)(m_i)}{\sum_{i=1}^{z} J_i\, m_i} \tag{17.4}$$

where

CR_i is reliability assigned to the ith cluster
z is the number of clusters
J_i is the event proximity for its cluster
m_i is the number of sensor nodes in the cluster

If J_i = 1, then the reliability is distributed among all the clusters based on their member density. By assigning a higher value of J_i, the sink can acquire more number of packets from the clusters closer to the event. The event proximity parameter, J_i, varies from cluster to cluster from a minimum value of 0 to a maximum value of 1.

The sink will vary the reliability values for the clusters if the event propagates to other areas. Their sensors will also form clusters based on the values of the measured event parameters. This idea of *dynamic reliability adaptation* at the sink (i.e., server) is helpful in terms of obtaining maximum information of the event.

17.3.2 MASN Routing Performance

Energy consumption: A major concern in MASN networking design is energy consumption. Our experiments have shown that most of the sensor battery is consumed in radio communications instead of in local data processing (such as ECG compression) or sensing (see Figure 17.7). Therefore, any MASN networking protocols (such as finding the optimal route) should have low complexity to save energy consumption.

Throughput: For better observation of a patient's health condition, a sensor can send out data at a high reporting frequency, and then use a high data rate to send out the large amount of sensed data wirelessly. Figure 17.8 shows the packet reception ratio (the number of received packets divided by the number of transmitted packets) for different sending rates (number of network packets per second). We can see that the MASN performance drops sharply if the sending rate is higher than 25 packets/s. Thus, it is important to use a reasonable reporting frequency in each medical sensor.

Scalability: We have investigated the MASN performance with the increasing number of sensors (it also means more patients if each patient carries one sensor). Our MASN system can still maintain good performance (reception

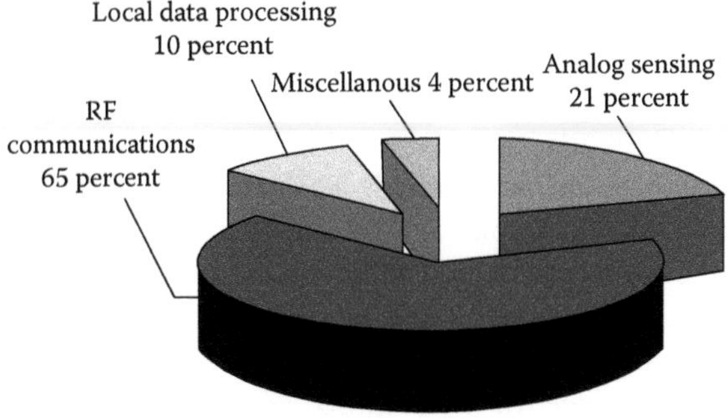

Figure 17.7 Energy consumption of MASN.

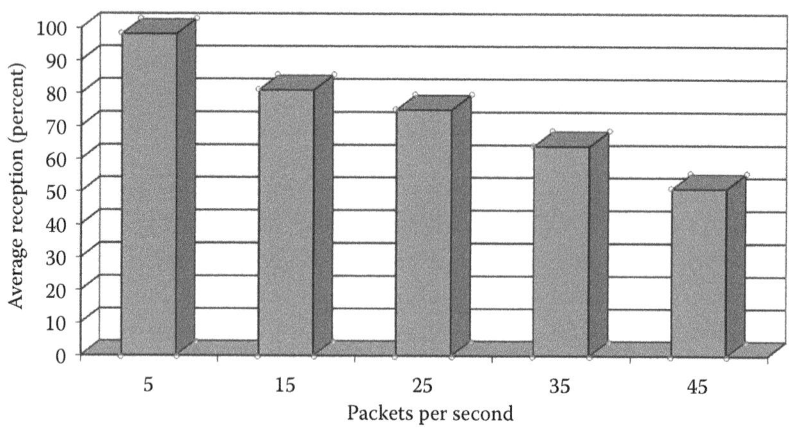

Figure 17.8 Reception ratio ~ sending rate.

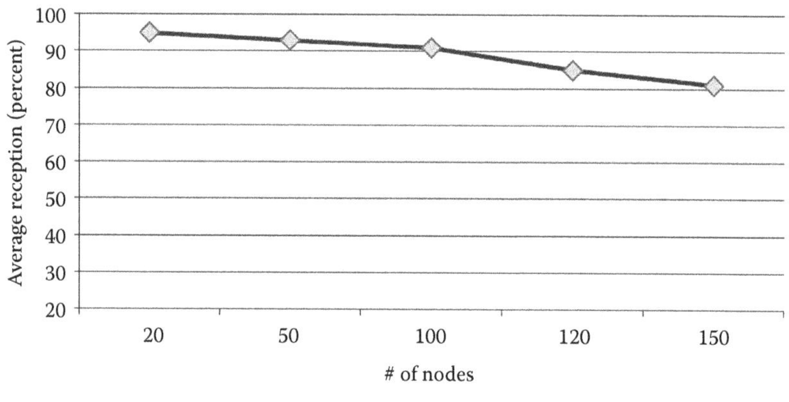

Figure 17.9 Reception ratio ~ number of nodes.

ratio >80 percent) even if there are a large number of patients (see Figure 17.9). It indicates that our MASN will be suitable to a large nursing home.

Mobility: We have tested the MASN delay performance under users' mobility behaviors. Currently, our system cannot achieve real-time data collection (delay >10 s) if the users move quickly (such as at 30 MPH) (see Figure 17.10). This is a future research topic.

Delay: We define aggregated packet delay as the time taken for the first aggregated event packet to reach the sink from the time an event is detected by the sensor nodes. This parameter represents the speed of reaction of the network to the event occurrence. In the proposed as well as Hybrid Energy-Efficient Distributed Clustering (HEED) [OYounis04] schemes, we consider that clusters are formed *on the fly* when the event occurs. In our experimental results

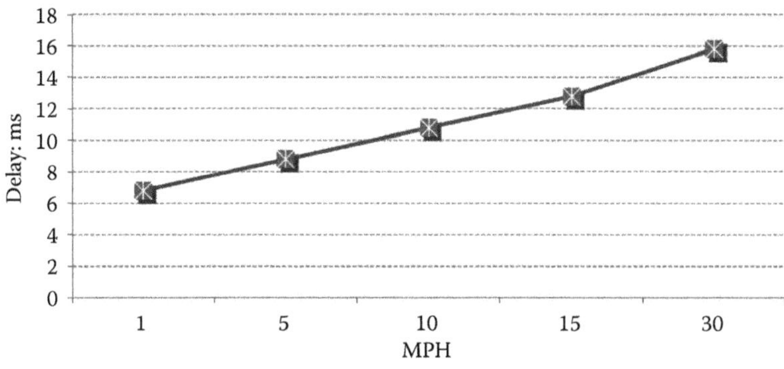

Figure 17.10 End-to-end delay ~ mobility.

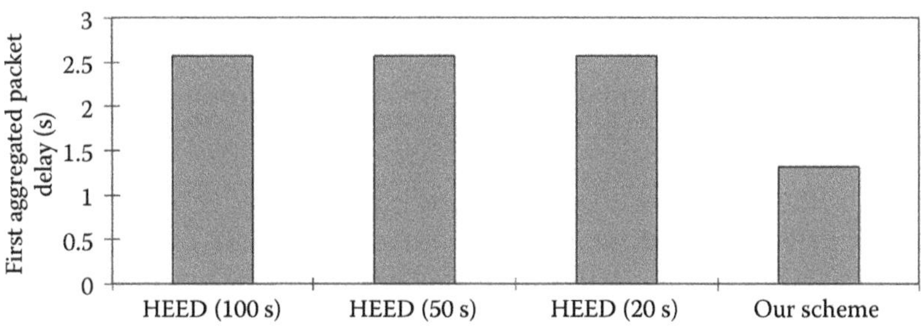

Figure 17.11 The first aggregated packet delay from the CH to the sink.

(see Figure 17.11), the HEED scheme needs more time for the first aggregate data packet to reach the sink due to the setup phase. In this phase, no packets are reported to the sink and clusters are formed with the help of overhead messages. In the proposed scheme, the event packets are transmitted to the sink even as the clusters are being formed.

17.4 MASN Software Design

17.4.1 ECG Sensor Mote Wireless Communication Software

All of the MASN RF mote control software runs in a special operating system called TinyOS [TinyOS07]. In the medical server that receives all patients' ECG data, we can monitor the entire MASN network topology. If two patients are close enough, a radio link will be shown between them to indicate the possibility of transmitting the ECG data between them (Figure 17.12).

In our software, we can remotely control the ECG sensors' performance parameters (such as ECG detection threshold) through the over-the-air command

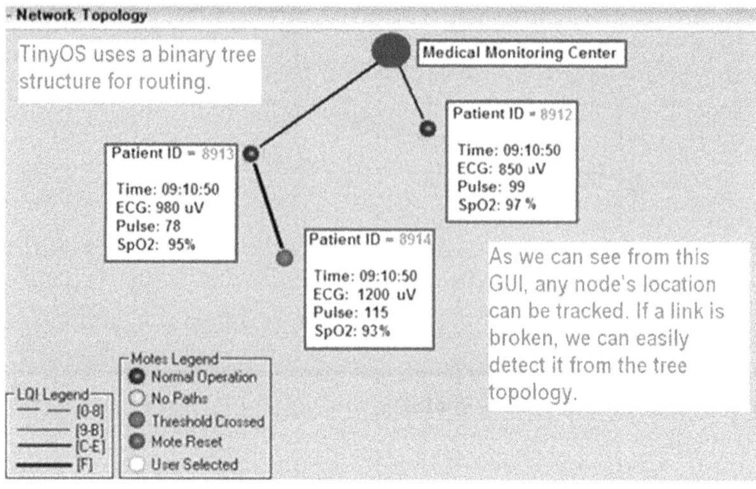

Figure 17.12 Cardiac-monitoring software for a simple nursing home with three cardiac patients. (From Hu, F. et al., *IEEE J. Sel. Areas Commun.*, 27(4), 450, 2009a)

transmission from the server to any ECG sensor. As shown in Figure 17.13, the ECG server (i.e., the MASN workstation) control parameters can be issued to a sensor to change its detection frequency (i.e., how many ECG values we should collect in each second).

VitalDust Plus [CodeBlue06] is used to display the data. It has two modules, the TinyOS software for the mobile platforms to sample and transmit vital sign data over the radio, and a Java GUI application to display the vital signs in a graphical form.

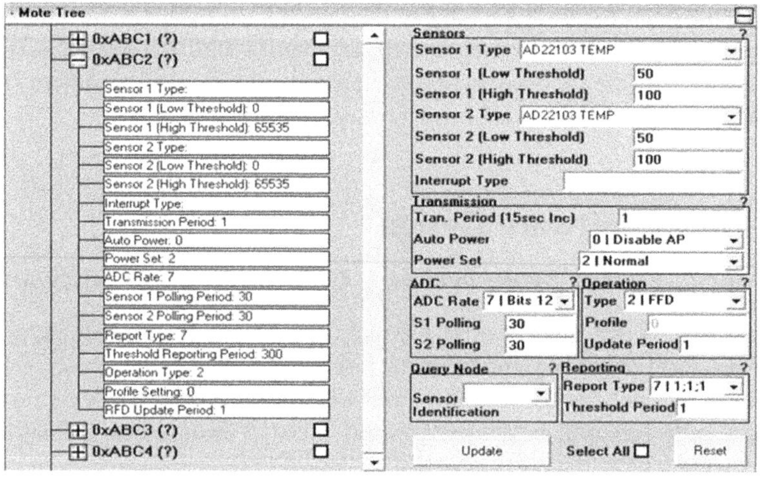

Figure 17.13 Remote control software to adjust ECG sensor parameters.

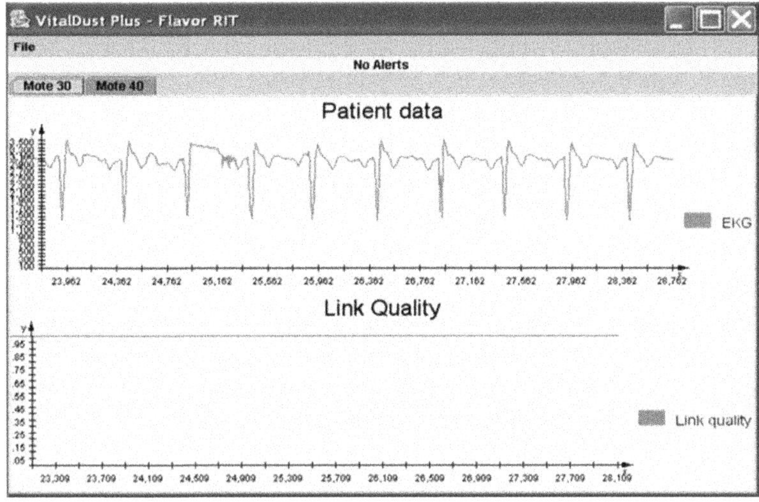

Figure 17.14 Enhanced VitalDust Plus.

Figure 17.14 is a screenshot of our enhanced software. It shows that a server is receiving patient data from two separate mobile platforms: mote30 and mote40. The patient data field is displaying the ECG waveform associated with the selected mobile platform. Only data from the currently selected mobile platform is sent to MATLAB® for signal processing. The link quality field shows the quality of the wireless signal also associated with the selected mobile platform.

17.5 Integration of RFID and Wearable Sensors [Laura07]

An RFID (radio frequency identification) system consists of a reader and some tags. The reader contains an antenna and a transceiver. It reads the tag's information and transfers the information to a processing device. The tag, or transponder, is an integrated circuit containing the RF circuitry and information to be transmitted.

RFID has been used to replace the universal product code (UPC) in supply-chain/object mobility monitoring applications in many organizations, such as Wal-Mart and the Depart of Defense [Wang06]. Industry and tele-healthcare corporations have seen the success and usefulness of RFID and are now beginning to incorporate it into healthcare scenarios to alleviate errors and to cut down costs. For example, a location-based medicare service (LBMS) was implemented in the Taipei Medical University Hospital that used RFID tags to locate both patients and hospital assets with successful results [Wang06]. Exavera's eSheperd has used RFID over a Wi-Fi network to track patients, staff, and supplies, including medication dispensed to patients by the staff [Exavera07]. En-Vision America has created a new way to provide prescription information to the user using RFID with ScripTalk [EnVision07]. When a patient using a ScripTalk reader submits a prescription, the pharmacy software prints and programs an auxiliary smart label using a dedicated, small-footprint printer. The smart label, which stores prescription information, is placed onto the prescription container by the pharmacist. At home, the patient uses a hand-held ScripTalk reader that speaks out the label information using a speech synthesis technology.

Why is the integration of RFID and sensors required? As it can be seen from the above discussions, sensors and RFID have different application scenarios. On the other hand, they represent two complementary technologies, and there would be a big advantage if such two technologies could be merged together. The following lists some benefits of the integration of RFID and wireless sensors.

1. RFID has the capability of tracking patients, which is a good complement to wireless sensors if used in disabled people tracking. Wherever they go, the RFID tags located in different places can tell us whether or not they are in a dangerous situation (for instance, near a road with a sharp slope). RFID can also help them recognize different medicines to be taken.
2. MSN sensors can provide various medical-condition-sensing capabilities that RFID cannot provide. More importantly, these wireless sensors have a CPU, which can run data processing and communication software. While RFID readers do not have such an intelligent processing capability, they can utilize MSN sensors to send out disabled people's tracking information to a control center. Thus, MSN makes RFID achieve remote transmission.
3. RFID is a single-hop wireless system, that is, an RFID reader can only communicate with tags that are one hop away (typically <3 m). Through the integration with sensor networks, the RFIDs can utilize the multi-hop, advanced mesh network protocols in MSNs to handle an arbitrary number of RFID readers and their complex communication issues.
4. RFID is typically a closed system, that is, current commercialized RFID readers do not allow customers to change its internal control software except some simple parameter configurations. By integrating with wireless sensors, the programmers can upload software codes to the sensor memory to indirectly process the RFID tags' data. For instance, a sensor program may store the disabled people's tracking information in a database for patient motion analysis.

In our work, we have successfully integrated wireless medical sensors (EKG/EMG) with RFID readers into one circuit board, and we have also created an integrated software to control RFID and sensor behaviors. Moreover, we ensure that different RFID/sensor boards communicate with each other without collisions.

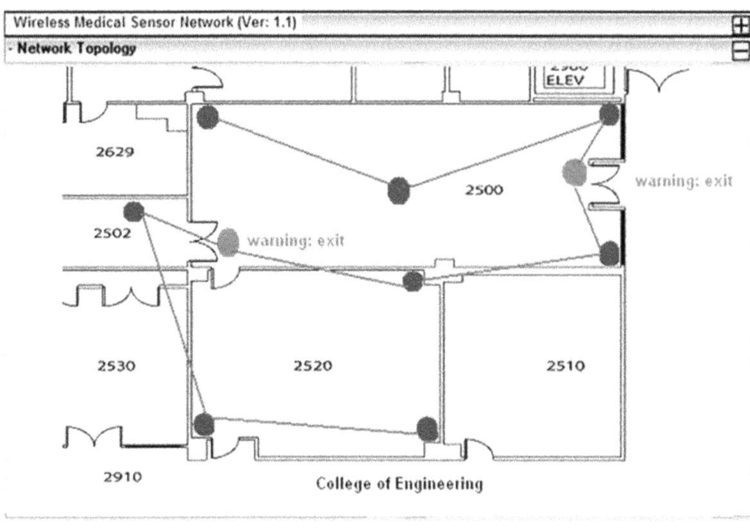

Figure 17.15 RFID for patient tracing.

RFID for road guide: The RFID can be used to keep track of disabled people. If a person gets closer to a place without a disability road, the patient's RFID reader can trigger a warning signal (such as making a sound in the patient's sensor). The system's software automatically draws the trajectory wherever the patient goes (as long as there is a tag in each location). Figure 17.15 shows the result of such a tracing.

RFID for medicine-taking guide: The developed RFID software allows a programmer to fill out all prescription information on an RFID tag, which is destined to be applied on a medication bottle. The tag will contain the patient's name, the name of the prescription, the quantity of medication in the bottle, the dose size, the doses needed per day, and the programmed node (reader) ID, which would be printed on the unit if this system were to be manufactured. A screenshot of the RFID application can be seen in Figure 17.16. The programmer places the RFID tag over the reader, fills in all the fields with the previously mentioned information, and hits the "Write Tag" button. If the programmer would like to check whether all information was appropriately entered, all he or she needs to do is press the "Read Tag" button, and the fields will be filled with the data he or she previously entered. If it is found that a mistake was made after reading back the tag information, the programmer can simply correct the appropriate field and rewrite to the tag. The status box above the buttons informs the programmer whether the read or write has failed or completed successfully.

RFID database: Following the above medicine-taking application, behind the scene, when the "Write Tag" button is pressed, a new entry will be placed in the database with all the information supplied by the programmer, as well as

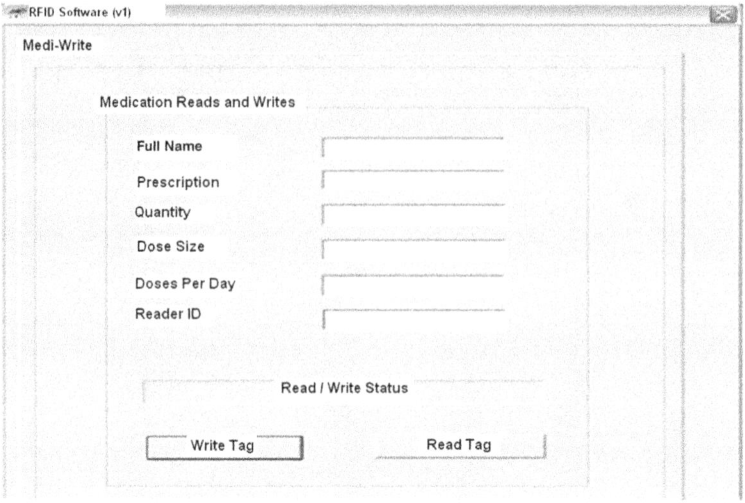

RFID Software (v1)

Medi-Write

Medication Reads and Writes

Full Name

Prescription

Quantity

Dose Size

Doses Per Day

Reader ID

Read / Write Status

Write Tag

Read Tag

Figure 17.16 RFID for patient medicine-taking management.

the RFID tag's ID. The tag ID is stored under the database field named tagID. There is also a field in the database, doseToday, that determines how many doses of that medication were taken for the current day. This is set to 0 when a new entry is added. A screenshot of the current database contents can be seen in Figure 17.17.

In the above database, the field names in the database have the following meaning (see Figure 17.18).

There are several situations in which the patient and healthcare personnel will receive an alert. These include the given medication not being in the database; a patient attempting to take medication that is not his or her own; if the patient is out of pills; and, finally, if the patient has not taken all the required doses of that medication for the day. For each of these errors, a pop-up will be displayed containing a time stamp, the patient who is incorrectly taking the medication, and the reason why the application was incorrect.

tagID	name	rx	QTY	doseSize	doseDay	doseToday	readerID
E005400001132A6C	Mary Woo	Prozac	36	2	2	0	3
E005400001132C43	Linda Doolittle	Zoloft	40	2	2	0	4
E005400001133032	Al Barr	Tylenol	40	2	2	0	2
E005400001132B72	Lou Smith	Aspirin	36	2	2	0	1

Figure 17.17 RFID database screenshot.

Database Field Name	Meaning
Name	Full name
Rx	Prescription
QTY	Quantity
doseSize	Dose size
doseDay	Doses per day
readerID	Reader ID

Figure 17.18 Database field names.

It is possible to determine if a medication is in the database by looking up the RFID tag ID under the tagID field in the database. If it does not exist in the database, it was never entered into the system, and an alert should be sent. To check if a patient is taking medication that is not his or her own, the software system could look up the tag ID in the database.

If this entry does not have a value in the *readerID* field of the database that matches the patient's mote ID, that is, the patient is attempting to take medication that is not his or her own, an alert must be sent. It can easily be determined if a patient is out of pills by checking the QTY field in the database for the corresponding tag ID received. If the value in the QTY is 0, an alert should be sent, so that the prescription may be refilled.

Finally, and, perhaps, most importantly, we should check to ensure that a patient is not about to overdose on his or her medication. The *doseToday* field in the database should be queried for the corresponding tag ID. If the value in this field is equal to the value in the *doseDay* field, the patient should not be taking anymore medication. If an attempt is made, an alert must be sent to keep the patient from overdosing. An example pop-up message for each of the previously described situations can be seen in Figures 17.19 through 17.22.

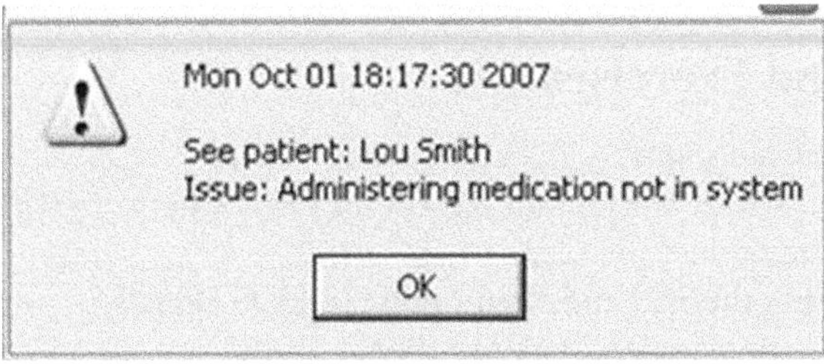

Figure 17.19 The given medication is not in the database.

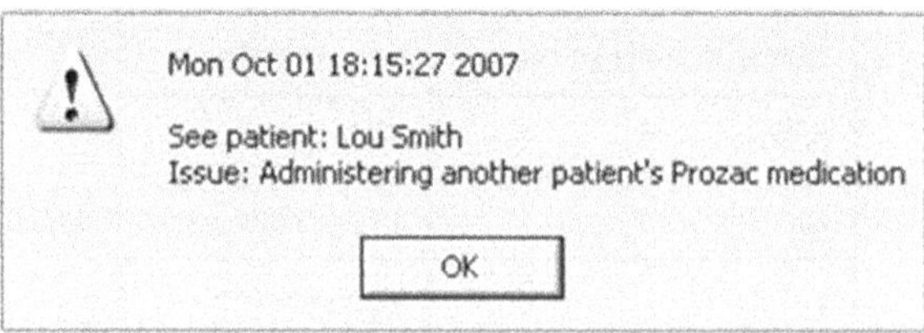

Figure 17.20 Patient attempting to take medication that is not his or hers.

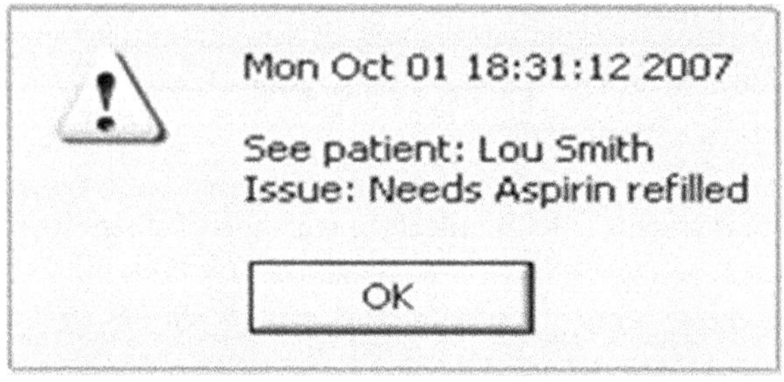

Figure 17.21 Patient is out of pills.

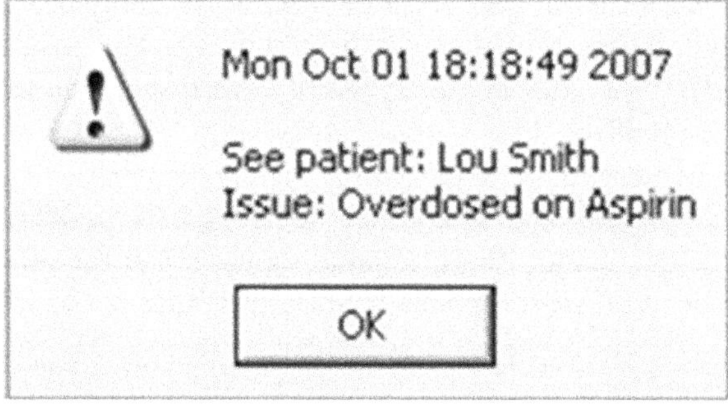

Figure 17.22 Patient is about to overdose.

Problems and Exercises

17.1 Multi-choice questions

1. An ECG sensor without an RF circuit cannot perform which of the following functions?
   a. Filtering thermal and circuit noise in ECG signals
   b. Converting analog signals to digital format
   c. Compressing the ECG stream by removing normal ECG patterns
   d. Selecting the nearest CH to forward the data

2. The ECG data could be damaged during wireless transmission due to some factors. Which of the following factors is not the main concern?
   a. Wireless transmission errors accumulate in each wireless hop.
   b. Radio signals are interfered with by some obstacles.
   c. The relay sensor's CPU processing unintentionally changes some data.
   d. The network attackers can falsify the ECG data.
3. The benefits of designing a customized communication board instead of using Crossbow motes include which of the following?
   a. Lower unit cost
   b. Longer RF communication range
   c. Easy change of software
   d. All of the above
4. The advantages of RFID (compared to general product bar codes) include which of the following?
   a. Longer product code reading distance
   b. Richer product information can be read
   c. Both a and b
   d. Possibility of running wireless network protocols
5. In the discussed work, the patient's ECG signals and medicine information could be read simultaneously in the same network packet because of which of the following?
   a. The integration of the RFID reader and the RF mote on the same board
   b. The sensed data from both the ECG sensor and the medicine sensor
   c. The voice signals from the patient on the medicine name
   d. None of the above

17.2 Some telemedicine systems use cell phones to send out medical data. Some others use a wireless LAN deployed in the building to send out data. Compared to these two approaches, what are the advantages of the sensor network–based telemedicine system? (*Hint*: Try to think of some scenarios where the traditional approaches do not work well.)

17.3 Conduct some Web research to summarize RFID systems' operation principles, design challenges, and application examples.

17.4 Some companies have designed special RFID products to monitor medicine-taking procedures. For instance, RFID tags may be put near each pill to detect the medicine dose. Please conduct some Web research to list a few examples of such applications.

17.5 In this chapter, we have proposed an enhanced cluster-based sensor network routing scheme. What advantages does it have compared to the LEACH scheme?

17.6 Besides ECG sensors, other medical sensors are also under research and development now. Can you conduct some independent research on diabetes patients by monitoring through glucose sensors and insulin pumps and summarize their operation principles?

Chapter 18

Case Study 2: Light Control

In this chapter, we introduce an interesting WSN application: light control. This chapter is based on the excellent work described in [Hamin06, Heemin07].

18.1 Introduction

It would be useful to collect the live light information from light sensors. Real-time data accounts for how characteristics (such as light intensity) change due to filament aging, supply voltage variation, changes in fixture position, color filters, etc. It is important to perform real-time measurement of light as it will take much time and effort to maintain desired intensities of lights for certain areas across many venues and over long time periods. Although we can measure light intensities through the currently available handheld manual light meters [Sek, Kon], these devices do not support *automatic* light control. They must be *manually* moved through different points in space. Cameras can provide only *reflected* light intensity, so it is important to study *incident* light to have measurements that are independent of surfaces and materials.

An intelligent light control system, called the *Illuminator*, has been developed in [Hamin06]. It can detect and control the best light actuation profiles using incident light measurements by light sensors and user requirements. The Illuminator can help media production staff to characterize, control, and set up lights in performance and filmmaking using WSN technologies. The Illuminator has three tasks (given a light setup and user constraints): (1) recommend the optimized sensor deployment, (2) collect the lights' characteristics, and (3) manage the best light actuation profiles satisfying user constraints. These constraints represent requirements about the aesthetic effect of lighting and include desired light intensities of the field or a high-level description of lighting conditions.

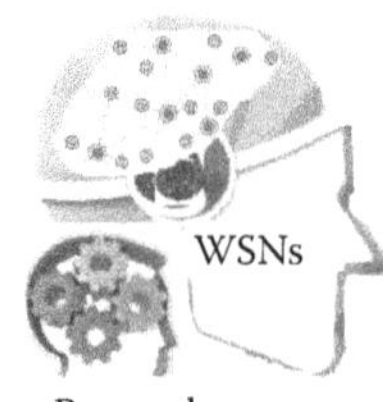

Although this book provides different WSNs applications, all systems have very similar networking issues such as sensor deployment, topology control, routing protocols, and congestion reduction. However, those systems have very different "analog sensor" design and corresponding sensor data analysis software. Therefore, you may put more learning focus on the specific analog sensor hardware part and its interface to an RF board.

This system of Heemin [Hamin06] (we call it *Heemin* in this chapter) assumes that lights have a fixed position over the time of Illuminator's control; but Heemin does not require knowledge of characteristics and locations of lights. Tracking and spotlighting using pan-tilt mounts is a well-known technology and can be implemented easily [Spo]. Heemin does allow *mobile* stage elements, equipment, and actors lit through these fixed lighting instruments, using *mobile tags*. Tag is a single entity that can sense light intensity and know its location.

To generate desired light levels at specific locations, we need to know the projection pattern of lights and brightness according to dimmer level. We call such information the *light characteristics*, and call the process of capturing this information as *light characterization*. The characterization process is done by turning on each light one by one at each dimmer level and by measuring the incident light intensities using wireless light sensors.

The Illuminator system could detect a light level with the best usage based on the user's requirements as well as the data from the light that has already been found. The Illuminator system also reconstructs other similar lighting effects in a different kind of physical light. To obtain this reconstruction, it requires a re-characterization of each of the lights within the current setup. An example is when the same performance needs to be done in different places or at different times. If the setup differs in any way, the setup will vary even though the film crew will attempt to set up the system of lights in the same way as before. Heemin uses Illuminator to document the results of the setup of the lighting (not just the physical setup and assignment of the equipment).

Figure 18.1 shows typical usage scenario of the Heemin Illuminator system. Based on user constraints and available light sensors, the Illuminator system recommends sensor placement. Then, a user deploys sensors based on the Illuminator's recommendation. The Illuminator automatically characterizes lights using deployed

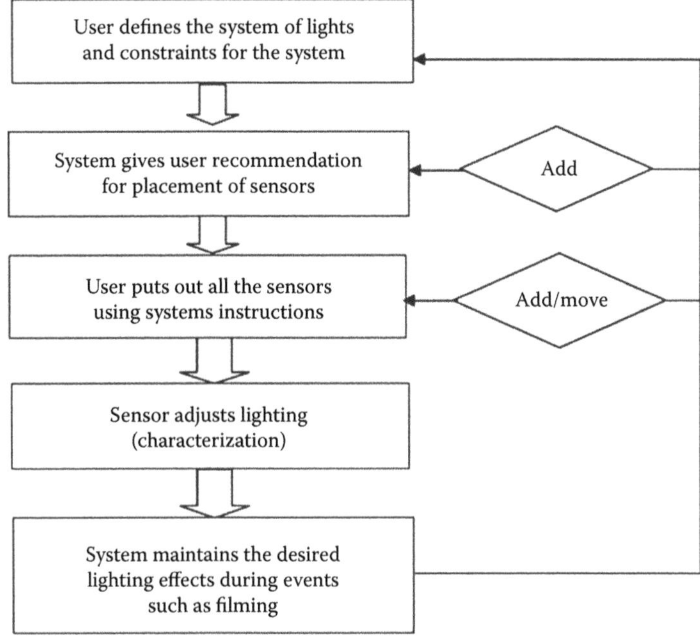

Figure 18.1 Usage scenario of the Illuminator system. (Adapted from Park, H., Design and implementation of a wireless sensor network for intelligent light control, PhD dissertation, Department of Electrical Engineering, UCLA, Los Angeles, CA, 2006.)

sensors. Once the light characterization process is done, light sensors can be removed from the set except for ones that are used for consistent illumination or tracking. In the rehearsal process, the Illuminator controls lights by online light actuation profile generation. A user may want to improve lighting design as rehearsal iterates. Improvement can be done by changing user constraints, adding more sensors for light characterization, and adding or moving light sensors for better illumination results. For example, if a user found that the characterization is insufficient for some area because of obstacles, he or she would want to deploy more sensors at the area of the obstacles.

WSNs can help with continuity management of lighting. The order of a film's sequence of events that an audience views is very different from the order in which they are produced. Film shots are generated based on the order that minimizes cost and makes best use of actors, crew, and locations. Note that the footage captured in different times should appear without difference when they are shown in a consecutive way, or, the system must be able to control the differences for creative show purposes. Therefore, we need to monitor and replicate the quality of light (illuminance and color) in each shot, so that footage captured at different times/locations does not show unexpected differences, which may not be perceived by the human eye but could affect the film stock. Heemin [Hamin06] has focused on lighting instrumentation as the first component of our *advanced technology for cinematography* (*ATC*) [WMB02].

Case study

Heemin [Hamin06] provides an interesting example on the importance of automatic light control: in the *Lord of the Rings* trilogy, the footage captured in three different movies has vastly different release dates and schedules. Although this movie staff comprised over 2400 people, maintaining continuity was remarkably difficult as notes had to be taken by hand and conditions were constantly changing. Therefore, continuity management is required for props, scenery, actors, and camera information, as well as lighting, though the term is typically applied to the management of nontechnical elements.

The Illumimote supports three different light sensing modalities: incident light intensity, color intensities, and incident light angle (the angle of ray arrival from the strongest source). It should also support two situational sensing modalities: attitude and temperature. The Illumimote has comparable performance to commercial light meters, and also meets the size and energy constraints imposed by its application in WSNs.

Design criteria for the Illumimote include the following capabilities: (1) *light intensity and color temperature sensing,* (2) *robustness against infrared energy,* (3) *wide dynamic range,* (4) *fast response time,* and (5) *high accuracy.* Illumimote is compatible with the Mica mote from Crossbow, a common platform in wireless sensor network research and development.

18.2 Illumimote's Sensors

The Illumimote's data acquisition is based on the three basic attributes of illumination: Signal strength (intensity), frequency (color), and transmission vector (incident light angle and sensor attitude). The Illumimote includes the following sensors: (1) *Incident light intensity sensor*: It detects incident light intensity with the precision of a commercial light meter (such as ekonic L558Cine [Sek] light meter). (2) *Color intensity sensors*: They detect red, green, and blue colors to calculate color temperature [WS82]. (3) *Incident light angle sensors*: They are to determine the angle to the strongest incident light source. (4) *Situational sensors*: Some additional sensors are included onboard to provide richer proprioceptive information [BFM06]. They include a gravity-based attitude sensor (accelerometer) for earth-plane relative transformation in the event that the sensor is not oriented parallel to the ground. A temperature sensor is used to detect overheating conditions that might occur under high intensity lighting.

18.3 System Architecture

The overall Heemin Illumimote system architecture diagram is shown in Figure 18.2. It only shows one light sensor channel out of the eight channels. The number of allocated light sensor channels depends on the number of detector circuits required to capture the illumination attribute. For example, the color temperature unit requires three channels for red, green, and blue luminosity. Signals from the eight light acquisition units and four situational units are multiplexed via the channel selection unit, and are sent to the ADC for conversion into a 10-bit digital signal. Its output data is conveyed to the sensor motes (Heemin used MicaZ motes) via either the I2C data bus or a direct 16550A compatible UART link that uses line-level (rail-to-rail) output. The operation of the Illumimote's units can be controlled directly from the mote via the I2C bus or locally by an onboard Atmel Atmega48 microprocessor.

18.4 Calibration

To convert the digitized sensor values to light intensity (lux), Heemin used linear transformation with two coefficients (i.e., $y = ax + b$, where y is the converted lux value, x is the ADC reading, and a and b are calibration coefficients). Specifically, Heemin used three steps to find the optimal coefficients:

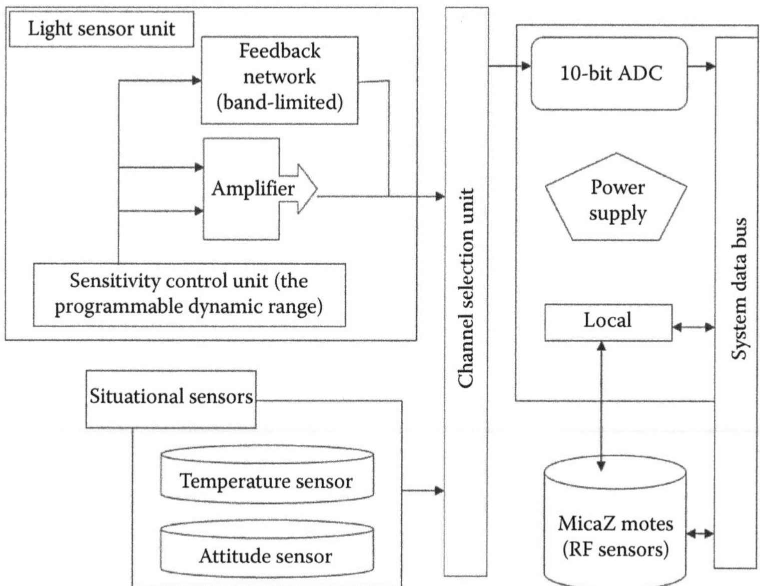

Figure 18.2 Architecture of the Illumimote. (Adapted from Park, H., Design and implementation of a wireless sensor network for intelligent light control, PhD dissertation, Department of Electrical Engineering, UCLA, Los Angeles, CA, 2006.)

Step 1: Plot the Illumimote's ADC readings with respect to reference lux values measured by a commercial light meter on 2D plane.
Step 2: Use MATLAB's® polyfit command to find a linear line (i.e., $y = a'x + b'$) that best represents the plot of the ADC values.
Step 3: We then get the calibration coefficients a and b by projecting the linear line ($y = a'x + b'$) onto $y = 18$. The projection is done by $a = 1/a'$ and $b = -b'$. We then use the collected ADC output values and calibrated a and b for Illumimote's six sensitivity settings.

Good idea

MATLAB Polyfit function has been used in many linear regression and function interpolation calculations. Its basic idea is to come up with a polynomial function to fit a series of experimental data points.

Color temperature of a light source is the black-body radiator's temperature in Kelvin that matches the hue of the light source [WS82]. However, as many light sources (except incandescent light) do not emit radiation like a black body, Heemin instead used *correlated color temperature* (CCT) to represent the color temperature of the light source. Color temperature calibration can be achieved by setting the factors that convert the Red, Green, Blue (RGB) raw readouts into RGB relative light intensities.

18.5 System Evaluation

To evaluate the Illumimote performance, Heemin integrated a wireless sensing system with the Illumimote. Their experimental setup is shown in Figure 18.3. To set up a light source, they used a tungsten-balanced incandescent lamp that generates a color

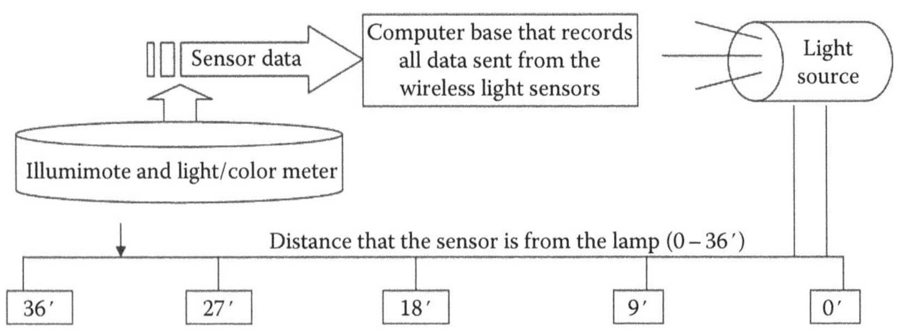

Figure 18.3 Experimental system setup. (Adapted from Park, H., Design and implementation of a wireless sensor network for intelligent light control, PhD dissertation, Department of Electrical Engineering, UCLA, Los Angeles, CA, 2006.)

temperature near 3200 K and can provide about 3000 lux brightness at a distance of 6 ft. It is a common light source in film sets, and has a well defined and very specific color temperature. To generate diverse brightness, they placed the Illumimote at 11 different points from 6 through 36 ft away from the light source in 3 ft each step.

Heemin developed three embedded software components for the experimental wireless sensing system. First, the sensitivity control software was downloaded to the Illumimote board. Second, Heemin has built programs in MicaZ mote for Illumimote driver and light-sensing applications. It uses SOS that is an OS for mote-class wireless sensor networks developed by NESL at UCLA [HKS05]. Finally, in the laptop (base station), a Java program was used to monitor and log the light measurements, and a visualization interface was used for real-time debugging and analysis. A Graphical User Interface (GUI) visualization interface is shown in Figure 18.4 to display the status of the Illumimote in real time. This GUI makes it easy to test and evaluate the Illumimotes visually and is a step toward designing the interface that could be used by a cinematographer.

The entire Illuminator system can be divided into three subsystems: sensor network, Illuminator core, and DMX controller and dimmer. Figure 18.5 shows the overall system connection of the Illuminator light control system.

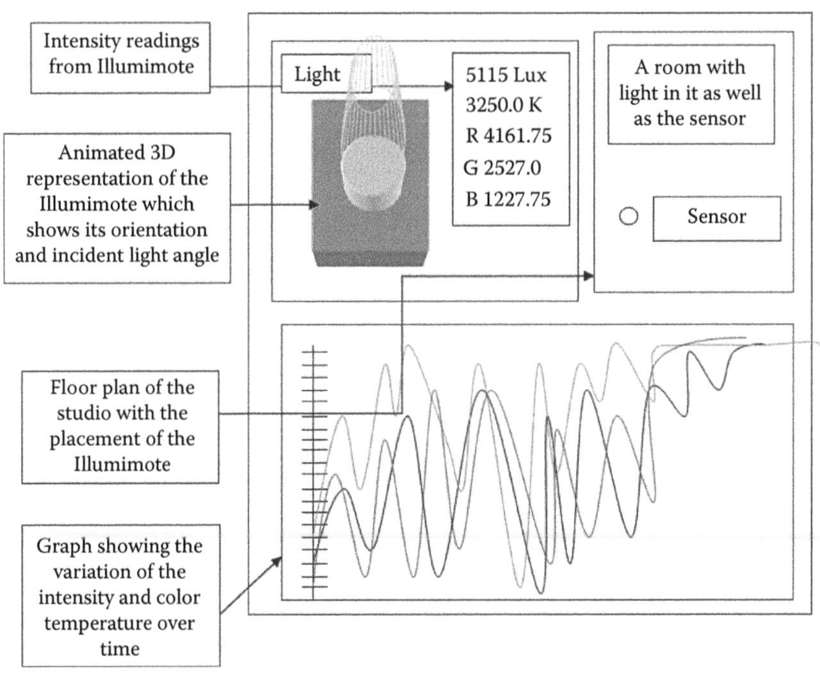

Figure 18.4 Screenshot of the real-time visualization interface (top) with a simplified version of interface (bottom). (Adapted from Park, H., Design and implementation of a wireless sensor network for intelligent light control, PhD dissertation, Department of Electrical Engineering, UCLA, Los Angeles, CA, 2006.)

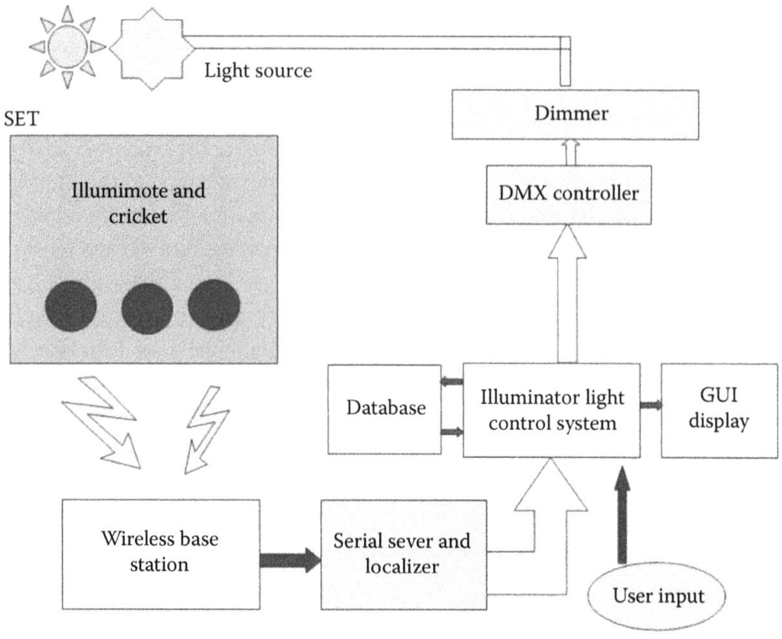

Figure 18.5 Illuminator light control system. (Adapted from Park, H., Design and implementation of a wireless sensor network for intelligent light control, PhD dissertation, Department of Electrical Engineering, UCLA, Los Angeles, CA, 2006.)

Heemin used a sensor network to measure the light intensities and sensor locations. It further consists of two subnetworks: one is the Cricket localization system and the other is single-hop MicaZ network with the Illumimote light-sensing board [PFG06]. Three Cricket [Priyantha05] nodes were used for beacon nodes that are precalibrated with their locations, and a Cricket was coupled with each Illumimote for localizing the light sensor module. To manage two sensor network platforms, two Java modules run concurrently: SerialServer for interfacing between the Illuminator core and sensor networks and Localizer for computing positions of the Cricket nodes based on the ultrasound range measurements.

Problems and Exercises

18.1 Multi-choice questions

1. The significance of this proposed light control sensor network includes
   a. Real-time data accounts for how characteristics like light intensity and color temperature change over time and deployments due to filament aging, supply voltage variation, changes in fixture position, color filters, etc.

   b. Through real-time measurement of light, we do not need to maintain desired intensities of lights for certain area across many venues and over long time periods.
   c. Current handheld manual light meters have not been incorporated in systems supporting automatic light control and must be manually moved through different points in space.
   d. All of the above.
2. Illuminator's roles do not include
   a. Light-to-light communications
   b. Characterize the lights
   c. Manage the best light actuation profiles satisfying user constraints
   d. Recommend sensor deployment
3. WSN is good for light management due to the following facts:
   a. We need to monitor and replicate the quality of light (illuminance and color) in each shot, so that footage captured at different times or in different locations does not show unexpected differences
   b. Maintaining continuity was remarkably difficult as notes had to be taken by hand and conditions were constantly changing
   c. Many large-budget feature films require significant postproduction digital image manipulation prior to release, which is quite expensive
   d. All of the above
4. Illuminator system could be useful for the following case(s):
   a. Entertainment and media production
   b. Achieving required illumination profiles in outdoor public venues
   c. Acoustic sensing in underwater imaging
   d. Both a and b
5. Illumimote's data acquisition capabilities include
   a. Signal strength (intensity)
   b. Frequency (color)
   c. Transmission vector (incident light angle and sensor attitude)
   d. All of the above

18.2 Why should we use sensor networks for light control?

18.3 How many capabilities does a light sensor have?

18.4 Explain each module of Illumimote's architecture.

18.5 Explain the principle of color temperature calibration.

References

[Abarroso05] A. Barroso, U. Roedig, and C. Sreenan, μ-MAC: An energy-efficient medium access control for wireless sensor networks, in *Proceedings of the Second IEEE European Workshop on Wireless Sensor Networks*, Istanbul, Turkey, January 2005, pp. 70–80.

[Abolhasan04] M. Abolhasan, T. Wysocki, and E. Dutkiewicz, A review of routing protocols for mobile ad hoc networks, *Ad Hoc Networks* (*Elsevier*), 2, 1–22, January 2004.

[Achandra00] A. Chandra, V. Gummalla, and J.O. Limb, Wireless medium access control protocols, *IEEE Surveys and Tutorials*, 3(2), 2–15, Second Quarter, 2000.

[ADoucet01] A. Doucet, N. Freitas, and N. Gordon, *Sequential Monte Carlo Methods in Practice*. Springer-Verlag, New York, 2001.

[Aelhoiydi04] A. El-Hoiydi and J.-D. Decotignie, WiseMAC: An ultra low power MAC protocol for the downlink of infrastructure wireless sensor networks, *IEEE Computers and Communications*, 1, 244–251, July 2004.

[Akan05] Ö.B. Akan and I.F. Akyildiz, Event-to-sink reliable transport in wireless sensor networks, *IEEE/ACM Transactions on Networking*, 13(5), 1003–1016, October 2005.

[Akcan06] H. Akcan, V. Kriakov, H. Brönnimann, and A. Delis, GPS-Free node localization in mobile wireless sensor networks, in *Proceedings of the Fifth ACM International Workshop on Data Engineering for Wireless and Mobile Access* (*MobiDE '06*), Chicago, IL, June 25, 2006, ACM, New York, pp. 35–42.

[Akyildiz02] I.F. Akyildiz, W. Su, Y. Sankarasubramaniam, and E. Cayirci, Wireless sensor networks: A survey, *Computer Networks* (*Elsevier*), 38, 393–422, March 2002.

[Akyildiz04] I.F. Akyildiz and I.H. Kasimoglu, Wireless sensor and actor networks: Research challenges, *Ad Hoc Networks* (*Elsevier*), 2, 351–367, October 2004.

[Akyildiz04a] I.F. Akyildiz, D. Pompili, and T. Melodia, Challenges for efficient communication in underwater acoustic sensor networks, *SIGBED Review*, 1(2), 3–8, July 2004.

[Akyildiz07] I.F. Akyildiz, T. Melodia, and K.R. Chowdhury, A survey on wireless multimedia sensor networks, *Computer Networks* (*Elsevier*), 51(4), 921–960, March 2007.

[Ahuja93] R.K. Ahuja, T.L. Magnanti, and J.B. Orlin, *Network Flows: Theory, Algorithms, and Applications*. Prentice Hall, Englewood Cliffs, NJ, February 1993.

[AManjeshwar01] A. Manjeshwar and D.P. Agrawal, TEEN: A routing protocol for enhanced efficiency in wireless sensor networks, in *Proceedings of the 15th IEEE International Parallel and Distributed Processing Symposium*, San Francisco, CA, April 2001, pp. 2009–2015.

[Anderson02] A. Mainwaring, J. Polastre, R. Szewezyk, D. Culler, and J. Anderson, Wireless sensor networks for habitat monitoring, in *ACM International Workshop on Wireless Sensor Networks and Applications* (*WSNA '02*), Atlanta, GA, September 2002.

[AMD03] AMD, AM49DL640BG Stacked Multi-Chip Package (MCP) Flash Memory and SRAM. 2003: http://www.amd.com/usen/assets/content_type/white_papers_and_tech_docs/26090a.pdf

[APerrig00] A. Perrig, R. Canetti, J. Tygar, and D. Song, Efficient authentication and signing of multicast streams over lossy channels, in *IEEE Symposium on Security and Privacy*, Oakland, CA, 2000.

[APerrig01] A. Perrig et al., SPINS: Security protocols for sensor networks, in *Proceedings of ACM MOBICOM*, Rome, Italy, 2001.

[APerrig02] A. Perrig, R. Canetti, J.D. Tygar, and D. Song, The TESLA broadcast authentication protocol, *CryptoBytes*, 5(2), 2–13, Summer/Fall 2002.

[ASavkin03] A. Savkin, P. Pathirana, and F. Faruqi, The problem of precision missile guidance: LQR and H1 control frameworks, *IEEE Transactions on Aerospace and Electronic Systems*, 39(3), 901–910, July 2003.

[ASavvides01] A. Savvides, C.-C. Han, and M. Srivastava, Dynamic fine-grained localization in ad-hoc networks of sensors, in *Proceedings of the Seventh ACM International Conference on Mobile Computing and Networking* (*Mobicom*), Rome, Italy, July 2001, ACM, New York, pp. 166–179.

[ASPNES03] J. Aspnes and G. Shah, Skip graphs, in *Fourteenth Annual ACM-SIAM Symposium on Discrete Algorithms*, Baltimore, MD, January 12–14, 2003, pp. 384–393.

[ASrinivasan06] A. Srinivasan, J. Teitelbaum, and J. Wu, DRBTS: Distributed reputation-based beacon trust system, in *Second IEEE International Symposium on Dependable, Autonomic and Secure Computing* (*DASC '06*), Indianapolis, IN, 2006, pp. 277–283.

[ASrinivasan08] A. Srinivasan and J. Wu, A survey on secure localization in wireless sensor networks, in *Encyclopedia of Wireless and Mobile Communications*, B. Furht, Ed., CRC Press/Taylor & Francis Group, Boca Raton, FL, 2008 (accepted for publication).

[Atmel01] Atmel Corporation, Atmega103(L) Datasheet. 2001, Atmel Corporation: http://www.atmel.com/atmel/acrobat/doc0945.pdf

[Atmel08] Atmel Corporation, http://www.atmel.com. 2008.

[AWoo01] A. Woo and D. Culler, A transmission control scheme for media access in sensor networks, in *Proceedings of the Seventh Annual International Conference on Mobile Computing and Networking* (*MOBICOM '01*), Rome, Italy, July 2001, pp. 221–235.

[AWood02] A. Wood and J. Stankovic, Denial of service in sensor networks, *IEEE Computer*, 35(10), 54–62, October 2002.

[AWood03] A. Wood, J. Stankovic, and S.H. Son, Jam: A jammed-area mapping service for sensor networks, in *Real-Time Systems Symposium*, Cancun, Mexico, 2003.

[Bao01] L. Bao and J.J. Garcia-Luna-Aceves, A new approach to channel access scheduling for ad hoc networks, in *Seventh Annual International Conference on Mobile Computing and Networking*, Rome, Italy, 2001, pp. 210–221.

[Bchen02] B. Chen, K. Jamieson, H. Balakrishnan, and R. Morris, Span: An energy-efficient coordination algorithm for topology maintenance in ad hoc wireless networks, *ACM Wireless Networks*, 8(5), 481–494, September 2002.

[Bdavid02] D. Braginsky, Estrin rumor routing algorithm for sensor networks, in *Proceedings of the First ACM International Workshop on Wireless Sensor Networks and Applications*, Atlanta, GA, September 2002, ACM, New York, pp. 22–31.

[BFM06] J. Burke, J. Friedman, E. Mendelowitz, H. Park, and M.B. Srivastava, Embedding expression: Pervasive computing architecture for art and entertainment, *Journal of Pervasive and Mobile Computing*, 2(1), 1–36, February 2006.

[Bharghavan93] V. Bharghavan, A. Demers, S. Shenker, and L. Zhang, MACAW: A media access protocol for wireless LAN's, in *Proceedings of ACM SIGCOMM Conference (SIGCOMM '94)*, London, U.K., August 1994, pp. 212–225.

[BHW97] B.H. Wellenhoff, H. Lichtenegger, and J. Collins, *Global Positions System: Theory and Practice*. Springer-Verlag, New York, 1997.

[Bkrap00] B. Krap and H.T. Kung, GPSR: Greedy perimeter stateless routing for wireless networks, in *Proceedings of MobiCom 2000*, Boston, MA, August 2000, pp. 243–254.

[BKusy07] B. Kusy, G. Balogh, A. Ledeczi, and M.M.J. Sallai, in-Track: High precision tracking of mobile sensor nodes, in *Fourth European Workshop on Wireless Sensor Networks (EWSN '07)*, Delft, the Netherlands, January 2007.

[Blom85] R. Blom, An optimal class of symmetric key generation systems, in *Advances in Cryptology: Proceedings of EUROCRYPT '84*, T. Beth, N. Cot, and I. Ingemarsson, Eds. Lecture Notes in Computer Science, Vol. 209, pp. 335–338, Springer-Verlag, Berlin, Germany, 1985.

[Blundo93] C. Blundo, A.D. Santis, A. Herzberg, S. Kutten, U. Vaccaro, and M. Yung, Perfectly secure key distribution for dynamic conferences, in *Advances in Cryptology—CRYPTO '92*, E. Brickell, Ed., Lecture Notes in Computer Science, Vol. 740, pp. 471–486, Springer-Verlag, Berlin, Germany, 1993.

[Bulusu00] N. Bulusu, J. Heidemann, and D. Estrin, GPS-less low cost outdoor localization for very small devices, *IEEE Personal Communications Magazine*, 7(5), 28–34, October 2000.

[Bulusu05] N. Bulusu, C. Chou, W. Hu, S. Jha, A. Taylor, and V. Tran, The design and evaluation of a hybrid sensor network for cane-toad monitoring, in *Proceedings of Information Processing in Sensor Networks*, Los Angeles, CA, April 2005.

[BWarneke01] B. Warneke, M. Last, B. Liebowitz, and K.S.J. Pister, Smart dust: Communicating with a cubic-millimeter computer, *IEEE Computer*, 34(1), 44–51, 2001.

[BYCHKOVSKIY03] V. Bychkovskiy, S. Megerian, D. Estrin, and M. Potkonjak, A collaborative approach to in-place sensor calibration, in *Proceedings of IPSN '03*, Palo Alto, CA, 2003.

[CardioNet08] CardioNet Inc. has developed an integrated technology and service–mobile cardiac outpatient telemetry (MCOT)—Which enables heartbeat-by-heartbeat, ECG monitoring, analysis and response, at home or away, 24/7/365. On CardioNet project details, please see: http://www.cardionet.com/

[Carlos04] C. Pomalaza-Ráez, Wireless sensor networks energy efficiency issues, (Lecture notes), Fall 2004, University of Oulu, Oulu, Finland.

[CcEnz04] C.C. Enz, A. El-Hoiydi, J.-D. Decotignie, and V. Peiris, WiseNET: An ultralowpower wireless sensor network solution, *IEEE Journal*, 37(8), 62–70, August 2004.

[CERPA01] A. Cerpa, J. Elson, D. Estrin, L. Girod, M. Hamilton, and J. Zhao, Habitat monitoring: Application driver for wireless communications technology, in *Proceedings of ACM SIGCOMM Workshop on Data Communications in Latin America and the Caribbean*, San Jose, Costa Rica, 2001.

[Chi06] C. Ma and Y. Yang, Battery-aware routing for streaming data transmissions in wireless sensor networks, *Mobile Networks and Applications*, 11, 757–767, 2006.

[Chieh-Yih05] C.-Y. Wan, A.T. Campbell, Member, IEEE, and L. Krishnamurthy, Pump-slowly, fetch-quickly (PSFQ): A reliable transport protocol for sensor networks, *IEEE Journal on Selected Areas in Communications*, 23(4), 862–872, April 2005.

[Chipcon08] On the Chipcon Inc. RF transceiver products, please see http://www.chipcon.com, Visited in June 2008.

[CIntanagonwiwat00] C. Intanagonwiwat, R. Govindan, and D. Estrin, Directed diffusion: A scalable and robust communication paradigm for sensor networks, in *Proceedings of the Sixth Annual International Conference on Mobile Computing and Networking (MobiCOM '00)*, Boston, MA, August 2000, ACM Press, New York, pp. 56–67.

[CKarlof03] C. Karlof and D. Wagner, Secure routing in sensor networks: Attacks and countermeasures, *Ad Hoc Networks*, Special issue on *Sensor Network Applications and Protocols (Elsevier)*, 1(2–3), 293–315, September 2003.

[Chang04] J.-H. Chang and L. Tassiulas, Maximum lifetime routing in wireless sensor networks, *IEEE/ACM Transactions on Networking*, 12(4), 609–619, August, 2004.

[Chehri06] A. Chehri, P. Fortier, and P.-M. Tardif, Application of ad-hoc sensor networks for localization in underground mines, in *Proceedings of the IEEE Annual Wireless and Microwave Technology Conference (WAMICON '06)*, Melbourne, FL, December 4–5, 2006, pp. 1–4.

[CMUcam08] The CMUcam2. http://www-2.cs.cmu.edu/ cmucam/cmucam2/index.html

[CodeBlue06] M. Welsh and B. Chen, CodeBlue: Wireless sensor networks for medical care, Division of Engineering and Applied Sciences, Harvard University, Cambridge, MA, 2006.

[CORMEN01] T.H. Cormen, C.E. Leiserson, R.L. Rivest, and C. Stein, *Introduction to Algorithms*, 2nd edn. The MIT Press, Cambridge, MA, 2001.

[CPERKINS00] C. Perkins, *Ad Hoc Networks*. Addison-Wesley, Reading, MA, 2000.

[CSavarese02] C. Savarese, Robust positioning algorithms for distributed ad hoc wireless sensor networks, Master's thesis, University of California at Berkeley, Berkeley, CA, 2002.

[Cschurgers01] C. Schurgers and M.B. Srivastava, Energy efficient routing in wireless sensor networks, in *Proceedings of IEEE MILCOM '01*, Vienna, VA, October 2001, Vol. 1, pp. 357–361.

[Crossbow08] On all wireless sensor network products (including motes, sensor boards, gateway, etc.) from Crossbow Inc., please see: http://www.xbow.com, Visited in June 2008.

[CYWan02] C.Y. Wan, A.T. Campbell, and L. Krishnamurthy, PSFQ: A reliable transport protocol for wireless sensor networks, in *Proceedings of the ACM WSNA*, Atlanta, GA, September 2002, pp. 1–11.

[DAI 04] H. Dai, M. Neufeld, and R. Han, ELF: An efficient log-structured flash file system for micro sensor nodes, in *SenSys '04: Proceedings of the Second International Conference on Embedded Networked Sensor Systems*, Baltimore, MD, 2004, ACM Press, New York, pp. 176–187.

[Dallas08] Dallas Semiconductor, DS2401 Silicon Serial Number: http://pdfserv.maximic.com/arpdf/DS2401.pdf

[Ddclark90] D.D. Clark and D.L. Tennenhouse, *Architectural Considerations for a New Generation of Protocols*, 20(4), 200–208, September 1990, ACM.

[DELIN00] K.A. Delin and S.P. Jackson, Sensor web for in situ exploration of gaseous biosignatures, in *Proceedings of the IEEE Aerospace Conference*, Big Sky, MT, 2000.

[DFox99] D. Fox, W. Burgard, F. Dellaert, and S. Thrun, Monte Carlo localization: Efficient position estimation for mobile robots, in *AAAI 1999*, Orlando, FL, 1999, pp. 343–349.

[DLiu05] D. Liu, P. Ning, and W. Du, Detecting malicious Beacon nodes for secure location discovery in wireless sensor networks, in *25th IEEE International Conference on Distributed Computing Systems* (*ICDCS '05*), Columbus, OH, 2005, pp. 609–619.

[DLiu05a] D. Liu, P. Ning, and W. Du, Attack-resistant location estimation in sensor networks, in *Proceedings of the Fourth International Conference on Information Processing in Sensor Networks* (*IPSN '05*), Los Angeles, CA, April 2005, pp. 99–106.

[DLM91] D.L. Mills, Internet time synchronization: The network time protocol, *IEEE Transactions on Communications*, 39(10), 1482–1493, October 1991.

[DLM92] D.L. Mills, Network time protocol (version 3): Specification, implementation, and analysis, Technical Report, Network Information Center, SRI International, Menlo Park, CA, March 1992.

[DOOLIN05] D. Doolin and N. Sitar, Wireless sensors for wildfire monitoring, in *SPIE Symposium on Smart Structures and Materials*, San Diego, CA, March 2005.

[Doyle93] M. Doyle, T.F. Fuller, and J. Newman, Modeling of galvanostatic charge and discharge of the lithium/polymer/insertion cell, *Journal of the Electrochemical Society*, 140(6), 1526–1533, 1993.

[DSchmidt07] D. Schmidt, M. Krämer, T. Kuhn, and N. Wehn, Energy modelling in sensor networks, *Advances in Radio Science*, 5, 347–351, 2007. See http://www.adv-radio-sci.net/5/347/2007/

[DSR] D. Johnson, D. Maltz, and J. Broch, The dynamic source routing protocol for multihop wireless ad hoc networks, in *Ad Hoc Networking*, C. Perkins, Ed., Addison-Wesley, Boston, MA, 2001.

[DSPComm08] DSPComm. available: www.dspcomm.com, Visited in 2008.

[Dulman03] S. Dulman, T. Nieberg, J. Wu, and P. Havinga, Trade-off between traffic overhead and reliability in multipath routing for wireless sensor networks, *IEEE WCNC*, New Orleans, LA, March 2003.

[DuW03] W. Du, J. Deng, Y.S. Han, and P.K. Varshney, A pairwise key pre-distribution scheme for wireless sensor networks, in *Proceedings of the 10th ACM Conference on Computer and Communications Security* (*CCS '03*), Washington DC, October 27–30, 2003, ACM, New York, pp. 42–51.

[DuW05] W. Du, J. Deng, Y.S. Han, P.K. Varshney, J. Katz, and A. Khalili, A pairwise key predistribution scheme for wireless sensor networks, *ACM Transactions on Information and System Security*, 8(2), 228–258, May 2005.

[Eelopez06] E.E. Lopez, J. Vales-Alonso, A.S. Martínez-Sala, J. García-Haro, P. Pavón-Mariño, and M.V.B. Delgado, A wireless sensor networks MAC protocol for real time applications, *Personal and Ubiquitous Computing*, 12(2), 111–122, January, 2008, ACM.

[Elnahrawy2003] E. Elnahrawy and B.R. Badrinath, Cleaning and querying noisy sensors, in *Proceedings of the Second ACM International Conference on Wireless Sensor Networks and Applications*, San Diego, CA, September 19, 2003.

[Ember08] On the RF and CPU chips from Ember Inc., see http://www.ember.com, Visited in June 2008.

[EnVision07] En-Vision America, ScripTalk, http://www.envisionamerica.com/scriptalk/scriptalk.php, downloaded 22 August 2007.

[EShih01] E. Shih et al., Physical layer driven protocol and algorithm design for energy-efficient wireless sensor networks, in *Proceedings of the ACM MOBICOM*, Rome, Italy, July 2001, pp. 272–286.

[ESouto04] E. Souto et al., A message-oriented middleware for sensor networks, in *Proceedings of the Second International Workshop Middleware for Pervasive and Ad-Hoc Computing* (*MPAC '04*), Toronto, Ontorio, Canada, October 2004, ACM Press, New York, pp. 127–134.

[Exavera07] Exavera Technologies, eShepherd overview, http://www.exavera.com/healthcare/eshepherd.php, downloaded 22 August 2007.

[FCristian89] F. Cristian, Probabilistic clock synchronization, *Distributed Computing*, 3, 146–158, 1989.

[Feng05] W. Feng, E. Kaiser, W.C. Feng, and M.L. Baillif, Panoptes: Scalable low-power video sensor networking technologies, *ACM Transactions on Multimedia Computing, Communications, and Applications*, 1(2), 151–167, May 2005.

[Finn87] G. Finn, Routing and addressing problems in large metropolitan-scale internet-works, Technical Report, ISI/RR-87-180, USC/ISI, March 1987.

[Fkuhn03] F. Kuhn, W. Roger, and Z. Aaron, Worst-case optimal and average-case efficient geometric ad-hoc routing, in *Proceedings of International Symposium on Mobile Ad Hoc Networking & Computing*, Annapolis, MD, June 2003.

[Fli06] F. Li, Y. Li, W. Zhao, Q. Chen, and W. Tang, An adaptive coordinated MAC protocol based on dynamic power management for wireless sensor networks, in *Proceedings of the 2006 International Conference on Wireless Communications and Mobile Computing*, Vancouver, British Columbia, Canada, July 2006, ACM, New York, pp. 1073–1078.

[FSimjee06] F. Simjee and P.H. Chou, Everlast: Long-life, supercapacitor-operated wireless sensor node, *ISLPED*, Tegernsee, Germany, 2006.

[Ftobagi75] F. Tobagi and L. Kleinrock, Packet switching in radio channels, Part II: Hidden-terminal problem in carrier sense multiple access and the busy-tone solution, *IEEE Transactions on Communications*, 23(12), 973–977, December 1975.

[FYe02] F. Ye, H. Luo, J. Cheng, S. Lu, and L. Zhang, A two-tier data dissemination model for large-scale wireless sensor networks, in *Eighth Annual International Conference on Mobile Computing and Networking* (*ACM Mobicom '02*), Atlanta, GA, September 2002, ACM Press, New York, pp. 148–159.

[Fye01] F. Ye, S. Lu, and L. Zhang, A scalable solution to minimum cost forwarding in large sensor networks, in *Proceedings of IEEE INFOCOM '01*, Anchorage, AK, April 2001, pp. 304–309.

[GANESAN03] D. Ganesan, B. Greenstein, D. Perelyubskiy, D. Estrin, and J. Heidemann, An evaluation of multi-resolution storage in sensor networks, in *Proceedings of the First ACM Conference on Embedded Networked Sensor Systems* (*SenSys*), Los Angeles, CA, 2003.

[GANESAN03a] D. Ganesan, D. Estrin, and J. Heidemann, Dimensions: Why do we need a new data handling architecture for sensor networks? *SIGCOMM Computer Communication Review*, 33(1), 143–148, January 2003.

[Ganesan01] D. Ganesan, R. Govindan, S. Shenker, and D. Estrin, Highly resilient, energy-efficient multipath routing in wireless sensor networks, *ACM SIGMOBILE Mobile Computing and Communication Review*, 5(4), 11–25, 2001.

[Gay03] D. Gay, P. Levis, R.V. Behren, M. Welsh, E. Brewer, and D. Culler, The nesC language: A holistic approach to networked embedded systems, in *Proceedings of SIGPLAN '03*, 2003.

[GGolub96] G. Golub, *Matrix Computations*. The Johns Hopkins University Press, Baltimore, MD, 1996.

[Girod01] L. Girod and D. Estrin, Robust range estimation using acoustic and multimodal sensing, in *IEEE/RSJ International Conference on Intelligent Robots and Systems* (*IROS 2001*), Maui, HI, October 2001.

[GloMoSim] X. Zeng, R. Bagrodia, and M. Gerla, GloMoSim: A library for parallel simulation of large-scale wireless networks, in *Proceedings of the 12th Workshop on Parallel and Distributed Simulations, PADS '98*, May 26–29, 1998, Banff, Alberta, Canada.

[Glu04] G. Lu, B. Krishnamachari, and C.S. Raghavendra, An adaptive energy-efficient and low-latency MAC for data gathering in wireless sensor networks, in *Proceedings of the IEEE 18th International Parallel and Distributed Processing Symposium*, Santa Fe, NM, April 2004, pp. 224–231.

[Hamin06] H. Park, Design and implementation of a wireless sensor network for intelligent light control, PhD dissertation, Department of Electrical Engineering, UCLA, Los Angeles, CA, 2006. Also see: http://nesl.ee.ucla.edu/fw/documents/journal/2006/Sensors_Illumimote_HeeminPark.pdf

[HanC05] C. Han, R. Kumar, R. Shea, E. Kohler, and M. Srivastava, A dynamic operating system for sensor nodes, in *Proceedings of the Third international Conference on Mobile Systems, Applications, and Services* (*MobiSys '05*), Seattle, WA, June 6–8, 2005, ACM, New York, pp. 163–176.

[Hartung06] C. Hartung, R. Han, C. Seielstad, and S. Holbrook, FireWxNet: A multi-tiered portable wireless system for monitoring weather conditions in wildland fire environments, in *Proceedings of the Fourth International Conference on Mobile Systems, Applications and Services* (*MobiSys '06*), Uppsala, Sweden, June 19–22, 2006, ACM, New York, pp. 28–41.

[HARVEY03] N. Harvey, M.B. Jones, S. Saroiu, M. Theimer, and A. Wolman, Skipnet: A scalable overlay network with practical locality properties, in *Proceedings of the Fourth USENIX Symposium on Internet Technologies and Systems* (*USITS '03*), Seattle, WA, March 2003.

[Hawkins80] D.M. Hawkins, *Identification of Outliers*. Chapman and Hall, New York, 1980.

[Heemin07] H. Park, J. Burke, and M.B. Srivastava, Design and implementation of a wireless sensor network for intelligent light control, in *Proceedings of the Sixth International Conference on Information Processing in Sensor Networks* (*IPSN '07*), Cambridge, MA, April 25–27, 2007, ACM, New York, pp. 370–379.

[Heinzelman02] W. Heinzelman, A. Chandrakasan, and H. Balakrishnan, An application-specific protocol architecture for wireless microsensor networks, *IEEE Transactions on Wireless Communications*, 1, 660–670, October 2002.

[HKS05] C.-C. Han, R. Kumar, R. Shea, E. Kohler, and M. Srivastava, A dynamic operating system for sensor nodes, in *Proceedings of the Third International Conference on Mobile Systems, Applications, and Services* (*MobiSys '05*), Seattle, WA, 2005, ACM Press, New York, pp. 163–176.

[Hojung07] H. Cha et al., Resilient, expandable, and threaded operating system for wireless sensor networks, in *IPSN '07*, Cambridge, MA, April 25–27, 2007.

[HOLMAN03] R. Holman, J. Stanley, and T. Ozkan-Haller, Applying video sensor networks to nearshore environment monitoring, *IEEE Pervasive Computing*, 2(4), 14–21, 2003.

[Honeywell08] Honeywell, 101 Columbia Road, Morristown, NJ 07962 USA. See: http://www.honeywell.com.

[Horn86] B.K.P. Horn, *Robot Vision*, 1st edn. The MIT Press, Cambridge, MA, 1986.

[Hschulzrinne96] H. Schulzrinne, S. Casner, R. Frederick, and V. Jacobson, RTP: A transport protocol for real-time applications. RFC1889, January 1996.

[HU03] F. Hu and S. Kumar, Multimedia query with QoS considerations for wireless sensor networks in telemedicine, in *Proceedings of Society of Photo-Optical Instrumentation Engineers—International Conference on Internet Multimedia Management Systems*, Orlando, FL, September 2003.

[Hu08] F. Hu, M. Jiang, L. Celentano, and Y. Xiao, Robust medical ad hoc sensor networks (MASN) with wavelet-based ECG data mining, *Ad Hoc Networks Journal* (*Elsevier*), 6(7), 986–1012, September 2008.

[Hu2009a] F. Hu, Y. Xiao, and Q. Hao, Congestion-aware, loss-resilient bio-monitoring sensor networking, *IEEE Journal on Selected Areas in Communications* (*JSAC*), 27(4), 450–465, anuary 2009.

[Hu2009b] F. Hu, S. Lakdawala, Q. Hao, and M. Qiu, Low-power, intelligent sensor hardware interface for medical data pre-processing, *IEEE Transactions on Information Technology in Biomedicine*, 13(4), 656–663, May 2009.

[Hu2009c] F. Hu, M. Jiang, M. Wagner and D. Dong, Privacy-preserving tele-cardiology sensor networks: Towards a low-cost, portable wireless hardware/software co-design, *IEEE Transactions on Information Technology in Biomedicine*, 11(6), 617–627, November 2007.

[Hu2009d] F. Hu, L. Celentano, and Y. Xiao, Error-resistant RFID-assisted wireless sensor networks for cardiac tele-healthcare, *Wireless Communications and Mobile Computing* (*Wiley*), 9, 85–101, February 2009.

[Hu2009e] F. Hu, P. Tilgman, S. Mokey, J. Byron, and A. Sackett, Secure, low-cost prototype design of underwater acoustic sensor networks, *Journal of Circuits, Systems, and Computers* (*World Scientific*), 17(6), 1203–1208, 2008.

[Hu2009f] F. Hu, Q. Hao, M. Qiu, and Y. Wu, Low-power electroencephalography sensing data RF transmission: Hardware architecture and test, in *ACM MobiHoc 2009—The First ACM International Workshop on Medical-grade Wireless Networks* (*WiMD '09*), New Orleans, LA, 2009.

[Huang07] T. Huang, K. Hou, H. Yu, E.T. Chu, and C. King, LA-TinyOS: A locality-aware operating system for wireless sensor networks, in *Proceedings of the 2007 ACM Symposium on Applied Computing* (*SAC '07*), Seoul, Korea, March 11–15, 2007, ACM, New York, pp. 1151–1158.

[Hui07] H. Song, Secure wireless sensor networks: Building blocks and applications, PhD dissertation, Department of Computer Science and Engineering, The Pennsylvania State University, University Park, PA, 2007.

[Hwendi00] W. Heinzelman, A. Chandrashekaran, and H. Balakrishnan, Energy efficient communication protocol for wireless microsensor networks, in *Proceedings of 33rd Hawaii International Conference on Systems Sciences*, Cambridge, MA, January 2000.

[HXia96] H. Xia, An analytical model for predicting path loss in urban and suburban environments, in *Proceedings of the Personal Indoor Radio Communication* (*PIRMC '96*), Taipei, Taiwan, 1996.

[IBorg97] I. Borg and P. Groenen, *Modern Multidimensional Scaling Theory and Applications*. Springer, New York, 1997.

[IEEE07] IEEE Standard for Information technology—Telecommunications and information exchange between systems—Local and Metropolitan area networks—Specific requirements, *Part II: Wireless LAN Medium Access Control* (*MAC*) *and Physical Layer* (*PHY*) *Specifications*, pp. 120–121, July 2007.

[Iglewicz93] B. Iglewicz and D.C. Hoaglin, *How to Detect and Handle Outliers*, ASQC Basic References in Quality Control, ASQC Quality Press, Milwaukee, WI, 1993.

[IKhalil05] I. Khalil, S. Bagchi, and N.B. Shroff, Analysis and evaluation of SECOS, a protocol for energy efficient and secure communication in sensor networks, *Ad Hoc Networks Journal (ADHOC)*, 5(3), 360–391, 2007.

[Intel02] Intel Corp, Intel Press Release: Intel Builds World's First One Square Micron SRAM Cell. 2002: http://www.intel.com/pressroom/archive/releases/20020312tech.htm.

[Internet07] (post date: 7-18-07). http://robotics.eecs.berkeley.edu/~roosta/SIRI2006.pdf

[INTERSEMA. 2002] INTERSEMA. 2002. MS5534A barometer module, Technical Report (October). Go online to http://www.intersema.com/pro/module/file/da5534.pdf

[IPetersen99] I. Petersen and A. Savkin, *Robust Kalman Filtering for Signals and Systems with Large Uncertainties*. Birkhäuser, Boston, MA, 1999.

[IPSec] Security architecture for the Internet Protocol. RFC 2401, November 1998.

[Irhee06] I. Rhee, A. Warrier, J. Min, and L. Ki, DRAND: Distributed randomized TDMA scheduling for wireless ad-hoc networks, in *Proceeding of IEEE MobiHoc*, Florence, Italy, May 2006, pp. 190–201.

[Irhee08] I. Rhee, A. Warrier, M. Aia, J. Min, and M.L. Sichitiu, Z-MAC: A hybrid MAC for wireless sensor networks, *IEEE/ACM Transactions on Networking*, 16(3), 511–524, June 2008.

[Issa06] I. Khalil, Mitigation of control and data traffic attacks in wireless ad-hoc and sensor networks, PhD thesis, Purdue University, West Lafayette, IN, 2006.

[Istepanian04] R.S.H. Istepanian, E. Jovanov, and Y.T. Zhang, Guest editorial introduction to the special section on M-health: Beyond seamless mobility and global wireless health-care connectivity, *IEEE Transactions on Information Technology in Biomedicine*, 8(4), 405–414, 2004.

[Jason03] J.L. Hill, System architecture for wireless sensor networks, PhD dissertation, Department of Computer Science, University of California at Berkeley, Berkeley, CA, Spring 2003.

[Jaein07] J. Jeong, X. Jiang, and D. Culler, Design and analysis of MicroSolar power systems for wireless sensor networks, Technical Report No. UCB/EECS-2007-24, http://www.eecs.berkeley.edu/Pubs/TechRpts/2007/EECS-2007-24.html

[JBeutel99] J. Beutel, Geolocation in a picoradio environment, Master's thesis, ETH Zurich, Zurich, Canton of Zurich, Switzerland, 1999.

[JElson02] J. Elson, L. Girod, and D. Estrin, Fine-grained network time synchronization using reference broadcasts, in *Proceedings of the Fifth Symposium on Operating Systems Design and Implementation (OSDI 2002)*, Boston, MA, December 2002, pp. 147–163.

[Jennifer08] J. Yick, B. Mukherjee, and D. Ghosal, Wireless sensor network survey, *Computer Networks*, 52(12), 2292–2330, August 22, 2008.

[JGProakis01] J.G. Proakis, E.M. Sozer, J.A. Rice, and M. Stojanovic, Shallow water acoustic networks, *IEEE Communications Magazine*, 39(11), 114–119, November 2001.

[John06] J.A. Stankovic, Wireless sensor networks, Department of Computer Science, University of Virginia, Charlottesville, VA, 2006. See: http://www.cs.virginia.edu/~stankovic/psfiles/wsn.pdf

[Johnson05] J. Johnson, J. Lees, M. Ruiz, M. Welsh, and G. Werner-Allen, Monitoring volcanic eruptions with a wireless sensor network, in *Proceedings of the Second European Workshop Wireless Sensor Networks (EWSN '05)*, Istanbul, Turkey, January 2005.

[Jonathan08] J. Bachrach and C. Taylor, Localization in sensor networks, computer science and artificial intelligence laboratory, Massachusetts Institute of Technology, Cambridge, MA; http://people.csail.mit.edu/jrb/Projects/poschap.pdf; Visited in 2008.

[JUANG02] P. Juang, O. Hidenkazu, M. Martonosi, L. Peh, D. Rubenstein, and Y. Wang, Energy-efficient computing for wildlife tracking: Design tradeoffs and early experiences with ZebraNet, *ASPLOS X*, San Jose, CA, October 2002.

[JZhao03] J. Zhao, R. Govindan, and D. Estrin, Computing aggregates for monitoring wireless sensor networks, in *Proceedings of the IEEE ICC Workshop Sensor Network Protocols Applications*, Anchorage, AK, May 2003, pp. 139–148.

[Jai04] J. Ai, J. Kong, and D. Turgut, An adaptive coordinated medium access control for wireless sensor networks, in *Proceedings of the Ninth IEEE International Symposium on Computer and Communications 2004*, Alexandria, Egypt, July 2004, Vol. 1, pp. 214–219.

[Jkulik02] K. Joanna, W. Heidemann, and H. Balakrishnan, Negotiation-based protocols for disseminating information in wireless sensor networks, *ACM Wireless Networks*, 8(2/3), 169–185, March–May 2002.

[Jli04] J. Li and G.Y. Lazarou, A bit-map-assisted energy-efficient MAC scheme for wireless sensor networks, in *Proceedings of the Third International Symposium on Information Processing in Sensor Networks*, Berkeley, CA, April 2004, ACM, New York, pp. 55–60.

[JNal-karaki04] J.N. Al-Karaki, R. Ul-Mustafa, and A.E. Kamal, Data aggregation in wireless sensor networks—Exact and approximate algorithms, in *Proceedings of IEEE Workshop on High Performance and Routing 2004*, Ames, IA, April 2004, pp. 241–245.

[Joseph05] J. Polastre, R. Szewczyk, and D. Culler, Telos: Enabling ultra-low power wireless research, in *Proceedings of the Fourth International Symposium on Information Processing in Sensor Networks 2005* (*IPSN 2005*), Los Angeles, CA, April 15, 2005, pp. 364–369.

[JPolastre04] J. Polastre, Interfacing Telos to 51-pin sensorboards, October 2004, http://www.tinyos.net/hardware/telos/telos-legacy-adapter.pdf

[Jpolastre04] J. Polastre, J. Hill, and D. Culler, Versatile low power media access for wireless sensor networks, in *Proceeding of Second International Conference on Embedded Networked Sensor Systems*, Baltimore, MD, October 2004, ACM, New York, pp. 95–107.

[JRice00] J. Rice et al., Evolution of seaweb underwater acoustic networking, in *Proceedings of the MTS/IEEE OCEANS*, Providence, RI, September 2000, Vol. 3, pp. 2007–2017.

[Karlof04] C. Karlof, N. Sastry, and D. Wagner, TinySec: A link layer security architecture for wireless sensor networks, in *Proceedings of the Second International Conference on Embedded Networked Sensor Systems* (*SenSys '04*), Baltimore, MD, November 3–5, 2004, ACM, New York, pp. 162–175.

[Karthikeyan] K. Vaidyanathan, S. Sur, S. Narravula, and P. Sinha, Data aggregation techniques in sensor networks, Technical Report OSU-CISRC-11/04-TR60, Department of Computer Science and Engineering, The Ohio State University, Columbus, OH, downloadable from: http://citeseerx.ist.psu.edu/viewdoc/summary?doi=10.1.1.60.937, Visited in 2009.

[Kavek04] K. Pahlavan and P. Krishnamurthy, *Principles of Wireless Networks: A Unified Approach*, 1st edn. Prentice Hall, Englewood Cliffs, NJ, 2004, ISBN: 8178086468.

[KAY93] S. Kay, *Fundamentals of Statistical Signal Processing, Volume I: Estimation Theory*. Prentice Hall, Upper Saddle River, NJ, 1993.

[Keoliver05] K.E. Oliver, Introduction to automatic design of wireless networks, *CrossRoads ACM Student Magazine*, 11(4), 1–4, 2005.

[Kjamieson03] K. Jamieson, H. Balakrishnan, and Y.C. Tay, Sift: A MAC protocol for Event-driven wireless sensor networks, in *Proceedings of the Third European Workshop on Wireless Sensor Networks*, Zurich, Switzerland, Lecture Notes in Computer Science, Vol. 3868, pp. 260–275, Springer Link, New York, May 2003.

[KOkeya05] K. Okeya and T. Iwata. Side channel attacks on message authentication codes, in *Second European Workshop on Security and Privacy in Ad Hoc and Sensor Networks*, Visegrad, Hungary, July 2005.

[Kon] K. Minolta, Minolta Color Meter IIIf. http://konicaminolta.com. 2008.

[KRamakrishnan90] K. Ramakrishnan and R. Jain, A binary feedback scheme for congestion avoidance in computer networks, *ACM Transactions on Computer Systems*, 8(2), 158–181, May 1990.

[KSanzgiri02] K. Sanzgiri, B. Dahill, B.N. Levine, C. Shields, and E. Belding-Royer, A secure routing protocol for ad hoc networks, in *Proceedings of the 10th IEEE International Conference on Network Protocols* (*ICNP*), Paris, France, 2002, pp. 78–87.

[Ksarvakar08] K. Sarvakar and P.S. Patel, An efficient hybrid MAC layer protocol utilized for wireless sensor networks, in *Proceedings of Fourth IEEE Conference on Wireless Communication and Sensor Networks '08*, Allahabad, India, December 2008, pp. 22–26.

[Ksohrabi00] K. Sohrabi, J. Gao, V. Ailawadhi, and G.J. Pottie, Protocols for self-organization of a wireless sensor network, *IEEE Personal Communications*, 7(5), 16–27, October 2000.

[Kyasanur03] P. Kyasanur and N.H. Vaidya, Detection and handling of MAC layer misbehavior in wireless networks, in *Proceedings of the International Conference on Dependable Systems and Networks* (*DSN '03*), San Francisco, CA, 2003, pp. 173–182.

[LAShort98] L.A. Short and E.H. Saindon, Telehomecare rewards and risks, *Caring*, 17(42), 36–40, 1998.

[Laura07] L.J. Celentano, RFID-assisted wireless sensor networks for cardiac tele-healthcare, MS thesis, Advisor: Dr. F. Hu, Department of Computer Engineering, Rochester Institute of Technology, New York, October 2007.

[Lcampelli07] L. Campelli, A. Capone, M. Cesana, and E. Ekici, A receiver oriented MAC protocol for wireless sensor networks, in *Proceedings of Mobile Ad Hoc and Sensor Systems 07*, Pisa, Italy, October 2007, pp. 1–10.

[LEWIS86] F.L. Lewis, *Optimal Estimation: With an Introduction to Stochastic Control Theory*. John Wiley & Sons, Inc., New York, 1986.

[Legg] G. Legg, ZigBee: Wireless technology for low-power sensor networks. TechOnline, May 2004.

[Levis06] P. Levis et al., TinyOS: An operating system for sensor networks, in *Ambient Intelligence*, W. Weber, J. Rabaey, and E. Aarts, Eds., Springer-Verlag, Berlin, Germany, 2004.

[LHu04] L. Hu and D. Evans, Using directional antennas to prevent wormhole attacks, in *Network and Distributed System Security Symposium* (*NDSS*), San Diego, CA, 2004.

[LHu04a] L. Hu and D. Evans, Localization for mobile sensor networks, in *Proceedings of the 10th Annual International Conference on Mobile Computing and Networking* (*MobiCom*), Philadelphia, PA, 2004, pp. 45–57.

[Linear04] Linear Technology. LTC1540: Nanopower comparator with reference. Datasheet, 7 December 2004. http://www.linear.com/pc/downloadDocument.do?navId=H0,C1,C1154,C1004,C1139,P1593,D1777

[LLazos04] L. Lazos and R. Poovendran, SeRLoc: Secure range-independent localization for wireless sensor networks, in *ACM WiSe*, Philadelphia, PA, 2004, pp. 21–30.

[Lli01] L. Li and J.Y. Halpern, Minimum-energy mobile wireless networks revisited, in *Proceedings of IEEE International Conference on Communications*, Helsinki, Finland, June 2001, Vol. 1, pp. 278–283.

[Lsubramanian00] L. Subramanian and R.H. Katz, An architecture for building self-configurable systems, in *Proceedings of MobiHoc 2000*, Boston, MA, November 2000, pp. 63–73.

[Macwilliams77] F. Macwilliams and N. Sloane, *The Theory of Error-Correcting Codes*. Elsevier Science, New York, 1977.

[MADDEN02a] S. Madden, M.J. Franklin, J.M. Hellerstein, and W. Hong, TAG: A Tiny Aggregation service for ad-hoc sensor networks, in *Proceedings of OSDI*, Boston, MA, 2002a.

[Manish06] M. Raghuvanshi, Implementation of wireless sensor mote, MTech thesis, Department of Nuclear Engineering and Technology, Indian Institute of Technology, Kanpur, India, 2006, see http://home.iitk.ac.in/~ynsingh/mtech/manish2006.pdf

[Marati02] A. Manjeshwar and D.P. Agarwal, APTEEN: A hybrid protocol for efficient routing and comprehensive information retrieval in wireless sensor networks, in *Proceedings of 15th IEEE Parallel and Distributed Processing Symposium*, Fort Lauderdale, FL, April 2002, pp. 195–202.

[Mark07] M. Luk, G. Mezzour, A. Perrig, and V. Gligor, MiniSec: A secure sensor network communication architecture, in *Proceedings of the Sixth International Conference on Information Processing in Sensor Networks* (*IPSN 2007*), Cambridge, MA, April 2007.

[Martin00] T. Martin, E. Jovanov, and D. Raskovic, Issues in wearable computing for medical monitoring applications: A case study of a wearable ECG monitoring device, in *Proceedings of the International Symposium on Wearable Computers* (*ISWC*), Atlanta, GA, 2000, pp. 43–50.

[Masoomeh07] M. Rudafshani and S. Datta, Localization in wireless sensor networks, in *IPSN'07*, Cambridge, MA, April 25–27, 2007.

[Mateusz07] M. Malinowski, M. Moskwa, M. Feldmeier, M. Laibowitz, and J.A. Paradiso, CargoNet: A low-cost MicroPower sensor node exploiting quasi-passive wakeup for adaptive asychronous monitoring of exceptional events, in *SenSys '07*, Sydney, Australia, November 6–9, 2007.

[MELEXIS02] MELEXIS, INC. 2002. MLX90601 infrared thermopile module, Technical Report (August). Go online to http://www.melexis.com/prodfiles/mlx90601.pdf

[Melodia05] T. Melodia, D. Pompili, and I.F. Akyildiz, On the interdependence of distributed topology control and geographical routing in ad hoc and sensor networks, *Journal of Selected Areas in Communications*, 23, 520–532, March 2005.

[Melodia07] T. Melodia, D. Pompili, V.C. Gungor, and I.F. Akyildiz, Communication and coordination in wireless sensor and actor networks, *IEEE Transactions on Mobile Computing*, 6(10), 1116–1129, October 2007. On Melodia's underwater sensor network papers: D. Pompili, T. Melodia, and I. Akyildiz, Three-dimensional and two-dimensional deployment analysis of underwater acoustic sensor networks, *Ad Hoc Networks* (*Elsevier*), 7(4), 778–790, June 2009; I.F. Akyildiz, D. Pompili, and T. Melodia, State of the art in protocol research for underwater acoustic sensor networks, *ACM Mobile Computing and Communication Review* (Invited Paper), October 2007.

[MGHunink97] M.G. Hunink et al., The recent decline in mortality from coronary heart disease, 1980–1990. The effect of secular trends in risk factors and treatment, *Journal of the American Medical Association*, 277, 535–542, 1997.

[Miaomiao08] M. Wang, J. Cao, J. Li, and S.K. Das, Middleware for wireless sensor networks: A survey, *Journal of Computer Science and Technology*, 23(3), 305–326, 2008.

[Min07] M.K. Park and V. Rodoplu, UWAN-MAC: An energy-efficient MAC protocol for underwater acoustic wireless sensor networks, *IEEE Journal of Oceanic Engineering*, 32(3), 710–720, July 2007.

[MMaroti04] M. Maroti, B. Kusy, G. Simon, and A. Ledeczi, The flooding time synchronization protocol, in *Proceedings of the Second International ACM Conference on Embedded Networked Sensor Systems* (*SenSys*), Baltimore, MD, 2004, ACM Press, New York, pp. 39–49.

[Mohamed02] M.G. Gouda, E.N. Elnozahy, C.-T. Huang, and T.M. McGuire, Hop integrity in computer networks, *IEEE/ACM Transactions on Networking*, 10(3), 308–319, June 2002.

[Moore04] D. Moore, J. Leonard, D. Rus, and S. Teller, Robust distributed network localization with noisy range measurements, in *Proceedings of the Second International Conference on Embedded Networked Sensor Systems*, Baltimore, MD, November 03–05, 2004.

[Newsome04] J. Newsome, E. Shi, D. Song, and A. Perrig, The sybil attack in sensor networks: Analysis & defenses, in *Proceedings of the Third international Symposium on Information Processing in Sensor Networks (IPSN '04)*, Berkeley, CA, April 26–27, 2004, ACM, New York, pp. 259–268.

[Ngajaweera08] N. Gajaweera and D. Dias, FAMA/TDMA hybrid MAC for wireless sensor networks, in *Proceedings of Fourth IEEE International Conference on Information and Automation for Sustainability '08*, Colombo, Sri Lanka, December 2008, pp. 67–72.

[Njamal04] N.A. Jamal and A.E. Kamal, Routing techniques in wireless sensor networks: A survey, *IEEE Wireless Communications*, 11(6), 6–28, December 2004.

[NPriyantha05] N. Priyantha, H. Balakrishnan, E. Demaine, and S. Teller, Mobile-assisted topology generation for auto-localization in sensor networks, in *Proceedings of Infocom*, Miami, FL, 2005.

[OYounis04] O. Younis and S. Fahmy, Distributed clustering in ad-hoc sensor networks: A hybrid, energy-efficient approach, in *Proceedings of the IEEE INFOCOM*, Hong Kong, China, March 2004.

[PBonnet01] P. Bonnet, J.E. Gehrke, and P. Seshadri, Towards sensor database systems, in *Proceedings of the Second International Conference on Mobile Data Management* (*MDM '01*), Hong Kong, China, January 2001, pp. 314–810.

[PDutta06] P. Dutta et al., Trio: Enabling sustainable and scalable outdoor wireless sensor network deployments, *IEEE SPOTS*, Nashville, TN, 2006.

[Peter05a] P. Desnoyers, D. Ganesan, and P. Shenoy, TSAR: A two tier sensor storage architecture using interval skip graphs, in *SenSys '05*, San Diego, CA, November 2–4, 2005.

[PFG06] H. Park, J. Friedman, P. Gutierrez, V. Samanta, J. Burke, and M.B. Srivastava, Illumimote: Multi-modal and high fidelity light sensor module for wireless sensor networks, *IEEE Sensors Journal*, 7(7), 996–1003, 2007.

[Philip03] P. Levis, N. Lee, M. Welsh, and D. Culler, TOSSIM: Accurate and scalable simulation of entire TinyOS applications, in *SenSys '03*, Los Angeles, CA, November 5–7, 2003.

[Pkarn90] P. Karn, MACA—A new channel access method for packet radio, in *ARRL/CRRL Amateur Radio Ninth Computer Networking Conference*, Montreal, Qubec, Canada, September 1990, pp. 1–5.

[PLevis02] P. Levis and D. Culler. Mate: A tiny virtual machine for sensor networks, in *Proceedings of the 10th International Conference on Architectural Support for Programming Languages and Operating Systems* (*ASPLOS-X*), San Jose, CA, 2002, ACM Press, New York, pp. 85–95.

[Plin04] P. Lin, C. Qiao, and X. Wang, Medium access control with a dynamic duty cycle for sensor networks, in *Proceedings of Wireless Communications and Networking Conference*, Piscataway, NJ, March 2004, Vol. 3, pp. 1534–1539.

[Pompili06] D. Pompili, T. Melodia, and I.F. Akyildiz, Routing algorithms for delay-insensitive and delay-sensitive applications in underwater sensor networks, in *Proceedings of the 12th Annual International Conference on Mobile Computing and Networking* (*MobiCom '06*), Los Angeles, CA, September 23–29, 2006, ACM, New York, pp. 298–309.

[Pompili09] D. Pompili, T. Melodia, and I.F. Akyildiz, Three-dimensional and two-dimensional deployment analysis for underwater acoustic sensor networks, *Ad Hoc Networks*, 7(4), 778–790, June 2009.

[Priyantha00] N.B. Priyantha, A. Chakraborty, and H. Balakrishnan, The cricket location-support system, in *Proceedings of the Sixth Annual ACM International Conference on Mobile Computing and Networking* (*MobiCom '00*), Boston, MA, August 2000, pp. 32–43.

[Priyantha05] N. B. Priyantha, The cricket indoor location system, PhD thesis. Computer Science and Engineering, Massachusetts Institute of Technology, MA, June 2005. Available at: http://nms.lcs.mit.edu/papers/bodhi-thesis.pdf

[PSikka06] P. Sikka, P. Corke, P. Valencia, C. Crossman, D. Swain, and G. Bishop-Hurley, Wireless ad hoc sensor and actuator networks on the farm, *IEEE SPOTS*, Nashville, Tennessee, 2006.

[Pubudu05] P.N. Pathirana, N. Bulusu, A.V. Savkin, and S. Jha, Node localization using mobile robots in delay-tolerant sensor networks, *IEEE Transactions on Mobile Computing*, 4(3), 285–296, May/June 2005.

[Purushottam07] P. Kulkarni, SensEye: A multi-tier heterogeneous camera sensor network, PhD thesis, Department of Computer Science, University of Massachusetts, Amherst, MA, February 2007.

[PXie05] P. Xie, J.-H. Cui, and L. Li, VBF: Vector-based forwarding protocol for underwater sensor networks, UCONN CSE Technical Report, UbiNet-TR05-03 (BECAT/CSETR-05-6), February 2005.

[PZhang04] P. Zhang, C.M. Sadler, S.A. Lyon, and M. Martonosi, Hardware design experiences in zebranet, in *ACM Sensys*, Baltimore, MD, 2004.

[Qfang03] Q. Fang, F. Zhao, and L. Guibas, Lightweight sensing and communication protocols for target enumeration and aggregation, in *Proceedings of MobiHoc 2003*, Annapolis, MD, June 2003, pp. 165–176.

[Qli01] Q.Li, J. Aslam, and D. Rus, Hierarchical power-aware routing in sensor networks, in *Proceedings of the DIMACS Workshop on Pervasive Networking*, Piscataway, NJ, April 2001, pp. 1–5.

[Radu05] R. Stoleru, T. He, J.A. Stankovic, and D. Luebke, A high-accuracy, low-cost localization system for wireless sensor networks, in *SenSys '05*, San Diego, CA, November 2–4, 2005.

[Rahimi05] M. Rahimi et al., Cyclops: In situ image sensing and interpretation in wireless sensor networks, in *Proceedings of the Third International Conference on Embedded Networked Sensor Systems* (*SenSys '05*), San Diego, CA, November 2–4, 2005, ACM, New York, pp. 192–204.

[Rahimi03] M. Rahimi, H. Shah, G. Sukhatme, J. Heidemann, and D. Estrin, Studying the feasibility of energy harvesting in a mobile sensor network, in *Proceedings of the IEEE International Conference on Robotics and Automation*, Taipai, Taiwan, May 2003, pp. 19–24.

[Rakhmatov03] D. Rakhmatov and S. Vrudhula, Energy management for battery-powered embedded systems, *ACM Transactions Embedded Computing Systems*, 2(3), 277–324, August 2003.

[Rappaport96] T.S. Rappaport, *Wireless Communication*: *Principles and Practices*. Prentice-Hall PTR, Upper Saddle River, NJ, 1996.

[RATNASAMY01] S. Ratnasamy et al., Data-centric storage in sensornets, in *ACM First Workshop on Hot Topics in Networks*, Princeton, NJ, 2001.

[RATNASAMY02] S. Ratnasamy et al., GHT—A geographic hash-table for data-centric storage, in *First ACM International Workshop on Wireless Sensor Networks and Their Applications*, Atlanta, GA, September 2002.

[Rcshah02] R.C. Shah and J.M. Rabaey, Energy aware routing for low energy ad hoc sensor networks, in *Proceedings of the IEEE WCNC '02*, Orlando, FL, March 2002, Vol. 1, pp. 350–355.

[Rkannan03] R. Kannan, K. Ram, S.S. Iyengar, and V. Kumar, Energy and rate based MAC protocol for wireless sensor network, in *Proceedings of the ACM SIGMOD 2003*, 32(4), 60–65, December 2003.

[Rramanathan97] S. Ramanathan, A unified framework and algorithms for (T/F/C) DMA channel assignment in wireless networks, in *Proceedings of IEEE INFOCOM*, San Francisco, CA, April 1997, Vol. 2, pp. 900–907.

[RShah03] R. Shah, S. Roy, S. Jain, and W. Burnette, Datamules: Modeling a three-tier architecture for sparse sensor networks, *Journal of Ad Hoc Networks* (*Elsevier*), 1(2–3), 215–233, 2003.

[RSivakumar99] R. Sivakumar, P. Sinha, and V. Bharghavan, CEDAR: A core-extraction distributed ad hoc routing algorithm, *IEEE Journal on Selected Areas in Communications*, Special issue on *Ad Hoc Networks*, 17(8), 1454–1465, August 1999.

[RSzewczyk04] R. Szewczyk, A. Mainwaring, J. Polastre, and D. Culler, An analysis of a large scale habitat monitoring application, in *Proceedings of the Second ACM Conference on Embedded Networked Sensor Systems* (*SenSys*), Baltimore, MD, November 2004.

[Rwheinzelman99] R.W. Heidemann, K. Joanna, and H. Balakrishnan, Adaptive protocols for information dissemination in wireless sensor networks, *ACM Mobicom '99*, Seattle, WA, August 1999, pp. 174–185.

[SAM06] T. Kuhn and P. Becker, A simulator interconnection framework for the accurate performance simulation of SDL models, in *System Analysis and Modeling*: *Language Profiles*, Lecture Notes in Computer Science, Vol. 4320, Springer, Berlin, Germany, 2006, ISBN 3-540-68371-2.

[Samuel02] R.S. Madden, M.J. Franklin, J.M. Hellerstein, and W. Hong, TAG: A Tiny AGgregation service for ad-hoc sensor networks, in *Fifth Symposium on Operating Systems Design and Implementation* (*OSDI 2002*), Boston, MA, 2002.

[SCapkun03] S. Capkun, L. Buttyán, and J.-P. Hubaux, SECTOR: Secure tracking of node encounters in multi-hop wireless networks, in *Proceedings of the First ACM Workshop on Security of Ad Hoc and Sensor Networks (SASN '03)*, Fairfax, VA, 2003, pp. 21–32.

[Sek] Sekonic, Sekonic L-558Cine DualMaster. http://www.sekonic.com/Products/L-558Cine.html. 2008.

[SENSIRION02] SENSIRION. 2002. SHT11/15 relative humidity sensor. Tech. rep. (June). Go online to http://www.sensirion.com/en/pdf/Datasheet_SHT1x_SHT7x_0206.pdf

[SensorSim] S. Park, A. Savvides, and M.B. Srivastava, SensorSim: A simulation framework for sensor networks, in *Proceedings of the 3rd ACM International Workshop on Modeling, Analysis and Simulation of Wireless and Mobile Systems*, Boston, MA, August 20, 2000. MSWIM '00, ACM, New York, pp. 104–111.

[Seth00] S. Edward-Austin Hollar, COTS Dust, MS thesis, Mechanical Engineering, University of California at Berkeley, Berkeley, CA, Fall 2000.

[Seung-Jong08] S.-J. Park, R. Vedantham, R. Sivakumar, and I.F. Akyildiz, GARUDA: Achieving effective reliability for downstream communication in wireless sensor networks, *IEEE Transactions on Mobile Computing*, 7(2), 214–230, February 2008.

[SFloyd93] S. Floyd and V. Jacobson, Random early detection gateways for congestion avoidance, *IEEE/ACM Transactions on Networking*, 1(4), 397–413, August 1993.

[SGaneriwal03] S. Ganeriwal, R. Kumar, and M.B. Srivastava, Timing-sync protocol for sensor networks, in *Proceedings of the First International ACM Conference on Embedded Networked Sensor Systems (SenSys)*, Los Angeles, CA, 2003, ACM Press, New York, pp. 138–149.

[Shanmugasundaram04] J. Shanmugasundaram, Querying peer-to-peer networks using P-trees, Technical Report TR2004-1926, Cornell University, Ithaca, NY, 2004.

[SLi03] S. Li, S. Son, and J. Stankovic, Event detection services using data service middleware in distributed sensor networks, in *Proceedings of the Second International Workshop Information Processing in Sensor Networks (IPSN '03)*, Palo Alto, CA, April 22–23, 2003, pp. 502–517.

[Slindsay02] S. Lindsay and C.S. Raghavendra, PEGASIS: Power-efficient gathering in sensor information systems, in *Proceedings of Aerospace Conference*, Big Sky, Mont, June 2002, Vol. 3, pp. 1125–1130.

[SLindsey02] S. Lindsey and C.S. Raghavendra, PEGASIS: Power efficient gathering in sensor information systems, in *Proceedings of the 2002 IEEE Aerospace Conference*, Big Sky, Mont, March 2002, pp. 1–6.

[Sony08]Sony SNC-RZ30N Camera driver. http://cvs.nesl.ucla.edu/cvs/viewcvs.cgi/CoordinatedActuation/Actuate/

[Sorber05] J. Sorber, N. Banerjee, M.D. Corner, and S. Rollins, Turducken: Hierarchical power management for mobile devices, in *Proceedings of MOBISYS*, Seattle, WA, 2005, pp. 261–274.

[SPalChaudhuri03] S. PalChaudhuri, A. Saha, and D.B. Johnson, Probabilistic clock synchronization service in sensor networks, Technical Report TR 03-418, Department of Computer Science, Rice University, Houston, TX, 2003.

[Sparton08] Sparton SP3003D Digital Compass. 2008. http://www.sparton.com/

[SRM05] S.R. Madden, M.J. Franklin, J.M. Hellerstein and W. Hong, TinyDB: An acquisitioned query processing system for sensor networks, *ACM Transactions Database Systems*, 30(1), 122–173, 2005.

[Spo] Spotlight. Website: http://www.spotlight.it

[SRoundy03] S. Roundy, B.P. Otis, Y.-H. Chee, J.M. Rabaey, and P. Wright, A 1.9ghz rf transmit beacon using environmentally scavenged energy, in *IEEE International Symposium on Low Power Electronics and Devices*, Seoul, Korea, 2003.

[SSL] OpenSSL. http://www.openssl.org

[Stargate08] Stargate platform. http://www.xbow.com/Products/XScale.htm. 2008.

[Stemm97] M. Stemm and R.H. Katz, Measuring and reducing energy consumption of network interfaces in hand-held devices, *IEICE Transactions on Communications*, E80-B(8), 1125–1131, August 1997.

[Stockdon00] H. Stockdon and R. Andholman, Estimation of wave phase speed and near-shore bathymetry from video imagery, *Journal of Geophysical Research*, 105(9), 22015–22033, September 2000.

[Sundararaman05] B. Sundararaman, U. Buy, and A. Kshemkalyani, Clock synchronization for wireless sensor networks: A survey, *Ad Hoc Networks* (*Elsevier*), 3, 281–323, May 2005.

[Sunil08] Z. Feng, S. Kumar, F. Hu, and Y. Xiao, E^2SRT: Enhanced event-to-sink reliable transport for wireless sensor networks, *Wireless Communications and Mobile Computing* (*Wiley*), November 2008 (accessible online). DOI: 10.1002/wcm.705.

[Sunil08a] S. Kumar, K.K.R. Kambhatla, B. Zan, F. Hu, and Y. Xiao, An energy-aware and intelligent cluster-based event detection scheme in wireless sensor networks, *International Journal of Sensor Networks* (*InderScience*), 3(2) 123–133, February 2008.

[Szhou07] S. Zhou, R. Liu, D. Everitt, and J. Zic, A^2-MAC: An application adaptive medium access control protocol for data collections in wireless sensor networks, in *Proceedings of IEEE ISCIT07*, Sydney, Australia, October 2007, pp. 1131–1136.

[Tanya06] T. Roosta, S.P. Shieh, and S. Sastry, Taxonomy of security attacks in sensor networks and countermeasures, in *First IEEE International Conference on System Integration and Reliability Improvements*, Hanoi, Vietnam, December 2006.

[TAOS] TAOS, INC. 2002. TSL2550 ambient light sensor, Technical Report (September). Go online to http://www.taosinc.com/images/product/document/tsl2550.pdf

[TCamp02] T. Camp, J. Boleng, and V. Davies, A survey of mobility models for ad hoc network research, *Wireless Communications and Mobile Computing*, 2(5), 483–502, 2002.

[Ti08] One of the largest chip production company—Texas Instruments, see http://www.ti.com, Visited in June 2008.

[Tian04] T. He, B.M. Blum, J.A. Stankovic, and T. Abdelzaher, AIDA: Adaptive application-independent data aggregation in wireless sensor networks, *Transactions on Embedded Computing System*, 3(2), 426–457, May 2004.

[TinyOS07] On TinyOS operating system, see http://www.tinyos.net, Visited in June 2007.

[Tmote06] Tmote invent user's manual, Technical Report, Moteiv, Inc., San Francisco, CA, February 2006.

[Transducer08] International Transducer Corporation. Available: www.itc-transducer.com. Visted in 2008.

[Tsai87] R.Y. Tsai, A versatile camera calibration technique for high-accuracy 3D machine vision metrology using off-the-shelf TV cameras and lenses, *IEEE Journal of Robotics and Automation*, RA-3(4), 323–344, August 1987.

[The03] H. Tian, J.A. Stankovic, C. Lu, and T. Abdelzaher, SPEED: A stateless protocol for real-time communication in sensor networks, in *Proceedings of Distributed Computing Systems 2003*, Providence, RI, May 2003, pp. 46–55.

[TVon92] T. von Eicken, D.E. Culler, S.C. Goldstein, and K.E. Schauser, Active messages: A mechanism for integrating communication and computation, in *Proceedings of the 19th Annual International Symposium on Computer Architecture*, Gold Coast, Australia, May 1992, pp. 256–266.

[Tvdam03] T. Van Dam and K. Langendoen, An adaptive energy-efficient MAC protocol for wireless sensor networks, in *Proceedings of First International Conference on Embedded Networked Sensor Systems*, Los Angeles, CA, November 2003, ACM, New York, pp. 171–180.

[TYlonen96] T. Ylonen, SSH—Secure login connections over the internet, in *Proceedings of the Sixth USENIX Security Symposium*, San Jose, CA, 1996.

[VJacobson88] V. Jacobson, Congestion avoidance and control, in *Proceedings of the ACM SIGCOMM Symposium*, Stanford, CA, August 1988.

[Victor04] V. Shnayder, M. Hempstead, B. Chen, G.W. Allen, and M. Welsh, Simulating the power consumption of large scale sensor network applications, in *SenSys '04*, Baltimore, MD, November 3–5, 2004.

[Virantha04] V. Ekanayake, C. Kelly IV, and R. Manohar, An ultra low-power processor for sensor networks, *ASPLOS '04*, Boston, MA, October 7–13, 2004.

[Vrajendran05] V. Rajendran, J.J. Garcia-Luna-Aceves, and K. Obraczka, Energy-efficient, application-aware medium access for sensor networks, in *Proceedings of IEEE Mobile Adhoc and Sensor Systems '05*, Washington, DC, November 2005, pp. 630–637.

[Vrajendran06] V. Rajendran, K. Obraczka, and J.J. Garcia-Luna-Aceves, Energy-efficient, collision-free medium access control for wireless sensor networks, in *Proceedings of the First International Conference on Embedded Sensor Systems (SenSys '03)*, Los Angeles, CA, February 2006, ACM, New York, Vol. 12, No. 1, pp. 63–78.

[VRaghunathan05] V. Raghunathan, A. Kansal, J. Hsu, J. Friedman, and M. Srivastava, Design considerations for solar energy harvesting wireless embedded systems, *IEEE SPOTS*, Los Angeles, CA, 2005.

[Vrodoplu99] V. Rodoplu and T.H. Meng, Minimum energy mobile wireless networks, *IEEE Journal on Selected Areas in Communications*, 17(8), 1333–1344, August 1999.

[Wan03] C. Wan, S.B. Eisenman, and A.T. Campbell, CODA: Congestion detection and avoidance in sensor networks, in *Proceedings of the First International Conference on Embedded Networked Sensor Systems (SenSys '03)*, Los Angeles, CA, November 5–7, 2003, ACM, New York, pp. 266–279.

[Wang06] S. Wang, W. Chen, C. Ong, L. Liu, and Y. Chuang, RFID application in hospitals: A case study on a demonstration RFID project in a Taiwan hospital, in *Proceedings of the 39th Annual Hawaii International Conference on System Sciences (HICSS '06)*, Kauai, HI, January 4–7, 2006, Vol. 8, p. 184a.

[Wang08] M.M. Wang, J.N. Cao, J. Li, and S. Das, Middleware for wireless sensor networks: A survey, *Journal of Computer Science and Technology*, 23(3), 305–326, May 2008.

[Ward97] A. Ward, A. Jones, and A. Hopper, A new location technique for the active office, *IEEE Personal Communications*, 4(5), 42–47, October 1997.

[WBHeinzelman02] W.B. Heinzelman, A.P. Chandrakasan, and H. Balakrishnan, An application-specific protocol architecture for wireless microsensor networks, in *Proceedings of the IEEE Transactions on Wireless Communications*, 1(4), 660–670, October 2002.

[WBHeinzelman04] W.B. Heinzelman et al., Middleware to support sensor network applications, *IEEE Network*, 18(1), 6–14, 2004.

[WINS] Wireless integrated network systems(wins). http://wins.rsc.rockwell.com/. 2008.

[WMB02] F. Wagmister, B. McDonald, J. Brush, J. Burke, and T. Denove, Advanced Technology for Cinematography, 2002. Website: http://hypermedia.ucla.edu/projects/atc.php

[WS82] W. Gunter and W.S. Stiles, *Color Science: Concepts and Methods, Quantitative Data and Formulae*, 2nd edn. John Wiley & Sons, New York, 1982.

[Wstallings04]W. Stallings, IEEE 802.11 Wireless LANs: From a to n, *IT Proceedings*, 6, 32–37, September–October 2004

[Wye02] W. Ye, J. Heidemann, and D. Estrin, An energy-efficient MAC protocol for wireless sensor networks, in *Proceedings of IEEE INFOCOM*, New York, June 2002, Vol. 3, pp. 1567–1576.

[Wye04] W. Ye, J. Heidemann, and D. Estrin, Medium access control with coordinated adaptive sleeping for wireless sensor networks, *IEEE/ACM Transactions on Networking*, 12(3), 453–506, July 2004.

[WSu05] W. Su and I.F. Akyildiz, Time-diffusion synchronization protocol for sensor networks, *IEEE/ACM Transactions on Networking*, 13(2), 384–398, 2005.

[XHong99] X. Hong, M. Gerla, G. Pei, and C.-C. Chiang, A group mobility model for ad hoc wireless networks, in *MSWiM '99*, Seattle, WA, 1999, ACM Press, New York, pp. 53–60.

[Xiang04] X. Ji, Localization algorithms for wireless sensor network systems, PhD thesis, Department of Computer Science and Engineering, The Pennsylvania State University, Philadelphia, PA, 2004.

[XJiang05] X. Jiang, J. Polastre, and D. Culler, Perpetual environmentally powered sensor networks, *IEEE SPOTS*, Los Angeles, California, April 2005.

[Yyan01] Y. Yan, R. Govindan, and D. Estrin, Geographical and energy aware routing: A recursive data dissemination protocol for wireless sensor networks, Technical Report UCLA-CSD TR-010023, August, 2001.

[YangXiao07] Y. Xiao, V.K. Rayi, B. Sun, X. Du, F. Hu, and M. Galloway, A survey of key management schemes in wireless sensor networks, *Computer Communications Journal* (*Elsevier*), Special issue on *Security on Wireless Ad Hoc and Sensor Networks*, 30(11–12), 2314–2341, September 2007.

[YCHu03] Y.-C. Hu, A. Perrig, and D.B. Johnson, Packet leashes: A defense against wormhole attacks in wireless networks, in *IEEE INFOCOM*, San Francisco, CA, 2003.

[YIyer05] Y. Iyer, S. Gandham, and S. Venkatesan, STCP: A generic transport layer protocol for sensor networks, in *Proceedings of 14th IEEE International Conference on Computer Communications and Networks*, San Diego, CA, October 2005.

[YTirta06] Y. Tirta, B. Lau, N. Malhotra, S. Bagchi, Z. Li, and Y.-H. Lu, Controlled mobility for efficient data gathering in sensor networks with passively mobile nodes, in *Sensor Network Operations*. Wiley-IEEE Press, Hoboken, NJ, 2006.

[Yuan06] H. Yuan, H. Ma, and H. Liao, Coordination mechanism in wireless sensor and actor networks, in *Proceedings of the First International Multi-Symposiums on Computer and Computational Sciences* (*IMSCCS '06*), April 20–24, 2006, Vol. 2, Zhejiang, China, pp. 627–634.

[Yxu01] Y. Xu, J. Heidemann, and D. Estrin, Geography informed energy conservation for ad hoc routing, in *Proceeding of MobiCom 2001*, Rome, Italy, July 2001, pp. 70–84.

[YXu01] Y. Xu, J. Heidemann, and D. Estrin, Geography informed energy conservation for ad hoc routing, in *Proceedings of the Seventh Annual International Conference on Mobile Computing and Networking*, Rome, Italy, July 2001.

[Zhang08] P. Zhang and M. Martonosi, LOCALE: Collaborative localization estimation for sparse mobile sensor networks, in *Proceedings of the 2008 International Conference on Information Processing in Sensor Networks* (*IPSN '08*), St. Louis, MO, April 22–24, 2008, pp. 195–206.

[Zigbee08] On Zigbee wireless communication standard please see http://www.zigbee.org, Visited in June 2007.

[ZLi05] Z. Li, W. Trappe, Y. Zhang, and B. Nath, Robust statistical methods for securing wireless localization in sensor networks, in *Proceedings of IPSN '05*, Los Angeles, CA, 2005.

Index

H

I

J

L

O

P

R

S

Z

Milton Keynes UK
Ingram Content Group UK Ltd.
UKHW050306161024
449569UK00033B/453